Handbook of Food Structure Development

Food Chemistry, Function and Analysis

Series editors:
Gary Williamson, *University of Leeds, UK*
Alejandro G. Marangoni, *University of Guelph, Canada*
Juliet A. Gerrard, *University of Auckland, New Zealand*

Titles in the series:
1: Food Biosensors
2: Sensing Techniques for Food Safety and Quality Control
3: Edible Oil Structuring: Concepts, Methods and Applications
4: Food Irradiation Technologies: Concepts, Applications and Outcomes
5: Non-extractable Polyphenols and Carotenoids: Importance in Human Nutrition and Health
6: Cereal Grain-based Functional Foods: Carbohydrate and Phytochemical Components
7: Steviol Glycosides: Cultivation, Processing, Analysis and Applications in Food
8: Legumes: Nutritional Quality, Processing and Potential Health Benefits
9: Tomato Chemistry, Industrial Processing and Product Development
10: Food Contact Materials Analysis: Mass Spectrometry Techniques
11: Vitamin E: Chemistry and Nutritional Benefits
12: Anthocyanins from Natural Sources: Exploiting Targeted Delivery for Improved Health
13: Carotenoid Esters in Foods: Physical, Chemical and Biological Properties
14: Eggs as Functional Foods and Nutraceuticals for Human Health
15: Rapid Antibody-based Technologies in Food Analysis
16: DNA Techniques to Verify Food Authenticity: Applications in Food Fraud
17: Advanced Gas Chromatography in Food Analysis
18: Handbook of Food Structure Development

How to obtain future titles on publication:
A standing order plan is available for this series. A standing order will bring delivery of each new volume immediately on publication.

For further information, please contact:
Book Sales Department, Royal Society of Chemistry, Thomas Graham House, Science Park, Milton Road, Cambridge, CB4 0WF, UK
Telephone: +44 (0)1223 420066, Fax: +44 (0)1223 420247,
Email: booksales@rsc.org
Visit our website at www.rsc.org/books

Handbook of Food Structure Development

Edited by

Fotis Spyropoulos
University of Birmingham, UK
Email: f.spyropoulos@bham.ac.uk

Aris Lazidis
University of Birmingham, UK
Email: a.lazidis@bham.ac.uk

and

Ian T. Norton
University of Birmingham, UK
Email: i.t.norton@bham.ac.uk

ROYAL SOCIETY OF CHEMISTRY

Food Chemistry, Function and Analysis No. 18

Print ISBN: 978-1-78801-216-4
PDF ISBN: 978-1-78801-615-5
EPUB ISBN: 978-1-78801-905-7
Print ISSN: 2398-0656
Electronic ISSN: 2398-0664

A catalogue record for this book is available from the British Library

The Royal Society of Chemistry is a charity, registered in England and Wales, Number 207890, and a company incorporated in England by Royal Charter (Registered No. RC000524), registered office: Burlington House, Piccadilly, London W1J 0BA, UK, Telephone: +44 (0) 20 7437 8656.

Visit our website at www.rsc.org/books
Printed in the United Kingdom by CPI Group (UK) Ltd, Croydon, CR0 4YY, UK

Preface

The importance of food structure for the development of products that meet consumer expectations in terms of being tasty, nutritious, pleasurable, and in many cases healthy, is nowadays well established. This book aims to provide an extensive and thorough review of our current understanding on food structure. It showcases how the latter can be developed through the interplay of formulation and processing elements, tailored to possess desirable performances, measured and/or characterized. In terms of content balance, it focuses on the basic principles relating to food structure development but, where necessary, progresses to greater depths of detail to also encompass more specialized technological features.

We strongly feel that the target audience for this book includes both formulators/scientists working within the food industry as well as researchers/academics primarily concerned with exploring this area from a more fundamental perspective. The book is therefore envisaged to act as a useful tool for product developers aiming to understand and address the challenges associated with current food structures and/or their performance, as well as those seeking inspiration for the creation of novel products and/or functionalities. At the same time, it will be of great assistance to early career academics with scientific interests in this area and also to their more established colleagues who, despite being already research active within the food science and engineering arena(s), wish to be kept well connected with current industrial needs and the research developments within this continuously evolving scientific field.

In order to meet its ambitious mandate, this book is structured in four discreet parts/themes. The first part focuses on the role of key components such as hydrocolloids, proteins, and oils/fats and that of fundamental colloidal microstructures (emulsions, foams) in structure development across a wide range of food categories. The interplay between processing

Food Chemistry, Function and Analysis No. 18
Handbook of Food Structure Development
Edited by Fotis Spyropoulos, Aris Lazidis and Ian T. Norton

Published by the Royal Society of Chemistry, www.rsc.org

routes and formulation elements is then further explored for confectionery and cereal food products. The use of starch and proteins to ultimately provide the aerated solid microstructure of bread serves as a good example of the journey from specific ingredients to the development of a final food structure. The second part of this book centers on the significant link between food structure and product performance/functionality. Here, specific attention is given to structure development for rheological, tribological, and mechanical performance(s), along with the role of food's microstructural design in terms of providing functionalities such as a desirable sensorial experience and the delivery of a range of species of nutritional/dietary significance. The important relationship between food performance/ functionality and food structure heavily relies on our capacity to accurately, reliably, and repeatably assess such complex microstructures. This is the focal point of the third part of this book, which examines in detail the most important approaches/methods for analyzing, characterizing, and predicting food structure (and its development) across a broad spectrum of length- and timescales. In its final part, this book attempts to examine some of the specific challenges that the food industry is presently faced with and the opportunities that these challenges are creating in terms of rethinking or rebalancing conventional food structure development. While some of these challenges have been long-standing (yet still very much relevant) items of scrutiny within the industrial and academic arenas (for example, the development of healthy or healthier food structures), others have more recently emerged (or are still emerging) to encompass themes ranging from the sustainability and flexibility of food processing approaches to food structure design that addresses the dietary needs of specific population groups.

To end, we would like to acknowledge all our friends and colleagues who have contributed with passion and expertise to this book. We are grateful for their tremendous efforts and humbled by the wealth of knowledge that they have so generously trusted upon us. Needless to say, this book would not have been realized without them. Finally, we would also like to thank the editorial staff in the Royal Society of Chemistry for their great support and endless patience throughout the preparation of this book.

FS would like to dedicate this book to Jasbir, Maiya and Solon, and AL to Elena and Alê.

Fotis Spyropoulos
Aris Lazidis
Ian T. Norton

Contents

Part A: Food Structure Development: The Interplay Between Processing Routes and Formulation Elements

Chapter 1 The Role of Hydrocolloids in the Development of Food Structure 3
H. Douglas Goff and Qingbin Guo

1.1 Introduction and Hydrocolloid Applications in Food Systems 3
1.2 Hydrocolloid Functionality in Food Systems 5
1.2.1 Solubility 6
1.2.2 Viscosity 8
1.2.3 Gelation 11
1.3 Molecular Structures of Hydrocolloids and Structure–Function Relationship 13
1.3.1 Molecular Weight and Molecular Weight Distribution 13
1.3.2 Charged and Neutral Polysaccharides 14
1.3.3 Branched and Linear Polysaccharides 15
1.3.4 Functional Groups 16
1.3.5 Molecular Conformation 17
1.4 Unique Examples 18
1.4.1 Sodium/Calcium Alginate 18
1.4.2 High-/Low-methoxyl Pectin 21
1.4.3 Methyl Cellulose 22
1.4.4 Carrageenan (Milk Reactivity) 23

Food Chemistry, Function and Analysis No. 18
Handbook of Food Structure Development
Edited by Fotis Spyropoulos, Aris Lazidis and Ian T. Norton

Published by the Royal Society of Chemistry, www.rsc.org

1.4.5 Gum Arabic 24
1.4.6 Xanthan 26
1.5 Conclusions and Future Perspectives 26
References 27

Chapter 2 The Role of Proteins in the Development of Food Structure 29
Valérie Gagnaire, Valérie Lechevalier, Marie-Hélène Famelart, Thomas Croguennec and Saïd Bouhallab

2.1 Introduction 29
2.2 Characteristics of Milk and Egg White Proteins 32
2.2.1 Caseins 33
2.2.2 Globular Proteins 34
2.3 Diversified Supramolecular Assemblies from Processed Globular Proteins and Potential Applications in Food 35
2.3.1 Assemblies Obtained by Tuning the Protein Environment 35
2.3.2 Assemblies Obtained on Heating Protein Solution 37
2.4 Examples of Food Matrices 41
2.4.1 Egg White Protein-based Gels 41
2.4.2 Casein-based Gels: Case of Yogurt 44
2.4.3 Casein-based Gels: Case of Cheeses 47
2.5 Conclusion and Future Trends 52
References 53

Chapter 3 Food Structure Development in Emulsion Systems 59
Ernesto Tripodi, Aris Lazidis, Ian T. Norton and Fotis Spyropoulos

3.1 Introduction 59
3.2 The Emulsion Microstructure 60
3.2.1 Conventional Emulsion Microstructure 60
3.2.2 Complex Emulsion Microstructures 66
3.3 Thermodynamic Aspects 69
3.4 Processing Routes for Emulsion Structure Development 69
3.4.1 Mechanical Methods 70
3.4.2 Nonmechanical Methods 77
3.5 Emulsion Microstructure Characterization 77
3.5.1 Visualization of the Emulsion Microstructure 77

3.5.2 Characterization of the Emulsion Microstructure and Its Components 78
3.5.3 Assessing Emulsion Flow and Thermal Behavior 79
3.6 Emulsion Microstructure and Performance 79
3.6.1 Emulsion Microstructure and Rheology 80
3.6.2 Emulsion Microstructure and Oral Processing 80
3.6.3 Emulsion Microstructure for Encapsulation and Release 81
3.6.4 Fat, Salt, and Sugar Reduction 82
3.7 Conclusions and Future Perspective 84
References 84

Chapter 4 The Role of Bubbles in the Development of Food Structure 93
Arpita Mondal and Keshavan Niranjan

4.1 Introduction 93
4.2 Bubble Structures in Milk 94
4.2.1 Process Parameters 95
4.2.2 Foam Processing Methods 98
4.3 Bubble Structures in Chocolate 101
4.3.1 Bubble Inclusion Methods 102
4.3.2 Influential Factors in Bubble Inclusion 103
4.3.3 Operational Variables 105
4.4 Bubble Structures in Dough and Bread 105
4.4.1 Bubble Inclusion Steps 106
4.4.2 Physical and Chemical Controlling Factors 108
4.4.3 Characterization of Bubbles in Bread 109
4.5 Conclusions 111
References 112

Chapter 5 Food Structure Development in Oil and Fat Systems 115
R. A. Nicholson and A. G. Marangoni

5.1 Introduction 115
5.2 Characterizing the Structure of Fats 116
5.2.1 Wide-angle X-ray Scattering and Polymorphism 116
5.2.2 Small-angle X-ray Scattering and Scherrer Analysis 117
5.2.3 ^{1}H Spin Diffusion NMR 119

5.2.4 Cryogenic Transmission Electron Microscopy 119
5.2.5 Microstructure 120
5.2.6 Ultra-small-angle X-ray Scattering 121
5.2.7 Fractal Dimension 124
5.3 Factors Affecting the Fat Structure 126
5.3.1 TAG Composition 126
5.3.2 Crystallization Temperature and Cooling Rate 127
5.3.3 Crystallization Under Shear 128
5.3.4 Addition of Emulsifiers 129
5.4 Relating Structure to Functional Properties 130
5.4.1 Oil-binding Capacity 130
5.4.2 Rheological Properties 131
5.5 Conclusions 132
Acknowledgements 132
References 132

Chapter 6 Food Structure Development in Chocolate 134
Emma J. McLeod and Peter J. Fryer

6.1 Introduction 134
6.2 Developments in Chocolate Formulation 135
6.3 Chocolate Manufacturing and Structure Control 137
6.3.1 Liquid Chocolate Mass Making 137
6.3.2 Tempering and Forming 139
6.4 Polymorphism in Cocoa Butter 139
6.4.1 Chocolate Cooling Model 141
6.5 Mechanisms for Bloom Formation 145
6.6 Conclusions and Future Needs 148
References 149

Chapter 7 Food Structure Development in Cereal and Snack Products 151
G. Della Valle, H. Chiron, A. Le-Bail and L. Saulnier

7.1 Introduction 151
7.2 Products and Materials 153
7.2.1 Wheat Components 153
7.2.2 State Diagrams 156
7.3 Processing 159
7.3.1 Operations and Development of Cellular Structure 159

7.3.2 Rheological Properties 162
7.3.3 Modeling Approaches 164
7.4 Conclusion and Main Prospects 167
Acknowledgements 169
References 169

Part B: Food Structure Development for Performance and Functionality

Chapter 8 Food Structure Development for Rheological/Tribological Performance 175
H. M. Shewan and J. R. Stokes

8.1 Introduction 175
8.2 Rheology Theory and Methods 176
8.2.1 Rheology Theory 176
8.2.2 Dynamic Rheology Measurement 177
8.2.3 Oscillatory Rheology Including Large and Small Oscillatory Shear 179
8.2.4 Thin Film Rheology 181
8.3 Tribology Theory and Methods 182
8.3.1 Stribeck Curves 183
8.4 Food Structure Design 185
8.4.1 Control of Viscosity 185
8.4.2 Tuning of the Yield Stress 190
8.4.3 Control of Fat Lubrication 191
8.5 Using Rheology and Tribology as Tools for Food Structure Development in Relation to Sensory Perception 192
8.6 Conclusion and Future Prospects 195
Acknowledgements 195
References 196

Chapter 9 Developing Food Structure for Mechanical Performance 199
I. K. Mohammed, C. G. Skamniotis and M. N. Charalambides

9.1 Introduction 199
9.2 Particulate Composites, Colloids, and Gels 200
9.2.1 Structure: Property Studies on Real Food Products 201
9.2.2 Structure: Property Studies for Gels and Filled Gels 205

9.3 Cellular Structures 209
9.3.1 Manufacturing Methods of Edible Foam Food Products 209
9.3.2 Mechanical Testing of Baked and Extruded Foam Foods 210
9.3.3 Microstructural Characterization of Foam Foods Using X-ray Micro Tomography 212
9.3.4 Finite Element Modeling of the Fracture of a Brittle Baked Wafer 213
9.4 Short Fiber Composites 214
9.4.1 Adding Fibers in Foods 214
9.4.2 Mechanical Characterization of Fiber Food Composites 215
9.4.3 Using Mechanical Properties to Simulate Food Oral Breakdown 218
9.5 Conclusion and Future Trends 219
Acknowledgements 221
References 221

Chapter 10 Design Structures for Optimal Sensory Performance 225
Marine Devezeaux de Lavergne and Benjamin le Révérend

10.1 Introduction—What Is Sensory Performance? 225
10.2 How Do Humans Sense and Process Foods in the Oral Cavity? 226
10.2.1 Sensory Physiology (*S*) 226
10.2.2 Olfaction 227
10.2.3 Taste 228
10.2.4 Trigeminal and Texture 228
10.2.5 Food Oral Processing Physiology (*P*) 231
10.3 How to Reverse Engineer Sensory Performance? 234
10.3.1 Quantify Sensory Perceptions and Product Performance 235
10.4 Foods as Time-dependent Stimuli for the Sensory System 239
10.4.1 Foods as Time-dependent Soft Matter Systems Under Mechanical Stresses 239
10.4.2 Foods as Time-dependent Soft Matter Systems Under Biochemical Stresses 240
10.4.3 Foods as Time-dependent Soft Matter Systems Under Thermal Stresses 241

10.5 Application of the Understanding of Food Sensing to Design Complex Structures Modulating Chemosensory Perception 241
10.6 Application of the Understanding of Food Sensing to Design Complex Structures Modulating Mechanical Perception 244
10.7 Conclusions and Future Avenues for Research in Food Perception 246
References 248

Chapter 11 The Development of Food Structures for the Encapsulation and Delivery of Bioactive Compounds 259
Nicolaas Jan Zuidam and Krassimir P. Velikov

11.1 Introduction 259
11.2 Food Products as a Vehicle for Bioactive Compounds 260
11.2.1 Iron-fortified Bouillon Cube 261
11.2.2 Food Products as Regulators of Digestion 261
11.2.3 Food as an Aid to Extract Bioactive Ingredients from Diet 262
11.3 Delivery Systems and Encapsulation Structures for Bioactive Compounds 262
11.3.1 Basic Structures 263
11.3.2 Homogeneous Core Systems 264
11.3.3 Matrix-type Encapsulates 272
11.3.4 Core-shell-type Encapsulates 276
11.4 Conclusions 280
Acknowledgements 280
References 280

Part C: Food Structure Analysis, Characterisation, and Modelling

Chapter 12 Food Structure Analysis Using Light and Confocal Microscopy 287
Camilla Öhgren, Patricia Lopez-Sanchez and Niklas Lorén

12.1 Introduction 287
12.2 Characterization Methods of Food Structures at the Micrometer Level 288
12.2.1 Light Microscopy 288
12.2.2 Confocal Laser Scanning Microscopy 290

12.3 Imaging and Dynamics of Foods 293
12.3.1 Cereal-based Porous and Heterogeneous Foods 293
12.3.2 Plant-based Foods 297
12.3.3 Fat-rich Foods 301
12.4 Conclusions and Future Perspective 305
Acknowledgements 306
References 306

Chapter 13 Food Structure Characterisation Using Small-angle Scattering Methods 309
Marta Martinez-Sanz, Amparo Lopez-Rubio and Elliot Gilbert

13.1 Introduction 309
13.2 Scattering Techniques 310
13.3 Small-angle Scattering Methods 312
13.4 Application of Small-angle Scattering to Characterise Food-relevant Materials 317
13.4.1 Polysaccharides 317
13.4.2 Proteins 328
13.4.3 Lipids 334
13.4.4 Complex Systems 336
13.4.5 Whole Foods 339
13.4.6 Other Food-related Applications 340
13.5 Conclusions and Future Perspectives 341
References 342

Chapter 14 Applications of Scanning Electron Microscopy and Atomic Force Microscopy to Food Structure Characterisation 361
Bryony James and Jenny Malmström

14.1 Introduction 361
14.2 Components of the Scanning Electron Microscope 362
14.3 The Compromises of High-resolution Imaging 362
14.4 Electron-specimen Interactions and Image Formation 365
14.5 Lens Aberrations 367
14.6 Overcoming the Limitations Imposed by Vacuum 367
14.6.1 Drying 368
14.6.2 Cryo-SEM 368
14.6.3 Environmental SEM 369
14.6.4 Examples 371

14.7 Atomic Force Microscopy 375
14.7.1 Polysaccharides 376
14.7.2 Lipids 377
14.7.3 Proteins 378
14.7.4 Plant-based Foods: Cell Wall Structure 378
14.8 Conclusions and Future Perspective 378
References 379

Chapter 15 Modelling and Computer Simulation Approaches to Understand and Predict Food Structure Development: Structuring by Gelation and Self-association of Biomolecules 383
Georgios Dalkas and Stephen R. Euston

15.1 Introduction 383
15.2 Basics of Computer-based Molecular Methods 384
15.2.1 Molecular Modelling 384
15.2.2 The MM Force Fields 385
15.2.3 Most Used Existing Force Fields 386
15.2.4 Docking Simulations 386
15.2.5 Molecular Dynamics Simulations 387
15.3 Structure Formation in Triglyceride Systems 388
15.3.1 Molecular Modelling and Computer Simulation Approaches to Characterise the Structure and Complexation of Oleogels 389
15.4 Molecular Modelling and Computer Simulation Approaches to Study Carbohydrate Gelation 393
15.5 Molecular Modelling and Computer Simulation Approaches to Study Protein Gelation 396
15.6 Conclusions and Future Perspective 397
Acknowledgements 398
References 398

Part D: Future Opportunities and Challenges in Food Structure Development

Chapter 16 Development of Food Structure *via* Sustainable Processing Systems 405
James Atuonwu, Savvas Tassou and Tim Foster

16.1 Introduction 405
16.2 Sustainable Food Processing Systems: An Overview 406

16.3 Sustainable Food Processing Technologies: Principles and Applications 407
16.3.1 High-pressure Processing HPP 407
16.3.2 Radiofrequency and Microwave Processing 408
16.3.3 Ohmic Heating 409
16.3.4 Pulsed Electric Field Processing (PEF) 410
16.3.5 Membrane Food Processes 411
16.4 Food Structure Development in Sustainable Processes: Current Status, Future Research Directions 412
16.4.1 HPP and Food Structure 412
16.4.2 RF/Microwave Processing and Food Structure 414
16.4.3 Ohmic Heating and Food Structure 416
16.4.4 Pulsed Electric Field Processing and Food Structure 416
16.4.5 Membrane Processing and Food Structure 417
16.5 Concluding Remarks and Future Perspectives 417
References 418

Chapter 17 Food Structure Development/Production Through Flexible Processes: The Use of Electric Fields to Enable Food Manufacturing 422
Ricardo N. Pereira, António A. Vicente and José A. Teixeira

17.1 Electric Field Processing: An Overview 422
17.2 Flexible Processes: Thermal and Nonthermal Processing 425
17.2.1 High-pulsed Electric Technologies 425
17.2.2 Ohmic Heating 428
17.3 Novel Perspectives Toward Food Structure 432
17.4 Research Needs 434
17.5 Conclusions 434
Acknowledgements 435
References 435

Chapter 18 Development of Healthy Food Structures: Reduction of Sugar, Salt, and Fat 439
Bettina Wolf

18.1 Introduction 439

18.2 Sugar 441
18.2.1 Biscuits and Cakes 442
18.2.2 Chocolate 445
18.2.3 Ice Cream 445
18.3 Salt 446
18.3.1 Cheese 446
18.3.2 Processed Meat 447
18.3.3 Baked Products (Bread) 447
18.4 Fat 448
18.5 Microstructure Approaches to Sugar, Salt, and Fat Reduction 448
18.5.1 Particles with Designed Morphology 448
18.5.2 Bubbles as Tastant-excluding Fillers 450
18.5.3 Droplets 450
18.5.4 Gelled Particles 453
18.5.5 Oleogels 453
18.6 Conclusions 455
References 455

Chapter 19 Food Structure Development for Specific Population Groups **459**
W. A. Alsanei and J. Chen

19.1 Introduction 459
19.2 Mechanisms Controlling Eating and Swallowing 460
19.2.1 Tongue Capability 460
19.2.2 Orofacial Muscle Activities 461
19.2.3 Temporo-spatial Coordination between Breathing and Swallowing Tracks 461
19.2.4 Bolus Formation Prior to Swallowing 461
19.3 Swallowing Disorders (Dysphagia) 462
19.4 Assessments and Grading of Eating and Swallowing Capability 462
19.4.1 Videofluoroscopy 462
19.4.2 Surface Electromyography 463
19.5 Texture Standardization of Diets for Specific Population Groups 463
19.5.1 Needs of Texture-modified Diet 463
19.5.2 Published National Standards 464
19.6 Functional Ingredients for Texture Modification for Specific Population Groups 470
19.6.1 Whey Proteins 470

19.6.2 Polysaccharides 471
19.6.3 Nanostructured Carriers 471
19.6.4 Incorporation of microbubbles 472
19.7 Conclusions 472
References 473

Subject Index **480**

Part A
Food Structure Development: The Interplay Between Processing Routes and Formulation Elements

CHAPTER 1

The Role of Hydrocolloids in the Development of Food Structure

H. DOUGLAS GOFF*[a] AND QINGBIN GUO[b]

[a] Department of Food Science, University of Guelph, Guelph, ON N1G 2W1, Canada; [b] State Key Laboratory of Food Nutrition and Safety (Tianjin University of Science & Technology), Ministry of Education, Tianjin, 300457, China
*Email: dgoff@uoguelph.ca

1.1 Introduction and Hydrocolloid Applications in Food Systems

Hydrocolloids are a class of food ingredients that are widely used in the development of food structure. Generally speaking, we can define *hydrocolloids* most simply as water-soluble polymers that contribute viscosity and gelation in solution.[1] "Gums" and "mucilages" are often used as synonyms. Most hydrocolloids are polysaccharides, but some proteins can also fall into this definition, *e.g.* gelatin and some milk-, egg-, and vegetable-derived protein isolates. However, not all food polysaccharides are hydrocolloids since many of the insoluble polysaccharides, such as unmodified cellulose, do not interact with water. Also, not all water-soluble polysaccharides contribute viscosity due to their molecular weight or molecular structure, *e.g.* the various oligosaccharides or the highly branched dextran, and thus are not considered as hydrocolloids in the conventional use. Granular starch

Food Chemistry, Function and Analysis No. 18
Handbook of Food Structure Development
Edited by Fotis Spyropoulos, Aris Lazidis and Ian T. Norton

Published by the Royal Society of Chemistry, www.rsc.org

may not act as a true hydrocolloid, but gelatinized and other modified starches certainly do. For the sake of clarity, we should also distinguish hydrocolloids from other food colloids. A *colloid* is defined in physical chemistry as a system made up of continuous and dispersed phases in which there is a discrete change in physical properties (*e.g.* density, optical properties) at the phase boundary.[1] Most food colloids are emulsions or foams. A solution or dispersion of hydrocolloids is not necessarily a colloid by this definition, although hydrocolloids are often added to food colloids for functionality or stability. A final distinction that can be made is the availability and use of hydrocolloids as an extracted ingredient, *e.g.* pectin or carrageenan, compared to that same component *in situ* within its source, such as pectin within a whole lemon or carrageenan within its seaweed source. Hydrocolloids are usually thought of as extracted and refined or semirefined food ingredients added to formulated food systems; hence functionality (*i.e.* viscosity, gelation) is also part of the definition. Often they are extracted from by-product streams of food processing (*e.g.* pomace from juice processing, okra from tofu processing, or bran or hulls from cereal and oilseed processing), hence adding value to the commodity. Therefore, it is important to distinguish hydrocolloids from colloids, to recognize which food macromolecules fall into this food ingredient category and why, as well as to understand that there are many overlaps.

Many hydrocolloids are available to the food industry, and as new sources are found from natural products or by-product streams, the list keeps growing. Traditional hydrocolloids are plant-derived ingredients such as pectin, modified starches, modified celluloses, guar gum, locust bean gum, and konjac mannan, along with exudate gums, such as gum arabic, gum ghatti, and tragacanth, and seaweed-derived ingredients, such as agar, alginates, and carrageenan. Gelatin from animal-derived collagen is also a traditional hydrocolloid. Some of these are more familiar in some parts of the world than in others. Newer hydrocolloid ingredients include gums from fermentation, *e.g.* xanthan, gellan, bacterial celluloses, curdlan, or dextran. Cereal β-glucans, arabinoxylans, and xyloglucans are being extracted and modified from hemicellulose to render them as hydrocolloids, and much of that interest is derived from their potential health benefits. Other examples include agricultural by-products such as soluble soybean polysaccharide from okra, flaxseed gum, yellow mustard gum, psyllium, tara gum, fenugreek, and corn fiber gum.

Hydrocolloids are used in foods for two main reasons, either for the physical functionality they add or for their nutritional benefits. Physical functionality is usually either to add viscosity or gelation to a food system. These properties are a manifestation of the chemical structure and conformation of the polymer in solution and in turn are used to deliver functionality such as the development of structure, modification of textural properties of a food system, or shelf-life extension (in which case the term "stabilizers" is often used to refer to the hydrocolloid ingredients). Food applications include soups, sauces, gravies, jams, jellies, baked goods, confections, pie fillings, puddings, ice cream, yogurt, foams and emulsions, and many others. Hydrocolloids are also used as edible barrier-layer films and as encapsulating materials for bioactive compounds. From a nutritional viewpoint, all of the soluble nonstarch polysaccharides and digestion-resistant starch fall into the category of soluble (or viscous) dietary fiber, and this makes up a large component of the hydrocolloid ingredients. These nondigestible carbohydrates can contribute many health benefits, depending on the quantities consumed. Some of the health benefits include reduction in gastroesophageal reflux, enhancement of satiety, reductions in postprandial blood sugar level and LDL cholesterol, and risk reduction of several chronic diseases, including Type II diabetes, obesity, cardiovascular disease, and colorectal cancers.[2–4] Many of the soluble dietary fibers/hydrocolloids are also fermentable, leading to numerous prebiotic functions and enhanced generation of short-chain fatty acids in the colon.[5] Also, some very specific bioactivities are currently being associated with nonstarch polysaccharides.[6] There can also be negative implications from consumption, including mineral binding and hence reduction in mineral absorption, as well as upset in colon functioning leading to irregular stool movement and stool water content and also perhaps enhanced flatulence. Individuals with diseases such as irritable bowel syndrome (IBS) or Crohn's disease need to monitor their intake of nondigestible polysaccharides carefully.

The study and application of food hydrocolloids are a well-developed research and technology area within food science. There are two major biennial conferences on the subject: Gums and Stabilisers for the Food Industry, hosted by the Food Hydrocolloids Trust and held its 20th conference in 2019, and the International Hydrocolloids Conference, approaching its 15th conference in 2020. Proceedings of the Gums and Stabilisers for the Food Industry conferences have been published under the same title after every conference by the Royal Society of Chemistry. A highly recognized journal in the field, *Food Hydrocolloids*, is now at Volume 93 in 2019. Several available textbooks cover this area; some notable titles include Cui,[7] Stephen *et al.*,[8] Biliaderis and Izydorczyk,[6] Morris,[9] Phillips and Williams,[10,11] Imeson,[12] Laaman,[13] and Eliasson.[14] In the remainder of this chapter, we will cover hydrocolloid functionality for structure development in food applications, discuss molecular structure–function relationships, and then present several examples of individual hydrocolloids that contribute unique food functionalities due to particular aspects of their structure (calcium alginate, methylcellulose, high- and low-methoxy pectin, κ-carrageenan, gum Arabic, and xanthan).

1.2 Hydrocolloid Functionality in Food Systems

Hydrocolloids have been extensively used in different food products as thickeners (soups, gravies, salad dressings, sauces, and toppings), water retention agents, stabilizers, emulsifiers, and gel-forming agents (jam, jelly, marmalade, restructured foods, and low-sugar per calorie gels). They also can be applied for inhibiting ice and sugar crystal formation in ice cream and the controlled release of flavors, *etc.* Some of the commonly used hydrocolloids are listed in Table 1.1: *e.g.* modified starch, gelatin, and pectin

Table 1.1 Examples of hydrocolloids used in various food categories.

Food category	Hydrocolloids used
Salad dressing	Xanthan gum, propylene glycol alginate (PGA), modified starch, microcrystalline cellulose (MCC), guar gum
Muscle foods	Modified starch, carrageenan, konjac glucomannan, alginate
Bakery products	Carboxymethyl cellulose (CMC), fenugreek gum, guar gum, konjac gum, xanthan gum
Bakery fillings	Locust bean gum, guar gum, pectin, alginate, PGA, cellulose derivatives, konjac gum, xanthan gum, carrageenan, agar, gellan gum
Frozen dairy desserts	Guar gum, locust bean gum, carrageenan, xanthan gum, alginate, cellulose derivatives, pectin, gelatin
Culture dairy products	Modified starch, locust bean gum, guar gum, gelatin, carrageenan, xanthan gum
Restructured foods	Alginate

can be used in yogurt; xanthan, pectin, and modified starch are used in salad dressing; xanthan and HPMC can be used in whipped topping, *etc.* To understand how the hydrocolloids contribute to the texture, appearance, and mouthfeel of the food products, three main functionalities are discussed further: solubility, viscosity, and gelation.

1.2.1 Solubility

As most hydrocolloids perform their function in aqueous solution, understanding the solubility of polysaccharides is very important, as it allows for the optimization of design and processing conditions of the food products. Application of hydrocolloids in most food products begins by fully dissolving. The dissolution of hydrocolloids is associated with a continuous hydration process, during which the intermolecular binding of the hydrocolloids is gradually converted to molecule–water binding. To achieve the full dissolution of the hydrocolloids, factors including particle size, temperature, and cations all should be considered. For example, compared to a fine powder (particle size higher than 100 mesh), coarse hydrocolloids with a mesh size less than 60 normally take a longer time to dissolve due to the longer time for water penetration. However, hydrocolloids with finer particle sizes may lump together when dissolving due to the formation of so-called fish eyes, *i.e.* particles with a hydrated surface but still undissolved powder in the center. To solve this problem, high agitation mixing and/or premixing with another dry ingredient such as sugar/salt are normally used in industrial applications. Recently, a granulation process has been adopted for hydrocolloid powders that also avoids lumping and helps with quick solubilization.[15] High temperature normally leads to quick dissolution of hydrocolloids; however, a temperature above 80 °C should be avoided for possible degradation of the hydrocolloid molecules. In addition, higher temperature also results in higher energy consumption, leading to increased cost for product manufacturing. Cations also should be considered for the dissolution of some hydrocolloids.

For example, calcium should be avoided for the dissolution of sodium alginate and pectin, as gelation could take place,[13] which prevents the full dispersion of the hydrocolloid molecules in water solution.

Most hydrocolloids are high-molecular-weight polymers and perform their functions as thickeners and stabilizers at concentrations less than 1 wt%. For example, only 0.25% xanthan gum in a pourable French dressing; 0.8% carrageenan in cured hams; 0.7% gums mixture in basic egg pasta. However, some gum-based emulsifiers such as gum arabic require concentrations of even up to 10% to achieve their function. Therefore, understanding the solubility of the gums is critically important for their food application. Different hydrocolloids display distinct solubilities due to their structural and conformational differences; *e.g.* starch is soluble only in hot water, while pullulan and gum arabic are readily dissolved in cold water. Some polysaccharides, such as cellulose, are not water soluble at all, although unmodified cellulose does not technically belong to the family of hydrocolloids. It has been summarized[16] that any structural feature hindering the intermolecular association leads to higher solubility, such as in branching structure or charged groups (carboxylate group, sulfate, or phosphate groups). On the other hand, structural characteristics that promote intermolecular association result in poor solubility, such as in linear chains, large molecular weight, and other regular structural features.

The solubility of some other hydrocolloids is summarized in Table 1.2. Gum arabic for example, has a highly branched molecular structure with excellent water solubility, up to 30% at room temperature, and has been commercially used as an emulsifier. Its detailed structure and functional relationships are discussed further in Section 1.4 of this chapter.

Table 1.2 Solubility of some typical polysaccharides and their structural characteristics. Adapted from ref. 16 [https://www.intechopen.com/books/solubility-of-polysaccharides] under the terms of a CC BY 3.0 license [https://creativecommons.org/licenses/by/3.0/].

Solubility groups	Polysaccharide	Monomer	Special/ charged group	Branching situation	Conformation
Poor solubility	Cellulose	4-β-D-GlcP	No	Linear	Ribbon
	Xylan	4-β-D-XlyP	No	Linear	Ribbon
	Mannan	4-β-D-ManP	No	Linear	Ribbon
Intermediate	Cereal β-glucan	4-β-D-GlcP, 3-β-D-GlcP	No	Linear	Random coil
	Konjac glucomannan	4-β-D-GlcP, 4-β-D-ManP	Acetyl group	Linear	Random coil
	Locust bean gum (galactomannan)	4-β-D-ManP, T-α-D-GalP	No	Less and short branched	Random coil
Easily soluble	Gum arabic per gum ghatii	3,4,6-, 3,6- Gal*p*, T-Ara*f*, *et al.*	Glucuronic acid	Highly branched	Spherical
	Pullulan	4-α-D-GlcP, 6- α-D-GlcP	No	Branched	Random coil

1.2.2 Viscosity

As an important functionality, the thickening properties that hydrocolloids contribute in food systems are mainly due to their viscosity. *Viscosity* can be expressed as the ratio of stress to strain (shear rate). Hydrocolloids can impart viscosity to a solution due to intermolecular entanglements, which lead to the resistance to flow under shearing forces.

Viscosity can be characterized by different techniques, some are very simple such as using Bostwick flow methods to determine the flow speed of a fluid food product under gravity down a defined slope for a specific time. However, *apparent viscosity*, which is the shear stress applied to a fluid divided by the shear rate, can be measured using more advanced equipment such viscometers or rheometers, where the shear stress/shear rate can be varied and the relationships between apparent viscosity and shear rate can be used to describe rheological properties. The rheological properties of hydrocolloids can be classified as shear thinning (pseudoplastic), shear thickening (dilatant), or Newtonian flow behavior, in which apparent viscosity is decreased, increased, and constant, respectively, with the applied shear rate (Figure 1.1).

Factors including solvent, temperature, concentration, pH, salt, and molecular structure all can affect the apparent viscosity of hydrocolloid solutions or dispersions. For example, higher temperature normally leads to a decrease in viscosity as it promotes disentanglement of the intermolecular chains. The Arrhenius equation reflects the relationships between apparent viscosity (η) and temperature (T) for solutions with Newtonian flow behavior:

$$\text{Log}\ \eta = \text{Log}\ A + \frac{E}{R}\frac{1}{T} \tag{1.1}$$

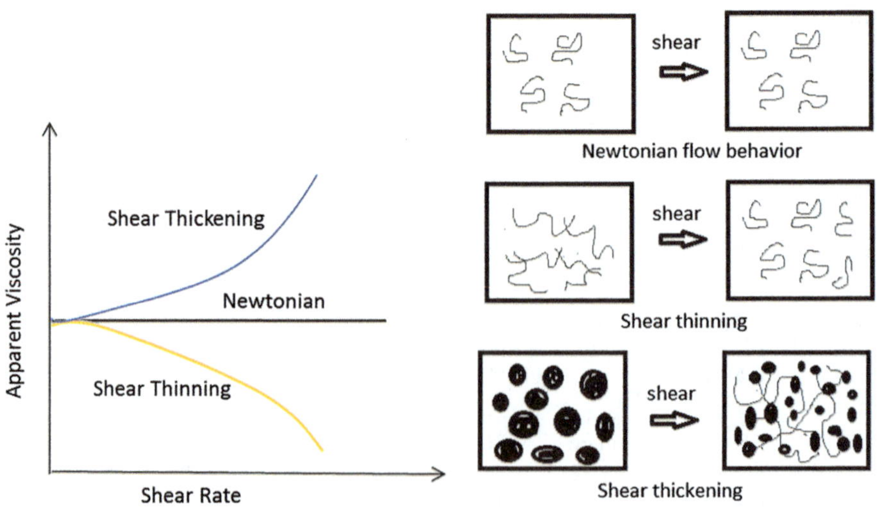

Figure 1.1 Shear flow behaviors of hydrocolloid solutions and the corresponding molecular interactions.

where E is the activation energy for viscous flow, R is the (ideal) gas constant, and A is a constant.

Viscosity is also highly affected by the concentration of the hydrocolloid. For dilute solutions, in which molecular interactions are rare, viscosity increases linearly with concentration but with a low slope. When concentration reaches a specific point where the molecules start to contact/interact with one another (referred to as the *critical concentration* of the hydrocolloid), the viscosity exhibits a step change, beyond which a higher rate of viscosity increase with concentration can be seen (*i.e.* slope increases) due to molecular entanglement, as shown in Figure 1.2.

To compare the viscosity of hydrocolloids in different solvents or under different conditions, intrinsic viscosity (also referred to as inherent viscosity) is generally used. Definitions such as relative viscosity, specific viscosity, and intrinsic viscosity are commonly used with the equations as follows:

$$\eta_{\text{rel}} = \eta_{\text{sol}}/\eta_{\text{solvant}} \tag{1.2}$$

$$\eta_{\text{sp}} = \eta_{\text{rel}} - 1 \tag{1.3}$$

The relationships of specific viscosity (η_{sp}), relative viscosity (η_{rel}), and intrinsic ($[\eta]$) viscosity could be expressed using the Huggins and Kramer equations:[18,19]

$$\eta_{\text{sp}} = [\eta]c + K'[\eta]^2c^2 \tag{1.4}$$

$$\ln(\eta_{\text{rel}}) = [\eta]c + (K' - 0.5)[\eta]^2c^2 \tag{1.5}$$

The intrinsic viscosity $[\eta]$ can be obtained by extrapolating concentration c to zero in either of these equations (Figure 1.2a) and represents molecular volume occupancy (volume per weight, *e.g.* mL g^{-1}). Intrinsic viscosity is proportional to the molecular weight according to the Mark–Houwink equation:[20,21]

$$[\eta] = kM_v{}^{\alpha} \tag{1.6}$$

The critical concentration (C*) for different hydrocolloids varies and is directly related to their individual intrinsic viscosities. The onset of entanglement for a wide range of neutral and charged polysaccharides is found to occur when $C[\eta] \approx 4$ (*i.e.* $C^* \approx 4/[\eta]$). In other words, the higher the intrinsic viscosity of the polysaccharide, the lower the concentration at which C* is exceeded (Figure 1.2b–d).

As can be seen from Table 1.1, mixtures of hydrocolloids are commonly used to impart improved rheological characteristics to food products, and, if the blend gives rise to synergistic interactions, an added incentive is a reduction in total hydrocolloid concentration and hence costs. Classic examples include galactomannan with xanthan gum or seaweed gum. For example, the synergistic effects of guar gum with xanthan, locust bean gum with xanthan, tara gum with xanthan, and locust bean gum with carrageenan have all been previously reported.[22]

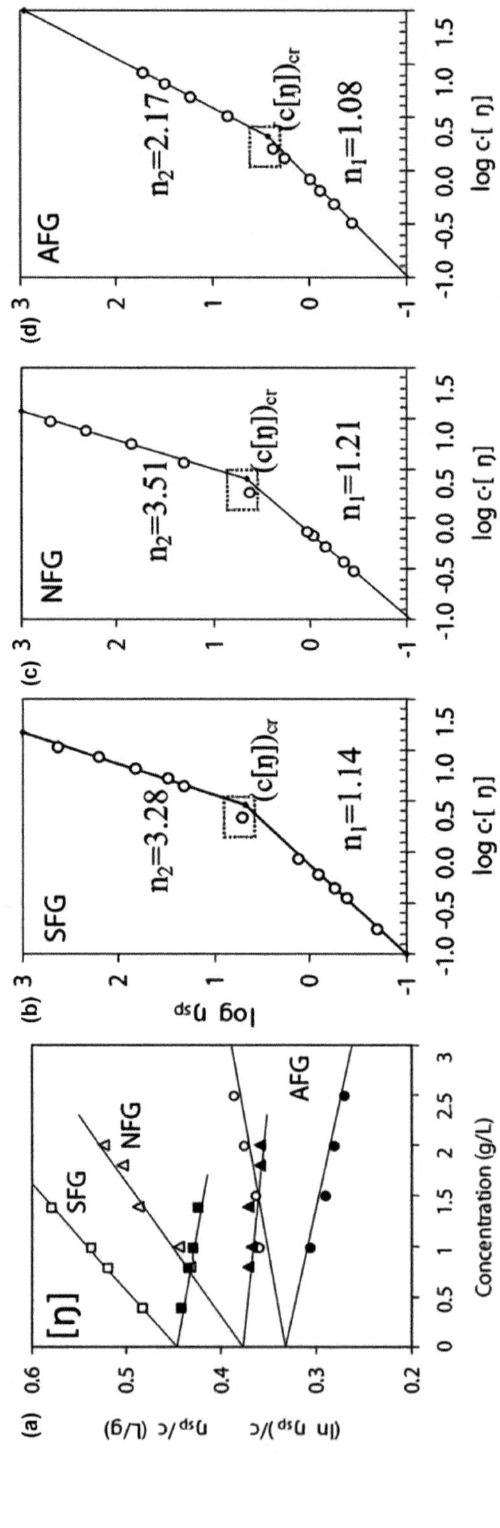

Figure 1.2 The intrinsic viscosity (a) and the dependence of viscosity on concentration (b–d) for three types of flaxseed gum. SFG: soluble flaxseed gum; NFG and AFG: neutral and acidic fraction gum. Adapted from ref. 17 with permission from Elsevier, Copyright 2012.

1.2.3 Gelation

Specific hydrocolloids can function as gel-forming agents in food products. The most commonly used gel-forming agents include the protein gelatin and the polysaccharides alginate, pectin, carrageenan, gellan, agar, modified starch, methyl cellulose, and hydroxypropyl methylcellulose (Table 1.3). Gel formation is the phenomenon involving the association or cross-linking of the polymer chains to form a three-dimensional network that traps or immobilises the water and other additives such as solutes and pigments within it. The associated regions, known as junction zones, may be formed by two or more polymer chains (Figure 1.3). The gelation process is essentially the formation of these junction zones.[23] The intermolecular interactions that contribute to

Table 1.3 Major hydrocolloid gelling agents and their characteristics. Adapted from ref. 22 with permission from Springer Nature, Copyright 2010.

Hydrocolloid as a gelling agent	Characteristics	Food application
Modified starch	Thermally irreversible opaque gel formed upon cooling	Dairy desserts
Agar	Thermally reversible gel formed upon cooling	Bakery products, jellies
κ-, ι-carrageenan	Thermally reversible gel formed upon cooling	Pudding, milkshakes, tofu
High methoxy pectin	Thermally reversible gel formed upon cooling at low pH	Jams, jellies, glazes, milk-based desserts
Low-methoxy pectin	Thermally reversible gel formed upon cooling at high pH	Low-sugar jams, jellies
Gellan gum	Thermally reversible transparent gel formed upon cooling	Water-based fruit-flavored jellies, lemon jelly
Alginate	Thermally reversible gel that does not melt on heating	Restructured foods, cold-prepared bakery creams
Methyl and hydroxypropylmethyl cellulose	Thermally reversible gel that does not melt on heating	Salad dressings, cake batters, beverages, whipped topping

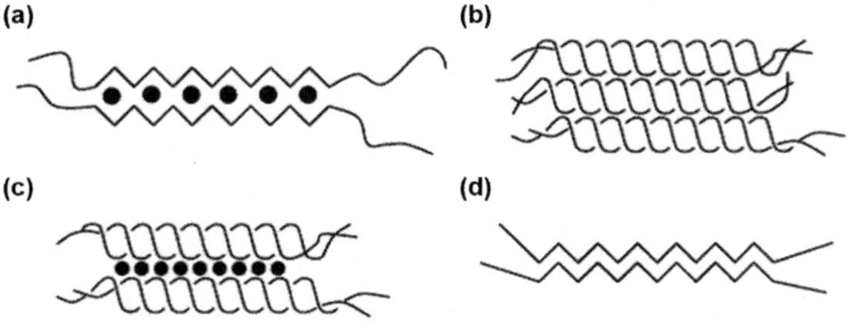

Figure 1.3 Examples of junction zones models for polysaccharides gels: (a) egg box junction, (b) aggregated double helical junction, (c) cation-promoted association of double helices, (d) association of extended ribbon-like structure. Adapted from ref. 24 with permission from Taylor & Francis Group, LLC, Copyright 2005.

junction zone formation include hydrogen bonding, ionic or ion–dipole bonding, van der Waals attraction, and hydrophobic interactions. The physical arrangement of these junction zones within the network can be affected by various parameters such as temperature, the presence of ions, and the inherent structure of the hydrocolloid. It also should be noted that the formation of junction zones by themselves can lead to molecular aggregation and precipitation of hydrocolloids if the zone of interaction is too long. Therefore, a *structure breaker* in the junction zone is also critical for gel formation. The structure breaker is responsible for limiting the length of one junction zone and allowing for the formation of another junction zone elsewhere in the same molecule, with differing molecules (Figure 1.3). This fills the three-dimensional space with the polymer and allows for the trapping and holding of a high degree of water. For example, (1→4)-β-D-glucans interact and lead to the precipitation of the molecules; they are therefore water insoluble and reluctant to form gels. However, mixed-link (1→3)(1→4)-β-D-glucans are somewhat water soluble due to the kinks of β 1→3 linked glycosidic bonds. Xylan structure is water insoluble, while arabinoxylans are water soluble and form gels due to the structure breaker of arabinose as a side chain. Rhamnose "kinks" in pectin are also structure breakers.

The junction zones formed by most gelling agents can be disrupted through heating and reformed upon cooling, with such species referred to as thermally reversible gels; however, for some other gelling agents, the molecular interactions are thermally irreversible.

To induce gelation, polysaccharides first need to be well dissolved or dispersed in solution and then exposed to a controlled change in environmental conditions that will lead to the formation of the three-dimensional structure (the junction zone). Gelation can be induced in three ways: ionotropic gelation, cold-set gelation, and heat-set gelation. For ionotropic gelation, the hydrocolloid (mostly negatively charged polysaccharides) could gel in the presence of ions (mostly cations). Typical examples are alginate, carrageenan, and low-methoxy pectin with Ca^{2+}. In cold-set gelation, first hydrocolloid powders are dissolved in warm/boiling water to achieve full dispersion and hydration, then the enthalpically stabilized interchain association can be formed upon cooling, leading to a three-dimensional network. Most of the hydrocolloids form gels by this mechanism; agar and gelatin are two typical examples. For heat-set gelation, hydrocolloids – such as curdlan, konjac glucomannan, methyl cellulose, hydroxypropylmethyl cellulose, starch, and globular proteins – first need to be heated. Heating results in the unfolding of their molecular structures, which are then rearranged into a network.[22] Increases in the temperature of methyl cellulose and hydroxypropylmethyl cellulose solutions will first lead to a minor decrease in viscosity, followed by an eventual sharp increase when the gelling temperature is reached (52 °C for methyl cellulose and 63–80 °C for hydroxypropylmethyl cellulose).

Hydrocolloids as gelling agents have been applied in many food products. For example, agar is used in water dessert gels, aspics, confectionery jellies, canned meats, icings, piping gels, and flan desserts. Agar is extracted from

Table 1.4 Typical water jelly formulation.

Ingredients	Quantity (%)
Sugar	12–15
Agar	0.8–0.85
Color	As required
Flavor	As required
Water or fruit juice	Up to 100

red seaweed (*Rhodophyceae*), is insoluble in cold water, and hydrates when boiled. Cooling an agar solution below 40 °C produces very firm brittle gels. A water jelly formulation is shown in Table 1.4.

1.3 Molecular Structures of Hydrocolloids and Structure–Function Relationship

As discussed in the Introduction, most hydrocolloids are polysaccharides, which are inherently heterogeneous species in terms of chemical structure and molecular weight distribution. The diversity in structural features, such as monosaccharide composition (simple *versus* complex), linkage patterns, chain shapes (linear *versus* branched), functional groups, and conformations, is linked directly to the physical properties of the hydrocolloids, including solubility, rheological properties (viscosity), and gelling properties. It can be generalized that any polysaccharide structure that hinders intermolecular association usually leads to higher solubility, such as branching or charged groups (carboxylate, sulfate, or phosphate groups); on the other hand, structural characteristics that promote intermolecular association result in poor solubility, such as linear chains, large molecular weight, and other regular structural characteristics. In terms of viscosity, normally higher molecular weight and molecules with rigid conformation result in higher viscosity. For gelation, any structure that enhances the formation of junction zones tends to form a gel.

1.3.1 Molecular Weight and Molecular Weight Distribution

Polysaccharides are polydisperse in molecular weight (M_w), which is referred to as molecular weight distribution. Molecular weight and molecular weight distribution play a critical role for the solubility, viscosity, and gelation of polysaccharides. High molecular weight molecules usually have a large excluded volume [eqn (1.7)], which promotes intermolecular interaction of polysaccharides and suppresses solubility.

$$V_h = \frac{[\eta]M}{\gamma N_A} \tag{1.7}$$

where V_h is the hydrodynamic volume, M is molar mass, N_A is Avogadro's number, and γ is Simha's parameter related to the shape of a particle. Almost all carbohydrate polymers with degrees of polymerization (DP) less than 20 are soluble in water.[25] Solubility decreases with the increase of molecular weight. However, polysaccharides with larger molecular weights normally generate higher viscosities under the same concentration, as such species tend to exhibit intermolecular associations. For example, the viscosity of cellulose gum is determined largely through controlling cellulose chain length or DP.

Molecular weight is also important for gelation. Intermolecular associations of polysaccharides, the prerequisite for gelation to occur, are stable only when the molecular chain length is long enough, typically with a DP value above 20.[24] Normally the gelling ability of the polysaccharides increases with the increase of M_w. The elastic modulus (G′, also referred to as storage modulus, reflecting the strength of the gel) increases with increasing molecular weight up to a certain point (then becoming independent of M_w), whereas rupture strength continues to rise with increasing molecular weight. It is noteworthy that for some neutral polysaccharides such as (1→3)(1→4)-β-D-glucans, lower molecular weight samples form stronger gels than higher molecular weight samples. To some extent, the gelation rate is reported to be inversely proportional to the molecular weight of the polysaccharide. When M_w is above ~500 kD, gel formation is inhibited. The mechanism proposed is as follows: Low-molecular-weight (1→3)(1→4)-β-D-glucans exhibit higher mobility in solution, which increases the probability of interactions between the molecules and hence the formation of ordered structures (junction zones).[24]

1.3.2 Charged and Neutral Polysaccharides

Some polysaccharides carry charged groups (mostly negatively charged, although some are positively charged) onto their molecular structure. The charged groups help with the solubility of polysaccharides in two ways: (1) increasing the molecular affinity to water and (2) preventing intermolecular associations due to the electrostatic effects posed by the charged group.

A relatively higher viscosity could be obtained for charged polysaccharides due to the chain extension caused by the repulsion of the charged group (*e.g.* carboxylic groups) within the molecule. Increasing the ionic strength of the solution could shield these charge effects, thus compromising the extension of the chain and therefore decreasing viscosity. However, when ionic strength reaches a critical value, the viscosity increases again due to the solvent environment change and increase of the intermolecular cross-links as well. Decreasing the pH value normally leads to a viscosity increase with negatively charged polysaccharides due to intermolecular association, and sometimes gel formation could be induced. One typical positively charged polysaccharide is chitosan, which is derived from the deacetylation of chitin. The positively charged groups (from the protonation of free amino groups) are the key to its water solubility. Chitosan is insoluble in basic environments due to the neutralization of the positive charge. However, in acidic environments, protonation of the amino groups increases the degree of water solubility. Due to this property, chitosan has been widely used for drug delivery, *e.g.* the transport of insulin.[16]

1.3.3 Branched and Linear Polysaccharides

Linear polysaccharides with highly regular conformations are mostly insoluble as they can form crystalline or partial crystalline structures due to strong intermolecular associations. Polysaccharides with branching structures demonstrate better solubility because (1) the branching structure could weaken intermolecular association due to steric effects, and (2) molecules with a branched structure exhibit smaller excluded volume and critical concentration, when compared to linear polysaccharides with the same molecular weight. However, in terms of gelling ability, molecules with a high degree of branching are somewhat prevented from forming junction zones and therefore are less likely to form gels.

Taking cellulose as an example, unmodified cellulose is essentially insoluble in aqueous media. Its solubility, however, can be highly increased by decreasing M_w and introducing either charged or branched groups to the molecule. The most commonly used modified celluloses (Figure 1.4) include methyl cellulose (MC), hydroxyl propyl cellulose (HPC), hydroxyl propyl methyl cellulose (HPMC), and carboxy methyl cellulose (CMC). CMC is both cold water and hot water soluble, while MC, HPMC, and HPC can dissolve only in cold water. MC and HPMC form gels with increasing temperature; HPC cannot form a gel while heating; instead, it precipitates when the temperature is increased above 45 °C.

Guar gum and locust bean gum both belong to the galactomannan family. The degree of branching for guar gum (galactose to mannose of 1 : 2) is higher than that of locust bean gum (galactose to mannose of about 1 : 4). In the case of guar, the higher degree of branching prevents the strong cohesion of the main backbones of different neighboring molecules, so that no extensive crystalline regions can be formed. However, locust bean gum is easier to form gels due to its lower degree of branching, which favors the formation of junction zones.

Carrageenans are a family of linear sulfated polysaccharides that are extracted from red edible seaweeds. There are three types of commercial carrageenans, including κ-carrageenan, λ-carrageenan, and ι-carrageenan, which exhibit different structural features as a result of the number and position of sulfate groups in the repeating unit of the polysaccharide (as shown in

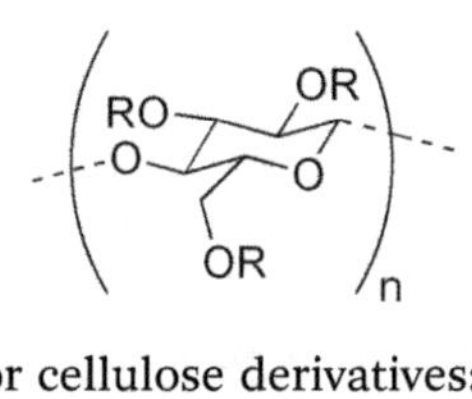

Figure 1.4 Schematic chart for cellulose derivatives: R = H or CH_3 for methyl cellulose; R = H or CH_2COOH for CMC; R = H or CH_3 or $CH_2CH(OH)CH_3$ for HPMC; R = H or $CH_2CH(OH)CH_3$ for HPC.
Adapted from ref. 16 [https://www.intechopen.com/books/solubility-of-polysaccharides] under the terms of a CC BY 3.0 license [https://creativecommons.org/licenses/by/3.0/].

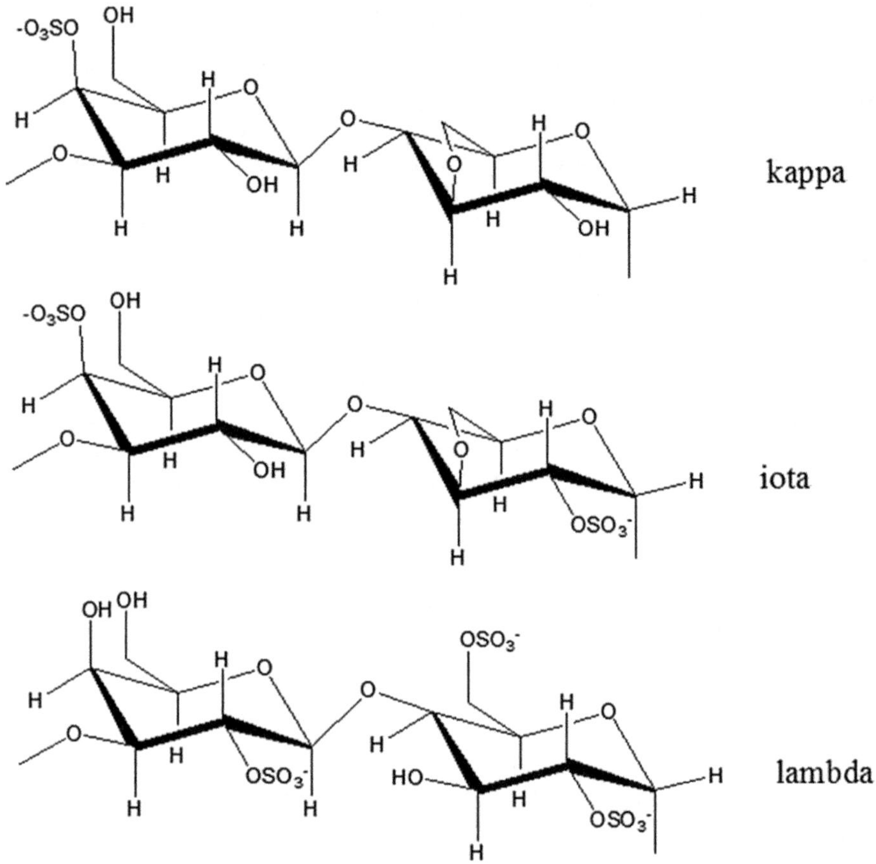

Figure 1.5 Molecular structural comparison for three carrageenan categories.

Figure 1.5). The higher levels of ester sulfate in λ-carrageenan (three sulfate groups per repeating unit) contribute to gel inhibition and result in good water solubility. With the decrease of sulfate groups in the repeating unit, carrageenan starts to form a soft gel (ι-carrageenan, two sulfate groups per repeating unit) and subsequently a rigid gel (κ-carrageenan, one sulfate group per repeating unit).

Starch contains two typical polysaccharides: amylose (mostly linear) and amylopectin (highly branched). During the gelatinization process upon heating, both amylose and amylopectin are released in solution. When cooling, molecules of amylose are much easier to cross-link with each other and form a three-dimensional network than those of amylopectin, which take a much longer time to gel; this process is called *retrogradation*.

1.3.4 Functional Groups

Similar to branching effects, the presence of some hydrophobic groups, *e.g.* *O*-acetyl (*O*-Ac) could also affect the solubility of polysaccharides. *O*-Ac substituents are present on many cell wall polymers including various hemicelluloses, the pectic polysaccharides, and the polyphenol lignin. For example, xylan, with a degree of acetylation (DA) of 2.0, only dissolves in nonpolar solvents like chloroform or polar aprotic solvents such as dimethyl sulfoxide, while the weakly acetylated xylan (DA ∼ 0.5) is totally water soluble, and the nonacetylated xylan (DA ∼ 0) is only partially soluble in hot water, due to spontaneous intramolecular hydrogen bonding.[16]

Some other special groups also highly affect the gelation of polysaccharides. For example, arabinoxylan, an important hemicellulose in cereal bran, and psyllium husk can easily form gels in the presence of ferulic acid. The gel-forming ability of arabinoxylan is highly reduced after alkaline treatment, as ferulic acid can be easily removed by alkaline treatment or extraction.[26]

1.3.5 Molecular Conformation

Conformation refers to the way that polymer chains align themselves in solution to adopt an orientation with a lower or the lowest energy. Two general types of conformation for polysaccharides, governed by the regularity of their molecular structure, can be identified: ordered conformation and disordered conformation. In aqueous solution, most of the nonstarch polysaccharides with heterogeneous structure demonstrate disordered conformation, including random coil, rigid, and spherical conformation.

If the values of the torsion angles are fixed by cooperative interactions between sugar residues, such as in solid or gel states, an ordered conformation can be adopted. Two general ordered conformations are the ribbon-like and helix conformations. Polysaccharides with ribbon-like conformation are most easily aligned and closely packed through numerous hydrogen bonds and van der Waals forces. The resultant compact structures essentially prevent solvent penetration and remain insoluble in water. Polysaccharides such as cellulose, xylan, and mannan, which contain the zig-zag types of linkages, all belong to this type. Another ordered conformation is the hollow helix conformation, in which the glycosidic bonds adopt a U-turn form, as in 3-β-D-Glc*P* (curdlan) and 4-α-D-Glc*P* (amylose from starch). Compared to the ribbon-like conformation, the hollow helix conformation has better solubility. Yet this is still not comparable to the solubility of polysaccharides with a disordered conformation (random coil); as an example, amylose, which adopts the helix conformation, is able to dissolve and form a gel in water.

Alginic acid is a copolymer of the building blocks of β-D-mannuronic acid (M) and α-L-guluronic acid (G) linked through 1–4 glycosidic bonds. The molecule contains G-blocks, M-blocks, and MG mixed regions. Due to conformational differences, molecules with a high ratio of G-block regions favor the formation of junction zones in the presence of Ca^{2+}. Therefore, high D-mannuronic acid alginates form turbid gels with low elastic moduli; in contrast, high L-guluronic acid alginates form transparent, stiffer, and more brittle gels. Detailed information regarding the gelation of alginate is provided in Section 1.4 of the present chapter.

In summary, the balance between molecule–molecule interactions and molecule–water interactions is the key to understanding polysaccharide solubility, viscosity, and gelation. In dilute solutions of polysaccharides with low M_w (intrinsic viscosity), the interactions between polysaccharide and water molecules are dominant, leading to low viscosity and normally a Newtonian flow behavior. When hydrocolloids reach the critical concentration or the molecular weight increases (intrinsic viscosity increases), molecules start to interact with one another, viscosity sharply increases, and solutions normally exhibit a pseudoplastic flow behavior. For some specific hydrocolloids with the involvement of ions, temperature change, pH change, or addition of other solutes, the interactions between polymer segments are dominant and could induce gelation when ordered molecular structures, *e.g.* junction zones, are formed. In other words, gelation occurs due to the formation of intra- and intermolecular associations, where hemiacetal oxygen and hydroxyl or methyl groups of the sugar residues of the polysaccharides contribute to hydrogen bonding or van der Waals forces of attraction.[27]

1.4 Unique Examples

As previously discussed, the many and varied molecular structures and interactions among polysaccharides can lead to unique functionalities for very specific applications.[7,8,10,14] In this section, a few of these unique functionalities and applications are presented as examples of the use of hydrocolloids in food structure development.

1.4.1 Sodium/Calcium Alginate

Alginates are structure-forming components in marine brown algae. Alginates are linear, unbranched block copolymers of (1–4)-linked β-D-mannuronic acid (M) and α-L-guluronic acid (G), distributed as long homopolymeric regions of M-blocks or G-blocks or heteropolymeric MG-, MGG- or MMG-structures, the frequency of which depends highly on source. The guluronate residues are in the $^1C_4$ conformation, while the mannuronate residues have the $^4C_1$ conformation. Since the $^1C_4$ conformation of the guluronate residues makes the glycosidic linkage diaxial in G-blocks, these have a very distinctive "eggbox" type of structure (Figure 1.6a). On the other hand, the M-block regions are characterized by diequitorial glycosidic linkages, leading to linear and extended structures (Figure 1.6a). The carboxylic acid group of both mannuronate and guluronate leads to ionized, negative charges at neutral pH, usually bound with Na^+. These two factors taken together are responsible for the strong ionic cross-linking of divalent cations with G-blocks and give rise to the well-known and highly utilized reaction between alginates (specifically the G-block regions) and calcium ions to create strong gels (Figure 1.6b).

The industrial formation of alginate gels is characterized by the control of the ionic cross-linking process. Typically, Ca^{2+} ions are introduced into

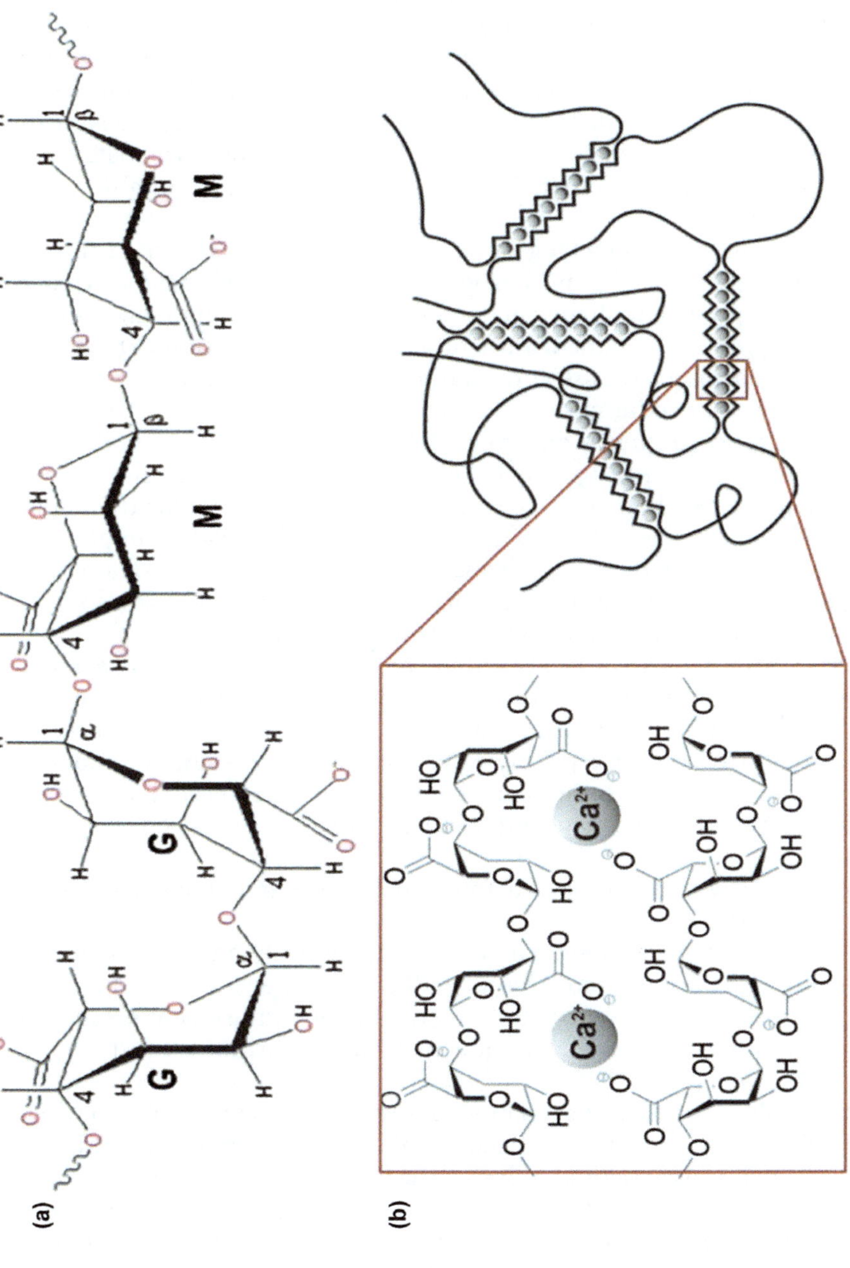

Figure 1.6 The structural units of alginate: (a) G: α-L-guluronic acid; M: (1–4)-linked β-D-mannuronic acid; (b) gel formation of alginate with calcium.
Adapted from ref. 28 with permission from Elsevier, Copyright 2015.

solutions of sodium alginate, where sodium has shielded the charges, and ion exchange occurs, leading to cross-linking of G-block regions (junction zone formation), while M-block regions or heteropolymeric regions act as structure breakers, leading to development of a three-dimensional macromolecular superstructure. The process has to be properly controlled due to the very rapid and irreversible binding reaction. Factors affecting gel formation kinetics and gel strength include the concentration, molecular structure (M : G ratio and distribution), and molecular weight of the alginate molecule being used; the source of the Ca^{2+} ion (chloride, carbonate, *etc.*); and the presence of other modifying factors such as chelators or glucono-σ-lactone (GDL).

The two primary methods for gelation are the diffusion set method and the internal set method. Diffusion setting gives rise to alginate bead or ribbon formation or specifically shaped gel particles. Typically, sodium alginate solution is dripped or streamed through a nozzle into solutions of calcium chloride. Rapid gelling occurs at the surface and diffuses inward as calcium ions exchange with sodium ions, leading to G-block cross-linking. The elasticity of the gel corresponds directly to G-block length. This method of gelation is used to create restructured beaded products like artificial berries, caviar, or gelled beads that can encapsulate other materials for the manufacture of products like fish feed. It is also used for the gelation of pimento strips, often used to stuff pitted olives, and for the formation of onion rings formulated from a slurried preparation of onions. Lately, it has also been used in molecular gastronomy to produce unique culinary dishes.[29]

Restructured foods are a good example to show how gelation is achieved. Restructured foods involve a complex assembly of raw materials and ingredients and a number of texture and structure processes. A restructured food is held together with a gelling system so that it can stand up under its own weight, for which sodium alginate is a good choice. A typical restructuring process consists of three parts:

1. A given raw material is disassembled and cut into smaller pieces of similar size.
2. The disassembled material is reassembled by intimately mixing it with sodium alginate or a solution of sodium alginate.
3. The mixture is then subjected to a treatment with a calcium salt. The reaction normally occurs at ambient temperature and pressure.

Table 1.5 shows a typical formulation for restructured onion rings, for which 1.1% sodium alginate and 3–5% calcium chloride hexahydrate are used.

The internal setting method relies on the slow release of the cross-linking ion. Either calcium carbonate or Ca^{2+} ions complexed in a chelating agent such as EDTA could be used. In either case, the release of Ca^{2+} ions is triggered by a gradual change in pH, often through a slow-release acid such as GDL. Internal set gels can be made in variously shaped molds for unique gel shapes or, once formed, can be further cut into pieces.

Table 1.5 Typical formulation of restructured onion ring. Adapted from ref. 13 with permission form John Wiley & Sons, Copyright © 2011 Blackwell Publishing, Ltd. and Institute of Food Technologists.

Phase	Ingredients	wt%
Alginate/onion	Minced, dehydrated onions	20
	Flour	14
	Salt	0.1
	Sodium alginate	1.1
	Water	64.80
	Total	100
Setting bath	Water	95–97
	Calcium chloride hexahydrate	3–5
	Total	100

1.4.2 High-/Low-methoxyl Pectin

Pectins, extracted for food use from the cell wall middle lamella of fruit waste (*e.g.* apple pomace or citrus peel), are defined as high-molecular-weight heteropolymers containing at least 65% galacturonic acid units, either free or naturally esterified with methanol. In their natural state, pectins also contain various neutral sugars, including rhamnose in the main chain and galactose, arabinose, rhamnose, and others as complex, neutral sugar side chains. The uronic acids and sugars are distributed as linear, unbranched blocks of α-1-4–linked homopolygalacturonan (HG; smooth regions) and as highly branched blocks comprised of rhamnogalacturonans (RG; hairy regions), although many of the side chains will be hydrolyzed under the hot acidic conditions of extraction. In the food industry, pectins are normally used for gelation, which is both pH and sugar-content dependent. A wide range of pectins is commercially available. Extracted pectins are normally 67–73% methyl esterified, referred to as high-methoxyl (HM) pectins, although the degree of esterification (DE) is easily reduced with processing. HM pectins require a high soluble solids (sugar) content and low pH to gel. Pectins with DE ~60% are slow-setting HM pectins, enabling gels to be prepared at higher sugar contents; these are used, for example, in sugar confectionery. Reduction in DE to <50% produces low-methoxyl (LM) pectins, which in terms of gelation are less reliant on soluble solids content and pH but more dependent on the presence of divalent cations (particularly calcium). Since pectins are widely used for food structure formation, and the resulting food microstructures are highly dependent on molecular structure and conformation, the gelation of HM *vs.* LM pectin will be described in more detail as another example of hydrocolloid use in food structuring.

With HM pectins, approximately 65% sugar and pH < 3.4 (rapid-set HM pectin) or 3.2 (slow-set HM pectin) are required for gelation. The most common applications for HM pectins are in fruit jams and jellies. The high solute content achieves a low-water-activity, hydrophobic environment that minimizes pectin-solvent interactions and promotes pectin–pectin interactions. The H^+ ions (at the low pH conditions) achieve sufficiently low ionization of the unesterified carboxyl groups to minimize electrostatic repulsions (charge neutralization). This results in junction zone formation in the HG regions and disorder in the RG regions, leading to the formation of a supermolecular three-dimensional gel structure.

With LM pectins, gelation is mainly controlled through interactions of unesterified (hence charged) galacturonic acid residues with neighbouring molecules through divalent cations (*e.g.* Ca^{2+}), forming junction zones similar in some respects to the gelation of alginates but perhaps with less well-defined "eggbox" cavities. LM pectins can gel over a wide range of sugar content, as low as 10%, and within a wide pH range, >pH 5.0, making them very suitable for low-sugar jam and jellies.

1.4.3 Methyl Cellulose

Cellulose, an insoluble and indigestible β-1-4–linked glucose polymer, can be chemically modified to make a wide range of functional food ingredients through addition of substituent groups to anhydroglucose monomers. The properties and functionality of modified celluloses depend entirely on the type and concentration of the substituent groups. Probably the most common example is carboxymethyl cellulose or cellulose gum, modified to provide water solubility and viscosity. One very unique cellulose derivative is methyl cellulose (MC) or its closely related molecule hydroxypropylmethyl cellulose (HPMC). Formation of MC replaces hydroxyl groups or the hydroxyl of carboxyl groups on the anhydroglucose monomers, with a methyl group (*i.e.* –OH converted to $-OCH_3$, or $-CH_2OH$ converted to $-CH_2OCH_3$). MC and HPMC are both soluble in cold water, implying they will hydrate well during low-temperature mixing of food formulations. However, they will both gel reversibly upon heating to above their gelation temperature (which is dependent on methyl substitution but typically >50 °C for MC and 65–80 °C for HPMC). MC produces firm gels, whereas HPMC (slightly more hydrophilic) produces softer gels. Thermogelation can be explained through dehydration of the polymer at elevated temperatures as a result of weak and reducing attractions between water and unsubstituted hydroxyl groups, followed by hydrophobic interactions between molecules.[30] Gel strength increases with heating time and temperature.

This thermogelation gives rise to the important functionalities of MC and HPMC. They serve well as binding and shape retention agents in hard-to-heat food composites that do not otherwise have good binding properties, *e.g.* reformed soy and plant-based protein products as meat analogues. In addition, they can act as barrier agents to reduce oil uptake in fried food batters, such as in onion rings or chicken nuggets. MC can also be used in bakery fillings, sauces, or other fillings within pastries, where boil-out (loss of filling due to boiling when the composite material is otherwise baked or fried, *e.g.* donut or pot pie fillings) needs to be controlled. In such cases, once thermogelation occurs, filling loss is minimized. In all of these examples, the texture upon consumption is not greatly affected by the thermogelation since it is reversible by the time the product has cooled for eating or within the mouth, although it does show considerable hysteresis from the heating curve.

1.4.4 Carrageenan (Milk Reactivity)

Carrageenans, as discussed in Section 1.3.3, are linear, negatively charged, sulfated ($-SO_3^-$) galactan polysaccharides with differing degrees of sulfation and 3,6-anhydrogalactose content (Figure 1.5). These polysaccharides are similar in backbone structure to agarose, although the latter is neutral and nonsulfated. This differing composition translates into varying functionality. κ- and ι-carrageenans are able to form gels, whereas λ-carrageenan cannot. One of the unique functionalities of κ-carrageenan specifically is its so-called milk reactivity, referring to its interaction with casein micelles, specifically κ-casein, in dairy systems. One aspect of this milk reactivity of κ-carrageenan in nongelled systems is its ability to inhibit visual phase separation between casein micelles and polysaccharides, which occurs readily due to biopolymer incompatibility. Because of the light-scattering effect of casein micelles, phase separation leads to a clear, watery, casein-depleted serum layer forming either above or below (depending on fat content and density) the decreasing-volume protein phase, which is white due to the presence of the casein micelles. Polysaccharide stabilizers often need to be added to dairy products for enhanced product functionality or shelf life. All that is required to stabilize a 4% milk protein per 0.14% locust bean gum system (typical of soft serve ice cream mix) from phase separating is 0.015–0.02% κ-carrageenan, and no other hydrocolloid has demonstrated similar functionality. Since this is a very specific, unique, and highly industrially relevant reaction, owing to the molecular structures and conformations of the two molecules involved and their interactions, it will be described in detail here as another example of hydrocolloids in food structure development.

There are two potential interactions by which κ-carrageenan could stabilize casein micelle systems containing hydrocolloids. The first proposes that the negatively charged κ-carrageenan interacts with a positively charged region of κ-casein (residues 97–112), thus adsorbing to the surface of the casein micelle. However, the surface of the micelle is overall negatively charged and stabilized sterically by the caseino-macropeptide portion of κ-casein (residues 106–169), the so-called hairy layer, which protrudes approximately 5 nm out from the micelle. Thus, for both steric and electrostatic reasons, the prevention of phase separation by direct absorption of κ-carrageenan to the micellar surface seems highly improbable. The second concept therefore proposes that phase separation is prevented *via* the formation of a weak κ-carrageenan gel, which holds the casein micelles suspended, even though concentrations of κ-carrageenan required for stability are below its critical gelling concentration.[31–33]

At elevated temperatures (>50 °C, depending on salt concentration), κ-carrageenan exists in solution as a random coil. However, when the temperature is sufficiently lowered, the polymer undergoes a transition from a coil to a helix, and κ-carrageenan gelation occurs as a result of helical aggregation. Ions also play a vital role in the formation of the κ-carrageenan helix and gelation. Cations, such as potassium and calcium, affect the transition temperature as well as aiding in helical aggregation, and certain anions, such as iodide, affect gelation by interacting with the helix and inhibiting helical aggregation, thus preventing gel formation. Within the ionic environment of milk (10.2 mM calcium and 33.3 mM potassium), κ-carrageenan is in an environment favorable for gelation, provided the polymer concentration is high enough.

Research work performed in the authors' laboratory suggests that both κ-carrageenan adsorption to casein micelle surfaces and κ-carrageenan helix–helix aggregation are required for the polymer to be effective in preventing casein micelle macroscopic phase separation from polysaccharides.[31–33] Electron microscopy and light scattering data suggest that κ-carrageenan in the helix form interacts with the surface of the casein micelle, perhaps because of its degree of sulfation and increase in charge density when in the helical form. Light and confocal microscopy showed that carrageenan-containing systems that did not exhibit visual phase separation were nevertheless phased-separated into water-in-water-type emulsions at the microscopic level. Phase separation experiments in 13% milk solids (not-fat), 0.14% locust bean gum mixed systems with κ-carrageenan in the coil state (>60 °C), κ-carrageenan in the helical state but with helix aggregation blocked (addition of NaI), λ-carrageenan, guar (all at 0.01–0.03%) and agarose (0.01–0.05%) were used to understand the conditions at which stabilization of phase separation occurred. The phase separation experiments suggest that the interaction between casein micelles and κ-carrageenan is not sufficient (by itself) to prevent macroscopic phase separation. By blocking helix aggregation, the stabilizing ability of the κ-carrageenan was reduced. In addition, to support the NaI experiments, replacing κ-carrageenan with nongelling λ-carrageenan resulted in visual phase separation even though the latter also adsorbs to the casein micelle. Conversely, agarose did not adsorb to the casein micelle, nor did it inhibit phase separation, but it did form a weak gel. Thus, this research suggested that both κ-carrageenan–casein micelle interaction and κ-carrageenan helix aggregation are required to prevent casein micelle–polysaccharide macroscopic phase separation and provide visual stability. Based on these results, it is believed that κ-carrageenan functions at the periphery of microscopically phase-separated discrete domains of casein micelles, stabilizing these (microdomains) and preventing them from coalescing and forming visual protein-enriched and protein-depleted phases.

1.4.5 Gum Arabic

Gum arabic is a good example of the use of a hydrocolloid as an emulsifying agent. Its specific use in this regard is for stabilization of flavor (citrus or cola) oil emulsions for carbonated beverage applications, since it is stable in acidic conditions, and protein emulsifiers typically are not. The polymer is able to inhibit flocculation and coalescence of oil droplets in both the concentrated flavor oil emulsion and in the diluted carbonated beverage over many months, giving it unique stabilizing properties for this application. This structuring property is again owing to its molecular structure and conformation.

Gum arabic is an exudate gum from *Acacia senegal* or *Acacia seyal trees* from Sudan, Nigeria, and Chad. The gum contains two main fractions, arabinogalactan (AG) and arabinogalactan protein (AGP), and it is the protein component that is the key to its emulsifying properties. The polysaccharide component of both fractions is comprised of a β-1-3-linked D-galactose backbone with extensive branching of mostly arabinose, galactose, and some rhamnose and glucuronic acid (~15%).[34] AG takes on the conformation of a thin disk-like structure. AGP has a protein content of ~10%, and its backbone is a polypeptide chain of ~250 amino acids. The AG-like polysaccharide disks of the AGP molecule, with a molecular weight of ~40 000 each, are attached to this protein backbone (Figure 1.7). The entire structure takes on a very compact spherical conformation. Therefore, solutions of gum arabic are very low in viscosity; a 30% gum arabic solution has a viscosity of ~300 mPa s, which is less than that of a 1% sodium carboxymethyl cellulose solution. When used for emulsification, the protein backbone is thought to adsorb to the interface of the oil droplet while the polysaccharide disks stick out into the aqueous phase, providing stability through both steric and electrostatic repulsion (from the glucuronic acids). Emulsification is achieved at fairly high gum arabic concentration due to the low protein content; for example, ~12% gum arabic is required to stabilize a 20% oil emulsion. However, the high surface coating of the oil droplet by the hydrophilic polysaccharide disks gives rise to a very stable emulsion microstructure. While some other gums have shown to provide a reduction in interfacial tension, this is usually associated with some coextracted protein content. However, in the case of gum arabic, the protein is covalently linked to the polysaccharide, making it unique and giving rise to its exceptional food structuring ability.

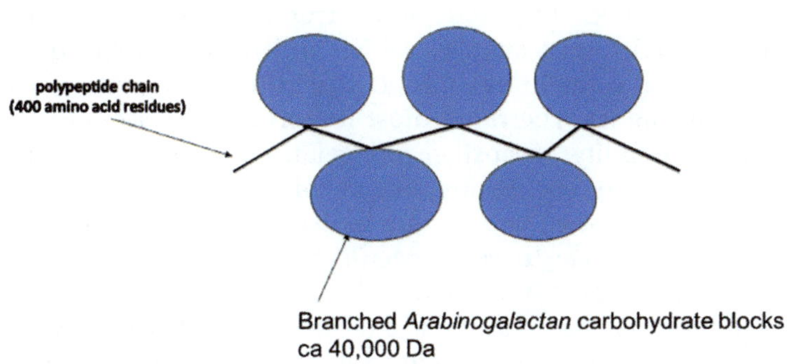

Figure 1.7 Schematic illustration of the Wattle blossom model of gum arabic.

1.4.6 Xanthan

A final example of connecting molecular structure closely to functionality and food structure is the high degree of pseudoplasticity exhibited by xanthan gum. Xanthan is a bacterial polysaccharide consisting of a cellulose backbone substituted at C-3 on alternating anhydroglucose units with a trisaccharide side chain containing two mannose units and a glucuronic acid unit, although there are several substituents on all three side chain residues. The molecular conformation is what gives xanthan its unique properties. It shows a fivefold right-handed helix such that the trisaccharide chains are aligned with the cellulosic backbone and stabilize it primarily through hydrogen bonding. This gives the molecule the structure of a rigid rod when in solution.[35] As a result, xanthan demonstrates high viscosity at low shear rates, low concentrations, and a high degree of pseudoplasticity and thixotropy. A 0.5% xanthan gum solution can exhibit a reduction of 4–5 log cycles in viscosity within a range of shear rates from 10^{-2} to 10^{3} s^{-1}. Xanthan solutions can appear almost gel-like but pour readily, thus providing long-term stability to colloidal systems. A common and demonstrable application is in salad dressings, where its unique functionality gives rise to the ability to be shaken and poured easily but show high holdup on the salad. It can also inhibit sedimentation of particulates in many low-viscosity fluids. The viscosity of xanthan solutions is stable over a wide temperature and pH range. Simplistically, it may be considered that the initially entangled rigid xanthan rods are encouraged to progressively align under the applied shear fields, which gives rise to such rapid drops in viscosity with increasing shear rates. However, the association between molecular structure and solution rheology may be more complex, in that xanthan dispersions have been shown to contain weakly associated microgels or xanthan aggregates due to incomplete conformational ordering of helical sequences,[36] and these aggregates may give rise to a dispersion more akin to a weak gel when at rest.

1.5 Conclusions and Future Perspectives

Hydrocolloids have been important food structuring agents for many years in many traditional foods, such as jellies and pies, and continue to play important and vital roles in novel food structure development, such as in films and encapsulants. The three most important hydrocolloid functional properties are solubility, viscosity, and gelation, which result from their molecular structures and conformations in solution, as was demonstrated in Section 1.3. It is very often the unique molecular structures of specific hydrocolloids that give rise to very specific food applications, as has been demonstrated in Section 1.4.

One of the big challenges for the use of hydrocolloids in today's food and consumer environment is the demand for natural, clean-label ingredients. While most of the hydrocolloids presently in use are from a natural stock, many of them are extracted or modified by industrial means that do not meet clean-label perceptions. Future developments in this area will likely produce an expanded array of hydrocolloid food ingredients, either from new and novel, untapped sources, many of these from agricultural by-product streams or from process modifications of existing products to render them more natural and/or environmentally friendly. Considerations for future hydrocolloid ingredients will include functionality, cost, and human safety, as has always been the case, but in addition the source and supply, including its carbon footprint, environmental impact, and "naturalness," will be increasingly scrutinized by discerning customers who want to know where their food comes from and how it is processed.

It is also worth noting that most nonstarch polysaccharide hydrocolloids are also dietary fibers, and a great deal of new interest and knowledge surrounds the bioactivity and health-promoting properties of dietary fiber. It is well recognized that most diets in the developed world do not contain enough specific fibers to obtain their full potential, so fiber-fortified foods are increasingly being developed and sought after. Ideal specific hydrocolloids will deliver both biological and physical functionality to food systems for appropriate structure development – enough of one without compromising the other.

References

1. E. Dickinson, *An Introduction to Food Colloids*, Oxford University Press, Oxford, 1992.
2. I. A. Brownlee, *Bioact. Carbohydr. Diet. Fibre*, 2014, **4**, 155.
3. P. I. Chater, M. D. Wilcox, J. P. Pearson and I. A. Brownlee, *Bioact. Carbohydr. Diet. Fibre*, 2015, **6**, 117.
4. S. Gill, P. I. Chater, M. D. Wilcox, J. P. Pearson and I. A. Brownlee, *Bioact. Carbohydr. Diet. Fibre*, 2018, **16**, 62.
5. K. N. Prasad and S. C. Bondy, *Bioact. Carbohydr. Diet. Fibre*, 2019, **17**, 100170.
6. C. G. Biliaderis and M. S. Izydorczyk, *Functional Food Carbohydrates*, CRC Press, Baco Raton, FL, 2007.
7. S. W. Cui, *Food Carbohydrates*, Taylor & Francis, Boca Raton, FL, 2005.
8. A. M. Stephen, G. O. Phillips and P. A. Williams, *Food Polysaccharides and Their Applications*, CRC Press, Baco Raton, FL, 2nd edn, 2006.
9. V. J. Morris, in *Understanding and Controlling the Microstructure of Complex Foods*, ed. D. J. McClements, CRC Press, Baco Raton, FL, 2007, 1, 1.
10. G. O. Phillips and P. A. Williams, *Handbook of Hydrocolloids*, CRC Press, Baco Raton, FL, 2nd edn, 2009.
11. G. O. Phillips and P. A. Williams, *Handbook of Food Proteins*, Woodhead Publishing Ltd., Oxford, UK, 2011.
12. A. Imeson, *Food Stabilisers, Thickeners, and Gelling Agents*, Wiley-Blackwell, Ames, IA, 2010.

13. T. R. Laaman, *Hydrocolloids in Food Processing*, Wiley-Blackwell, Ames, IA, 2011.
14. A.-C. Eliasson, *Carbohydrates in Food*, CRC Press, Boca Raton, FL, 3rd edn, 2016.
15. V. De Simone, A. Dalmoro, G. Lamberti, D. Caccavo, M. d'Amore and A. A. Barba, *J. Drug Delivery Sci. Technol.*, 2019, **49**, 513.
16. M. Q. Guo, X. Hu, C. Wang and L. Ai, in *Solubility of Polysaccharides*, ed. Z. Xu., IntechOpen, Winchester, UK, 2017, p. 2.
17. K. Y. Qian, S. W. Cui, Y. Wu and H. D. Goff, *Food Hydrocolloids*, 2012, **28**, 275.
18. M. L. Huggins, *J. Amer. Chem. Soc.*, 1942, **64**, 2716.
19. E. O. Kraemer, *Ind. Eng. Chem.*, 1938, **30**, 1200.
20. H. Mark, in *Der festeKörper*, ed. R. Sänger, Hirzel, Leipzig, 1938, pp. 65–104.
21. R. Houwink, *J. Prakt. Chem.*, 1940, **157**, 15.
22. D. Saha and S. Bhattacharya, *J. Food Sci. Technol.*, 2010, **47**, 587.
23. D. Oakenfull, *Crit. Rev. Food Sci. Nutr.*, 1987, **26**, 1.
24. W. Qi and S. W. Cui, in *Food Carbohydrates: Chemistry, Physical Properties, and Applications*, ed. S. Cui, CRC Press, Boca Raton, FL, 2005, pp. 161–218.
25. R. L. Whistler, in *Carbohydrates in Solution*, American Chemical Society, 1973, pp. 242–255.
26. J.-Y. Yin, S.-P. Nie, J. Li, C. Li, S. W. Cui and M.-Y. Xie, *J. Agric. Food Chem.*, 2012, **60**, 7981.
27. M. Tako, *Adv. Biosci. Biotechno.*, 2015, **6**, 22–36.
28. M. Bruchet and A. Melman, *Carbohydr. Polym.*, 2015, **131**, 57.
29. N. Caporaso and D. Formisano, *Food Rev. Int.*, 2016, **32**, 417.
30. S. A. Arvidson, J. R. Lott, J. W. McAllister, J. Zhang and F. S. Bates, *et al.*, *Macromolecules*, 2013, **46**, 300.
31. P. A. Spagnuolo, D. G. Dalgleish, H. D. Goff and E. R. Morris, *Food Hydrocolloids*, 2005, **19**, 371.
32. C. Vega and H. D. Goff, *Int. Dairy J.*, 2005, **15**, 249.
33. C. Vega, D. G. Dalgleish and H. D. Goff, *Food Hydrocolloids*, 2005, **19**, 187.
34. S.-P. Nie, C. Wang, S. W. Cui, Q. Wang, M.-Y. Xie and G. O. Phillips, *Food Hydrocolloids*, 2013, **31**, 42.
35. H. Habibi and K. Khosravi-Darani, *Biocatal. Agric. Biotechnol.*, 2017, **10**, 130.
36. H. R. Badwaik, T. K. Giri, K. T. Nakhate, P. Kashyap and D. K. Tripathi, *Curr. Drug Delivery*, 2013, **10**, 587.

CHAPTER 2

The Role of Proteins in the Development of Food Structure

VALÉRIE GAGNAIRE, VALÉRIE LECHEVALIER, MARIE-HÉLÈNE FAMELART, THOMAS CROGUENNEC AND SAÏD BOUHALLAB*

UMR 1253, STLO, INRA, Agrocampus Ouest, 65 rue de Saint-Brieuc, F-35042, Rennes, France
*Email: Said.bouhallab@inra.fr

2.1 Introduction

Proteins are known to have major structural and catalytic functions in many biological processes. In the food sector, beyond their roles in nutrition, bioactivity, and health, food proteins are among the essential dietary components with a significant effect on food structure and quality. They play a key role as food additives and ingredients in the structuration/texturing of many food products by virtue of their ability to form a variety of gels and to stabilize multiphasic systems such as emulsions and foams. The ability of a given protein to express such functional properties depends on its specific spontaneous or induced interactions with water, other proteins, or other food components.

Proteins comprise specific chains of amino acids that fold into unique native structures arranged in specific spatial conformations driven by a number of covalent and noncovalent interactions with various energies: *e.g.* hydrogen bonding (2–30 kcal mol^{-1}), electrostatic (1–20 kcal mol^{-1}), van der Waals forces (0.1–1 kcal mol^{-1}), hydrophobic packing (1–3 kcal mol^{-1}),

Food Chemistry, Function and Analysis No. 18
Handbook of Food Structure Development
Edited by Fotis Spyropoulos, Aris Lazidis and Ian T. Norton

Published by the Royal Society of Chemistry, www.rsc.org

disulfide bonds (80–100 kcal mol^{-1}), *etc.* Among the electrostatic interactions, the coulombic interactions between opposite charges, cation–π interactions, and van der Waals interactions play important roles in defining the stability of proteins and their subsequent assemblies. The hydrophobic effect and hydrogen bonds are also of prime importance, and water is a major factor that contributes to protein structure, denaturation, and stability, as well as to the aggregation pathways.[1] Hydrogen bonding occurs when two electronegative atoms, such as nitrogen and oxygen, interact with the same hydrogen. The water-binding capacity of food proteins is a measure of the total amount of water that can be absorbed per weight of a protein powder. This property is based on the direct interaction of protein molecules with water. The ability of food proteins to entrap and retain water is generally improved after thermal protein denaturation, despite the fact that this may cause a loss in terms of solubility.[2]

Protein interaction with water is important not only for determining protein solubility but also because it governs the ability of these species to form network structures, such as gels or interfacial films that stabilize foam bubbles.[3] Water is therefore fundamental because of its role in defining hydrophobic attractions. Hence, the noncovalent forces that drive the properties of food proteins are strongly dependent on the water molecules that form the solvating medium in most foods. Covalent interactions, *i.e.* S–S bonds, generally occur after induced protein denaturation (*e.g.* as a response to heat treatment or high pressure), leading to irreversible associations between the unfolded protein molecules.

According to the balance between these interactions and depending on pH, ionic strength, and intermolecular interactions, protein molecules may self-associate to form dimers and higher-order oligomers and supramolecular structures with various shapes and sizes. The amino acid sequences generally remain unchanged during denaturation, except under drastic conditions, but changes in the secondary and tertiary structures increase protein interactions and favor aggregation due to the exposure of hydrophobic amino acids on the protein surface.[4]

Owing to their versatile properties, there is an increasing demand for a better understanding of food protein assemblies and their subsequent relationships with the functional properties of the formulated food product. The development of food structure is a compromise of balanced interactions among proteins, protein–water, protein–other components, all driven by competition or displacement phenomena. Competition and displacement are thermodynamically under the control of relative affinities among the involved components, including water. In other words, the contribution of a given protein to food structure development depends on its structural properties and on how these properties are affected, modulated, and oriented under the influence of interactions with other components during food processing. These interactions between proteins and surrounding environmental components, in particular water, can be shifted by chemical modifications that change the size, conformation, and functional group and charges of the proteins. The global scheme behind the process-structure–property interplays is given in Figure 2.1.

Proteins are classified as structured globular proteins, membrane proteins, and unstructured or fibrous proteins. Many recent works relate the roles of proteins of various origins in the development of food structure, including plant and cereal proteins,[6,7] soy proteins,[8] meat and muscle proteins,[9] and seafood proteins.[10] This chapter will focus on proteins from milk (globular proteins and unstructured caseins) and from egg white (globular proteins). These proteins constitute a good example to illustrate the role of proteins in the development of food structure as a whole. They are extensively and commonly used not only to stabilize oil-in-water emulsions and

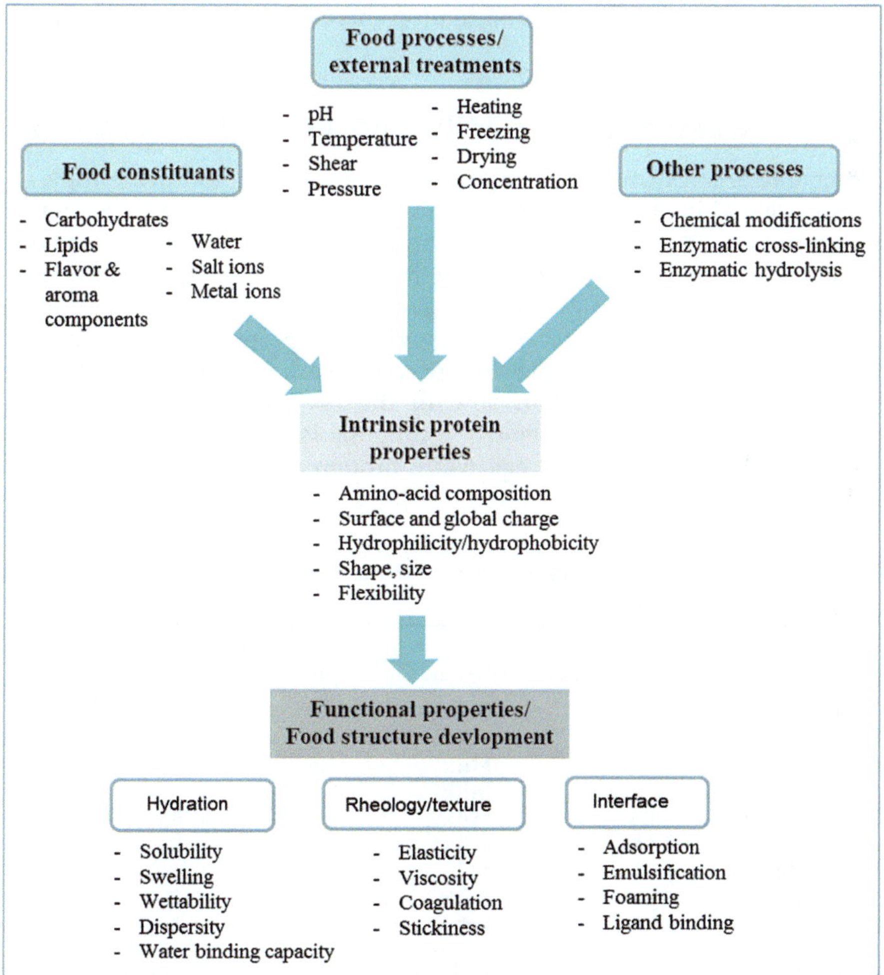

Figure 2.1 Factors affecting the functional properties of proteins and their subsequent contribution to food structure.
Adapted from ref. 5 with permission from Elsevier, Copyright 2018.

foams, given their great adsorption ease at the oil–water and air–water interfaces (covered in more detail in other chapters in this book), but also to control and diversify food texture.

We will first describe the current knowledge on the physicochemical data of milk and egg white proteins and then the subsequent diversity of assemblies that can be obtained during their processing and controlled structuration. In a final section, through the three examples provided, we will focus on protein gels, illustrating how the sol–gel transition of processed proteins contributes to the development of "real" food. Gelation is a protein functionality of great importance in foodstuffs as it affects not only their elasticity and texture but also their overall characteristics, including organoleptic ones. Protein gelation can be induced by the addition of heat, chemicals, or enzymes[11] or a combination thereof, as we will see in the next section.

2.2 Characteristics of Milk and Egg White Proteins

The characteristics of the major identified proteins from bovine milk and chicken egg white are presented in Table 2.1. The total protein content of bovine milk is 32–35 $g L^{-1}$, and its natural pH is 6.5–6.7. Milk is also an emulsion of fat globules that will not be considered in this chapter. Milk proteins are divided into two distinct groups: caseins, which belong to the unstructured protein group, and whey globular proteins. Egg white is an aqueous, viscous solution of globular proteins, carbohydrates, and minerals. Proteins constitute more than 90% of the dry matter of the egg white solids. The total number of egg white proteins is not precisely known, and until recently, only the main egg white proteins have been identified[12–14] (Table 2.1).

Table 2.1 Physicochemical characteristics of the main protein from bovine milk, milk serum, and egg white.

Protein	Abbreviation	Origin	Isoelectric point	Molecular weight, M_w (kDa)
α_s-Casein	α_s-CN	Milk	5.0	24
β-Casein	β-CN	Milk	5.4	24
κ-Casein	κ-CN	Milk	5.6	19
Lactoferrin	LF	Milk	8.5	83
α-Lactalbumin	α-LA	Milk	4.7	14.2
β-Lactoglobulin	β-LG	Milk	5.2	2×18.3
Serum albumin	BSA	Milk	5.0	66.3
Osteopontin	OST	Milk	3.6	36
Ovalbumin	OVA	Egg white	4.5	45.5
Ovomucoid	—	Egg white	5.1	28
Ovomucin	—	Egg white	4.5–5.0	>230
Ovotransferrin	Ovo	Egg white	6.7	76
Lysozyme	LYS	Egg white	10.7	14.3
Avidin	AVI	Egg white	9.5	4×14.3

2.2.1 Caseins

The four main caseins (α_{s1}-, α_{s2}-, β-, and κ-caseins, abbreviated to CN hereafter) account for 80% of milk proteins and are present at ~10, 2.6, 9.3, and 3.3 $g L^{-1}$, respectively.[15] Some of their characteristics are given in Table 2.1. They are responsible for most of the milk transformations to dairy gels, such as cheeses and yogurts. Caseins are unstructured proteins that contain some secondary organization but mainly random coil structures. Due to their high number of proline residues in the casein sequence (17, 10, 35, and 20 for α_{s1}-, α_{s2}-, β-, and κ-CN, respectively), they are not well-structured proteins and cannot therefore be crystalized. Part of the serine residues are phosphorylated.[16] The α_{s2}- and κ-CN both contain 2 cysteines, mostly connected in intramolecular disulfide bonds.[17] κ-CN contains glycans in its C-terminal section, among them the negatively charged sialic acid. In milk, caseins are natively assembled in an original quaternary suprastructure, confusingly called *casein micelle*, and are negatively charged. The average size of the casein micelle is 130–160 nm.[16] Its structure has long been a subject of controversy, with different proposed models.[18–20] Caseins are phosphorylated proteins that interact strongly with multivalent cations. Moreover, as they present some hydrophobic patches, they also self-assemble, depending on ionic strength, negative charge, and consequently pH conditions.

The casein micelles are spherical, highly hydrated (4 g water per gram protein), and sponge-like objects with a large diameter size distribution (20–600 nm). They are stable colloidal macromolecular structures in milk with continuous and dynamic exchange with the soluble phase. Their integrity is based on many kinds of interactions involving hydrophobic effect, steric repulsions, ionic (calcium bridges) and electrostatic interactions, hydrogen and disulfide bonds. As hydrophobic interactions decrease at low temperature, caseins (mainly β-CN) can diffuse in the serum phase of cooled milk. However, caseins are mainly present in the casein micelles, although a small part of them are present as soluble casein assemblies. κ-CN is mainly located at the surface of the micelle, with its negatively charged C-term chain protruding as a 5- to 10-nm hairy layer that is responsible for the electrostatic and steric stabilization of the overall casein micelles. Milk minerals (mainly calcium, magnesium, and phosphate) and citrate also partition between whey and the colloidal phase according to complex equilibria, depending on complex salt solubility, association constants, and milk physicochemical conditions (pH, ionic strength, temperature, among the most prevalent parameters). Some calcium is directly bound to phosphoserine residues, and part of the calcium phosphate of milk also interacts with phosphoserines as mineral nanoclusters, called *colloidal calcium phosphate* (CCP).[21]

Casein micelles are able to form acid gels, *i.e.* yogurts, enzymatic rennet gels, and cheeses; see examples in Section 2.4 of the present chapter. The ability of casein micelles to form gels is largely dependent on the various processes used and gives the micelles a high propensity to reorganize in order to sustain such changes. These processes induce the destabilization of the overall micellar structure by (1) physicochemical changes, when pH drops during acidification and temperature changes and mechanical action are applied; (2) enzymatic cleavage of the κ-casein, when coagulation enzymes are used such as chymosin or rennet; and (3) a combination of both. When the milk is acidified, the CCP is progressively solubilized, which is complete at pH 5.2.[22] The caseins are further released below pH 5.1, when electrostatic and steric repulsion between the micelles is no longer sufficient to inhibit aggregation and this in turn leads to precipitation or gelation and the complete destructuring of the casein micelles into soluble caseins named caseinate.[23] The strength of the electrostatic and hydrophobic interactions between caseins is governed by the residual charges on the casein molecule according to the processes applied, the type of caseins, and the temperature of the process used.[24]

Casein micelles within the range of 25–160 $g\,L^{-1}$ are able to gel between pH 5.2 and 6.7 above a critical temperature (T_{crit}) that decreases with decreasing pH value and increasing casein concentration.[23] Above this T_{crit}, the rate of gelation no longer depends on the temperature of gelation. The stiffness of the gel increases with the increase in the casein concentration but does not depend on the pH and the temperature used to induce the gelation. The mechanism is different for the caseinate gelation, as the rate of gelation is much lower and increases with decreasing pH due to weaker electrostatic repulsions between proteins. Gelation begins above 60 °C and does not exhibit a T_{crit} value. The stiffness of the gels increases with decreasing pH values. These casein gels are probably formed by hydrophobic interactions, as their rate of formation increases with increasing gelation temperatures, but hydrogen bonds are also shown to be involved in their strengthening. Differences between micellar casein and caseinate gelation are thought to be due to either the micellar structure or the presence of calcium phosphate in the micelle.

2.2.2 Globular Proteins

Whey proteins represent 20% of total proteins in milk; the main ones are β-lactoglobulin (β-LG, 3.2 $g\,L^{-1}$) and α-lactalbumin (α-LA, 1.2 $g\,L^{-1}$).[15] These two proteins are acidic and highly important as they drive the functional properties of whole whey protein concentrates and isolates used in food formulations. Lactoferrin is the major basic protein present in bovine milk (0.1–0.2 $g\,L^{-1}$) (Table 2.1).[25,26]

Egg white proteins have an acidic isoelectric point apart from lysozyme and avidin. They are all glycoproteins except for lysozyme and are rich in sulfur amino acids. Many of these proteins are very heat sensitive and/or highly susceptible to surface denaturation, thus contributing to the remarkable technofunctional qualities of egg white. Eight of the eleven major globular proteins from milk and egg white exhibit an acidic character and consequently are negatively charged at physiological pH. Processing of these proteins in pure or mixed systems allows for the production of a diversity of protein assemblies that contribute to the development and enhancement of food macrostructures.

2.3 Diversified Supramolecular Assemblies from Processed Globular Proteins and Potential Applications in Food

Depending on the physicochemical conditions and processing forces, several supramolecular assemblies are obtained from processing food globular proteins. The structure (shape, size, density) of such assemblies may be tuned by controlling parameters such as protein concentration (below the critical concentration of gelation), temperature and heating time, pH, and the presence of small solutes.[27] Figure 2.2 illustrates the main protein assemblies that are described in literature, but presently only few of them are encountered in food products (microparticulated aggregates, fractal aggregates, microgels). The processing conditions required for their formation and their potential food applications are summarized in Table 2.2.

2.3.1 Assemblies Obtained by Tuning the Protein Environment

Globular proteins, native or modified in their structure (hydrolyzed[36]), are able to interact and form various supramolecular structures by changing the chemical conditions of the protein solutions (pH, type and concentration of salts).

2.3.1.1 Heteroprotein Complex Coacervates

Heteroprotein complex coacervates (HPCC) are formed within a very narrow pH range, at low ionic strength, and for specific protein stoichiometry through the interaction and spontaneous assembly between two oppositely charged proteins (Figure 2.2I). Complex coacervation is a liquid–liquid phase separation that starts by the formation of a microphase separation with the formation of droplets of about 3–5 μm, followed by their coalescence, resulting in a macrophase separation. In contrast to protein assemblies induced by drastic conditions, the mild conditions involved in this case are advantageous as they give rise to the possibility of producing reversible supramolecular assemblies, since only weak noncovalent interactions are involved. This offers a real opportunity for the control of both the assembly process of proteins into nano- and microstructures and of the

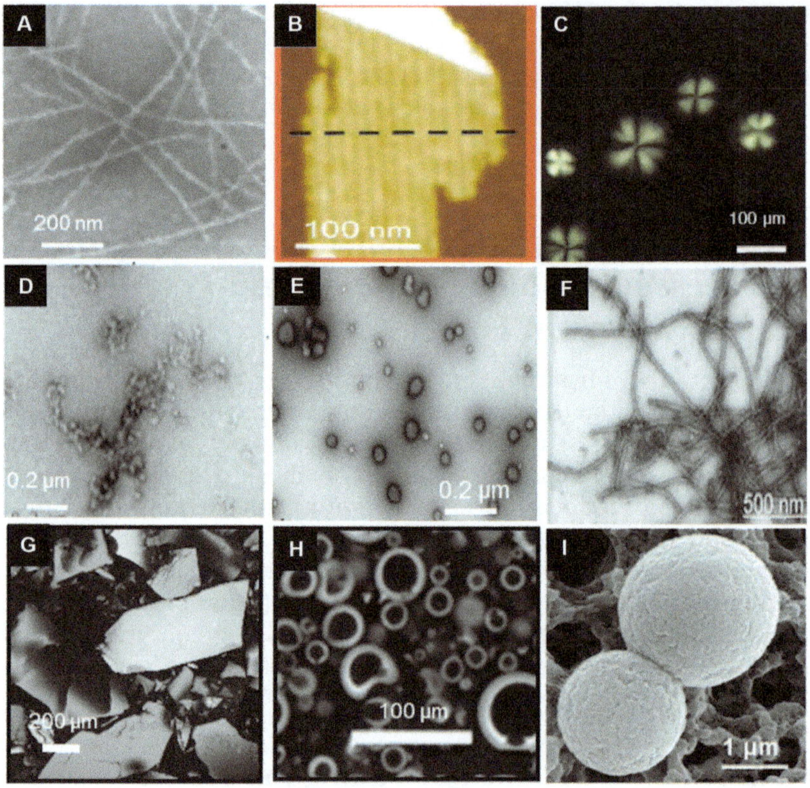

Figure 2.2 Diversity of supramolecular structures reported in the literature after processing of milk or egg proteins: (A) fibrils from heated mixture of β-LG and α-LA at pH 2;[28] (B) ribbons from heated β-LG at pH 2;[29] (C) spherulites of β-LG;[30] (D) fractal aggregates from heated β-LG at neutral pH;[31] (E) microgels from heated β-LG at pH 5.8 (personal data); (F) nanotubes from hydrolyzed α-LA;[32] microparticles of freeze-dried (G) or spray-dried (H) whey protein isolate after dry heating at alkaline pH;[33,34] (I) heteroprotein complex coacervates of lysozyme/α-LA.[35] (A, D, E, F): transmission electron microscopy; (B): atomic force microscopy; (C): polarized light microscopy; (G, H): confocal laser scanning microscopy; (I): scanning electron microscopy.

disassembly process. The formation of HPCC has been described for various mixtures of oppositely charged milk and egg proteins, including, LF/β-LG, LYS/α-LA, OVA/LYS, and AVI/OVA.[37] Electrostatic attraction between the two proteins, associated with an entropic contribution resulting from the release of counter-ions and water molecules, constitutes the main driving force for the initial contacts leading to coacervation.

Because of their characteristics, heteroprotein complex coacervates have also been considered for the encapsulation and the targeted delivery of bioactive compounds.[38,39] The complex coacervation is also a means of preparing translucent and highly concentrated protein solutions

Table 2.2 Process-driven aggregation of food proteins: processing conditions and potential applications for food structure development.

Supramolecular shape	Process conditions	Potential use for food structuring
Fractal aggregates	Heating, pH 7	Texture; water retention; interfacial stabilization
Microgels	Heating, pH 5.8	Texture/water retention
Fibrils	Heating, pH 2	Viscoelasticity
Nanotubes	Enzymatic hydrolysis	Encapsulation
Whey protein microparticles	Freeze-drying/dry heating/ pH > 9	Texture; water retention
Whey protein microspheres	Spray-drying/dry heating/ pH > 9	Encapsulation; texture
Complex coacervates	Intermediate pH between two opposite isoelectric points	Encapsulation

(~300 g L^{-1}). The coacervates exhibit a much higher viscosity than a globular protein solution, at a similar total concentration.

2.3.1.2 Nanotubes

Tubular strands of about 20 nm in diameter and several micrometers in length, called *nanotubes*, were successfully formed from partially hydrolyzed α-LA molecules, the second most abundant whey protein of bovine milk (Figure 2.2F). The nanotubes were formed in a narrow range of protein concentrations and require precise solution conditions: neutral pH, calcium concentration, hydrolysis conditions.[32] A shift of the optimal conditions and the presence of other proteins inhibit the self-assembly process into nanotubes and rather leads to the formation of random aggregates.

The nanotubes exhibit features that could lead to promising future food applications, namely their hollow elongated structure and their stability under processing; they could, for example, withstand conditions similar to pasteurization. Because of their elongated structure, these nanotubes form strong gels with a high storage modulus even at a weight protein fraction as low as 3%. The reversibility and disassembly properties of these nanotubes provide other potential applications, such as encapsulation and controlled release of bioactive components.[32]

2.3.2 Assemblies Obtained on Heating Protein Solution

Globular proteins are particularly sensitive to heat. Depending on the heating conditions below the critical concentration of gelation, solutions of proteins are able to generate various supramolecular structures: fractal aggregates, microgels, fibrils, multistranded ribbons, and spherulites. Protein heating in a dry state is another means to create original structures, called *protein microparticles*. Microparticulated proteins are classically obtained by a combination of heat and shear treatments.

2.3.2.1 Microgels

Globular proteins (BSA, β-LG, LYS, *etc.*) are able to form irreversible, well-defined, particulate aggregates of several tenths to hundreds of nm when heated close to their pI (pH = 5.7–5.9)[40] or in the presence of salts ($CaCl_2$).[41] Hydrophobic interactions constitute the main driving force for protein assembly in this case. These particulate aggregates self-assemble when the electrostatic repulsions are too low for their stabilization in solution.[42] Particulate aggregates are also formed by heating a mixture of β-LG and α-LA (ratio 80 : 20) at pH 5.8 (Figure 2.2E). At this pH, both proteins are negatively charged, but it was suggested that the presence of α-LA in the aggregates reduces the hydrophobic attractive interactions for the formation of round and dense particles. A suspension of narrowly spherical aggregates, with a core–shell structure of diameter around 100 nm, is also obtained by heating a mixture of oppositely charged proteins (OVA/LYS or Ovo/LYS).[43] Stable spherical nanoparticles of LYS and β-CN are obtained by heating the protein mixture under specific pH conditions and in appropriate ratios.[44] The low electrostatic repulsion between proteins during the heat-induced assembly process favors the growth of the aggregates into spherical structures.[45] Then particulate aggregates are formed close to the protein pI or in the presence of salts due to the screening of exposed ionized groups. pH, ionic strength, hydrophilic/hydrophobic balance in the polypeptide chain, protein concentration, heating temperature ramp. and heating time are all factors tuning the size of the formed particulate aggregates.[42,46]

The microgels are stable to sedimentation and can withstand further heat treatments even at high protein concentration at various physicochemical conditions, while the heating of the same proteins in their "native state" leads to a gel. This is of particular interest for high protein drinks[47,48] that have to be heat-treated to ensure microbial safety. Moreover, the microgels irreversibly anchor at the interface of oil droplets and air bubbles. This gives exceptional emulsion stability against coalescence[49] and delays liquid drainage in foams as well as the disproportionation of air bubbles.[50] Microgels can form cold-gels with $CaCl_2$ addition, but gels show low stiffness, high heterogeneity, and syneresis. The $CaCl_2$ content influences the rate of gelation but does not affect much the stiffness of the final gels; nonetheless, gel stiffness was shown to decrease with an increase in the size of the microgels between 140 and 370 nm.[51]

2.3.2.2 Fractal Aggregates

Fractal aggregates (Figure 2.2D) of different sizes and density have been produced by heating β-LG at 80 °C in the pH range 6–6.8.[52] Experiments on β-LG indicate that proteins unfolded (by heat treatment) first aggregate into oligomers and then into small soluble aggregates of less than 50 nm in length and 10 nm in width, called *primary aggregates*.[53,54] Upon prolonged heating and when protein concentration is high enough, these primary aggregates[54] further associate, forming self-similar or fractal aggregates.[55] The size and the density of the fractal aggregates mainly depend on protein concentration, pH, and ionic strength. The density of the fractal aggregates increases by decreasing the pH from 6.8 to 6.0 or by increasing ionic strength, due to a higher degree of ramification. Suspensions of the fractal aggregates were transparent because smaller aggregates were embedded within the larger assemblies in a hierarchical manner, with local electrostatic repulsions.[56]

Fractal aggregates also form cold-gels with acid (HCl) or $CaCl_2$ addition.[51,57] For both gels, the speed of gelation increases strongly with increasing temperatures, but the stiffness of the final gels is unchanged with reference to the gelation temperature. Though the gelation mechanisms are different, the stiffness of both gels are of the same order of magnitude; however, acid-induced gels show more syneresis than $CaCl_2$-induced gels. The rate of acid-induced gelation increases strongly with increasing HCl content, but excessive amounts of HCl, corresponding to a net protein charge density (α) lower than -2.5, produce only turbid suspensions with a high degree of syneresis. The stiffness of the final gels was unchanged with aggregate sizes between 35 and 200 nm but significantly increased with the aggregate concentration between 30 and 60 $g\,L^{-1}$.[57] $CaCl_2$-induced gels show high stiffness, low heterogeneity, and syneresis. The content in $CaCl_2$ that is required to induce the sol–gel transition decreases with the size of aggregates, and increasing the content in $CaCl_2$ favors gel syneresis. As for microgels, increasing the content in $CaCl_2$ only increases the speed of gelation but does not change the final gel stiffness. The gel stiffness also increases with fractal aggregate content between 7 and 60 $g\,L^{-1}$ in a log–log relationship and independently of the aggregate size.[51]

Mixtures of sodium caseinate and fractal aggregates also form acid- or $CaCl_2$-cold gels.[58] The addition of caseinate does not change the microstructure or the stiffness of the gels but slows down the gelation rate due to competition between the two protein species for calcium binding.

Given their specific properties, fractal aggregates and microgels from whey proteins can be used as substitute texturing agents instead of polysaccharides for the development of dairy products.

2.3.2.3 Fibrils, Multistranded Ribbons, and Spherulites

Several milk and egg white globular proteins (*e.g.* β-LG, LYS), as well as unstructured caseins (*e.g.* κ-CN), are able to form fibrils – linear polymers of about 3 to 10 nm in width and hundreds of nm to μm in length (Figure 2.2A). Fibril formation from globular proteins, such as whey, egg white, and soy, requires a preliminary step of protein unfolding and/or hydrolysis in order to expose residues prone to establishing intermolecular β-sheets on the protein surface. Protein fibrils are extensively studied, and theoretical models have been proposed that explain and predict their behavior. Fibril formation follows a nucleation/growth mechanism. The nucleation step is usually the limiting step and is affected by shearing, addition of fibril seeds, or increasing heating temperature.[59] The growth of protein fibrils is hierarchical and unidirectional. This requires specific physicochemical conditions allowing the proteins to expose hydrophobic patches to the solvent and to conserve some net charge at the surface of the fibrils. For this reason, intrinsically unfolded proteins such as κ-CN are also well adapted to fibril formation.[60,61] In general, fibrils from globular proteins are obtained under elevated temperatures, low pH, and low ionic strength conditions.[29] The shape and size of the fibrils are tuned by ionic strength. The screening of the electrostatic charges by salts accelerates the aggregation step, resulting in shorter and more curved fibrils but with unchanged thicknesses of around 3.5 nm in the range of 10–100 mM of ionic strength.[62]

Protein fibrils can also be ordered into larger structures that are either linear (helical ribbons; Figure 2.2B) or spherical (spherulites; Figure 2.2C). Ribbons result from the lateral stacking of fibrils, observed only after the prolonged heating (at acidic pH) of globular proteins such as LYS or β-LG. Spherulites result from the radial arrangement of unbranched protein fibrils and are characterized by a semicrystalline structure.[63] The spherulites' diameter can reach several hundreds of micromillimeters. The subtle balance between electrostatic interactions and hydrophilic/hydrophobic effects has been assumed to govern the association of single fibrils into spherulites.

Fibrils could be used as thickening agents in food products as they increase bulk viscosity at relatively low protein concentrations. The thickening properties of the fibrils result from their large hydrodynamic diameter and their ability to form entangled networks that are progressively disrupted under increasing shearing, giving rise to a shear-thinning behavior.[64] Fibrillated protein solutions were also reported to have interesting foaming and emulsifying properties, but the contribution of molecular entities present within the fibrillated protein solutions (other than the fibrils themselves) toward this interfacial functionality is still a matter of debate.[65]

2.3.2.4 Protein Microparticles

Combined heat and shear treatments of whey proteins is a classic well-applied processing approach used by the dairy industry to produce microparticulated protein ingredients as fat replacer agents.[66] Heating under specific conditions of a sufficient whey protein concentration leads to aggregates that interact to form a three-dimensional network. However, if shearing is applied simultaneously during the heating step, individual microparticles are produced instead of a connected bulk protein network.[67]

Dry-heating (DH) is another way of processing that is largely used at industrial scale to improve the functional properties of egg white proteins. Egg white dry-heating allows soluble aggregates to form with improved foaming, emulsifying, and gelling properties (see Section 2.4). Several studies also report on the application of dry heating at neutral or acidic pH to modify whey proteins. Recent studies showed that DH at 100 °C of whey protein powders, conditioned at pH 9.5 with a water activity (a_w) of around 0.2, leads to the formation of whey protein microparticles (WPM).[33] Interestingly, the shape and size of the WPM depended on the upstream processing applied in order to dry the protein isolate or concentrate before the DH step. WPM obtained by DH of freeze-dried powders have the same angular shape as the powder flakes and are highly heterogeneous in size, ranging from 50 to 2000 μm (Figure 2.2G).[33] In contrast, WPM obtained from DH of spray-dried powders are spherical with an open hollow structure and a narrower size range centered around 60–90 μm (Figure 2.2H).[34] Dry-heated microparticles exhibit a very high water-binding capacity, ranging from 20 to 60 g of water per g of microparticles, depending on the upstream processing used. Hence spherical WPM leads to very high viscosity values. The cross-linked powder particles were able to swell when placed in an aqueous phase. Moreover, changing the pH of the microsphere suspension led to reversible changes of their size. Addition of lactose to the whey protein isolate powder promoted the speed of formation of these microparticles, leading to higher particle density and yield of conversion. This constitutes an excellent example of the possibility of controlling the size and shape of protein microparticles by controlling the size and shape of the initial powder.

2.4 Examples of Food Matrices

2.4.1 Egg White Protein-based Gels

The gelling properties of egg white are used in many food applications (charcuterie, fish or vegetable terrine, pastries, *etc.*). The quality of egg white gels is closely related to the physicochemical characteristics of its proteins, such as their mean hydrophobicity (Bigelow index), surface hydrophobicity, and net charge, which can be modified by environmental conditions, especially pH, ionic strength, and type of ions, as well as the technological treatments they were subjected to. Egg white gels are called *thermotropic gels* as the physicochemical process leading to the gel is the thermogelation of its proteins.

When heated, egg white loses its fluidity around 60 °C, but maximum firmness is reached above 70 °C. With the exception of ovomucin and ovomucoid, all egg white proteins coagulate with heat. However, they do not all react in the same way to heat treatment; for example, in egg white at pH 7, the denaturation temperatures of ovalbumin, lysozyme, and ovotransferrin are 85 °C, 74 °C, and 63 °C, respectively.[68] The latter is thus the most heat-sensitive protein, explaining why it is often regarded as the "initiator" of egg white protein gelation. However, its low denaturation temperature limits the firmness of egg white gels. The coagulation temperature of egg white indeed increases and the rheological properties of the gels are improved by selectively removing ovotransferrin from egg white or by binding metal ions (Al^{3+}, Fe^{3+}) onto this protein, thereby increasing its thermal stability.

The thermogelation of egg white proteins, resulting from a change in the balance between attractive and repulsive forces between proteins, corresponds to the long established model of heat-induced gelation of globular proteins. It involves two stages: the heat denaturation of proteins and the aggregation of denatured proteins.

The macroscopic characteristics of the egg white thermogels obtained (firmness, breaking strength, elasticity, water retention capacity, *etc.*) depend on the number and type (energy) of interactions among proteins and between proteins and water molecules. In thermotropic egg white gels, the interprotein bonds are mainly of low-energy (hydrophobic and electrostatic interactions, hydrogen bonds), but high-energy covalent bonds (disulfide bonds) also exist. Regardless of the type of interaction, their formation depends on protein conformation, *i.e.* the level of denaturation reached after the first stage (k_d), which in turn will determine the level of exposure of reactive groups or portions of the protein sequence. The formation of interprotein interactions also depends on the medium conditions that can facilitate or limit their establishment. This will either accelerate or slow down the aggregation stage (k_a) and thus restrict or increase the level of denaturation at the end of the first stage (reaction rate ratio of the two stages).

These mechanisms have been studied in the case of ovalbumin thermogelation and the impact of ionic strength and pH on the structure and properties of the resulting gels. Thermal denaturation results in an overall increase in the surface hydrophobicity of ovalbumin. When heat is applied at high ionic strength, the surface charges carried by the protein are screened, causing a decrease in electrostatic repulsion and favoring the formation of hydrophobic associations among denatured proteins. Under these conditions, random aggregates of partly denatured proteins quickly appear, resulting, above the critical gelation concentration, in opaque particulate gels with low water retention capacity, elasticity, and rigidity. Conversely, at low ionic strength, the electrostatic repulsions delay aggregation, thus promoting a greater level of denaturation (first stage). Moreover, this aggregation can occur only at certain contact points (hydrophobic patches), resulting in the formation of linear aggregates below the critical gelation concentration or stranded gels above the critical gelation concentration (Figure 2.3). Since ovalbumin plays a major role in the gelling properties of egg white, the impact of ionic strength as just described can be transposed to egg white.

Like ionic strength, which impacts egg white protein gelation by screening the protein charges, pH is also a major factor influencing the gelation of egg white by modifying the net charge of the proteins. Close to their pI, proteins tend to form random aggregates and particulate gels, similar to what has been described at high ionic strength. This results in low rheological properties for egg white at around pH 5. In contrast, egg white exhibits the best gelling properties away from the protein's pI, especially at basic pH (~pH 9). Under these conditions, strong electrostatic repulsions delay aggregation, increase denaturation, and promote the formation of linear aggregates and stranded gels (Figure 2.4). But the higher reactivity of thiol groups within this pH region may also be the cause for the improved gelling properties by

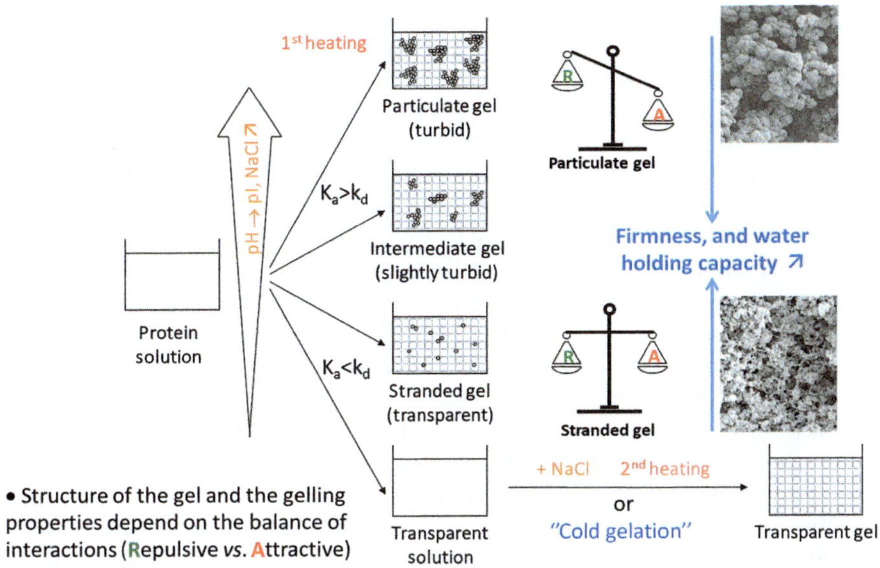

Figure 2.3 Effect of ionic strength and pH on the thermogelation of egg white proteins: mechanistic model[69,70] and electron microscopy images of typical egg white gels.[71]

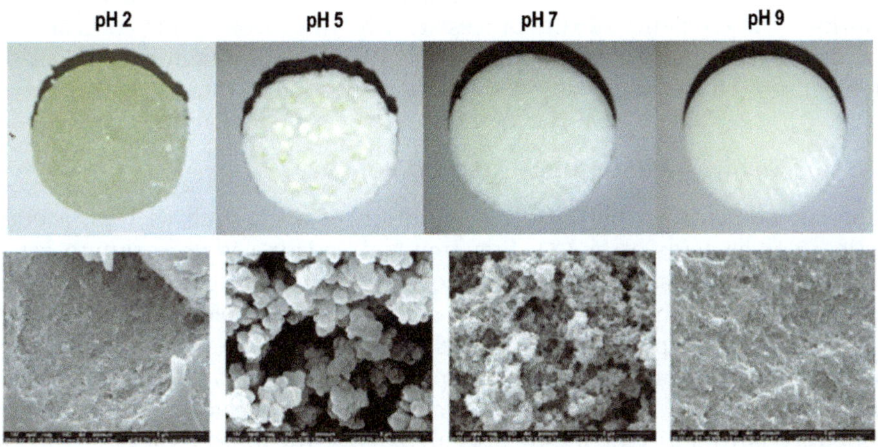

Figure 2.4 Macro- and microstructure (SEM) of egg white gels at: pH 2 (ionic strength 0.05 M); pH 5 (ionic strength 1 M); pH 7 (ionic strength 0.05 M); and, pH 9 (ionic strength 0.05 M).[76]

facilitating the formation of disulfide bonds. This is of special interest for the peelability of hard-boiled eggs, which is conversely weak for very fresh eggs. Indeed, egg white pH is around 7.5 just after laying but rapidly increases to 9.2 after 2 or 3 days due to a loss of CO_2 through the shell. To exhibit good peelability, egg white pH should be over 8.5.[72]

Egg white is mainly used as an ingredient in many complex food matrices. Its gelling properties and especially its gel strength are improved when mixing with either whey milk proteins[73,74] or caseins at pH over 7.[75]

A two-stage heating process was proposed to prepare transparent, very firm, and elastic egg white gels with exceptional water retention capacity.[77] The first stage consists of heating egg white, demineralized (by dialysis), to form linear polymers of denatured proteins in solution. The second stage involves a heat treatment, this time in the presence of salts, which induces the onset of interactions among the linear polymers of denatured proteins. These interactions do not imply the disintegration of the soluble polymers initially obtained, which explains why the gels stay transparent.[70] The soluble polymers obtained after the first heat treatment can also form a gel without any other heat treatment: This is called *cold gelation*.[78] This gelation is obtained by decreasing the electrostatic repulsions between the soluble polymers either by adding salts and/or by changing the pH toward the pI of the polymeric structures.[79]

To improve the gelling properties of egg white, Kato *et al.*[80] proposed an original approach consisting of the extensive denaturation of the proteins while preventing aggregation. Such conditions can be obtained by heating desugarized egg white powder at high temperatures (80 °C) for a long time (up to 10 days). This treatment increases the protein flexibility and the exposure of reactive groups that can further interact to strengthen the gel formed when previously dry-heated egg white is solubilized and heated in solution. The efficiency of this process can be improved by pH control.[81,82] The dry-heating process is today the basis for mass production practices of high-gel egg white powders.

2.4.2 Casein-based Gels: Case of Yogurt

Milk can be acidified by the direct addition of acid, such as HCl, acetic, lactic, or citric acids or by CO_2 injection.[83] As stirring is necessary to get a homogeneous pH value, a precipitate forms rather than a gel. This process is largely used in the industry to produce acid casein and caseinates. But if acid is added at 0–4 °C, precipitation is not observed. A further warming of milk forms a homogeneous acid gel.[84] The pH of milk can also be slowly reduced by fermentation to pH 4.6.[85] Another way to produce an acid gel in the laboratory is to add glucono-delta-lactose (GDL) to milk. GDL ($pK_a = 3.86$) hydrolyzes slowly in gluconate ions and releases protons to milk. The final pH value of milk decreases with the GDL content, the dilution of milk (as a high protein content increases the milk buffering), and temperature.

When milk is acidified, the negative charge of the casein micelle decreases to its pI, leading to a reduction in the electrostatic protection of casein micelles, to the collapse of their negatively charged hairy layer, and to micelle aggregation. Moreover, due to the protonation of phosphate and citrate in the serum phase of milk, calcium colloidal complexes displace to more ionized and soluble forms; CCP and calcium directly bound to

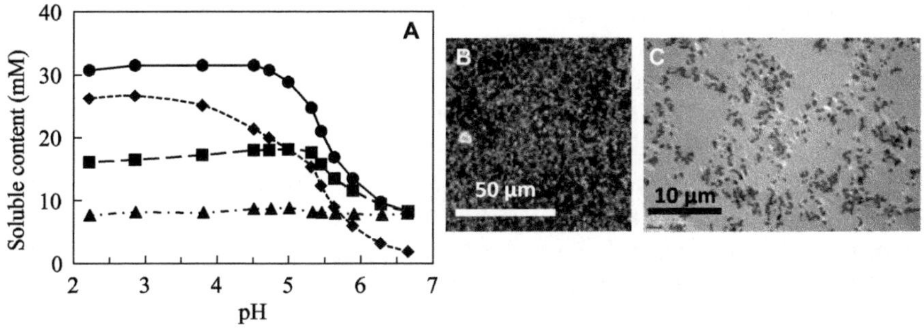

Figure 2.5 (A) Soluble content in calcium (●), inorganic phosphate (■), citrate (▲) and ionic calcium (◆) in milk as a function of pH.[86] Microstructure of acid gels observed by confocal laser scanning microscopy (B), or transmission electron microscopy (C) (Famelart, personal data).

phosphoserine are progressively dissolved (Figure 2.5A). CCP is completely released from the micelle at around pH 5.2, while calcium bound to phosphoserine continues to solubilize until pH 3.5. This mineral solubilization leads to the disruption of the casein micelle.

When the pH of milk approaches 5.5, a certain amount of total casein (and mainly β-CN) becomes soluble (10%, 30%, and 55% at 40 °C, 20 °C, and 4 °C, respectively[86]), but the micelle maintains its integrity until pH 5.2.[22] At pH 5.1, caseins start to aggregate, and then a gel is formed, as observed in micrographs (Figures 2.5B and C) and as measured by a viscosity increase. As micelle disintegration begins before gelation, the acid gelation of milk is not simply the aggregation of native micelles.

This hypothesis is confirmed by the abnormal behavior of the ζ-potential of casein.[22] A minimum is observed at pH 5.2, just before the onset of aggregation, followed by an increase and a subsequent decrease at the point of gel formation. It has been suggested that β-CN, predominantly found in the serum phase of milk, could begin to reassociate with the depleted and loose casein micelle at its pI (5.2) because of its positive charge. Particles in the acid gel are much larger than in the initial native casein micelle.[87]

2.4.2.1 *Yogurt Manufacture*

Yogurt is produced by fermentation with lactic acid bacteria. A preheating treatment of milk is required before bacteria inoculation to avoid the growth of contaminant microorganisms. In fact, a severe milk preheating is also applied for yogurt texture improvement (80 °C to 95 °C for 5–10 min[88]). This heat treatment induces the almost complete denaturation of whey proteins. The proteins unfold, and the hydrophobic and cysteine residues become exposed to solvent and react to form noncovalent (hydrophobic and ionic) and covalent (disulfide) interactions.[89] The free accessible thiol of β-LG initiates disulfide interchange reactions, while thiol–thiol oxidation is also observed in the presence of air.[90] It results in the formation of fractal complexes among β-LG, α-LA, and κ-CN (and incidentally α_{s2}-CN) that either stay attached to the micellar surface (micelle-bound complexes) or dissociate from the micelle to form new particles of sizes 50–100 nm (soluble complexes).[89,91–93]

The micelle-bound complexes (0.5–3.5 moles of whey proteins per mole of κ-CN) appear as either 100-nm-long or 20- to 100-nm-round appendages on the micelle surface. Soluble complexes (1–5 moles of whey proteins per mole of κ-CN) have a round shape of 20–50 nm or 100- to 200-nm noodle-like shape, as reviewed by Donato and Guyomarc'h.[89] Soluble complexes have a pI around 4.4–4.5 and bear a negative charge at the milk pH (from −15 to −20 mV). They are insoluble at acid pH values, and their surface hydrophobicity is higher than that of the native casein micelle.

While whey proteins, in their native form, remain soluble throughout the whole range of pH, they coaggregate with caseins during the acidification of preheated milk and form new interactions that contribute to gel formation. This substantially increases the firmness of the acid gel (Figures 2.6A) in relation to changes to the microstructure of the gels (Figures 2.6B and C). It has also been shown that the soluble complexes co-gel with the casein micelles at a higher pH value than in unheated milk and are colocalized within the gel network.[94] Whatever the location of complexes in the serum or colloidal phases, both contribute to the protein connectivity in the acid gel.[94] Microstructure of acid gels formed from the unheated milk system appeared more homogeneous, with smaller pore sizes and a lower connectivity between aggregates, than the gels from preheated milk. Casein is less solubilized at pH 5.5 during the acidification of preheated milk,[95] and particles in the gel formed from preheated milk appear smaller.

Properties of soluble complexes have been studied in order to understand the acid gel structure, either by collecting natural soluble complexes[89,96] or by using whey protein model complexes.[97–102] By changing the properties of

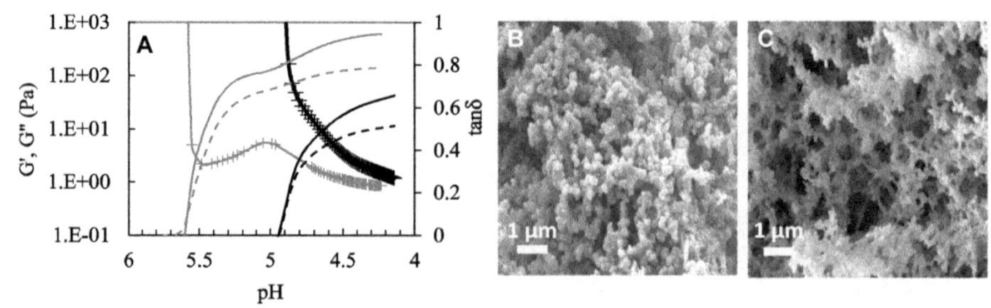

Figure 2.6 (A) Storage (solid line), loss (dashed line) modulus and loss tangent (+) as a function of time for yogurt fermented at 45 °C from unheated milk (black) and heated milk (90 °C–10 min) (gray). Scanning electron microscopy of yogurt made from (B) unheated and (C) pretreated milk (90 °C–10 min) (Famelart, personal data).

complexes, it has been concluded that complexes with high pI and high hydrophobicity allow for acid gel formation at a higher pH value, while only the latter (hydrophobicity) is able to dramatically increase the gel firmness.

2.4.2.2 Structure of Stirred Yogurt

While set-style yogurts are produced by distributing milk after the bacterial inoculation and conducting the fermentation in separated retail cartons, the fermentation of stirred yogurts is performed in large tanks. When the pH reaches a value close to 4.6, the gel is broken by stirring and sieving, pumped through a heat exchanger for cooling and distributed. During this step, the gel structure is disrupted into a fluid of very low viscosity consisting of a suspension of acid gel grains floating in the serum phase. But during storage, stirred yogurts show a recovery of texture[103,104] that highly depends on the intensity of the applied shear during breakdown.[105] The decrease in G' of set-style yogurt due to shear shows an asymptotic behavior with time, as does the recovery of stirred yogurt.[105] While acid gels do not form at very low temperature,[84] the storage temperature of stirred yogurt, in the range 4–50 °C, does not affect its texture recovery.[106] Even though hydrophobic interactions are essential for the buildup of set-style yogurts, they are not involved in structure recovery of the stirred gel.[106] A lower pH value at stirring (4.4, as compared to 4.7 or 5.0) increases structure recovery,[106] though another study does not report a large pH effect (4.5 or 4.25[105]). However, the effect of pH on the texture recovery is due only to the reduction in repulsive electrostatic interactions and not to attractive electrostatic interactions. Moreover, disulfide bonds are not involved in texture recovery.[104] Only a reduction of calcium content by citrate addition as a chelating agent promotes structure and texture recovery.[104] Explanations for this effect are still unknown.

The formation of yogurt structure is largely dependent on the suprastructure organization of the caseins in the micelle. As previously described in this chapter, the localization of κ-CN on the micelle surface has two main consequences for the formation of yogurts. First, the stabilization by the hairy layer disappears at pH values close to the pI of casein during the acidification of milk, allowing micelles to interact and form the acid gel. Second, heat-induced interactions between surface κ-CN and soluble whey proteins are formed, consistent with a dramatic increase in gel firmness. In fact, many other milk transformation processes take advantage of this original suprastructure, which also probably explains why its structure has attracted so much attention since the 1960s.

2.4.3 Casein-based Gels: Case of Cheeses

In this part of the chapter, we focus only on three flagship types of cheeses. Actually, there are several hundred types of final cheeses with various structural, organoleptic, nutritional, and biological characteristics. This is due to a large number of process conditions, among which are the ones relating to aging, that can vary in terms of duration, temperature of storage, and diversity of the microorganisms already present in the milk or added as starters, which hydrolyze, among other things, the caseins into peptides and free amino acids. First, we focus on mozzarella cheeses, produced through the application of specific temperature and mechanical processes to form stranded structures, evidenced from the micro- to macroscale. Second, we present cheeses with very different final structures and textures, due to the varied manufacturing and aging conditions they are exposed to, such as Swiss-type cheeses and Camembert cheese.

2.4.3.1 Mozzarella Cheeses

The mozzarella cheeses are rapidly consumed after manufacture (~two weeks of storage) and are mainly used for their technofunctional properties, such as melting and stretching in various cooking applications (pizza, pasta, grilled sandwiches, sauces, *etc.*). In order to avoid undesirable bacterial contamination, the process used has led to a peculiar gel structure. First, the milk is generally pasteurized prior to the addition of lactic acid bacteria (LAB) starters (*Streptococcus thermophilus* alone or with *Lactobacillus helveticus* or *L. delbrueckii*) and coagulant enzymes to form the paracasein gel. Then the gel is cut into pieces to expel the whey, and the resulting curd (paracasein network) is acidified by lactic acid produced by the bacteria until the pH drops to around 5.2 for bovine milk and 4.9 for buffalo milk. This crucial difference is because buffalo milk is richer in proteins and in colloidal calcium phosphate than bovine milk. The acidification rate is responsible for the extent of colloidal calcium phosphate release, the casein hydrolysis, the rheological properties of the cheeses, and their capability to melt and stretch.[107] The curd is then heated in hot water to around 70 °C, kneaded, and stretched, giving mozzarella its typical fibrous texture. During stretching, the casein network is able to maintain its integrity when the continuous stress is applied to the cheese. The casein molecules must interact with one another, release stress (applied to the cheese), and become pliable.[24] The protein matrix thus forms a bundle of long strands that contain columns of melted fat suspended in whey and bacteria.[108] The heat treatment is required to kill the LAB starters and other contaminant microorganisms. Second, it stops the acidification at a calcium concentration that is sufficient to sustain the interactions between calcium and the phosphorylated sites of the caseins, as well as to promote strand formation during the stretching step. Third, the coagulant enzyme is denatured, avoiding further casein hydrolysis during storage and damage of the formed strands. The curd is then molded, brined, and either stored in cheese whey or dry-packed. Some modifications in the manufacture process can be made to (1) produce cheeses with a lower moisture content, by adding food-grade acid (*e.g.* lactic acid, citric acid) instead of LAB starters and by heating the curd during whey drainage;[109] (2) standardize the proteins-to-fat ratio and calcium content by using microfiltration of the milk;[110] (3) homogenize the milk prior to manufacturing, so that fat and protein losses during the stretching step can be limited;[111] and (4) treat the milk by high pressure to increase the water-holding capacity of the paracasein matrix.[112]

As for renneted cheeses, the specificity of the coagulant enzyme in mozzarella is crucial not only to form the paracasein gel by hydrolysis of the κ-CN but also to control the extent of hydrolysis of the other caseins, especially α_{s1}-CN. Actually, to have high stretching properties means to have caseins with low level of hydrolysis that can interact with one another as well as with the calcium. The replacement of bovine chymosin by camel chymosin led to a higher proteolysis and a lower chewiness and firmness due to changes in the hydrolysis specificity on α_{s1}-CN.[113] When the coagulant enzyme hydrolyzes α_{s1}-CN at the Phe_{23}–Phe_{24} peptide bond, in various types of cheeses such as cheddar or mozzarella, the resulting paracasein gel has a weaker structure and lower fracture stress and firmness.[114] When another coagulant from the microbial *Cryphonectria parasitica* was used, the resultant mozzarella cheeses were softer in unmelted cheese texture, were more meltable during cooking, and had lower apparent viscosity and more free oil release upon melting than other cheeses, in which chymosin was used instead.[115] In contrast, in most mozzarella cheeses, plasmin, an endogenous milk enzyme that preferentially hydrolyzes β-CN, only weakly contributes to proteolysis. A balance has to be obtained to hydrolyze the caseins at the appropriate level to favor flowability and stretching, along with the release of the fat. The optimal proteolysis values are in the range from 5% to 11% of soluble nitrogen to total nitrogen content. As the temperature within the curd can vary between 58 °C and 68 °C, changes to the amount and the activity of the residual coagulant are obtained, as well as to the extent of the proteolysis during the cold storage. During the aging of mozzarella cheese, changes in paracasein network structure are more related to rearrangements of caseins than to proteolysis.

Fat, salt, and whey protein aggregates content are also among the parameters that can influence the structure of the paracasein network. The fat level in traditional mozzarella cheeses is generally ~20%. When the fat level is reduced, an increase in rigidity, elasticity, and hardness is observed, as well as a loss of meltability.[116] The salt content of the cheese (after brining) changes the overall water-holding capacity and increases the swelling of the strands of mozzarella cheeses, with the latter leading to a more homogeneous and continuous paracasein network, as observed using scanning electron microscopy.[109] The strands contain whey that becomes less expressible during storage due to the changes in water-holding capacity. The casein hydration enhances the displacement of neighboring planes of the protein matrix when the cheese is cooked on pizza and improves the flowability and stretchability of the heated cheese.[117] Therefore, increasing the salt-to-moisture ratio within the range of 0.5%–5.1% modifies the overall rheological properties of the cheeses and induces an increase in firmness, fracture stress, and sensory hardness.[118]

The denaturation of whey proteins during milk pretreatment above 65 °C impacts the clotting process, on the one hand, by slowing or inhibiting the primary phase of the rennet reaction because κ-CN–β-LG cross-linking reduces the susceptibility of κ-CN to hydrolysis by rennet. On the other hand, denatured whey proteins impair melting and stretching capacity and reduce the oiling-off as a result of the increased interaction between fat and proteins, changes in the protein matrix of the curd, or the binding of water and calcium by the denatured whey proteins.[119] However, in low-fat mozzarella cheese, the addition of denatured microparticulated whey proteins (with dimensions of 0.5–5 μm) has been successfully used as a fat replacement strategy. These microparticulated structures behave as a noninteracting filler within the cheese matrix and mimic the properties of the (similarly sized) native milk fat globules, ultimately providing a creamy mouthfeel.[120]

2.4.3.2 *Emmental or Swiss-type Cheeses*

The Swiss-type cheeses are classified as renneted and plasmin-type cheeses having the highest dry matter content, up to 62% (w/w) for Emmental cheeses, giving a dense curd microstructure. These cheeses are kept for long storage periods over months and even years, like parmigiano reggiano cheeses, another member of this hard-type cheese category. The Emmental cheese manufacture starts with the renneting step using chymosin or coagulant from *Cryphonectria parasitica*, which is more thermolabile than the chymosin. After coagulation, the curd is cut and cooked under stirring at 50–55 °C for 15–45 min to shrink the curd grains to a tenfold size reduction in order to expel the maximum whey and to completely inactivate the coagulant enzyme. The acidification by the lactic acid bacteria starters (*S. thermophilus* and *L. helveticus* or *L. delbrueckii*) occurs only when the cheese is molded and pressed, after the cooking step. This acidification takes place for several hours during the pressing step at a temperature that slightly decreases from 50 °C and leads to a final pH 5.2–5.35. When the draining of the whey is over (or almost over) during the pressing step, the minerals are mainly kept in the curd when the acidification occurs. The resulting high-calcium content of the Emmental cheeses leads to a soft and elastic texture. The cheese is then brined and stored under ripening room conditions that favor the development of propionibacteria (*P. freudenreichii*), a ripening flora that is responsible for the formation of holes (eyes) and flavor compounds. Actually, the firm and dry rind formed after brining reduces the loss of CO_2, produced during propionic acid fermentation, while the consistency of the cheese supports the opening of the cheese body and the forming of the eyes during ageing.[121] Cheese mechanical properties imply high resistance to breaking (fracture stress) and sufficient aptitude to flow (elongational viscosity) for eye formation at the proper ageing time, *i.e.* during the warm room period in which the temperature reaches 22–24 °C for 2–3 weeks.[122]

In Emmental cheeses, the proteolysis is one main parameter that induces changes in structure and texture. Proteolysis is indeed a continuous process in which plasmin and proteolytic enzymes from the starters, *i.e.* proteinases and peptidases, are the main responsible factors. The relatively high pH of the cheese allows for the continuous activity of plasmin throughout the ripening time (preferentially on β- and α_{s2}-CN), and the curd cooking step also enhances the conversion of plasminogen to plasmin by inactivating an inhibitor of an activator of plasminogen.[123] The cell envelope proteinases of lactobacilli, used as starters, have large substrate specificities, as well as numerous peptidases that are released (after bacteria death) in the cheese during aging, are able to give a specific peptide fingerprint per type of cheese.[124] The hydrolysis of both α_{s1}-CN (by LAB) and β-CN (by plasmin and LAB) weakens the number and strength of the protein–protein interactions between casein molecules in the paracasein network. Differences between the proteinases of *L. delbrueckii* spp. *bulgaricus* and *L. helveticus* strains are shown to influence functional properties such as hardness, cohesiveness, melting strength, and even stretching quality.[125] Additionally, soluble peptides, derived from the cleavage of α_{s1}-CN (f1-23) by different strains of lactobacilli, positively correlate with meltability and stretching characteristics of cheeses made with those strains. Cheeses made with lactobacilli were shown to be both more hydrolyzed and softer than the control cheeses made using streptococci alone.[125] Depending on the specificity of the lactobacilli cell-wall proteinases, the type and the number of peptides identified by MS/MS containing more than 20 amino acids from α_{s1}- and α_{s2}-CN were different.[124] These peptides, preferably hydrophobic, were more abundant in the cheeses with increased elastic properties and improved viscoelastic properties under cooking conditions, for culinary applications, such as mozzarella cheeses. It was therefore assumed that such peptides could be involved in the formation of fibers *via* hydrophobic bonds between native caseins and peptides upon heating.[126] More recently, an *in silico* method to establish relationships between a technofunctional property of cheese and a peptide profile has shown that not only the size of peptides but also their physicochemical and structural characteristics are involved in Emmental cheese stretchability.[127]

The uptake and diffusion of salt into Emmental cheeses during brining remains very modest in relation to the high weight of the cheeses (75–120 kg) and in the range of 0.5% to 1.6% (salt to moisture). A salt gradient exists between the core and the surface of the cheese that persists throughout the aging time. However, the cheese composition is hardly affected by these differences in terms of textural and functional properties.

2.4.3.3 *Camembert Cheeses*

Camembert cheese is a renneted and acidified cheese that has residual coagulant activity during ripening (in contrast to the two previous cheese types), since the temperature does not exceed 35 °C during cheese manufacture. The pH of the curd is in the pH 4.4–4.8 range and is responsible for the huge demineralization of the paracasein network and for the crumbly texture that develops during the first days of ripening. The whey is removed *via* a perforated mold containing the curd that is turned every 6 to 12 hours to allow even whey drainage. Thus, Camembert cheeses have a high moisture content (~54%). The LAB starters (*Lactococcus lactis*) grow in the inner part of the cheeses while a mold, *Penicillium camemberti*, and a yeast, *Geotrichum candidum*, progressively cover the surface of the cheese. Over time, the fusion of casein micelles inside and between the curd grains creates a more homogeneous structure with smaller serum cavities, which can lead to a very marked softening, almost liquefaction, of the body from the surface to the center by the end of the aging time.[117] Such a change in the casein network structure results mainly from the hydrolysis of α_{s1}-CN by the residual coagulant, the LAB starter proteolytic enzymes, and the increase in pH (pH ~7 at the cheese surface) due to the high proteolytic activity of *P. camemberti* and the production of ammonia.[128,129] These enzymatic and physicochemical variations induce, in turn, changes in the casein–casein interactions and the migration of calcium from the inner to the outer layer of the cheese, where it precipitates as calcium lactate or calcium phosphate. The increase in pH stimulates the plasmin activity that also plays an important role in the softening of Camembert cheeses by hydrolyzing β-CN.[130] In Camembert cheese, NaCl concentration has a major influence on proteolysis and pH changes, as well as on mold surface growth.[117] The higher the salt content is, the lower the proteolysis and the changes in overall softening will be.

2.5 Conclusion and Future Trends

The structure and properties of protein-based foods are governed by the intrinsic properties of the used proteins, the composition of the food matrix, and the processing conditions. In the previously mentioned protein-based food matrices, proteins in their "native state" are modified *in situ* by enzymes, temperature, shearing, *etc.*, in order to obtain the described food structures.

In the context of the growing demand for protein worldwide, milk-based and egg protein-based products are still promising, either as singular species or in mixtures with other protein sources, in order to diversify and optimize food offers in terms of food functionalities and nutrition. Recent knowledge supports the notion that food matrix microstructure modulates the efficiency of digestive processes, as well as various physiological functions, in the human body. This underscores the need to understand the local and mesoscale interactions involved in the development of the food microstructure *in situ* or under relevant experimental conditions.

New uses of the proteins, as "structuring blocks," are now emerging by taking advantage of the malleable properties of milk and egg proteins to build *ex situ* a large diversity of ordered nano- and microstructures (microgels, fibrils, complex coacervates, nanotubes, *etc.*), with novel properties that can be used as ingredients in order to design foodstuffs with new structures and properties. Whereas the structure of these nano- and microstructures is rather well characterized (see Section 2.3), most of their properties as well as their further applications in foods are still to be investigated.

The combined use of powerful biophysical techniques and concepts developed in soft matter physics offers a real opportunity for the food scientific community and the food industry to notably improve our understanding of the main structural elements in building food products. In this sense, further research on methodological transfer and adaptation are required for tailoring new protein assemblies and related functionalities for novel applications in food formulations.

References

1. Y. Levy and J. N. Onuchic, *Proc. Natl. Acad. Sci. U. S. A.*, 2004, **101**, 3325–3326.
2. H. Ma, P. Forssell, R. Partanen, J. Buchert and H. Boer, *J. Agric. Food Chem.*, 2011, **59**, 13246–13253.
3. S. Damodaran, in *Fennema's Food Chemistry*, CRC Press, Boca Raton, FL, 4th edn, 2007, pp. 217–330.
4. E. A. Foegeding and J. P. Davis, *Food Hydrocoll.*, 2011, **25**, 1853–1864.
5. E. C. Y. Li-Chan and I. M. E. Lacroix, in *Proteins in Food Processing*, ed. R. Y. Yada, Woodhead Publishing, 2018, pp. 1–25.
6. M. N. Riaz, in *Handbook of Food Proteins*, ed. G. O. Phillips and P. A. Williams, Woodhead Publishing, 2011, pp. 395–418.
7. N. Guerrieri and M. Cavaletto, in *Proteins in Food Processing*, ed. R. Y. Yada, Woodhead Publishing, 2018, pp. 223–243.
8. K. Nishinari, Y. Fang, T. Nagano, S. Guo and R. Wang, in *Proteins in Food Processing*, ed. R. Y. Yada, Woodhead Publishing., 2018, pp. 27–45.
9. Y. L. Xiong, in *Proteins in Food Processing*, ed. R. Y. Yada, Woodhead Publishing, 2018, pp. 127–148.
10. R. Tahergorabi, S. V. Hosseini and J. Jaczynski, in *Handbook of Food Proteins*, ed. G. O. Phillips and P. A. Williams, Woodhead Publishing, 2011, pp. 116–149.
11. A. N. A. Aryee, D. Agyei, and C. C. Udenigwe, in *Proteins in Food Processing*, ed. R. Y. Yada, Woodhead Publishing, 2018, pp. 27–45.
12. E. Li-Chan and S. Nakai, *Crit. Rev. Poult. Biol. USA*, 1989, **2**, 21–58.
13. C. Guérin-Dubiard, M. Pasco, D. Mollé, C. Désert, T. Croguennec, and F. Nau, *J. Agric. Food Chem.*, 2006, **54**, 3901–3910.
14. K. Mann, *Proteomics*, 2007, **7**, 3558–3568.
15. P. Walstra and R. Jenness, *Dairy Chemistry and Physics*, John Wiley and Sons, Inc., New York, 1984.
16. J. A. O'Mahony and P. F. Fox, in *Advanced Dairy Chemistry*, Springer, Boston, MA, 2013, pp. 43–85.

17. T. Huppertz, in *Advanced Dairy Chemistry*, Springer, Boston, MA, 2013, pp. 135–160.
18. A. Bouchoux, G. Gésan-Guiziou, J. Pérez and B. Cabane, *Biophys. J.*, 2010, **99**, 3754–3762.
19. D. G. Dalgleish, *Soft Matter*, 2011, **7**, 2265–2272.
20. C. Holt, J. A. Carver, H. Ecroyd and D. C. Thorn, *J. Dairy Sci.*, 2013, **96**, 6127–6146.
21. C. Holt, E. S. Sørensen and R. A. Clegg, *FEBS J.*, 2009, **276**, 2308–2323.
22. I. Heertje, J. Visser and P. Smits, *Food Microstruct.*, 1985, **4**, 267–277.
23. P. Thomar and T. Nicolai, *Colloids Surf. B Biointerfaces*, 2016, **146**, 801–807.
24. J. A. Lucey, M. E. Johnson and D. S. Horne, *J. Dairy Sci.*, 2003, **86**, 2725–2743.
25. K. Brew, in *Advanced Dairy Chemistry—1 Proteins*, Springer, Boston, MA, 2003, pp. 387–419.
26. L. Sawyer, in *Advanced Dairy Chemistry*, Springer, Boston, MA, 2013, pp. 211–259.
27. A. M. Donald, *Soft Matter*, 2008, **4**, 1147–1150.
28. S. G. Bolder, H. Hendrickx, L. M. C. Sagis and E. van der Linden, *J. Agric. Food Chem.*, 2006, **54**, 4229–4234.
29. C. Lara, J. Adamcik, S. Jordens and R. Mezzenga, *Biomacromolecules*, 2011, **12**, 1868–1875.
30. K. R. Domike and A. M. Donald, *Biomacromolecules*, 2007, **8**, 3930–3937.
31. T. Loiseleux, A. Rolland-Sabaté, C. Garnier, T. Croguennec, S. Guilois, M. Anton and A. Riaublanc, *Food Hydrocoll.*, 2018, **74**, 197–206.
32. J. F. Graveland-Bikker and C. G. de Kruif, *Trends Food Sci. Technol.*, 2006, **17**, 196–203.
33. M. H. Famelart, E. Schong and T. Croguennec, *J. Food Eng.*, 2018, **224**, 112–120.
34. E. Schong and M.-H. Famelart, *Food Res. Int.*, 2018, **113**, 210–220.
35. M. Nigen, T. Croguennec, M.-N. Madec and S. Bouhallab, *FEBS J.*, 2007, **274**, 6085–6093.
36. L. Lacou, J. Léonil and V. Gagnaire, *Food Hydrocoll.*, 2016, **57**, 187–199.
37. T. Croguennec, G. M. Tavares and S. Bouhallab, *Adv. Colloid Interface Sci.*, 2017, **239**, 115–126.
38. F. Diarrassouba, G. Remondetto, G. Garrait, P. Alvarez, E. Beyssac and M. Subirade, *Food Chem.*, 2015, **173**, 203–209.
39. A.-L. Chapeau, G. M. Tavares, P. Hamon, T. Croguennec, D. Poncelet and S. Bouhallab, *Food Hydrocoll.*, 2016, **57**, 280–290.
40. C. Schmitt, C. Bovay, A.-M. Vuilliomenet, M. Rouvet and L. Bovetto, *Food Hydrocoll.*, 2011, **25**, 558–567.
41. T. Phan-Xuan, D. Durand, T. Nicolai, L. Donato, C. Schmitt and L. Bovetto, *Food Hydrocoll.*, 2014, **34**, 227–235.
42. L. Donato, C. Schmitt, L. Bovetto and M. Rouvet, *Int. Dairy J.*, 2009, **19**, 295–306.
43. S. Yu, P. Yao, M. Jiang and G. Zhang, *Biopolymers*, 2006, **83**, 148–158.

44. X. Pan, S. Yu, P. Yao and Z. Shao, *J. Colloid Interface Sci.*, 2007, **316**, 405–412.
45. S. Mossa, F. Sciortino, P. Tartaglia and E. Zaccarelli, *Langmuir*, 2004, **20**, 10756–10763.
46. M. R. H. Krebs, K. R. Domike and A. M. Donald, *Biochem. Soc. Trans.*, 2009, **37**, 682–686.
47. D. Sağlam, P. Venema, E. van der Linden and R. de Vries, *Curr. Opin. Colloid Interface Sci.*, 2014, **19**, 428–437.
48. M. Chevallier, A. Riaublanc, C. Lopez, P. Hamon, F. Rousseau, J. Thevenot and T. Croguennec, *Food Hydrocoll.*, 2018, **76**, 164–172.
49. M. Destribats, M. Rouvet, C. Gehin-Delval, C. Schmitt and B. P. Binks, *Soft Matter*, 2014, **10**, 6941–6954.
50. C. Schmitt, C. Bovay and M. Rouvet, *Food Hydrocoll.*, 2014, **42**, 139–148.
51. A. Kharlamova, T. Nicolai and C. Chassenieux, *Food Hydrocoll.*, 2018, **79**, 145–157.
52. R. N. Zúñiga, A. Tolkach, U. Kulozik and J. M. Aguilera, *J. Food Sci.*, 2010, **75**, E261–E268.
53. M. Pouzot, T. Nicolai, R. W. Visschers and M. Weijers, *Food Hydrocoll.*, 2005, **19**, 231–238.
54. N. Mahmoudi, S. Mehalebi, T. Nicolai, D. Durand and A. Riaublanc, *J. Agric. Food Chem.*, 2007, **55**, 3104–3111.
55. T. Nicolai, M. Britten and C. Schmitt, *Food Hydrocoll.*, 2011, **25**, 1945–1962.
56. W. Inthavong, A. Kharlamova, C. Chassenieux and T. Nicolai, *Soft Matter*, 2016, **12**, 2785–2793.
57. A. Kharlamova, C. Chassenieux and T. Nicolai, *Food Hydrocoll.*, 2018, **81**, 263–272.
58. A. Kharlamova, T. Nicolai and C. Chassenieux, *Int. Dairy J.*, 2018, **86**, 110–119.
59. M. R. H. Krebs, D. K. Wilkins, E. W. Chung, M. C. Pitkeathly, A. K. Chamberlain, J. Zurdo, C. V. Robinson and C. M. Dobson, *J. Mol. Biol.*, 2000, **300**, 541–549.
60. H. M. Farrell, P. H. Cooke, E. D. Wickham, E. G. Piotrowski and P. D. Hoagland, *J. Protein Chem.*, 2003, **22**, 259–273.
61. J. Leonil, G. Henry, D. Jouanneau, M.-M. Delage, V. Forge and J.-L. Putaux, *J. Mol. Biol.*, 2008, **381**, 1267–1280.
62. L. N. Arnaudov and R. de Vries, *Biomacromolecules*, 2006, **7**, 3490–3498.
63. E. H. C. Bromley, M. R. H. Krebs and A. M. Donald, *Faraday Discuss.*, 2005, **128**, 13–27.
64. S. M. Loveday, J. H. Su, M. A. Rao, S. G. Anema and H. Singht, *J. Agric. Food Chem.*, 2012, **60**, 5229–5236.
65. M. Mohammadian and A. Madadlou, *Trends Food Sci. Technol.*, 2018, **75**, 115–128.
66. A. Y. Tamime, M. Kaláb, D. D. Muir and E. Barrantes, *Int. J. Dairy Technol.*, 1995, **48**, 107–111.
67. T. Spiegel, *Int. J. Food Sci. Technol.*, 1999, **34**, 523–531.

68. J. W. Donovan, C. J. Mapes, J. G. Davis and J. A. Garibaldi, *J. Sci. Food Agric.*, 1975, **26**, 73–83.
69. E. Doi, *Trends Food Sci. Technol.*, 1993, **4**, 1–5.
70. N. Kitabatake, Y. Tani and E. Doi, *J. Food Sci.*, 1989, **54**, 1632–1638.
71. T. Croguennec, F. Nau and G. Brulé, *J. Food Sci.*, 2002, **67**, 608–614.
72. W. M. Britton and D. L. Fletcher, *Poult. Sci.*, 1987, **66**, 453–457.
73. E. Matringe, R. P. T. Luu and D. Lorient, *J. Food Sci.*, 1999, **64**, 787–791.
74. K. J. Aryana, Z. Z. Haque and P. D. Gerard, *Int. J. Food Sci. Technol.*, 2002, **37**, 643–652.
75. N. Matsudomi, Y. Kanda and H. Moriwaki, *J. Agric. Food Chem.*, 2003, **51**, 7140–7145.
76. K. Nyemb, C. Guérin-Dubiard, S. Pézennec, J. Jardin, V. Briard-Bion, C. Cauty, S. M. Rutherfurd, D. Dupont and F. Nau, *Food Hydrocoll.*, 2016, **54**, 315–327.
77. N. Kitabatake, A. Shimizu and E. Doi, *J. Food Sci.*, 1988, **53**, 735–738.
78. C. M. Bryant and D. J. McClements, *Trends Food Sci. Technol. U. K.*, 1998, **9**, 143–151.
79. S. J. Choi, S. E. Lee and T. W. Moon, *J. Food Sci.*, 2008, **73**, C313–C322.
80. A. Kato, H. R. Ibrahim, H. Watanabe, K. Honma and K. Kobayashi, *J. Agric. Food Chem.*, 1989, **37**, 433–437.
81. Y. Mine, *Food Res. Int.*, 1996, **29**, 155–161.
82. Y. Mine, *J. Agric. Food Chem.*, 1997, **45**, 2924–2928.
83. F. Gaucheron, *Minéraux et Produits Laitiers*, Tec & Doc, 2003.
84. S. P. F. M. Roefs, A. E. A. de Groot-Mostert and T. Van Vliet, *Colloids Surf.*, 1990, **50**, 141–159.
85. W. J. Lee and J. A. Lucey, *Asian-Australas. J. Anim. Sci.*, 2010, **23**, 1127–1136.
86. O. Mekmene, Y. Le Graët and F. Gaucheron, *J. Agric. Food Chem.*, 2010, **58**, 4440–4447.
87. M.-H. Famelart, J. Tomazewski, M. Piot and S. Pezennec, *Int. Dairy J.*, 2004, **14**, 313–321.
88. T. Benezech and J. F. Maingonnat, *J. Food Eng.*, 1994, **21**, 447–472.
89. L. Donato and F. Guyomarc'h, *Dairy Sci. Technol.*, 2009, **89**, 3–29.
90. J. N. de Wit, *Trends Food Sci. Technol.*, 2009, **20**, 27–34.
91. S. G. Anema and Y. Li, *J. Dairy Res.*, 2003, **70**, 73–83.
92. F. Guyomarc'h, A. J. R. Law and D. G. Dalgleish, *J. Agric. Food Chem.*, 2003, **51**, 4652–4660.
93. A. J. Vasbinder, A. C. Alting and K. G. de Kruif, *Colloids Surf. B Biointerfaces*, 2003, **31**, 115–123.
94. F. Guyomarc'h, M. Jemin, V. Le Tilly, M. N. Madec and M. H. Famelart, *J. Agric. Food Chem.*, 2009, **57**, 5910–5917.
95. A. Laligant, M. H. Famelart, G. Brul, M. Piot and D. Paquet, *Le Lait*, 2003, **83**, 181–192.
96. K. Jean, M. Renan, M. H. Famelart and F. Guyomarc'h, *Int. Dairy J.*, 2006, **16**, 303–315.

97. R. Andoyo, F. Guyomarc'h, C. Cauty and M.-H. Famelart, *Food Hydrocoll.*, 2014, **37**, 203–212.
98. R. Andoyo, F. Guyomarc'h, A. Burel and M.-H. Famelart, *Food Hydrocoll.*, 2015, **51**, 118–128.
99. M. Morand, F. Guyomarc'h and M. H. Famelart, *Dairy Sci. Technol.*, 2011, **91**, 97–126.
100. M. Morand, F. Guyomarc'h, S. Pezennec and M. H. Famelart, *Int. Dairy J.*, 2011, **21**, 670–678.
101. M. Morand, F. Guyomarc'h, D. Legland and M.-H. Famelart, *Int. Dairy J.*, 2012, **23**, 9–17.
102. M. Morand, A. Dekkari, F. Guyomarc'h and M. H. Famelart, *Int. Dairy J.*, 2012, **25**, 103–111.
103. E. Rönnegård, J. Jiang and P. Dejmek, in *Developments in Food Engineering*, Springer, Boston, MA, 1994, pp. 90–92.
104. M. Renan, F. Guyomarc'h, V. Arnoult-Delest, D. Pâquet, G. Brulé and M. H. Famelart, *J. Dairy Res.*, 2008, **75**, 450–456.
105. E. Rönnegård and P. Dejmek, *Le Lait*, 1993, **73**, 371–379.
106. M. Renan, F. Guyomarc'h, V. Arnoult-Delest, D. Pâquet, G. Brulé and M.-H. Famelart, *Int. Dairy J.*, 2009, **19**, 142–148.
107. P. F. Fox and P. L. H. McSweeney, in *Cheese: Chemistry, Physics & Microbiology*, ed. P. L. H. McSweeney, P. F. Fox, P. D. Cotter and D. W. Everett, Boston, MA, 4th edn, 2017, pp. 5–21.
108. C. J. Oberg, W. R. McManus and D. J. McMahon, *Food Struct.*, 1993, **12**, 251–258.
109. B. M. Paulson, D. J. McMahon and C. J. Oberg, *J. Dairy Sci.*, 1998, **81**, 2053–2064.
110. R. L. Brandsma and S. S. H. Rizvi, *Int. J. Food Sci. Technol.*, 2001, **36**, 611–624.
111. M. A. Rudan, D. M. Barbano, M. R. Guo and P. S. Kinstedt, *J. Dairy Sci.*, 1998, **81**, 2065–2076.
112. C. E. O'Reilly, P. M. Murphy, A. L. Kelly, T. P. Guinee, M. A. E. Auty and T. P. Beresford, *Innov. Food Sci. Emerg. Technol.*, 2002, **3**, 3–9.
113. A. C. Moynihan, S. Govindasamy-Lucey, J. J. Jaeggi, M. E. Johnson, J. A. Lucey and P. L. H. McSweeney, *J. Dairy Sci.*, 2014, **97**, 85–96.
114. L. K. Creamer and N. F. Olson, *J. Food Sci.*, 1982, **47**, 631–636.
115. J. J. Yun, J. L. Kiely, P. S. Kindstedt and D. M. Barbano, *J. Dairy Sci.*, 1993, **76**, 3657–3663.
116. S. Henneberry, M. G. Wilkinson, K. N. Kilcawley, P. M. Kelly and T. P. Guinee, *Dairy Sci. Technol.*, 2015, **95**, 613–638.
117. T. P. Guinee and P. F. Fox, in *Cheese: Chemistry, Physics & Microbiology*, ed. P. L. H. McSweeney, P. F. Fox, P. D. Cotter and D. W. Everett, Boston, MA, 4th edn, 2017, pp. 317–375.
118. B. Ganesan, K. Brown, D. A. Irish, C. Brothersen and D. J. McMahon, *J. Dairy Sci.*, 2014, **97**, 1970–1982.
119. F. Guyomarc'h, *Le Lait*, 2006, **86**, 1–20.

120. D. J. McMahon, M. C. Alleyne, R. L. Fife and C. J. Oberg, *J. Dairy Sci.*, 1996, **79**, 1911–1921.
121. A. Thierry, F. Berthier, V. Gagnaire, J. R. Kerjean, C. Lopez and Y. Noël, in *Technology of Cheesemaking*, ed. B. A. Law and A. Y. Tamime, Wiley-Blackwell, Chichester, 2nd edn, 2010, pp. 360–383.
122. P. Walstra, *IDF Bull*, 1991, **268**, 16–25.
123. B. Ismail and S. S. Nielsen, *J. Dairy Sci.*, 2010, **93**, 4999–5009.
124. L. Sadat-Mekmene, R. Richoux, L. Aubert-Frogerais, M.-N. Madec, C. Corre, M. Piot, J. Jardin, S. le Feunteun, S. Lortal and V. Gagnaire, *J. Dairy Sci.*, 2013, **96**, 1455–1470.
125. B. S. Oommen, D. J. McMahon, C. J. Oberg, J. R. Broadbent and M. Strickland, *J. Dairy Sci.*, 2002, **85**, 2750–2758.
126. R. Richoux, L. Aubert, G. Roset and J. R. Kerjean, *Dairy Sci. Technol.*, 2009, **89**, 31–41.
127. L. Lacou, S. Le, S. Pezennec and V. Gagnaire, *Colloids Surf. – Physicochem. Eng. Asp.*, 2015, **475**, 44–54.
128. Y. Le Graët, A. Lepienne, G. Brulé and P. Ducruet, *Le Lait*, 1983, **63**, 317–332.
129. A. Noomen, *Neth. Milk Dairy J*, 1983, **37**, 229–232.
130. A. Delacroix-Buchet and S. Fournier, *Le Lait*, 1992, **72**, 53–72.

CHAPTER 3

Food Structure Development in Emulsion Systems

ERNESTO TRIPODI,*[a] ARIS LAZIDIS,[a,b] IAN T. NORTON[a] AND FOTIS SPYROPOULOS*[a]

[a] Chemical Engineering Department, University of Birmingham, UK;
[b] Nestlé Product Technology Centre, York, UK
*Emails: E.Tripodi@bham.ac.uk; F.Spyropoulos@bham.ac.uk

3.1 Introduction

Food emulsions represent structurally complex systems containing a multitude of constituents. These molecular species (*e.g.* water, fat, emulsifiers, carbohydrates, proteins, salt, sugar, *etc.*) contribute through their micro- and macroscale characteristics and interactions to the emulsion microstructure, which ultimately determines many of the physical properties and functionalities of the food they are part of. For this reason, the development of emulsion architectures with tailored features represents a critical step in optimizing or enhancing the features of existing products as well as designing novel, tastier, and healthier foods.

Controlling the performance of emulsions can be conducted by carefully choosing both the formulation and the processing characteristics. The cautious selection of the emulsion constituents and the understanding of their contribution to the product microstructure offer the opportunity to shape the properties of conventional emulsions and design structurally complex systems with enhanced bulk properties, as well as to ensure their stability. The selection of the most suitable manufacturing method represents an additional critical step, during which ingredients are brought

Food Chemistry, Function and Analysis No. 18
Handbook of Food Structure Development
Edited by Fotis Spyropoulos, Aris Lazidis and Ian T. Norton

Published by the Royal Society of Chemistry, www.rsc.org

together to create the desired microstructural features. Both micro- as well as macrocharacteristics of the product can be probed, measured, and quantified through the use of an ever expanding array of analysis techniques. The discussion that follows ultimately demonstrates that the formulation and processing aspects involved in the development of emulsion structure can be controlled to optimize or design systems with specific functionality and performance.

3.2 The Emulsion Microstructure

Whilst emulsions define, to a great extent, the rheology and texture of many food systems, controlling the development of their microstructure, its behavior, and stability can sometimes be challenging. In an effort to systematically address these challenges, one can design emulsion-based products by the careful choice of (1) the formulation and (2) the processing route. This section focuses on the former and aims to describe the basic constituents (and their contribution to the microstructure) of conventional emulsions but also to that of more complex systems.

3.2.1 Conventional Emulsion Microstructure

In its simplest definition, an oil-in-water (o/w) emulsion (Figure 3.1A) consists of (1) oil droplets (dispersed phase) surrounded by (2) water (continuous phase) and (3) the interfacial region (interface) that separates the two.[1] Water-in-oil (w/o) emulsions (Figure 3.1B) are similarly defined.[1] Each of these three emulsion structural features has a different contribution to the final architecture and thus to its bulk properties.[2]

3.2.1.1 The Continuous Phase

The bulk properties of the continuous phase influence a wide range of emulsion characteristics but most importantly its stability and texture. Texture is related to mouth perception and to attributes such as creaminess, thickness, and mouthfeel. Texture in relation to emulsions can typically be instrumentally characterized through the study of their rheological and tribological properties.[3–6] Dynamic viscosity, an important rheological parameter of the continuous phase of emulsions can be tuned in order to enhance mouthfeel perception as well as to extend product shelf life.[7] In o/w emulsions, increasing the viscosity of the continuous phase is usually achieved by the inclusion of thickening or bulking agents into the formulation.[8,9] Typical thickeners that are found in foods are polysaccharides (gums), starch, pectin, cellulose or its derivatives, as well as proteins or blends of any of the above.[10] In w/o emulsions, the continuous phase rheology is usually determined by the presence (or not) of a fat crystal network and its physical characteristics such as fat composition, solid fat content, crystal size/shape, and the strength of the network that entraps the water

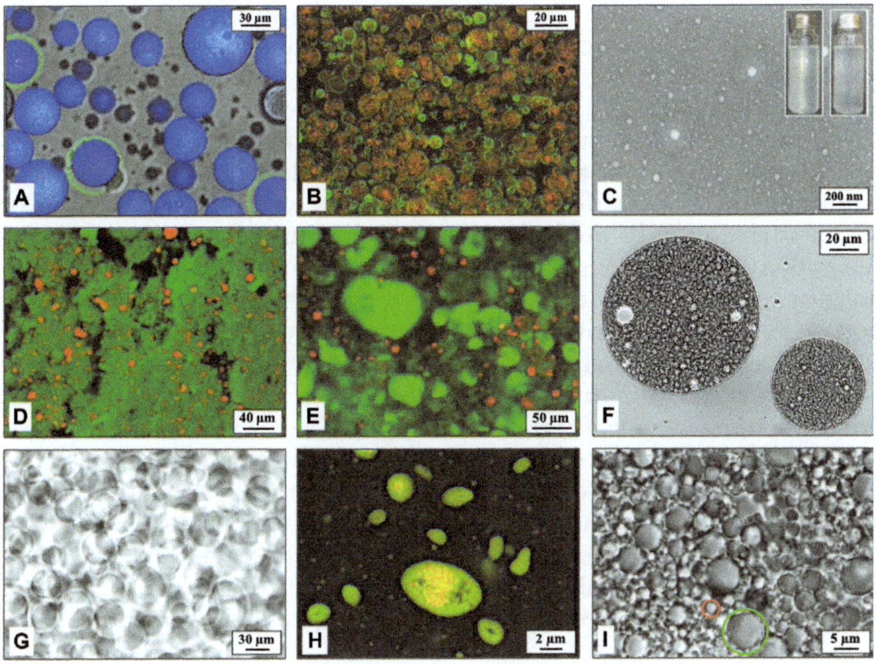

Figure 3.1 Simple and complex emulsion microstructures: (A) o/w emulsion stabilized by sodium caseinate/chitosan (NaCAS/CS) complexes; fluorescent emissions from the NaCAS/CS complexes (green) and the oil droplets (blue); (B) w/o emulsion stabilized by tripalmitin crystal nanoparticles; fluorescent emissions from the water droplets (red) and tripalmitin (green); (C) o/w nanoemulsion stabilized by Tween 80; insets are images of near transparent nanoemulsion samples, taken immediately (left) and 56 days (right) after formation; (D) o/w emulsion gel of whey protein isolate (thermally set); fluorescent emissions from the oil droplets (red) embedded within the WPI gel network (green); (E) o/w emulsion gel of Pea Protein (pH-set); fluorescent emissions from the oil droplets (red) embedded within the PP gel network (green); (F) $w_1/o/w_2$ double emulsion stabilized by PGPR (internal w_1/o interface) and Tween 20 (external o/w_2 interface); (G) gelatin/dextran w/w emulsion under mild shear; (H) sodium caseinate/pectin hydrogel particles containing o/w emulsion; fluorescent emissions from the oil droplets (red) and the protein (green) part of their hydrogel particle carriers; (I) a/o/w emulsions showing air cells (*e.g.* green circle) and oil droplets (*e.g.* red circle); air cells (~7 μm) are codispersed with oil droplets (~0.6 μm) within an aqueous medium to give the final triphasic emulsion microstructure. [A]. Adapted from ref. 71 with permission from American Chemical Society, Copyright 2018. [B] Image courtesy of Dr Richard Watson, University of Birmingham. [C]. Adapted from ref. 72 with permission from Elsevier, Copyright 2016. [D]. Adapted from ref. 73 with permission from Elsevier, Copyright 2019. [E]. Adapted from ref. 74 with permission from Elsevier, Copyright 2018. [F] Image courtesy of Mr. Mehul Shingadia, University of Birmingham. [G]. Adapted from ref. 75 with permission from AIP Publishing, Copyright 2003. [H]. Adapted from ref. 76 with permission from Elsevier, Copyright 2013. [I]. Adapted from ref. 77 with permission from Elsevier, Copyright 2009.

droplets.[11] Controlling and altering the fat composition and the processing conditions (such as thermal history, applied mechanical forces, *etc.*) offer a direct way to modify the network properties and thus the emulsion microstructure in this case.[12]

Whether the continuous phase is predominantly water (o/w) or oil (w/o) also significantly impacts on the emulsion's electrical conductivity. The monitoring of the electrical conductivity can be used to evaluate emulsion stability (creaming or phase inversion or, in a more extreme case, phase separation), as well as to assess both the droplet size and the distribution polydispersity.[13–15] The emulsion's capacity to be diluted is another important feature that is primarily controlled by the continuous phase and that can have a major impact in oral processing, when the emulsion is diluted with saliva and maximum flavor release is required, or during processing in order to improve emulsification efficiency.[16]

The relative solubility of the dispersed into the continuous phase has a major influence on emulsion stability. The onset and progression of Ostwald ripening increase at higher dispersed phase solubility.[17] The control of this phenomenon is of vital importance for beverages and drinks containing flavor oils (high-water solubility), whereas it has negligible impact in dairy-like emulsions usually containing triglyceride oils (low-water solubility).[18]

3.2.1.2 *The Dispersed Phase*

The dispersed phase has a dramatic impact on the final properties of emulsion products (texture, stability, appearance, delivery, and encapsulation of functional components, *etc.*) through features such as droplet size and distribution, volume fraction, charge, type, and physical state of the droplet constituents. The contribution of the dispersed phase to the overall product rheology is predominately linked to the droplet size, its polydispersity, and concentration/volume fraction.[19,20] For volume fractions (ϕ) below the maximum packing concentration ($\phi < 0.75$, for spheres), smaller droplets can result in pronounced variations in product viscosity and mechanical properties, such as viscoelasticity, that in turn can affect mouthfeel.[21] At values close to the maximum packing concentration ($0.50 < \phi < 0.75$), emulsion polydispersity plays a major role in final properties. When two monodisperse systems of different sizes are mixed together, they exhibit a minimum viscosity at a specific combination, a minimum not observed for trimodal (or more) dispersions.[22] For even more concentrated emulsions ($\phi > 0.75$), droplets deform by coming into contact with one another with the final system, exhibiting solid-like properties, such as a yield stress; mayonnaise is a classic example of such an effect.[22]

The charge of the dispersed phase plays a significant role in the physicochemical stability and in the organoleptic and nutritional properties of emulsions, mostly due to the large variety of components that may interact with the droplet interface. Emulsion stability's prevention of creaming and flocculation can dramatically decrease as a consequence of the presence of salts that act as charge screeners for ionic emulsifiers.[23] Due to their charge, droplets may attract metal ions present in solution, thus accelerating oxidative damage, or drops may also bind oppositely charged flavor molecules and affect their release performance; both effects can alter product taste.[24,25]

The dispersed phase of some emulsion-based food products may undergo partial crystallization, milk, margarine, butter, and ice cream being a few examples. Depending on shear and temperature, the presence of crystals in droplets can cause partial coalescence, which influences the physical stability, texture, and rheological features of the product.[26] Although partial coalescence is undesirable for applications requiring high stability against creaming and flocculation, it also represents an essential step in the production of foods, such as margarine, butter, and ice cream, improving product stiffness and stability.[27–29]

Finally, features such as droplet concentration, size, and refractive index have a large influence on the appearance of emulsion-based products (*e.g.* lightness, color, *etc.*), which in turn can influence consumer perception.[30] Lightness and color can be enhanced by increasing the difference in the refractive indices of the two phases.[31] Alternatively, transparent products can be obtained by matching the refractive indexes of both the continuous and dispersed phases or by significantly reducing droplet size below a certain threshold value (*e.g.* nanoemulsions).[32,33]

3.2.1.3 The Interface

The interfacial region of a food emulsion is mainly characterized by the presence of water, oil, and surface active molecules, with the latter, generally called emulsifiers, being used to extend emulsion longevity against thermodynamic instability.[34] During processing, emulsifiers are expected to quickly adsorb at the interface, reducing interfacial tension and thus favoring droplet breakup, at the same time limiting droplet contacts and coalescence; the latter performance is also required during storage in order to extend final product stability.[35,36] Typically, emulsifiers are divided into three broad categories:[37,38] low-molecular-weight surfactants; high-molecular-weight surfactants, or most commonly proteins; and nanoparticles.

3.2.1.3.1 Low-molecular-weight Surfactants. Low-molecular weight surfactants represent surface-active molecules consisting of a hydrophilic head group and a hydrophobic tail that, once at the oil/water interface, orientate their structure so that these two domains are largely present in the water and oil phases, respectively.[39] Depending on the relative size of the hydrophilic and hydrophobic parts, surfactants may be more suitable to stabilize o/w or w/o emulsions. The hydrophilic lipophilic balance (HLB) value represents an empirical method to evaluate this tendency;[40] surfactants with low HLB values (3–8) tend to form w/o emulsions, while those with high HLB values (8–20) tend to stabilize o/w emulsions. Surfactants are also categorized based on the charge of their head group.[41,42]

Nonionic surfactants possess no charge and find extensive use as stabilizers in food emulsions due to their low toxicity and irritability; examples are ethoxylated sorbitan esters (Tweens and Spans), polyoxyethylene ethers (POE), and sugar esters (sorbital monooleate, sucrose palmitate). *Ionic surfactants* can carry either a positive or a negative charge, but their application is restricted to products requiring low surfactant concentration, since they can cause irritation and have varying levels of toxicity; examples of negatively charged surfactants are citric acid ester of monoglycerides (CITREM), diacetyl tartaric acid monoglycerides (DATEM), and sodium laureth sulfate (SLS), while lauric arginate is an example of a positively charged surfactant that also exhibits high antimicrobial activity.[43] Finally, *zwitteronic surfactants* show both a positive and a negative charge, and so their net behavior depends on solution conditions (*e.g.* pH, ionic strength, *etc.*); common examples are represented by phospholipids, such as lecithins, which show improved stabilization performance when used in combination with other emulsifiers.

Surfactants are often used in mixtures where their synergistic behavior can enhance formulation performance (*e.g.* improved shelf life and product stability against temperature variations).[44] Surfactants quickly and reversibly adsorb at an emulsion interface by generating low interfacial tensions, thus favoring the formation of small droplets, although they do not usually form strong interfaces (compared to proteins or particles that form viscoelastic interfacial layers instead), and therefore droplet collisions in such systems may be prone to coalesce.[45] An important feature of low-molecular-weight surfactants is that they exist as monomers (with high surface activity) below a certain concentration but form aggregates (*i.e.* micelles) above this critical micelle concentration (CMC).[46] When CMC is exceeded, emulsion properties such as interfacial tension drastically change; for instance, o/w interfacial tension decreases by increasing the surfactant concentration while still below the CMC but remains relatively constant above it.[47]

In practice, food products rarely contain only surfactants, as these are used in combination with proteins, polysaccharides, and in many cases particles. Under these scenarios, emulsions may exhibit a completely different behavior than when the surfactant(s), protein(s), polysaccharide(s), or particles species are used alone.[48]

3.2.1.3.2 High-molecular-weight Emulsifiers: Proteins. Proteins are an important class of emulsifiers that determine the properties (texture, stability, and mouthfeel) of a wide range of products such as milk, whipped cream, and salad dressing, among others.[49] Compared to low-molecular-weight surfactants, proteins are also surface active but adsorb at the interface at a slower rate due to their larger hydrodynamic volume. Furthermore, upon adsorption, proteins lower the interfacial tension to a lesser extent due to the physical challenges in orienting their hydrophilic and lipophilic domains (which are spread across their more complex molecular structure) to be positioned within the water or oil phases, respectively.[50] Proteins reversibly adsorb at the interface, forming a viscoelastic layer of gel-like properties that greatly enhances droplet resistance to coalescence.[51,52] Features of the formed interfacial layer depend on protein configuration, which in turn is mainly determined by protein concentration and its structure (*e.g.* distribution of the hydrophobic and hydrophilic parts on the protein chain), and can be susceptible to solution conditions such as pH, ionic strength, and temperature.[53] Emulsion droplets tend to flocculate at pH values close to the isoelectric point of the protein or when the ionic strength in the system exceeds a critical value; in both cases, droplet repulsion *via* an adsorbed protein-driven electrostatic mechanism is not sufficient to prevent droplet attraction.[54] Furthermore, as a consequence of thermal variations, proteins may unfold at the interface, exposing reactive groups (*e.g.* sulfhydryl or nonpolar groups), which cause attraction between droplets and in turn promote their flocculation or coalescence.[55] Nevertheless, establishing a general protein functionality against these parameters is not trivial, since in food products, proteins are often present as mixtures.[50]

3.2.1.3.3 Nanoparticles. Emulsions stabilized by solid nanoparticles (Pickering emulsions; see Figures 3.1A and B) have received great attention due to their enhanced stability against coalescence and their capacity to provide surfactant-free formulations. These characteristics are particularly attractive for applications where high concentrations of surfactants are associated with toxicity or irritancy and/or the used surfactants originate from or are associated with unsustainable sources.[56] Particles can achieve effective stabilization if their size is at least one order of magnitude smaller than that of emulsion droplets. If successfully adsorbed at an interface, they can form a mechanical barrier against droplet coalescence.[38] Particles are generally not surface active; therefore their adsorption at the interface does not (significantly) lower interfacial tension.[57] Instead, the thermodynamic advantage associated with their interfacial adsorption resides with their capacity to reduce the effective surface area of oil/water contact.[58] Similarly, to surfactant HLB, particle wettability is the criterion used to establish the suitability of these species to stabilize o/w or a w/o emulsions. Wettability is expressed in terms of the contact angle (θ), which provides a measure of the degree of particle hydrophilicity/hydrophobicity.[48] For efficient stabilization, particles should be predominantly wetted by the continuous phase: $\theta < 90°$ for particles preferentially wetted by water (o/w emulsions) or $\theta > 90°$ for the opposite case. However, at the same time, θ should not be too close to either 0 or 180°, in which case the energy of particle desorption from the interface becomes much lower; the energy of desorption for particles of $\theta \sim 90°$ is found to be (depending also on their actual dimensions) in the order of 10^6 kT, compared to that for low-molecular-weight emulsifiers, which is in the region of 10 kT.[59] Due to the large energy needed for desorption, it is hypothesized that interfacial adsorption of particles is almost irreversible.[60]

Although superior in terms of stability, Pickering emulsions usually possess droplets with larger diameters compared to those delivered *via* an equivalent processing environment but in the presence of surfactants. This is a direct consequence of the inability of particles to (significantly) lower interfacial tension (*i.e.* promote droplet breakup) and of their much reduced interfacial adsorption rate.[61] These issues can be to an extent addressed by carefully controlling the processing environment used to generate these structures, in terms of both energy input and residence time.[62] However attractive, the use of particles for the stabilization of food emulsions is challenging. One of the main issues hindering the industrial application of these systems is the difficulty of sourcing suitable food-grade colloidal species to be used in commercial products or, similarly, constructing (from food-grade material) particulate structures within the appropriate size regime for Pickering functionality. In the last years, investigation of the stabilization performance of food-grade particles such as starch and quinoa granules, cellulose, and chitin nanocrystals, as well as flavonoid particles, has opened new pathways for the successful introduction of Pickering emulsions in the food sector.[63–67]

3.2.2 Complex Emulsion Microstructures

As The Consumer Demand For Healthier And More Functional products has increased during the last number of years, it has generated further academic and industrial interest in creating more complex emulsion microstructures capable of facilitating the development of such emulsion-based formulated foods.[68,69] As a consequence, a number of complex emulsion architectures (see Figures 3.1C–I), such as nano-, double, water-in-water, hydrogel particle–stabilized, and air-filled emulsions, have received a great deal of attention[70] and the present section aims to briefly introduce these systems and highlight their major advantages and drawbacks over conventional emulsions.

3.2.2.1 Nanoemulsions

Nanoemulsions (Figure 3.1C) are similar to conventional emulsions from a structural point of view, but with droplet dimensions spanning from 10 to 200 nm,[33] In the food industry, nanoemulsions have been receiving particular attention due to opportunities they offer over traditional emulsions such as optical transparency, improved encapsulation, delivery, and bioavailability of functional components; enhanced texture at lower dispersed phase concentration; and increased stability against creaming and flocculation.[78,79] Several challenges still limit the use of nanoemulsions in food products: (1) poor stability against Ostwald ripening and chemical degradation, (2) the requirement for a larger emulsifier concentration, and (3) the enhanced bioavailability of components whose (now higher) consumption may result in an increased risk of toxicity.[80]

3.2.2.2 Gelled Emulsions

Emulsion gels refer to a class of complex systems where it is possible to convert a biopolymer (either protein or polysaccharide) stabilized oil-in-water emulsion to a soft/solid-like material by controlling the mechanism of biopolymer gelation *via* common processing operations such as heating (Figure 3.1D), addition of ionic species (*e.g.* Ca^{2+}), acidification (Figure 3.1E), or enzymatic treatment.[10] In this case, emulsion droplets can be classified as active or inactive fillers to the external network. Active fillers are bound to the gel matrix and can either increase or decrease the gel stiffness depending on the ratio of the droplet/gel moduli, while inactive fillers do not interact with the gel network and always tend to decrease the matrix stiffness.[81] Besides constituting the bases of a wide variety of food products (*e.g.* creams, mayonnaise, yogurt, *etc.*), emulsion gels have shown promising applications in the controlled release of lipophilic components, extended resistance against lipid oxidation, and potential in the development of reduced oil/calorie emulsion-based products.[82–84]

3.2.2.3 Duplex Emulsions

Duplex or double emulsions (Figure 3.1F), such as water-in-oil-in-water ($w_1/o/w_2$) or oil-in-water-in-oil ($o_1/w/o_2$) emulsions, have been termed emulsions of emulsions: a primary (simple) emulsion (*e.g.* w/o) further emulsified into another aqueous phase to ultimately give water droplets (w_1) that are within oil droplets (o) that are, in turn, suspended in a water continuous phase (w_2).[85] Of particular interest to the food industry are $w_1/o/w_2$ emulsions due to their potential to be used for the design of low-fat products, for encapsulating and protecting water-soluble bioactive compounds (within their inner aqueous compartments), as well as controlling aroma and flavor release.[86] The major challenges that hinder the successful industrial application of double emulsions are mainly associated with (1) finding scalable and reliable manufacturing methods for their manufacture and (2) the stability issues that arise as a consequence of the osmotic pressure difference between the two water phases.[87]

3.2.2.4 Water-in-water Emulsions

Water-in-water (w/w) emulsions (Figure 3.1G) represent colloidal dispersions of water droplets within another immiscible water continuous phase, in the presence of a charged biopolymer, or a mixture of two biopolymers , or even a mixture of a biopolymer with a micelle-forming surfactant.[88] The w/w systems form due to phase separation phenomena taking place as a result of the thermodynamic incompatibility (or compatibility) of the water-soluble biopolymers (proteins or polysaccharides) included in the formulation; a typical example is the gelatin/dextran aqueous mixtures.[75] As a result, an emulsion-like microstructure can be produced at water contents approaching 99% but with dispersed phase fractions of even 50%.[89] The phase behavior and thus w/w microstructure are influenced by biopolymer physicochemical properties (*e.g.* molecular weight, charge, *etc.*) and solution conditions (pH, ionic strength, temperature, and applied shear).[90] These emulsions are characterized by extremely low values of interfacial tension (in the order of $\mu N\,m^{-1}$ rather than the $mN\,m^{-1}$ values measured for a typical oil/water interface) and have shown relevance in applications involving (1) texture modification through rheology, (2) encapsulation and delivery of (mainly) hydrophilic bioactive substances, (3) and the production of fat-reduced products.[91–93]

3.2.2.5 Filled-hydrogel Particle Emulsions

Filled hydrogel particle emulsions (Figure 3.1H) can be thought as a type of oil-in-water1-in-water2 ($o/w_1/w_2$) emulsions, in which oil droplets (o) are trapped within biopolymer particles (w_1), which in turn are suspended in a continuous phase (w_2). Again in this case, the two aqueous phases are immiscible as a result of phase separation phenomena (thermodynamic incompatibility), as with w/w emulsions.[94] The resulting particle properties (*e.g.* shape, structure, dimensions, appearance, *etc.*) can be controlled by varying operating conditions, such as heating, cooling, *etc.*[95] Filled-hydrogel particle emulsions have shown promise for applications aiming to extend product physical stability, the encapsulation and release of lipophilic bioactives, and the production of reduced-fat products.[96] With reference to the last two functionalities, the oil content present in these systems (in comparison to the absence of any oil in w/w emulsions) can be used to enclose lipophilic species and produce reduced-fat products of superior organoleptic properties. However, the additional processing steps and material required for their production still pose a serious challenge to their widespread application in food products.[96]

3.2.2.6 Air-filled Emulsions

Air-filled emulsions (Figure 3.1I) represent a three-phasic system where two dispersed phases, oil droplets and air bubbles (of comparable size), are embedded within an aqueous continuous phase.[97] The inclusion of microbubbles together with oil droplets allows the production of reduced-calorie emulsions with sensory and textural behavior that closely resembles that of their full-fat counterparts.[98] However, the stability of such three-phase systems is seriously jeopardized by instability phenomena encountered in both emulsions and foams (*e.g.* gravitational drainage, coalescence, and disproportionation). Specialty proteins (hydrophobins) have been shown to mitigate such instabilities, but their excessive cost, together with the challenges associated with the very production of the air-filled emulsions, have hindered the industrial application of these systems.[99,100]

3.3 Thermodynamic Aspects

Emulsions are inherently unstable systems, and the thermodynamic aspects of emulsion stability/instability phenomena have been extensively reviewed elsewhere (*e.g.* see ref. 2). This section briefly outlines the main mechanisms by which emulsion instability can develop and progress to, given enough time, complete phase separation. These include *creaming* (or *sedimentation*), the upward (or downward) movement of droplets under the influence of gravity because of a density difference between the continuous and dispersed phases;[101] *flocculation*, referring to the aggregation of two or more droplets into so-called flocs without loss to their individual integrity;[50] *coalescence*, the merger of two or more droplets into a single larger entity,[102] or *partial coalescence*, occurring in o/w emulsions containing an oil phase (often triglycerides) that undergoes partial crystallization;[103] *Ostwald ripening*, the diffusional mass transfer of dispersed phase from smaller to larger droplets due to the Laplace pressure difference that exists between such disparate in size entities;[104] and *phase inversion*, referring to the reversal between the dispersed and continuous phases in an emulsion (*e.g.* an oil-in-water emulsion inverting to a water-in-oil emulsion).[105] These types of instabilities are not limited to simple emulsions alone, with more complex microstructures also prone to the same or related destabilization phenomena. For example, double emulsions suffer from poor stability due to Laplace and/or osmotic pressure differences between their two aqueous phases,[106] a phenomenon that is in many ways similar to Ostwald ripening. In addition to physical instability phenomena, chemical destabilization processes, such as lipid oxidation, could also potentially lead to microstructural damage to emulsions.[26] Overall, it is important to note that emulsion instability is a rather complex process taking place *via* a number of well understood but almost always interrelated and interpromoting mechanisms.

3.4 Processing Routes for Emulsion Structure Development

In order to tailor the desired emulsion microstructure, a very important parameter is the choice of the most suitable manufacturing method.[107] Homogenization is the technique responsible for bringing together the two immiscible phases, along with surfactants and other stabilizers, to create the desired microstructure (*i.e.* average droplet size and size distribution) that ultimately determines the final properties of the product, such as texture, appearance, and shelf life.[107] This section deals with the processing aspects of creating an emulsion: from the theoretical concepts of emulsification to the description of conventional and unconventional processing techniques used nowadays for emulsion manufacturing at both the industrial and laboratory scales.

Emulsion microstructure, in terms of processing, strongly depends on the type of homogenization device employed and its operational principles.[108]

Homogenization techniques can be either mechanical or nonmechanical;[109] in the former category, emulsion formation occurs under the application of highly disruptive forces (high- or intermediate-energy methods) or spontaneously as a consequence of droplet detachment from a pore or microchannel edge (low-energy methods). On the other hand, nonmechanical methods exploit physical principles to generate emulsion microstructure.

3.4.1 Mechanical Methods

3.4.1.1 Droplet Breakup Versus Droplet Coalescence

In mechanical methods, the emulsion microstructure is formed by a balance between droplet disruption (breakup) and droplet coalescence.[109] *Droplet breakup* largely depends on the choice of the technique since the geometrical characteristics of the device, the flow regime, and the energy input determine the mode of droplet formation. Droplet disruption can take place only if (1) the external disruptive forces are able to overcome Laplace pressure and (2) these forces are applied for a period longer than the time required for droplet deformation.[110] Laplace pressure is the force responsible for maintaining the droplets' spherical shape, and it counteracts any external stresses. It is proportional to interfacial tension and inversely proportional to the Sauter mean diameter of the droplets. This means that the force resisting droplet breakup increases as the dimension of the droplet reduces; *i.e.* larger external stresses must be applied to break smaller droplets.

External stresses acting on emulsion droplets arise from the flow regime generated within the homogenization device. It is possible to distinguish between laminar viscous (LV), interfacial-driven (ID), turbulent, and cavitation flow regimes.[16,110] In the *laminar flow regime*, droplets are subjected to forces generated by simple shear, pure elongational stresses as well as a combination of both.[35] Elongational stresses are more effective than shear forces since these cause the rotation of the liquid inside a drop, thus dissipating part of the energy applied for droplet disruption.[111] During pure elongation, drop recirculation is inhibited, and a larger portion of the total energy input is actually used to cause droplet breakup.[111] In the laminar regime, the viscosity ratio (defined as the ratio of the viscosity of the dispersed phase over the viscosity of the continuous phase) has a fundamental importance because pure shear droplet disruption cannot occur for a viscosity ratio above 4, whereas in the case of pure elongation, droplet disruption is always possible for any value of viscosity ratio.[39] The *interfacial-driven* flow regime differs from conventional methods (*i.e.* where the interfacial tension counteracts disruptive forces through the Laplace pressure) because in this case the interfacial tension is the driving force causing the spontaneous droplet snap-off in correspondence to the edge of a pore (or microjunction).[110] In fact, when droplets approach the edge of a microchannel or pore, they have a noncircular shape usually with a higher interfacial energy than the corresponding spherical droplet of the same volume.

After detachment from the edge, the droplet will spontaneously adopt the lowest interfacial-energy configuration.[112,113] *Turbulent flow* is the regime most commonly encountered in traditional emulsification methods. In the turbulent regime, fluid flow is chaotic and is characterized by the presence of a spectrum of eddies within the fluid.[35] In this case, the smallest eddies are responsible for generating the energy dissipation responsible for droplet breakup.[110] From the Kolmogorov theory, it is possible to estimate the dimension of the smallest eddy within the turbulent environment based on the value of the energy dissipation rate imposed by the homogenization techniques, as well as the continuous phase flow and physicochemical properties (*i.e.* the viscosity and density, respectively).[114] Based on the comparison between the average emulsion droplet size and the smallest Kolmogorov eddy size, it is possible to distinguish between the turbulent inertial (TI) and turbulent viscous (TV) flow regimes.[115] In the TI-flow regime, the resulting droplet size is larger than the smallest eddy size predicted by the Kolmogorov theory. In this regime, the resulting droplet size does not depend on the viscosity of the continuous phase but is independent on the dispersed phase mass fraction.[116] In the TV-flow regime, the resulting droplet size is smaller than that predicted by Kolmogorov, which makes this regime more effective in producing smaller droplet sizes (in comparison to the TI one).[116] Finally, the *cavitation flow regime* is characterized by the presence of shock waves that propagate through the fluid, causing droplets to be disrupted, providing efficient mixing and resulting in the formation of very fine and highly stable emulsions.[117] This flow regime is of fundamental importance in high-pressure and ultrasonic homogenization processes.[118,119]

During each of these flow regimes, it is also possible to define the concept of maximum stable droplet diameter, which provides an estimation of the largest droplet diameter that can resist disruption under specific energy dissipation conditions.[120] Several studies have proposed a direct proportionality relationship between this maximum stable diameter and the mean emulsion droplet size.[121–123] If the constant of proportionality is known, the Sauter diameter can replace the maximum droplet size in these relationship, thus providing further ways to predict and optimize the process performance.[109] Nonetheless, it must be noted that to establish this relationship is not always possible and may change depending on the formulation and processing technique investigated.[124,125]

Emulsion droplet size reduction is clearly affected by the employed mechanical method as well as specific formulation elements such as viscosities of the two phases, dispersed phase volume fraction, and interfacial tension. However, droplet size decrease is also heavily hindered by *droplet coalescence* throughout the duration of an emulsification process. Droplet coalescence is mainly caused by droplet collisions during homogenization.[126] Emulsifiers are commonly associated with a decrease in interfacial tension (certainly in the case of surfactants) and thus can facilitate droplet breakup. Yet emulsifiers are also included in emulsion formulations in order to prevent droplets from coalescing.[127] An ideal emulsifier must adsorb at the interface as quickly as possible and create a strong boundary that is robust enough not to rupture upon droplet contacts/collisions.[127] The resistance of an interfacial layer to coalescence depends on the type as well as on the concentration of the emulsifier; surfactants quickly adsorb and form a fluid interface when compared to the more slowly adsorbing proteins or particles that instead form a viscoelastic interfacial layer.[128] Depending on the amount of emulsifier used during homogenization, one can distinguish between the surfactant-poor and -rich regimes.[128,129] Within the surfactant-poor regime, final droplet size is strictly dependent on the amount of surfactant used. In this case, the presence of surfactant at the interface is inadequate (partial interfacial coverage), and contacts between droplets will tend to result in coalescence with emulsion droplet size increasing (or similarly total interfacial area decreasing) until surfactant concentration is sufficient to allow full interfacial coverage.[128,129] In the surfactant-rich regime, final droplet size is primarily dependent on the type of homogenization process chosen and the flow regime induced, since the ample emulsifier content in the system can ensure full interfacial coverage of droplets and thus limited coalescence events.[128,129]

3.4.1.2 Energy Density and Energy Efficiency

Mechanical emulsification methods can be compared based on the concepts of energy density and energy efficiency.[2] Energy density, defined as the energy input per unit volume of emulsion, has been shown to be a suitable measure for evaluating the extent of droplet disruption during continuous mechanical emulsification.[109,130] In fact, data of Sauter mean droplet diameters as a function of corresponding energy density values have been shown (when plotted on a logarithmic scale for both axes) to give a linear relationship, with a slope that remains constant for the same type of mechanical process; *e.g.* a slope of 0.4 for high-shear mixing or a slope of 0.6 for high-pressure homogenization.[109,130] This relationship, also known as process function, provides a strong basis for product design of emulsion-based products, as it relates an emulsion microstructure element (mean droplet size) to the processing conditions utilized for its production (energy density), thus essentially allowing for the control of emulsion structure *via* the selection of appropriate process conditions.[109,130]

Emulsification techniques can also be compared based on the concept of energy efficiency, defined as the ratio of the theoretical energy required to form an emulsion (proportional to the interfacial tension and the difference in interfacial area) over the energy density (*i.e.* the actual amount of energy expended by the homogenization technique).[110] In order to form small droplets, homogenization techniques must create conditions where the actual energy expenditure is much larger than the theoretical one in order to overcome the significant pressure gradient present within droplets.[2,110] This is why, normally, emulsification processes are methods of relatively low energy efficiency, with typical energy efficiency values ranging from 10^{-3}% (high-energy methods) to 1% (intermediate-energy methods).[2,110]

3.4.1.3 *High-energy Versus Intermediate-energy Mechanical Methods*

Mechanical methods can be divided in high- and intermediate-energy approaches.[108] High-energy methods include well established techniques such as high-shear mixing, high-pressure homogenization, and sonication. Intermediate-energy methods, such as confined impinging jets, microfluidic devices, and membrane emulsification, have been developed much more recently. Although these approaches show promise in terms of emulsion structure formation at a much reduced energy density and with a significantly higher energy efficiency (in comparison to typical high-energy processing routes), their industrial adoption is currently hindered mainly due to manufacturing challenges associated with the relatively modest product throughputs that can be presently achieved.

3.4.1.3.1 High-energy Mechanical Methods. *High-shear mixers* are the most common devices used in food industry for emulsion manufacturing.[131] Within this unit operation, ingredients are blended together inside a mixing head that creates high-energy dissipation rates as a consequence of the movement of the rotor head and its proximity to the shear stator.[132,133] The intensity of the mixing can be adjusted by controlling the revolution of the rotating head and the distance (from the rotor) that the stator is placed together with its geometry.[131] They can operate both in batch or in continuous mode, process fluids of viscosity within the low to medium range, and produce emulsion droplets down to about 1 μm.[16] Essentially a variant of high-shear mixers, *colloid mills* also work on the rotor-stator principle.[134] The intensity of the shear stresses achieved in this case can be higher than those in simple high-shear mixing, and colloid mills can relatively easily handle high-viscosity material, producing emulsions with droplet sizes as small as 1 μm.[2,16] In *high-pressure homogenizers*, usually a coarse preemulsion is forced through a narrow gap where it is subjected to highly disruptive forces (a combination of turbulence and cavitation or laminar and elongational, depending on the geometry of the nozzle) caused by flow under large pressures (typical values of 50 to 1000 bar).[135,136] High-pressure homogenizers can handle fluids of low to medium viscosity and can produce droplets as small as 0.1 μm.[2,16] In *ultrasonic homogenization*, ultrasonic waves cause the generation of cavitation effects that propagate and disrupt droplets.[118,137] Although most commonly used at a bench scale, variants operating in a continuous mode and at a large scale do exist.[137] Depending on the configuration, nanoemulsions (droplet diameter $<$ 1 μm) can be produced using ultrasonic homogenization, which can process fluids of low to medium viscosity.[2,16]

3.4.1.3.2 Intermediate-energy Mechanical Methods. High-energy approaches to emulsification have been traditionally used due to their capability of delivering large product throughputs, to handle a large variety of raw material, and to produce considerably small droplets.[108] Nonetheless, these techniques suffer from (1) low energy efficiencies (a large part of energy input is wasted without being utilized for droplet size reduction purposes), (2) increases to product temperature during processing (as a result of energy being dissipated as heat), which can be detrimental to the quality of heat-sensitive species, (3) relatively low batch-to-batch reproducibility (in terms of droplet size and size distribution control) due to the large volumes of processed material.[138,139] In recent years, more sustainable, lower-energy alternatives have been proposed to overcome the limitations of high-energy manufacturing techniques. Confined impinging jets (Figure 3.2) and microfluidic and membrane emulsification (Figure 3.3) represent some examples of such intermediate-energy methods.

In *confined impinging jets* (*CIJs*), the collision of two fluids within a mixing chamber causes large levels of turbulence (Figure 3.2) over a confined volume of product, in turn causing droplet breakup and size reduction.[140] In this case, the process conditions can be tuned by adapting the flow rate of the two jets. At a lab scale, this technique has shown potential in delivering continuous operation, coupled with high product throughputs at low energy input, processing low-viscosity emulsions and obtaining droplet sizes of a few micrometers.[141,142] *Membrane emulsification* and *microchannel emulsification* represent techniques where emulsion formation occurs spontaneously as a consequence of droplet detachment from a membrane pore or at a microchannel junction.[139,143] Both methods rely on a bottom-up approach to emulsion formation (droplets are produced one at a time to give the required dispersed phase fraction) rather than the top-down approach of turbulent processing (comminution of one phase within another immiscible one). The performance of both processes can be controlled by adjusting the geometrical characteristics (*e.g.* pore/microchannel dimensions/shape and material of construction) or by changing their configuration (rotating and cross-flow membrane, or flow focusing and edge microchannel emulsification).[143,144] Although both techniques offer the major advantage of producing emulsions with highly controlled droplet sizes in a highly energy-efficient manner, they do suffer from (1) low product throughput that currently limits their use at bench-scale operation and (2) only being able to handle fluids of low viscosity.[2,16] Approaches currently explored in terms of increasing the throughput of membrane and microchannel emulsification methods include the use of membrane modules of much higher porosity and straight through pore geometry[145] (*e.g.* laser-drilled steel membranes) and surfactant delivery through the dispersed phase (as opposed to its conventional positioning within the continuous phase; see Figure 3.3)[146] and the development of microfluidic parallelization arrangements.[147]

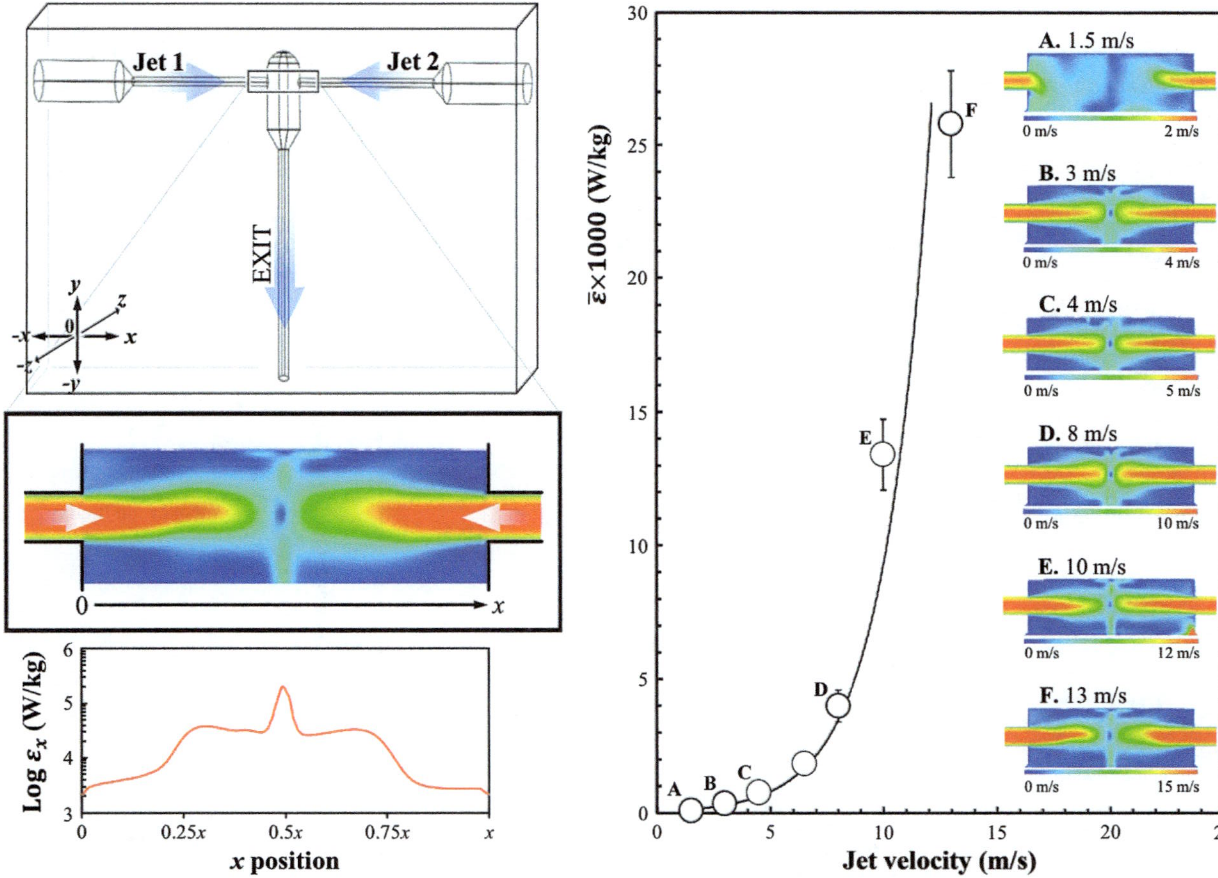

Figure 3.2 Top left: An example of a confined impinging jet (CIJ) geometry showing the two channels of jet entry. Middle left: A typical CFD velocity profile showing jet collision and associated energy dissipation values (ε_x) calculated along the jet collision path (x position; $0 \rightarrow x$) by the model (bottom left). Right: CFD-derived mean energy dissipation values ($\bar{\varepsilon}$) generated within the CIJ chamber as a function of jet velocity. Inset: CFD profiles are also provided for selected jet velocities.

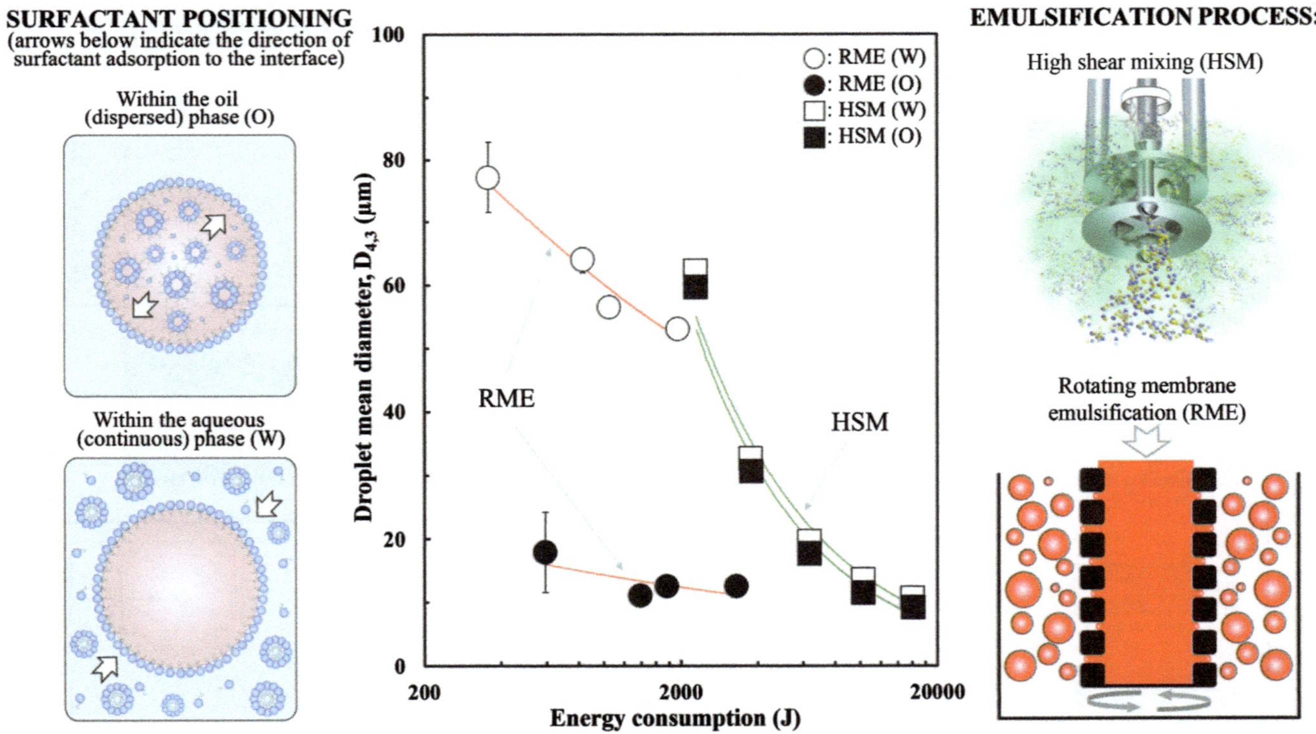

Figure 3.3 Effect of surfactant (Tween 20) positioning (dispersed *vs* continuous phase; see left-hand side) on the droplet size (μm) of o/w emulsions produced either by high shear mixing (HSM) or rotating membrane emulsification (RME) and associated energy consumption (J).
Data shown adapted from ref. 146. HSM image courtesy of Silverson (http://www.silverson.co.uk/ images/uploads/products/lab-how-it-works-stage-4.jpg).

3.4.2 Nonmechanical Methods

Nonmechanical emulsification methods exploit specific physical principles in order to drive emulsion formation, with the *phase inversion temperature* (PIT) method among the most heavily used technique.[148] The PIT method takes advantage of the change in the optimal curvature and solubility of nonionic surfactants, as a consequence of variations in temperature, to induce a transition from an o/w emulsion (below the surfactant PIT) to a w/o emulsion (above the surfactant PIT).[149,150] Through this triggered phase inversion phenomenon, the PIT method allows the formation of nanoemulsions with controlled microstructures by using a nonmechanical and low-cost approach to emulsification (compared to traditional emulsification techniques based on turbulence).[151,152] On the other hand, the higher surfactant-to-oil ratio required, coupled with the limited applicability of the technique to only a relatively small combination of water/oil/surfactant systems and experimental conditions, has hindered its industrial adoption.[153]

3.5 Emulsion Microstructure Characterization

The emulsion microstructure, as well as its individual structural constituents, can significantly affect, among other things, the processability, texture, sensory perception, and overall functionality, as well as the shelf life of the final food system. It is therefore critical for formulators to have the appropriate experimental tools to be able to characterize, evaluate, and measure a range of emulsion structural and bulk properties. Part C (Food Structure Analysis, Characterization and Modeling) of this book presents an extensive discussion on the currently available toolbox for the interrogation of food structure, which quite often is indeed relevant also to emulsion microstructure. So, in order not to replicate the discussion carried out elsewhere, the aim of the present section is to provide only a brief overview of the most important experimental techniques used to probe the emulsion microstructure and to allow the extraction of information on the overall emulsion architecture (*e.g.* droplet size) or its individual features (*e.g.* the interfacial layer), as well as on product bulk properties (*e.g.* flow behavior).[154]

3.5.1 Visualization of the Emulsion Microstructure

Light microscopy (*LM*) is suitable for the visualization of microstructural features with dimensions of ~1 μm and can provide information regarding droplet size distribution and the presence of flocculated or coalesced droplets.[2] *Electron microscopy* (*EM*) has a higher resolution that allows for microstructural elements smaller than 1 μm to be visualized, such as nanosized droplets, nanoparticles, micelles, or vesicles.[155] It is worth noting that sample preparation in both LM and EM could impact on the to-be-characterized microstructure, thus generating artifacts.[156,157] *Atomic force microscopy* (*AFM*) offers the advantage of significantly high resolution, allowing the visualization of structures and aggregates at the molecular level (*e.g.* biopolymers, surfactants), coupled with minimal specimen preparation.[157–159] One of the main disadvantages of AFM is that imaging soft structures can be challenging.[158]

3.5.2 Characterization of the Emulsion Microstructure and Its Components

Static light scattering (*SLS*) is commonly used to measure emulsion droplet sizes and droplet size distributions.[160,161] Droplets with diameters within the range of 100 nm–1000 μm can be in theory detected, but, in order to avoid multiscattering, excessive dilution is needed (<0.01%), which can affect the emulsion microstructure.[160] *Dynamic light scattering* (*DLS*) overcomes some of the limitations of the static configuration and allows for measurement of colloidal matter down to 1 nm. DLS exploits the continuous rearrangement of droplets due to Brownian motion that causes intensity fluctuations over time, which are converted to information regarding droplet size and droplet size distribution data.[162]

In *neutron magnetic resonance* (*NMR*), the emulsion is subjected to a static magnetic field gradient, and the hydrogen atoms are excited to higher energy levels, causing a signal that (1) is detectable from the NMR and (2) has an amplitude dependent on the movement of the nuclei in the sample.[163] Thus, by measuring the attenuation of the signal generated by the molecule movement in an emulsion, it is possible to derive information regarding the droplet size and droplet size distribution.[154] NMR is an useful tool in terms of non-intrusively determining the droplet size distribution of dilute as well as concentrated emulsions, emulsion stability, and fat content, with minimal sample preparation.[163] Nevertheless, despite its great potential, the wide application of NMR in industrial settings is somewhat limited by its cost.[2]

Several methods are available for the measurement of the oil–water *interfacial tension* in the presence of an emulsifier species, with direct measurement techniques using a microbalance (the Du Noüy ring and the Wilhelmy plate) and methods based on gravity-distorted drops (*Pendant drop*) among the most important.[101,164] The Pendant drop technique also offers the possibility of varying the drop volume in a controlled manner. The monitoring of droplet interfacial changes as a consequence of volume deformation can result in the determination of the interfacial tension and interfacial dilatational rheology.[2] Further characterization methods of interfacial layer properties are currently available, *e.g.* spinning drop, maximum bubble pressure, capillary rise, among others.[101,164]

Neutron scattering (NS) has been used for the determination of emulsion properties such as droplet shape, dimension, and distribution, as well as to probe the structure of the interface (*e.g.* thickness and distribution of components as a function of the distance from the surface).[165]

3.5.3 Assessing Emulsion Flow and Thermal Behavior

Rheological methods study how a material deforms/flows once subjected to a certain force or deformation, and (bulk) rheology can represent a useful tool to extract information on food systems relative to their processing, shelf life, and sensory perception.[166] Several experimental tests exist that can be categorized depending on the mode of deformation (transient or dynamic).[21] Transient tests evaluate time-dependent properties such as viscosity, relaxation stress, and compliance, whereas dynamic tests study properties in the frequency domain, and information regarding the sample viscoelasticity can be extracted.[21] *Interfacial rheology* techniques are available to characterize the relationship between the interfacial stress and the resulting interfacial layer deformation.[167,168] Depending on emulsifier type, the interface may show different viscoelastic properties in response to shear, dilation, or oscillation, equivalent to shear, elongation, and oscillation in bulk rheology.[167,168] *Tribology* (thin film rheology) provides information on texture and mouthfeel through the characterization of friction and lubrication properties of a thin film of product.[169] The understanding of the influence of factors such as the emulsion droplet size, stability to coalescence, fat content, among others, on the emulsion tribological properties can potentially allow for the design of reduced-fat products that provide a full-fat oral response.[170]

In food emulsions, *differential scanning calorimetry* (*DSC*) represents a useful tool to determine a wide range of properties (*e.g.* water and fat crystallization, emulsion stability and partial coalescence, lipid emulsifier crystallization, emulsion type, and droplet size distribution) as a consequence of thermal transitions mimicking processing and storage conditions (*e.g.* crystallization, melting, *etc.*).[171]

3.6 Emulsion Microstructure and Performance

The demand from consumers for healthier options across different processed food categories, following increased public awareness of the impact of diet on health, has presented formulators and industry with the need to design convenient and tasty but healthy foods with enhanced functionalities.[172] In Part B (Food Structure Development for Performance and Functionality) of this book, the relationship between food structure design and food system performance is explored in great detail, including, where appropriate, in-depth discussion relating to emulsion microstructures. The current section aims to provide a much more concise outline of recent applications where emulsion systems have been applied in foods in order to provide specific functionality/performance. The relationship between the design of an emulsion microstructure and the performance that this provides is a common thread in the segments that follow.

3.6.1 Emulsion Microstructure and Rheology

The understanding of the rheology of dispersions has been shown to be a useful tool in the prediction and control of processability (*i.e.* flowability) and sensory features of emulsion-based products such as creaminess, thickness, pourability, spreadability, *etc.*[21] All of these properties can be associated to the measured shear viscosity, which is ultimately influenced by a variety of structural factors (*e.g.* dispersed phase volume fraction, droplet size and size distribution, dispersed and continuous phase rheology, *etc.*).[173]

One of the major factors that drastically changes emulsion rheology is the *dispersed phase volume fraction* (ϕ). In dilute systems ($\phi < 5\%$) where colloidal forces and hydrodynamic interparticle interactions may be considered negligible, emulsion viscosity (η) depends only on the continuous phase viscosity (η_o) and linearly increases with ϕ. As the concentration of dispersed phase increases and, in the absence of colloidal interactions, emulsion viscosity (in addition to ϕ) is also influenced by the intrinsic viscosity $[\eta]$, which takes into account the shape of the droplets ($[\eta] = 2.5$ for hard spheres) and the maximum packing concentration (ϕ_c), which is the fraction of closely packed particles (in the range of 0.6–0.7 for spherical and noninteracting spheres). When colloidal interactions cannot be neglected, emulsion viscosity is more appropriately linked to the effective volume fraction, resulting from the presence of a steric or electrostatic layer around the droplets.[174] Above the maximum packing concentration, droplets are in such close proximity that they start deforming, and the emulsion exhibits a solid-like behavior characterized by an elastic modulus and a yield stress.[166]

Besides the concentration of the dispersed phase, the droplet size also plays a major role in determining the final emulsion rheology. For dispersed phase volume fractions below ϕ_c, smaller droplet sizes give rise to a larger viscosity increase (with increasing ϕ) and exhibit a more pronounced non-Newtonian behavior compared to emulsions with the same overall dispersed phase volume but of greater average droplet sizes.[20,166,175] For moderately concentrated dispersions ($\phi < \phi_c$), the emulsion polydispersity has a higher degree of importance over the average droplet size. Droplets with a bimodal size distribution can pack more efficiently, and the resulting emulsion viscosity is lower than either of the viscosities of the two size populations if present as monodisperse emulsions.[20,166,175] However, when emulsions with three or more different average droplet sizes are mixed together, this effect is lost, and the viscosity of the blend increases with the degree of emulsion polydispersity.[19] Above ϕ_c, emulsions exhibit a variegated rheological behavior, having a mechanical response both solid- and liquid-like depending on the applied load.[176]

3.6.2 Emulsion Microstructure and Oral Processing

Emulsion oral processing tends to be evaluated through a combination of rheological and tribological techniques.[172] Rheology best describes the initial stages of oral manipulation, while tribology corresponds to the squeezing and swallowing steps that follow.[177] Several studies have shown a good correlation between initial thickness perception, sliminess and stickiness and shear viscosity measured at 50 s^{-1}.[178] On the other hand, thin-film tribological properties demonstrated a more sensible relation to texture. For instance, comparisons between low- and full-fat emulsion samples (of matching rheological properties) revealed that tribology measurements were able to capture the distinct microstructural differences among the two systems that bulk rheology failed to recognize.[179]

Although rheological and tribological tests can correlate well with overall oral perception, changes to emulsion composition and microstructure in response to "environmental" conditions within the oral cavity (*e.g.* pH, temperature, *etc.*) and under the dynamic forces applied during oral manipulation must also be considered.[172] During oral processing, emulsions are prone to droplet coalescence as well as flocculation, with both phenomena linked to a creamy perception.[180–182] In gelled emulsions, with oil droplets trapped within the gel network, fat release is dependent on whether the globules act as active or inactive fillers, and sensory perception is related to the melting of the gel structure.[183]

3.6.3 Emulsion Microstructure for Encapsulation and Release

Emulsions represent an efficient medium where functional components can be encapsulated and then released at specific locations within the gastrointestinal (GI) tract in a controlled and gradual manner or in response to an environmental cue (*e.g.* change in pH, ionic strength, temperature, *etc.*).[184] *Conventional* as well as *nano*emulsions have been extensively studied for the encapsulation of lipophilic species (*e.g.* fatty acids, carotenoids, and antioxidants).[185] Their release performance can be tuned through manipulation of microstructural parameters such as droplet size, composition of the interfacial layer, and physical state of the lipid phase. In general, smaller droplet sizes allow for greater release rates.[70,96] Coating the interface of emulsion droplets with surface-active agents (*e.g.* phospholipids, proteins) has been shown to delay chemical degradation of the enclosed bioactives and slow down their release as a result of interfacial displacement of the original emulsifiers by the surface-active moieties secreted by the human body.[185,186] Finally, the presence of crystals within oil droplets has been shown to delay the release of lipophilic components.[187] Although both conventional and nanoemulsion microstructures have shown great promise for the encapsulation and release of lipophilic components, it is worth noting that a large proportion of bioactives with relevance to the food industry are primarily water soluble (*e.g.* minerals, vitamins, enzymes, proteins, *etc.*).[188]

More complex emulsion microstructures have also shown potential in terms of acting as suitable carrier systems. *Gelled emulsions* can be utilized for the controlled and/or triggered release of lipophilic components at specific locations in the GI tract by tuning changes to the gel structure in response to variations in temperature, pH, salt concentration, or the application of mechanical forces.[189] Droplets that are bound to the matrix (inactive fillers) are usually released more quickly than unbound droplets (active fillers).[189] *Double emulsions* have attracted attention as effective media for the encapsulation of both hydrophilic and lipophilic components, with $w_1/o/w_2$ microstructures being the main focus.[190] Hydrophilic components can be incorporated in the inner w_1 phase, whereas lipophilic bioactives are enclosed within the oil compartment of the double emulsion architecture. Oil droplets can also act as a barrier to the discharge of w_1-enclosed hydrophilic actives by increasing the diffusion timescales for their release.[190,191] Similarly to this, *hydrogel filled particle emulsions*[96] can operate as carriers of lipophilic components within their oil droplets, with the biopolymer particle network surrounding these, further regulating release performance.

Although currently more relevant to pharmaceutics, the capacity of emulsion microstructures to facilitate the release of multiple actives (corelease) is an extremely attractive proposition that can be easily envisaged to apply within the foods arena in the future. Microemulsions and double emulsions have been studied for the corelease of at least two actives. However, codelivery in both cases is practically limited to either simultaneous or sequential, rather than independent release profiles.[192–194] A simple o/w emulsion microstructure has also been shown to act in this capacity but to additionally offer the coencapsulation of two model actives in a segregated manner and to deliver these independently in response to disparate stimuli/triggers.[71]

3.6.4 Fat, Salt, and Sugar Reduction

Despite their significant contribution to the taste, aroma, and texture of food products, fat, salt, and sugar have been shown to influence long-term health status. Formulators, regulatory bodies, and more recently consumers are now increasingly aware of the diet-related health implications arising from the consumption of foods containing high amounts of these components, and the public in particular has become more conscious of their eating habits. In emulsion-based products, engineering the emulsion microstructure has been proposed as a possible vehicle to lower fat, salt, and sugar content without affecting significantly the organoleptic properties of products developed for a healthier lifestyle.

Fat reduction in emulsion systems is not a trivial operation since fats play a determining role in the overall texture and mouthfeel of these systems, which in turn make the product acceptable and enjoyable from a customer perspective.[195] Some strategies (*e.g.* interfacial engineering, addition of thickening agents, controlling the manifestation of instability phenomena) have been proposed in addition to the use of complex types of emulsions to achieve this target. In the effort to reduce lipid adsorption within the GI tract, interfacial properties of the oil droplets can be modified either by using an emulsifier able to create a viscoelastic interfacial layer (*e.g.* proteins), adopting Pickering stabilization or by coating droplets with non-digestible fibers (*e.g.* inulin-type fructans).[196–198] Textural and rheological features of full-fat products can be duplicated by controlling the network formation caused by thickening agents (*e.g.* agar, alginate, xanthan) or by using fat substitutes such as protein microparticles or starch granules as fat replacers.[199–202] The general drawback of these approaches is that their adoption may not be able to compensate for the loss of other emulsion food properties such as appearance, mouthfeel (associated with thin-film rather than bulk flow behavior), or flavor profile, due to the reduced fat content.[203,204] During oral processing, the occurrence of emulsion instability phenomena such as flocculation or partial coalescence can be induced to increase the viscosity of the product and thus mimic fat-related sensory attributes such as creaminess and mouth coating.[205,206] On the other hand, if not well controlled, coalescence or phase inversion instability phenomena can negatively impact product perception.[203,204] In *double emulsions* (*e.g.* $w_1/o/w_2$), fat reduction can be in theory achieved while still preserving both rheological and sensorial properties of the full fat because (1) the incorporation of w_1 droplets within the oil allows for less fat being used, while at the same time (2) the overall volume fraction of perceived oil globules in the system remains practically unchanged.[195,207] Similarly to double emulsions, the inclusion of air bubbles as replacers of fat globules in *air-bubble-filled emulsions* has shown the prospect of creating triphasic systems of reduced calorific content and similar rheological features as the corresponding simple o/w emulsions.[98,208] Finally, *hydrogel-particle-filled emulsions* have also shown potential in replicating the sensorial performance of traditional full-fat products. This can be achieved by manipulating the responsiveness of the gel particle to the oral stresses involved during oral processing, as well as by controlling how particle deformation would regulate the release/expulsion of the fat droplets it contains.[76]

For the food industry, *salt* and *sugar* reduction also represents a challenging task.[92] At the moment, the only successful applied approaches consist in the use of salt or sugar substitutes such as mineral salts in the former case or either natural (*e.g.* Steviol glycosides, stevioside) or artificial (*e.g.* aspartame, acelsufame-K) sweeteners in the latter case.[209,210] Both, however, can have major drawbacks in terms to off tastes (for both salt replacers and natural sweeteners) or perceived/actual health hazards (for artificial sweeteners).[209,210] Attempts for salt and sugar reduction have also been conducted through the manipulation of the emulsion microstructure, with double emulsions and gels filled with emulsion droplets among the main strategies. For instance, through the encapsulation of salt or sugars within the inner (w_1) droplets of a double emulsion ($w_1/o/w_2$), it is possible to regulate the release of both components and effectively optimize their detection by the taste receptors in the mouth. As a result, this approach can be employed to enhance the saltiness/sweetness perception of emulsions with reduced salt/sugar content to ideally match that of their full salt/sugar predecessors.[98,211] Another proposed strategy involves trapping sugar (or equivalently salt) molecules within a gel network filled with emulsion droplets.[212–214] By modulating the interaction between the droplets and the gel network, it is possible to tune the fracture behavior of these systems. The formation of more brittle gels, *i.e.* with a lower Young's modulus, allows (upon mastication) the rupture of the network into smaller fragments, in turn enhancing sweetness (or saltiness) perception.[212,214]

3.7 Conclusions and Future Perspective

The relationship between emulsion microstructure development, through the control of both formulation and processing aspects, and product performance has been the common theme throughout this chapter. Nonetheless the design and development of emulsion structures is a rather complex operation and one that is practically impossible to cover in full detail within the confines of a single book chapter. If nothing else, the discussion presented here serves to clearly demonstrate the significant formulation/processing challenges facing both industry and academia for the development of new, enjoyable, nutritious and inexpensive emulsion-based food microstructures that meet growing consumer needs. It is these very challenges that also present food formulators and engineers with exciting opportunities for further innovation.

The continuously increasing public awareness of the impact of diet on health, has increased the demand for healthy foods and thus the necessity to develop and design novel products having superior properties. Despite the great deal of interest that has been dedicated to this, the combination of food engineering and sensory sciences and human physiology, in order to correlate biological response (*e.g.* digestion, absorption) and organoleptic properties (*e.g.* mouthfeel, taste, aroma) to the emulsion microstructural features, remains a fascinating and still expanding research area. In addition, the production of convenient and tasty foods should strive to improve current manufacturing efficiencies and production costs either through the development of new emulsification methods or through the further optimization of existing methods. Successful scale-up and scale-down approaches are equally important in order to develop processing techniques that possess the flexibility to be adapted to large-scale industrial manufacturing, as well as to late customization/home finishing of food products.

References

1. M. M. Robins and P. J. Wilde, in *Encyclopedia of Food Sciences and Nutrition*, ed. B. Caballero, Academic Press, Oxford, 2003, pp. 1517–1524.

2. D. J. McClements, *Food Emulsions: Principles, Practices and Techniques*, CRC Press, Boca Raton, 2016.
3. Z. Ma and J. I. Boye, *Food Bioprocess Technol.*, 2013, 6, 648.
4. M. Akhtar, J. Stenzel, B. S. Murray and E. Dickinson, *Food Hydrocolloids*, 2005, **19**, 521.
5. S. Aben, C. Holtze, T. Tadros and P. Schurtenberger, *Langmuir*, 2012, **28**, 7967.
6. D. J. McClements, *Food Hydrocolloids*, 2000, **14**, 173.
7. D. Saha and S. Bhattacharya, *J. Food Sci. Technol.*, 2010, **47**, 587.
8. G. Sworn, in *Gums and Stabilisers for the Food Industry 12*, ed. P. A. Williams and G. O. Phillips, Royal Society of Chemistry, Cambridge, 2004, vol. 2, pp. 13–22.
9. J. de Vries, in *Gums and Stabilisers for the Food Industry 12*, ed. P. A. Williams and G. O. Phillips, Royal Society of Chemistry, Cambridge, 2004, vol. 3, pp. 23–31.
10. E. Dickinson, *Food Hydrocolloids*, 2012, **28**, 224.
11. F. R. Lupi, D. Gabriele, B. de Cindio, M. C. Sánchez and C. Gallegos, *J. Food Eng.*, 2011, **107**, 296.
12. S. S. Narine and A. G. Marangoni, *Food Res. Int.*, 1999, **32**, 227.
13. E. Tyrode, J. Allouche, L. Choplin and J.-L. Salager, *Ind. Eng. Chem. Res.*, 2005, **44**, 67.
14. K. Feitosa, S. Marze, A. Saint-Jalmes and D. J. Durian, *J. Phys.: Condens. Matter*, 2005, **17**, 6301.
15. E. M. Varka, C. Ampatzidis, M. Kostoglou, T. Karapantsios and V. Dutschk, *Colloids Surf., A*, 2010, **365**, 181.
16. P. Walstra and P. E. A. Smulders, in *Modern Aspects of Emulsion Science*, ed. B. P. Bink, Royal Society of Chemistry, Cambridge, 1998, vol. 2, pp. 56–99.
17. D. J. McClements, L. Henson, L. M. Popplewell, E. A. Decker and S. Jun Choi, *J. Food Sci.*, 2012, **77**, C33.
18. D. T. Piorkowski and D. J. McClements, *Food Hydrocolloids*, 2014, **42**, 5.
19. R. Pal, *AIChE J.*, 1996, **42**, 3181.
20. R. Pal, *Curr. Opin. Colloid Interface Sci.*, 2011, **16**, 41.
21. G. Tabilo-Munizaga and G. V. Barbosa-Cánovas, *J. Food Eng.*, 2005, **67**, 147.
22. R. Pal, *Food Hydrocolloids*, 2006, **20**, 997.
23. S. Ogawa, E. A. Decker and D. J. McClements, *J. Agric. Food Chem.*, 2003, **51**, 2806.
24. H. Singh, A. Ye and D. Horne, *Prog. Lipid Res.*, 2009, **48**, 92.
25. L. Mei, E. A. Decker and D. J. McClements, *J. Agric. Food Chem.*, 1998, **46**, 5072.
26. J. N. Coupland and D. J. McClements, *Trends Food Sci. Technol.*, 1996, **7**, 83.
27. K. Boode, P. Walstra and A. E. A. de Groot-Mostert, *Colloids Surf., A*, 1993, **81**, 139.
28. H. D. Goff, *Curr. Opin. Colloid Interface Sci.*, 2002, **7**, 432.

29. P. Buldo, U. Andersen and L. Wiking, *Food Biophys.*, 2013, **8**, 262.
30. D. J. McClements, *Curr. Opin. Colloid Interface Sci.*, 2002, **7**, 451.
31. W. Chantrapornchai, F. M. Clydesdale and D. J. McClements, in *Color Quality of Fresh and Processed Foods*, American Chemical Society, 2008, vol. 26, pp. 364–387.
32. J. Ziming Sun, M. C. E. Erickson and J. W. Parr, *Int. J. Cosmet. Sci.*, 2005, **27**, 355.
33. H. D. Silva, M. â. Cerqueira and A. A. Vicente, *Food Bioprocess Technol.*, 2012, **5**, 854.
34. P. J. Wilde, *Curr. Opin. Colloid Interface Sci.*, 2000, **5**, 176.
35. P. Walstra, *Chem. Eng. Sci.*, 1993, **48**, 333.
36. J. Palanuwech and J. N. Coupland, *Colloids Surf., A*, 2003, **223**, 251.
37. K. P. Das and J. E. Kinsella, in *Advances in Food and Nutrition Research*, ed. J. E. Kinsella, Academic Press, 1990, pp. 81–201.
38. E. Dickinson, *Curr. Opin. Colloid Interface Sci.*, 2010, **15**, 40.
39. P. Walstra, *Physical Chemistry of Foods*, CRC Press, New York, 2003.
40. R. C. Pasquali, M. P. Taurozzi and C. Bregni, *Int. J. Pharm.*, 2008, **356**, 44.
41. H. J. Butt, K. Graf and M. Kappl, *Physics and Chemistry of Interfaces*, Wiley-VCH, Weinheim, 2003.
42. D. J. McClements and J. Rao, *Crit. Rev. Food Sci. Nutr.*, 2011, **51**, 285.
43. Y. Chang, L. McLandsborough and D. J. McClements, *Food Chem.*, 2015, **172**, 298.
44. B. Kronberg, *Curr. Opin. Colloid Interface Sci.*, 1997, **2**, 456.
45. D. G. Dalgleish, *Food Hydrocolloids*, 2006, **20**, 415.
46. R. Zana, *Dynamics of Surfactant Self-Assemblies*, CRC Press, Boca Raton, 2005.
47. M. Stang, H. Karbstein and H. Schubert, *Chem. Eng. Process.*, 1994, **33**, 307.
48. R. Pichot, F. Spyropoulos and I. T. Norton, *J. Colloid Interface Sci.*, 2009, **329**, 284.
49. E. Dickinson, *Colloids Surf., B*, 2010, **81**, 130.
50. S. Damodaran, *J. Food Sci.*, 2005, **70**, R54.
51. P. Wilde, A. Mackie, F. Husband, P. Gunning and V. Morris, *Adv. Colloid Interface Sci.*, 2004, **108–109**, 63.
52. A. Dan, G. Gochev, J. Krägel, E. V. Aksenenko, V. B. Fainerman and R. Miller, *Curr. Opin. Colloid Interface Sci.*, 2013, **18**, 302.
53. E. Dickinson, *Colloids Surf., B*, 2001, **20**, 197.
54. L. M. C. Sagis and E. Scholten, *Trends Food Sci. Technol.*, 2014, **37**, 59.
55. H. Singh and A. Sarkar, *Adv. Colloid Interface Sci.*, 2011, **165**, 47.
56. Y. Chevalier and M.-A. Bolzinger, *Colloids Surf., A*, 2013, **439**, 23.
57. J. Xiao, Y. Li and Q. Huang, *Trends Food Sci. Technol.*, 2016, **55**, 48.
58. R. Aveyard, B. P. Binks and J. H. Clint, *Adv. Colloid Interface Sci.*, 2003, **100–102**, 503.
59. E. Dickinson, *Trends Food Sci. Technol.*, 2012, **24**, 4.

60. T. N. Hunter, R. J. Pugh, G. V. Franks and G. J. Jameson, *Adv. Colloid Interface Sci.*, 2008, **137**, 57.
61. E. Dickinson, *J. Sci. Food Agric.*, 2013, **93**, 710.
62. N. Zhang, L. Zhang and D. Sun, *Langmuir*, 2015, **31**, 4619.
63. I. Kalashnikova, H. Bizot, B. Cathala and I. Capron, *Langmuir*, 2011, **27**, 7471.
64. Z. Luo, B. S. Murray, A. Yusoff, M. R. A. Morgan, M. J. W. Povey and A. J. Day, *J. Agric. Food Chem.*, 2011, **59**, 2636.
65. A. Yusoff and B. S. Murray, *Food Hydrocolloids*, 2011, **25**, 42.
66. M. V. Tzoumaki, T. Moschakis, V. Kiosseoglou and C. G. Biliaderis, *Food Hydrocolloids*, 2011, **25**, 1521.
67. L. J. Duffus, J. E. Norton, P. Smith, I. T. Norton and F. Spyropoulos, *J. Colloid Interface Sci.*, 2016, **473**, 9.
68. I. Norton, S. Moore and P. Fryer, *Obes. Rev.*, 2007, **8**, 83.
69. J. M. Aguilera, *J. Food Eng.*, 2005, **67**, 3.
70. D. J. McClements, *Adv. Colloid Interface Sci.*, 2015, **219**, 27.
71. F. Spyropoulos, D. Kurukji, P. Taylor and I. T. Norton, *Langmuir*, 2018, **34**, 3934.
72. M. I. Guerra-Rosas, J. Morales-Castro, L. A. Ochoa-Martínez, L. Salvia-Trujillo and O. Martín-Belloso, *Food Hydrocolloids*, 2016, **52**, 438.
73. N. Luo, A. Ye, F. M. Wolber and H. Singh, *Food Hydrocolloids*, 2019, **92**, 19.
74. S. Ben-Harb, M. Panouillé, D. Huc-Mathis, G. Moulin, A. Saint-Eve, F. Irlinger, P. Bonnarme, C. Michon and I. Souchon, *Food Hydrocolloids*, 2018, **77**, 75.
75. B. Wolf and W. J. Frith, *J. Rheol.*, 2003, **47**, 1151.
76. C. Chung, B. Degner, E. A. Decker and D. J. McClements, *Innovative Food Sci. Emerg.*, 2013, **20**, 324.
77. F. L. Tchuenbou-Magaia, I. T. Norton and P. W. Cox, *Food Hydrocolloids*, 2009, **23**, 1877.
78. S. Goindi, A. Kaur, R. Kaur, A. Kalra and P. Chauhan, in *Emulsions*, ed. A. M. Grumezescu, Academic Press, New York, 2016, vol. 19, pp. 651–688.
79. T. Tadros, P. Izquierdo, J. Esquena and C. Solans, *Adv. Colloid Interface Sci.*, 2004, **108–109**, 303.
80. F. Donsì, in *Nanoemulsions*, ed. S. M. Jafari and D. J. McClements, Academic Press, New York, 2018, vol. 11, pp. 349–377.
81. G. Sala, F. van de Velde, M. A. Cohen Stuart and G. A. van Aken, *Food Hydrocolloids*, 2007, **21**, 977.
82. A. C. K. Sato, K. E. F. P. Moraes and R. L. Cunha, *Food Hydrocolloids*, 2014, **34**, 184.
83. L. Oliver, L. Berndsen, G. A. van Aken and E. Scholten, *Food Hydrocolloids*, 2015, **50**, 74.
84. M. E. Malone and I. A. M. Appelqvist, *J. Controlled Release*, 2003, **90**, 227.

85. N. Garti and C. Bisperink, *Curr. Opin. Colloid Interface Sci.*, 1998, **3**, 657.
86. G. Muschiolik, *Curr. Opin. Colloid Interface Sci.*, 2007, **12**, 213.
87. N. Garti and A. Benichou, in *Food Emulsions*, ed. S. Friberg, K. Larsson and J. Sjöblom, Marcel Dekker, New York, 2003, vol. 10, pp. 362–422.
88. I. Capron, S. Costeux and M. Djabourov, *Rheol. Acta*, 2001, **40**, 441.
89. D. M. A. Buzza, P. D. I. Fletcher, T. K. Georgiou and N. Ghasdian, *Langmuir*, 2013, **29**, 14804.
90. W. J. Frith, *Adv. Colloid Interface Sci.*, 2010, **161**, 48.
91. A. Pawlik, P. Fryer and I. T. Norton, in *Formulation Engineering of Foods*, ed. J. E. Norton, P. J. Fryer and I. T. Norton, Wiley-Blackwell, Oxford, 2013, vol. 5, pp. 82–107.
92. J. E. Norton, Y. Gonzalez Espinosa, R. L. Watson, F. Spyropoulos and I. T. Norton, *Food Funct.*, 2015, **6**, 663.
93. E. Dickinson, *Food Hydrocolloids*, 2016, **52**, 497.
94. A. Matalanis, U. Lesmes, E. A. Decker and D. J. McClements, *Food Hydrocolloids*, 2010, **24**, 689.
95. D. J. McClements, *Adv. Colloid Interface Sci.*, 2017, **240**, 31.
96. D. J. McClements, *Curr. Opin. Colloid Interface Sci.*, 2012, **17**, 235.
97. F. L. Tchuenbou-Magaia, PhD Thesis, University of Birmigham, 2012.
98. J. E. Norton, G. A. Wallis, F. Spyropoulos, P. J. Lillford and I. T. Norton, *Annu. Rev. Food Sci. Technol.*, 2014, **5**, 177–195.
99. F. L. Tchuenbou-Magaia, N. Al-Rifai, N. E. M. Ishak, I. T. Norton and P. W. Cox, *J. Cell. Plast.*, 2011, **47**, 217–232.
100. B. J. D. Le Révérend, I. T. Norton, P. W. Cox and F. Spyropoulos, *Curr. Opin. Colloid Interface Sci.*, 2010, **15**, 84.
101. T. Tadros, in *Encyclopedia of Colloid and Interface Science*, ed. T. Tadros, Springer, Berlin, Heidelberg, 2013, p. 635.
102. I. B. Ivanov, K. D. Danov and P. A. Kralchevsky, *Colloids Surf., A*, 1999, **152**, 161.
103. K. Boode and P. Walstra, *Colloids Surf., A*, 1993, **81**, 121.
104. A. Kabalnov, in *Modern Aspects of Emulsion Science*, ed. B. P. Binks, Royal Society of Chemistry, Cambridge, 1998, **vol.** 7, pp. 205–260.
105. B. W. Brooks, H. N. Richmond and M. Zerfa, in *Modern Aspects of Emulsion Science*, ed. B. P. Binks, Royal Society of Chemistry, Cambridge, 1998, vol. 6, pp. 175–204.
106. G. Muschiolik and E. Dickinson, *Compr. Rev. Food Sci. Food Saf.*, 2017, **16**, 532.
107. E. J. Windhab, M. Dressler, K. Feigl, P. Fischer and D. Megias-Alguacil, *Chem. Eng. Sci.*, 2005, **60**, 2101.
108. S. M. Jafari, E. Assadpoor, Y. He and B. Bhandari, *Food Hydrocolloids*, 2008, **22**, 1191.
109. H. Schubert and R. Engel, *Chem. Eng. Res. Des.*, 2004, **82**, 1137.
110. M. Rayner, in *Engineering Aspects of Food Emulsification and Homogenisation*, ed. M. Rayner and P. Dejmek, CRC Press, Boca Raton, 2015, vol. 2, pp. 3–32.
111. C. L. Tucker III and P. Moldenaers, *Annu. Rev. Fluid Mech.*, 2002, **34**, 177.

112. M. Yasuno, M. Nakajima, S. Iwamoto, T. Maruyama, S. Sugiura, I. Kobayashi, A. Shono and K. Satoh, *J. Member. Sci.*, 2002, **210**, 29.
113. I. Kobayashi, S. Mukataka and M. Nakajima, *J. Colloid Interface Sci.*, 2004, **279**, 277.
114. N. Vankova, S. Tcholakova, N. D. Denkov, I. B. Ivanov, V. D. Vulchev and T. Danner, *J. Colloid Interface Sci.*, 2007, **312**, 363.
115. N. Vankova, S. Tcholakova, N. D. Denkov, V. D. Vulchev and T. Danner, *J Colloid Interface Sci.*, 2007, **313**, 612.
116. S. Tcholakova, I. Lesov, K. Golemanov, N. D. Denkov, S. Judat, R. Engel and T. Danner, *Langmuir*, 2011, **27**, 14783.
117. A. Patist and D. Bates, *Innovative Food Sci. Emerging Technol.*, 2008, **9**, 147.
118. J. P. Canselier, H. Delmas, A. M. Wilhelm and B. Abismaïl, *J. Dispersion Sci. Technol.*, 2002, **23**, 333.
119. A. Håkansson, L. Fuchs, F. Innings, J. Revstedt, B. Bergenståhl and C. Trägårdh, *J. Food Eng.*, 2010, **100**, 504.
120. J. A. Boxall, C. A. Koh, E. D. Sloan, A. K. Sum and D. T. Wu, *Langmuir*, 2012, **28**, 104.
121. F. B. Sprow, *Chem. Eng. Sci.*, 1967, **22**, 435.
122. R. V. Calabrese, C. Y. Wang and N. P. Bryner, *AIChE J.*, 1986, **32**, 677.
123. J. Baldyga and W. Podgórska, *Can. J. Chem. Eng.*, 1998, **76**, 456.
124. G. Zhou and S. M. Kresta, *Chem. Eng. Sci.*, 1998, **53**, 2063.
125. A. W. Pacek, C. C. Man and A. W. Nienow, *Chem. Eng. Sci.*, 1998, **53**, 2005.
126. S. Tesch and H. Schubert, *J. Food Eng.*, 2002, **52**, 305.
127. S. Tesch, C. Gerhards and H. Schubert, *J. Food Eng.*, 2002, **54**, 167.
128. S. Tcholakova, N. D. Denkov, D. Sidzhakova, I. B. Ivanov and B. Campbell, *Langmuir*, 2003, **19**, 5640.
129. S. Tcholakova, N. D. Denkov and T. Danner, *Langmuir*, 2004, **20**, 7444.
130. H. Schubert, K. Ax and O. Behrend, *Trends Food Sci. Technol.*, 2003, **14**, 9.
131. J. Zhang, S. Xu and W. Li, *Chem. Eng. Process.*, 2012, **57–58**, 25.
132. R. V. Calabrese, T. P. K. Chang and P. T. Dang, *AIChE J.*, 1986, **32**, 657.
133. S. Hall, M. Cooke, A. El-Hamouz and A. J. Kowalski, *Chem. Eng. Sci.*, 2011, **66**, 2068.
134. K. Urban, G. Wagner, D. Schaffner, D. Röglin and J. Ulrich, *Chem. Eng. Technol.*, 2006, **29**, 24.
135. L. Lee and I. T. Norton, *J. Food Eng.*, 2013, **114**, 158.
136. C. Qian and D. J. McClements, *Food Hydrocolloids*, 2011, **25**, 1000.
137. O. Behrend, K. Ax and H. Schubert, *Ultrason. Sonochem.*, 2000, **7**, 77.
138. S. van der Graaf, C. G. P. H. Schroën and R. M. Boom, *J. Memb. Sci.*, 2005, **251**, 7.
139. G. T. Vladisavljević, I. Kobayashi and M. Nakajima, *Microfluid. Nanofluid.*, 2012, **13**, 151.
140. B. K. Johnson and R. K. Prud'homme, *AIChE J.*, 2003, **49**, 2264.

141. S. W. Siddiqui, W. A. F. Wan Mohamad, M. F. Mohd Rozi and I. T. Norton, *Ind. Eng. Chem. Res.*, 2017, **56**, 12833.
142. S. W. Siddiqui and I. T. Norton, *J. Colloid Interface Sci.*, 2012, **377**, 213.
143. S. M. Joscelyne and G. Trägårdh, *J. Memb. Sci.*, 2000, **169**, 107.
144. F. Y. Ushikubo, D. R. B. Oliveira, M. Michelon and R. L. Cunha, *Food Eng. Rev.*, 2015, **7**, 393.
145. R. D. Hancocks, F. Spyropoulos and I. T. Norton, *J. Memb. Sci.*, 2016, **497**, 29.
146. D. M. Lloyd, I. T. Norton and F. Spyropoulos, *J. Food Eng.*, 2015, **166**, 316.
147. T. Nisisako, T. Ando and T. Hatsuzawa, *Lab Chip*, 2012, **12**, 3426.
148. T. F. Tadros, in *Emulsion Science and Technology*, ed. T. F. Tadros, Wiley-VCH Verlag GmbH & Co. KGaA, Weinheim, 2009, vol. 1, pp. 1–56.
149. S. L. Ee, X. Duan, J. Liew and Q. D. Nguyen, *Chem. Eng. J.*, 2008, **140**, 626.
150. P. Fernandez, V. André, J. Rieger and A. Kühnle, *Colloids Surf., A*, 2004, **251**, 53.
151. P. Izquierdo, J. Esquena, T. F. Tadros, C. Dederen, M. J. Garcia, N. Azemar and C. Solans, *Langmuir*, 2002, **18**, 26.
152. P. Izquierdo, J. Esquena, T. F. Tadros, J. C. Dederen, J. Feng, M. J. Garcia-Celma, N. Azemar and C. Solans, *Langmuir*, 2004, **20**, 6594.
153. A. Perazzo, V. Preziosi and S. Guido, *Adv. Colloid Interface Sci.*, 2015, **222**, 581.
154. D. J. McClements, *Crit. Rev. Food Sci. Nutr.*, 2007, **47**, 611.
155. V. Klang and C. Valenta, *J. Drug Delivery Sci. Technol.*, 2011, **21**, 55.
156. A. Caillet, C. Cogné, J. Andrieu, P. Laurent and A. Rivoire, *Food Sci. Technol.*, 2003, **36**, 743.
157. H. Yang, Y. Wang, S. Lai, H. An, Y. Li and F. Chen, *J. Food Sci.*, 2007, **72**, R65.
158. D. Ricci and P. C. Braga, in *Atomic Force Microscopy*, ed. P. C. Braga and D. Ricci, Humana Press, New York, 2004, 1, 3–12.
159. A. P. Gunning, A. R. Mackie, P. J. Wilde and V. J. Morris, *Langmuir*, 2004, **20**, 116.
160. D. Lehner, G. Kellner, H. Schnablegger and O. Glatter, *J. Colloid Interface Sci.*, 1998, **201**, 34.
161. H. Lindner, G. Fritz and O. Glatter, *J. Colloid Interface Sci.*, 2001, **242**, 239.
162. S. K. Brar and M. Verma, *Trends Anal. Chem.*, 2011, **30**, 4.
163. B. Balinov, F. Mariette and O. Söderman, in *Food Emulsions*, ed. S. Friberg, K. Larsson and J. Sjöblom, Marcel Dekker, New York, 2003, vol. 15, pp. 601–640.
164. M. Korenko and F. Šimko, *J. Chem. Eng. Data*, 2010, **55**, 4561.
165. J. H. E. Hone, T. Cosgrove, M. Saphiannikova, T. M. Obey, J. C. Marshall and T. L. Crowley, *Langmuir*, 2002, **18**, 855.
166. S. R. Derkach, *Adv. Colloid Interface Sci.*, 2009, **151**, 1.
167. B. S. Murray, *Curr. Opin. Colloid Interface Sci.*, 2002, **7**, 426.

168. R. Miller, V. B. Fainerman, J. Krägel and G. Loglio, *Curr. Opin. Colloid Interface Sci.*, 1997, **2**, 578.
169. S. M. Goh, in *Encyclopedia of Tribology*, ed. Q. J. Wang and Y.-W. Chung, Springer US, Boston, MA, 2013, pp. 3871–3873.
170. D. M. Dresselhuis, H. J. Klok, M. A. C. Stuart, R. J. de Vries, G. A. van Aken and E. H. A. de Hoog, *Food Biophys.*, 2007, **2**, 158.
171. S. Miao and L. Mao, in *Differential Scanning Calorimetry*, ed. E. Chiavaro, CRC Press, Boca Raton, 2015, vol. 10, pp. 243–272.
172. E. Dickinson, *Food Hydrocolloids*, 2018, **77**, 372.
173. J. R. Mitchell and B. Wolf, in *Practical Food Rheology*, ed. I. T. Norton, F. Spyropoulos and P. W. Cox, Blackwell Publishing Ltd., Hoboken, New Jersey, 2010, vol. 8, pp. 173–192.
174. T. F. Tadros, *Colloids Surf., A*, 1994, **91**, 39.
175. P. Fischer and E. J. Windhab, *Curr. Opin. Colloid Interface Sci.*, 2011, **16**, 36.
176. S. Cohen-Addad and R. Höhler, *Curr. Opin. Colloid Interface Sci.*, 2014, **19**, 536.
177. J. R. Stokes, M. W. Boehm and S. K. Baier, *Curr. Opin. Colloid Interface Sci.*, 2013, **18**, 349.
178. M. E. Malone, I. A. M. Appelqvist and I. T. Norton, *Food Hydrocolloids*, 2003, **17**, 763.
179. J. R. Stokes, in *Food Oral Processing*, ed. J. Chen and L. Engelen, Blackwell Publishing Ltd., Hoboken, New Jersey, 2012, vol. 12, pp. 265–287.
180. A. Sarkar, K. K. T. Goh and H. Singh, *Food Hydrocolloids*, 2009, **23**, 1270.
181. R. Zhang, Z. Zhang, H. Zhang, E. A. Decker and D. J. McClements, *Food Hydrocolloids*, 2015, **45**, 175.
182. D. M. Dresselhuis, E. H. A. de Hoog, M. A. Cohen Stuart, M. H. Vingerhoeds and G. A. van Aken, *Food Hydrocolloids*, 2008, **22**, 1170.
183. G. A. van Aken, E. H. A. de Hoog and M. H. Vingerhoeds, in *Designing Functional Foods*, ed. D. J. McClements and E. A. Decker, Woodhead Publishing, Hoboken, New Jersey, 2009, pp. 481–501.
184. M. Noshad, M. Mohebbi, A. Koocheki and F. Shahidi, *J. Dispersion Sci. Technol.*, 2016, **37**, 56.
185. D. J. McClements and Y. Li, *Adv. Colloid Interface Sci.*, 2010, **159**, 213.
186. L. Sagalowicz and M. E. Leser, *Curr. Opin. Colloid Interface Sci.*, 2010, **15**, 61.
187. S. Abramov, P. Ruppik and H. P. Schuchmann, *Processes*, 2016, **4**, 25.
188. D. J. McClements, E. A. Decker, Y. Park and J. Weiss, *Crit. Rev. Food Sci. Nutr.*, 2009, **49**, 577.
189. L. Mao, Y. H. Roos, C. G. Biliaderis and S. Miao, *Crit. Rev. Food Sci. Nutr.*, 2017, **57**, 3173.
190. F. Jiménez-Colmenero, *Food Res. Int.*, 2013, **52**, 64.
191. L. Mao, D. Wang, F. Liu and Y. Gao, *Crit. Rev. Food Sci. Nutr.*, 2018, **58**, 770.

192. N. P. Aditya, S. Aditya, H. Yang, H. W. Kim, S. O. Park and S. Ko, *Food Chem.*, 2015, **173**, 7.
193. D. Chong, X. Liu, H. Ma, G. Huang, Y. L. Han, X. Cui, J. Yan and F. Xu, *Microfluid. Nanofluid.*, 2015, **19**, 1071.
194. S. P. Callender, J. A. Mathews, K. Kobernyk and S. D. Wettig, *Int. J. Pharm.*, 2017, **526**, 425.
195. C. Lobato-Calleros, J. Reyes-Hernández, C. I. Beristain, Y. Hornelas-Uribe, J. E. Sánchez-García and E. J. Vernon-Carter, *Food Res. Intern.*, 2007, **40**, 529.
196. P. J. Wilde and B. S. Chu, *Adv. Colloid Interface Sci.*, 2011, **165**, 14.
197. X. Wu, L. Zhang, X. Zhang, Y. Zhu, Y. Wu, Y. Li, B. Li, S. Liu, J. Zhao and Z. Ma, *Sci. Rep.*, 2017, **7**, 12079.
198. J. L. Buttriss and C. S. Stokes, *Nutr. Bull.*, 2008, **33**, 186.
199. M. Rayner, A. Timgren, M. Sjöö and P. Dejmek, *J. Sci. Food Agric.*, 2012, **92**, 1841.
200. R. N. Tharanathan, *Crit. Rev. Food Sci. Nutr.*, 2005, **45**, 371.
201. V. Aykan, E. Sezgin and Z. B. Guzel-Seydim, *Euro Fed Lipid*, 2008, **110**, 516.
202. E. T. Rolls, A. B. Norton, A. Lazidis, I. T. Norton and T. Mills, *Cereb Cortex*, 2018, **28**, 4080.
203. D. J. McClements, *Adv. Nutr.*, 2015, **6**, 338S.
204. C. Chung, G. Smith, B. Degner and D. J. McClements, *Crit. Rev. Food Sci. Nutr.*, 2016, **56**, 650.
205. Y. Mao and D. Julian McClements, *J. Food Sci.*, 2012, **77**, E144.
206. J. Benjamins, M. H. Vingerhoeds, F. D. Zoet, E. H. A. de Hoog and G. A. van Aken, *Food Hydrocolloids*, 2009, **23**, 102.
207. C. Lobato-Calleros, M. T. Recillas-Mota, T. Espinosa-solares, J. Alvarez-Ramirez and E. J. Vernon-Carter, *J. Texture Stud.*, 2009, **40**, 657.
208. I. T. Norton, P. W. Cox and F. L. Tchuenbou-Magaia, *U. S. Pat.*, 20110287150, 2010.
209. N. Chiu, L. Hewson, N. Yang, R. Linforth and I. Fisk, *Food. Sci. Technol.*, 2015, **63**, 65.
210. C. Belloir, F. Neiers and L. Briand, *Curr. Opin Clin. Nutr. Met. Care*, 2017, **20**, 279.
211. N. Chiu, A. Tarrega, C. Parmenter, L. Hewson, B. Wolf and I. D. Fisk, *Food Hydrocolloids*, 2017, **69**, 450.
212. G. Sala and M. Stieger, *Food Hydrocolloids*, 2013, **30**, 73.
213. T. Mills, F. Spyropoulos, I. T. Norton and S. Bakalis, *Food Hydrocolloids*, 2011, **25**, 107.
214. A. Koliandris, A. Lee, A.-L. Ferry, S. Hill and J. Mitchell, *Food Hydrocolloids*, 2008, **22**, 623.

CHAPTER 4

The Role of Bubbles in the Development of Food Structure

ARPITA MONDAL*[a] AND KESHAVAN NIRANJAN[b]

[a] Department of Service Sector Management, Sheffield Hallam University, Howard St, Sheffield S11WB, UK; [b] Department of Food and Bioprocessing Sciences, University of Reading, Reading RG6 6AP, UK
*Emails: afsniran@reading.ac.uk; arpi1981@gmail.com

4.1 Introduction

Bubbles add novel structures and distinctive organoleptic properties, depending on the net gas content and bubble size distribution in many food and drink products that include bread, chocolate, ice cream, whipped cream, and dairy-based beverages such as cappuccino. While bubbles do not add any calorific value to foods, they do change their texture and mouthfeel by influencing attributes such as foaminess in cappuccino, sponginess in bread, and brittleness or creaminess in aerated chocolate. Thus, most aerated products are well accepted and appreciated by the consumers due to their appearance, texture, sophisticated mouthfeel, volume, and the rate of release of aroma in case of dairy-based coffee beverages. Cognizance of the influence of different processing methods and conditions is an essential step toward understanding bubble formation, its growth and stabilization, and finally its effect on the cellular structure of food. This chapter discusses the role of bubbles in structure development with reference to three specific food products with widely differing viscosities: milk (low viscous), chocolate (intermediate viscous), and bread (viscoelastic). More specifically, this chapter addresses the incorporation of bubbles by different methods

Food Chemistry, Function and Analysis No. 18
Handbook of Food Structure Development
Edited by Fotis Spyropoulos, Aris Lazidis and Ian T. Norton

Published by the Royal Society of Chemistry, www.rsc.org

and highlights the effect of processing conditions on the development and stabilization of bubble-containing structures.

4.2 Bubble Structures in Milk

Milk foam may be considered to be a biphasic colloidal system where gas bubbles, normally air, are dispersed in a continuous milk phase.[1] Foaming properties of milk to form novel structures are important in the processing of many dairy-based products that include whipped cream, ice cream, or cappuccino-style beverages. Foamed milk adds significant character in the appearance and sensory attributes of milk-based coffee beverages (cappuccino, latte, mocha). For example, foams for cappuccinos are required to have a velvety shiny appearance, with small densely packed air bubbles. Milk foam can also be undesirable in case of the reconstitution of infant milk powder or formula. Milk foams are colloidal systems where the structure is created by energy input into milk that includes air bubbles, aqueous phase, surface-active proteins and fat. Energy input is required to overcome the interfacial free energy of the system, which increases during foaming because of enlargement of the surface area. Once formed, the coalescence of gas bubbles normally occurs rapidly in order to minimize surface energy and area. However, intrinsically active proteins and fat tend to form a cohesive viscoelastic film around the bubbles that decrease the interfacial tension and act as a barrier to prevent coalescence, which facilitates the formation of small bubbles.[2] The foaming properties of milk are usually expressed in terms of foamability and foam stability. Foamability is generally described by the total volume of foam, say in milliliters, obtained immediately after foaming under controlled conditions (gas flow/level of agitation, *etc.*) and time. The stability of foam is expressed in terms of half-life, which is the time (say, in minutes) taken for the foam to collapse to half its original volume normally under ambient conditions. Overrun is another foam characteristic parameter, which refers to the volume of air incorporated in the foam. Overrun is generally defined as the difference in density between the milk and the foam, divided by the density of the foam. Therefore, foam overrun increases with decreasing density of the foam, which is commonly associated with more voluminous foams.

The physical stability of foam is pivotal from a selling point of view, as the foam should remain stable throughout the consumption of the beverage. For frothed beverages like cappuccino, the foam should be stable at least for 10–15 min.[3] In these type of beverages, baristas often create latte art on the surface of the beverage by utilizing the contrast between the milk bubbles and the coffee. A small bubble size in the latte part ensures a bright foam, which leads to the sharp visual contrast between the foamed milk and the coffee. Bubble sizes and their distributions also play important roles in developing the final foam structure. Bubble size is mainly influenced by the method and intensity of foaming. Bubbles are generally small and spherical immediately after foaming; with time, they coalesce with other bubbles to

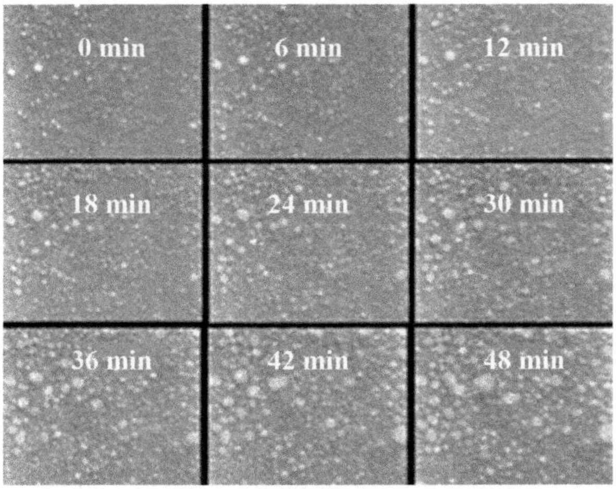

Figure 4.1 Changes occurring in skim milk foam upon storage at 25 °C for up to 48 min following foam formation by agitation.
Reproduced from ref. 3 with permission from John Wiley & Sons, Copyright © 2010 Society of Dairy Technology.

become larger, acquire irregular shapes, and finally burst (Figure 4.1). Foam structure generally disintegrates *via* three possible phenomena: drainage of liquid, coalescence, and disproportionation. The density difference between air in the bubbles and the continuous phase, along with gravitational force, causes drainage of liquid from a foam head into the liquid while air moves upward.[3] Coalescence of bubbles occurs because of liquid drainage from the lamella film as two bubbles approach each other, leading to film thinning and rupture.[4] Coalescence of gas bubbles leads to a decrease in the number of gas bubbles in the foam while increasing their sizes. The spreading of surface-active material also causes thinning and collapse of the liquid film, as a result of which coalescence may occur.[3] Disproportionation takes place as a result of the transport of gas from smaller to larger bubbles due to the concentration gradient caused by the difference in Laplace pressure between the bubbles of different sizes. Through this process, smaller bubbles shrink as the larger bubbles grow, until the former disappears. A smaller spread in the size distribution of bubbles can reduce the rate of disproportion.

Foam properties depend on the process conditions under which these are formed and the method employed to produce the foam. Using the same foaming method, it is possible to generate foams with diverse properties if the process conditions are changed.

4.2.1 Process Parameters

Foaming properties are mainly affected by the biochemical composition of milk, the heat treatment it receives, and the temperature during foaming.

Different manufacturing conditions and compositions of milk – for example, whole-fat milk, low-fat milk, reconstituted milk, long shelf life milk – may have different foaming properties.

4.2.1.1 *Milk Protein/Lipid*

Surfactants in milk, such as proteins (high-molecular-mass surfactants) and polar lipids (monoglycerides and diglycerides, free fatty acids, and phospholipids; low-molecular-mass surfactants), are determining factors of foamability and foam stability. Proteins stabilize foams by forming a viscoelastic film on the surface, whereas lipids stabilize foams *via* the Gibbs–Marangoni mechanism.[3] The Marangoni effect causes rapid adsorption of surfactants from the bulk phase to the areas of lower concentration in the adsorbed film. The effect allows free lateral diffusion of molecules in the surface layer, allowing rapid equilibration of concentration gradients. Since the stabilization mechanisms of proteins and polar lipids are mutually incompatible, systems containing both proteins and polar lipids (*e.g.* whole milk) may show weaker foam stability than systems stabilized by pure compounds (skim milk).

4.2.1.2 *Free Fatty Acid (FFA)*

Lipolysis of milk fat by milk lipoprotein lipase results in the release of surface active monoglycerides, diglycerides, and free fatty acids (FFAs), all of which are highly detrimental to the formation and stability of milk foams. FFAs in milk are low-molecular-weight surfactants, which compete with milk proteins for adsorption at the air–serum interface of the foams and are capable of displacing proteins from the interface. Adsorption of FFAs leads to disruption of protein–protein interactions, and, as a result, milk proteins are unable to form a stable viscoelastic film around the air bubbles. Weakening of the elastic film around the air bubbles leads to the coalescence of bubbles, thereby destabilizing the foam structure. As FFAs are more surface active than milk proteins, they dominate the air–serum interface at higher concentrations and displace the proteins, making the viscoelastic interface weaker. Therefore foams formed from milks with higher FFA content are unstable and coarsen more rapidly when compared to foams formed from milks containing lower FFA content. FFA levels as low as 1 μequiv mL^{-1} can have a detrimental effect on the steam foaming properties of milk.[5] Foam stability decreases with an increase in the FFA content. Lower FFA concentration results in the foam remaining smooth and creamy until it reaches its half-life. Good manufacturing practice from cow to consumer has a major impact on FFA concentration. In the dairy industry, lipolysis can be induced at any stage of milk processing starting from the farm, during transport, or in the factory due to agitation and pumping, homogenization, mixing raw and homogenized milk, and freezing/thawing of milk.[6] High-temperature short-time (HTST) pasteurization (72 °C for 15 s) effectively reduces lipolysis and the consequent release of FFA by inactivating milk lipase.[7]

4.2.1.3 Change of pH

Milk pH can influence foaming properties through various mechanisms. Generally milk pH stays in the "neutral" region (6.0–7.0), which has little effect on foamability, foam stability, overrun, and resistance to drainage and bubble diameter in the case of skim milk foam. If pH is adjusted to more than 7.0, it does not affect the overrun of milk foams; to the contrary, it improves the resistance of the foams to drainage, which may be related to the increased viscosity of milk at this pH because of the partial disruption of casein micelles.[8] Microbial growth induces changes in pH or degradation of fat and proteins. Reduced pH increases the level of nonmicellar caseins, calcium, and phosphate. When pH is reduced to 5.6, an increase in foamability takes place because of the dissociation, particularly of β-casein, from the casein micelles, which aids rapid and effective covering of the air–water interface.[9] As pH lessens further to 4.5–5.0, skim milk has very poor foaming properties because of the precipitation of caseins at this pH.[10] However, low-pH milk has been used only for research studies. In general, low pH can cause coagulation or curdling with a smell and taste of "sour" in liquid milk.

4.2.1.4 Skim Milk Versus Whole Milk

Casein predominates the foaming behavior of skim milk, which makes it best in producing a voluminous and stable foam. Skim milk foams are quite stable during storage even after 1 hr of producing the foam, where the volume of liquid underneath the foamy head steadily increases due to drainage of liquid from the head while foam volume stays the same. As a result, bubble size coarsens during storage.

In the case of whole milk, fat concentration and physical state affect foam capacity and stability because the fat can displace the protein from the interface, but when the foaming temperature is above the melting point of fat, it becomes less significant. The foam stability of whole milk can be reduced at low temperature (5–35 °C). This reduction is due to the recrystallization of globular milk fat. Fat crystals in the globules can pierce foam lamella, resulting in instability. However, the effect of the recrystallization of globular fat on foam stability decreases as the globular size decreases. Therefore, homogenization techniques, which generate small fat globules, can improve foam stability. The higher the homogenization pressure is, the higher the bubble count and the smaller the bubble sizes will be, in turn enhancing foaming properties.[3]

4.2.1.5 Foaming Temperature

Foaming temperature is the most important process parameter regulating foam structure. Foaming temperature has a proportional relationship with foamability, which means the foam volume increases with increasing temperature. It might be due to the fact that both surface tension and viscosity of skim milk decrease with increasing temperature.[11] This causes the proteins present to diffuse to the surface more rapidly and stabilize bubbles more effectively. The mean diameter of bubbles in the foam also decreases with increasing temperature. This is again related to a decrease in surface tension with increasing temperature. All these effects result in foams with higher stability. However, foam volume increases with temperature up to certain level. Foam is more stable at 45–50 °C compared to foam created at 70 °C. A low level of nonmicellar caseins and the lack of denaturation of whey protein at this temperature could lead to the formation of a strong interfacial layer consisting predominantly of casein micelles.

4.2.1.6 *Influence of Milk Heat Treatment*

Heat treatment causes denaturation of whey proteins in skim milk, which is supposed to affect the foaming properties; however, the results after denaturation do not seem to differ a lot. The foamability of pasteurized milk does not vary a lot from UHT-treated skim milk.[11] However, the overrun of skim milk foams may increase with increasing heat load applied to the milk due to availability of β-casein, differences in the ionic composition of milk, and low concentrations of surface-active peptides.[9] The foam stability of milk is largely unaffected by heat treatment during the first 30 min. During storage, the foam stability of heat-treated milk is lower than that of non-heat-treated milk.[3] Foam stability in pasteurized skim milk between 5–85 °C is higher than foams formed in UHT-treated skim milk, under otherwise similar conditions.

4.2.2 Foam Processing Methods

There are several milk foam processing methods: steam injection, cold aeration, agitation, supersaturation, and gas bubbling.[2] Steam injection and agitation are the most commonly employed methods. Compared to the effects of milk composition and processing conditions on the foaming properties of milk, foams prepared by different methods do not vary markedly.[12]

4.2.2.1 *Steam Injection*

Steam injection has been extensively used in the preparation of the traditional barista-style milk foams in coffee shops. In this method, steam is injected into the milk through small openings in a porous sheet, as a result of which gas bubbles are formed. Steam injection increases the temperature of the milk significantly. Injection into a given volume of milk for 50 s increases its temperature to >85 °C; however, the foam temperature does not exceed 65 °C and rapidly decreases to <50 °C within 4 min. Pressure and flow rates are also important factors in injection processes.[2] Figure 4.2

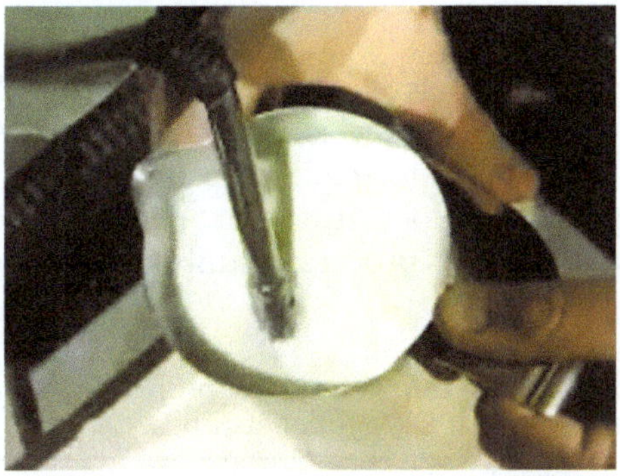

Figure 4.2 Traditional method of milk frothing by steam injection. Reproduced from ref. 2 with permission from C. A. Jimenez-Junca.

shows the traditional method of milk frothing by steam injection. Modern machines take advantage of an ejector-type device to restrict the passage of steam and generate the necessary pressure drop to suck air or a mix of air and milk in order to generate the foam.

4.2.2.2 Cold Aeration

Similar to steam injection, cold aeration of milk refers to processes where milk is foamed through the incorporation of air or another gas at near ambient temperature. In this process, air is forced into the milk through a nozzle or other type of perforated surface in order to create small air bubbles, which are rapidly stabilized by the surface-active proteins in milk. The duration of aeration, the gas pressure, and the temperature of the milk are the main determining factors of foam stability. This method is not commonly used for commercial processes; however, it has been widely used in scientific studies.[3]

4.2.2.3 Agitation

This method includes stirring, whipping, or beating, where the mechanical energy is transferred to the gas–liquid interface, as a result of which bubbles are formed in the liquid. Agitation leads to the initial formation of comparatively large air bubbles, which can be further broken into smaller ones. Some products like whipped cream or ice cream are prepared by this method.[3] Rotation speed, the intensity and duration of agitation, and the temperature of the milk are essential variables in the whipping process. Small milk whisking devices for home use and high-shear mixers and blenders used during the reconstitution of milk powders in the preparation of recombined milk or infant formula fall in this category. A certain angle of the milk foamer (about 50°) should be maintained as the tilt angle strongly influences the quality and volume of the foamed milk.

4.2.2.4 Supersaturation

This process is based on dissolving a gas in a liquid under pressure. Thus when the pressure is released, the solubility of the gas is reduced, and consequently the bubbles are formed. Aerosol-whipped cream and some milk foams are prepared by supersaturation.[2]

Table 4.1 shows the effect of process conditions on foam properties.

Table 4.1 Effect of process conditions on foam properties. Reproduced from ref. 2 with permission from C. A. Jimenez-Junca.

Process variable	Effect on foam properties
Temperature	Heating protein solutions between 60 °C and 90 °C increased protein denaturation and decreased the drainage, but bubbles were bigger.[13] Milk was better foamed at temperatures greater than 50 °C due to the fat being completely liquefied, although it depends on the fat content.[8,11,14] There was a direct relationship between foam stability and temperature (21–80 °C) for whey protein, but it had a detrimental effect in the case of sodium caseinate,[15] Preheating whey protein solutions between 70 °C and 100 °C showed that 80 °C is the optimal temperature to obtain more stable foams with high rigidity.[16]
Flow	Increasing the ratio of gas to protein solution flow rates improved the overrun and foam rigidity, but the drainage and bubble sizes were greater.[13] Lower flow of protein solution in a whipper increased the formation and stability of foams.[15] Levels of 20 mL min^{-1} gas flow rate produced foams with optimum properties in a sparging process using ovalbumin.[17] Foam drainage was faster and the extent of drainage was greater for larger bubbling flow rates for different protein solutions.[18]
Rotor speed	Higher foaming capacity and smaller bubble size was observed at higher whipper speed (5000–15 000 rpm).[15] Viscosity and mixer rotation speed (50–600 rpm) were the more important parameters affecting bubble sizes; higher speeds produce lower bubble sizes.[19] Raising the rotational speed (300–1200 rpm) decreased the bubble size and increased the elastic modulus (G') of different dairy emulsions recipes.[20]
Others	Geometry: Using 8 pairs of rotor-stator produced foams with better properties (smaller bubble size and higher overrun).[21] Container properties: Liquid drainage was lower in containers with hydrophobic walls, and it was higher when the diameter was increased.[22]

4.3 Bubble Structures in Chocolate

In this era of calorie-conscious consumers, aerated chocolates are well accepted and appreciated due to their sophisticated mouthfeel and perceived lower calorie. Bubble inclusion to create aerated chocolate has been practiced since 1935 when Rowntree-Mackintosh introduced aerated chocolate as Aero in the UK market. Later, Cadbury, Kraft Foods, and other brands also launched aerated chocolates. While bubbles do not add any caloric value to foods, they do change the texture and mouthfeel characteristics such as brittleness or creaminess of the aerated chocolate bar. Bubbles can also help to modulate the flavor and increase the novelty of the product. In contrast to research aiming to understand the role of bubbles in the structure development of low-viscosity products (*e.g.* carbonated beverages, milk-based coffee beverages, *etc.*) and high-viscosity products (*e.g.* bread dough), there are relatively fewer studies on intermediate viscous products such as chocolates and cake batter. The quality of aerated chocolates is measured by three parameters: density, gas holdup, and bubble sizes. Gas holdup is measured by comparing the density of aerated chocolate and gas-free density of the liquid or solidified chocolate. Bubble size can be measured by nondestructive, noninvasive X-ray tomography or by other means.[23] Bubble sizes play an important role in characterizing mouth-feel. In the context of aerated chocolates, consumer perception can be related to either the smooth mouthfeel of microaerated chocolate (characterized by the presence of smaller bubbles hardly visible to the unaided eye) or the crispness and brittleness of macroaerated chocolate (characterized by the presence of larger bubbles).[23] In practice, differences in bubble sizes and gas holdup can be created in chocolates by using different gases in the process. The methods employed to include gas bubbles in chocolate are gas sparging under pressure, vacuum application, extrusion, and reverse phase aeration.[24] Bubble inclusion using different gases under pressure or under vacuum occurs based on the balance between pressure and interfacial forces.[25] If the radius of a bubble is r, the outside pressure on the bubble is p_1, and the inside pressure is p_2 (as shown in Figure 4.3), then the external force on the bubble will be $\pi r^2 p_1$, and the inside force in the bubble will be $\pi r^2 p_2$. Surface tension (T) also produces the force on the circumference of the bubble, which can be expressed as $2\pi r\mathrm{T}$. In order for the bubble to be stable:

$$\pi r^2 p_1 + 2\pi r\mathrm{T} = \pi r^2 p_2 \tag{4.1}$$

which can be simplified to give the well-known Laplace equation:

$$p_2 - p_1 = 2\mathrm{T}/r \tag{4.2}$$

For a given chocolate, smaller bubbles will have higher internal pressure compared to larger bubbles. If two bubbles of different sizes come in contact, the gas will move from the high-pressure (*i.e.* smaller) to the low-pressure (*i.e.* larger) bubble, which will eventually cause coalescence of the bubbles.

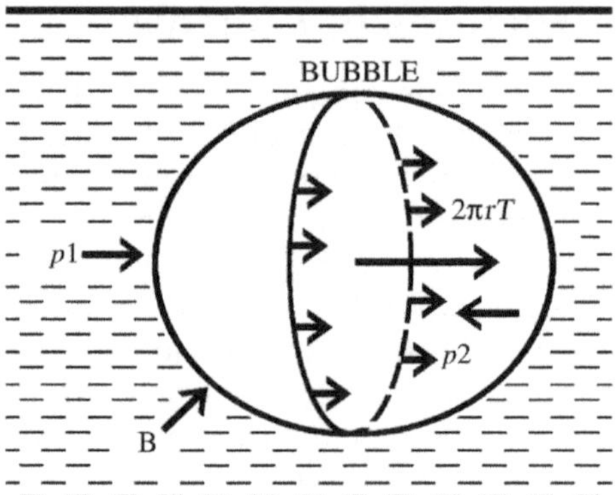

Figure 4.3 Pressure within a bubble inside chocolate.
Reproduced from ref. 25 with permission from the Royal Society of Chemistry.

4.3.1 Bubble Inclusion Methods

4.3.1.1 Gas Sparging Under Pressure

In this method, the gas is dissolved and mixed with tempered chocolate under pressure. Then the mixture is exuded through a nozzle when expansion of the mixture causes the gas to come out of the solution, thereby forming small bubbles. The chocolate is then deposited in molds and shaken vigorously to form larger bubbles due to coalescence. The chocolate is then cooled down to solidify and preserve the bubbles. This method produces a uniform "thin-walled" bubble structure, trapped in viscous chocolate foam, with bulk densities down to 0.4–0.7 $g\,ml^{-1}$.[26]

4.3.1.2 Vacuum Application

The vacuum application process consists of mixing gas into melted chocolate suspension, injecting the liquid chocolates into molds, expanding the bubbles under vacuum, and then cooling immediately to solidify the structure in which bubbles are trapped. The tempering condition (which is required to achieve the right type of crystals to stabilize the chocolate foam), the vacuum application process, and the subsequent cooling steps are crucial to get the right textured bubble-containing chocolate. It is essential that centers of the mold be properly cooled in order to avoid collapsing. It is obvious that the lower the pressure is, the higher the gas holdup and fraction of bubble in the chocolate foam will be. This leads to high levels of coalescence, which further results in larger bubbles. This process may produce an aerated center with a variable size and a reduced bulk density of 0.4–0.7 $g\,ml^{-1}$ from an initial chocolate density of 1.25–1.30 $g\,ml^{-1}$.[24] Vacuum pressure can vary from 0.68 bar to 0.95 bar but may also be lower.[26]

4.3.1.3 Extrusion

In this method, tempered chocolate is extruded through a die, which incorporates air holes down its length and by applying low-pressure air to maintain the structure. This process results in a honeycomb-like structure that gives a crisp texture. Extrusion is commonly carried out at ambient temperature where pressure varies from 10bar to 300 bar.[26] This process is more efficient compared to conventional melting process as the extruded product is capable of maintaining its shape, and there is no need for an extra cooling step.

4.3.1.4 Reverse Phase Aeration

This method consists of dissolving sugar in water and emulsifying liquid chocolate and milk into that sugar water.[24] This mix is then concentrated to adjust the moisture content to 10–12%, normally in a continuously scraped vacuum evaporator, then to extrude the mix in a vacuum chamber to form bubbles mainly in the aqueous phase. In this method, a stable aerated chocolate is achievable with very low bulk densities between 0.1 $\mathrm{g\,ml^{-1}}$ and 0.3 $\mathrm{g\,ml^{-1}}$.[26]

4.3.2 Influential Factors in Bubble Inclusion

Several factors may influence bubble formation in chocolate. These are classified as intrinsic, extrinsic, system, and operational variables.[27] Intrinsic factors are attributable to the basic "natural" ingredients of chocolate, whereas extrinsic factors are due to ingredients that are added to the chocolate. System factors are related to the liquid chocolate behavior (*e.g.* tempering and rheology), and operational variables (*e.g.* temperature, pressure, type of gas, *etc.*).

4.3.2.1 Intrinsic Factors

4.3.2.1.1 Percentage and Type of Fat. A high percentage of fat is beneficial for aerated chocolate as it can retain a higher amount of gas. Those types of fat, which start crystallizing rapidly and then slow down to allow more gas entrapment, are good for bubble inclusion. If the fat is highly polymorphic, such as cocoa butter, it is difficult to control bubble inclusion because the fat continues to crystallize irrespective of cooling rate. Generally, α and β crystals, especially the latter, facilitate inclusion of an ample amount of small bubbles.[28] Small β crystals are also able to arrange themselves at the interface, yielding a large surface area-to-volume ratio of fat. These crystals are also slow melting but with the fastest setting properties. However, large β crystals result in including larger and fewer bubbles. A high percentage of liquid fat is beneficial for higher gas holdup if crystallization progresses continuously. However, too much liquid fat can also cause bubbles to burst by spreading over the interface. As cocoa butter replacer, butterfat, palm kernel, or coconut oil is good for aerated chocolate as these can be crystallized in β form, which reduces densities from 1.3 to 0.2 $g\,cm^{-3}$.[24]

4.3.2.1.2 Sugar/Total Solid/Moisture. High-grade sugar can significantly vary foaming tendency; for example, beet sugar has higher foam value than cane sugar.[29] Also, refined sugar is preferable for better aeration. The total solid and moisture content of chocolate are also important intrinsic factors for bubble inclusion as these are able to influence the viscosity of chocolate considerably.

4.3.2.2 Extrinsic Factors

4.3.2.2.1 Emulsifiers. Emulsifiers influence the degree of aeration and rheological properties of chocolate by reducing the interfacial tension. They can also influence the crystal structure of fats and alter setting properties. A combination of polyglycerol polyricinoleate (PGPR) and lecithin are used for better bubble inclusion in chocolate.[30] Lecithin and glycerol monostearate (GMS) can reduce the density up to 0.2 $g\,ml^{-1}$ during bubble inclusion.[26]

4.3.2.2.2 Tempering. One of the most important steps in chocolate processing is tempering. Chocolate should be correctly tempered to form the crystals that help to stabilize the chocolate. This step is also significant as it affects the amount of gas dissolved in the chocolate and the viscosity that is required to hold the gas bubble.[24]

4.3.2.2.3 Type of Gas. The amount of gas dissolved in the fat phase of chocolate varies with the type of gas and pressure and mixing control. Air and most of the gases dissolved in the fat phase get released during cooling, except carbon dioxide, which remains dissolved even at low temperature.[31,32] Carbon dioxide and nitrous oxide are able to form larger bubbles with thin lamellae compared to nitrogen and argon.[23] Chocolates made with argon and nitrogen were perceived to be harder, less aerated, slow to melt in the mouth, and having a higher overall flavor intensity.[23] The mixing method employed, percentage of gas mixed, nozzle size, gas flow rate, and mixing head space are all equally important to get a good dispersion and uniform bubble size in good-quality bubble-containing chocolate.

4.3.2.3 System Factors

4.3.2.3.1 Viscosity. Cocoa butter exhibits Newtonian flow properties, but with the inclusion of cocoa solid, sugar, and milk solid, it becomes a non-Newtonian suspension, which exhibits a yield value and plastic viscosity.[25] Yield value influences bubble stability, especially when chocolate is made under vacuum, whereas plastic viscosity is more critical for bubble growth when aerating under pressure. The rheology of aerated and unaerated chocolates is also influenced by other important intrinsic and extrinsic factors (for example, temperature, moisture content, particle size distribution, conching time, fat or solid content, emulsifiers, vibration) as these factors influence yield value and plastic viscosity.

4.3.3 Operational Variables

4.3.3.1 Mixing Head and Speed in Continuous Process

Mixing speed and volume of mixing head play a vital role in creating a bubble-containing chocolate dispersion. The type of mixer influences the energy input. The higher the mixing speed is, the lower the bubble size will be; however, if the speed is low, the required texture of aerated chocolate will not be achieved.[33] The mixer should have high mixing efficiency but with sufficient cooling rate to decrease the possibility of detempering, which may take place due to viscous heat dissipation during mixing.[34]

4.3.3.2 Fluid and Gas Flow Rate

Bubble size may increase with increased fluid flow rate or decrease with stronger shear. Increased gas flow rate results in larger bubbles. Gas dissolves when it is introduced into the liquid inside a mixing head operating under pressure. When the pressure is released, bubble nuclei form and eventually coalesce to form larger bubbles. However, when gas flow is more than an optimum value, the mixer will be unable to disperse all the gas, which may result in the formation of heterogeneous foam.[19]

4.3.3.3 Effect of Vibration

Mechanical vibration plays an influential role in bubble-containing chocolate. It may either retard or enhance the bubble numbers. Low oscillation reduces chocolate viscosity, thereby increasing bubble rise velocity.[35] The frequency and amplitude of vibration control the plastic viscosity and yield value of liquid chocolate. For aerated chocolate, mechanical vibration is generally conducted in the horizontal direction in order to spread chocolate and to reduce the loss of trapped gas bubbles by applying shear force, which overpowers yield value and reduces plastic viscosity.

4.4 Bubble Structures in Dough and Bread

Gas bubbles contribute 70% of bread's final volume.[36] Bubbles, which are initially created during the mixing of ingredients by trapping air, contribute to bread's appeal through a number of distinctive quality characteristics such as its volume, brightness, shape, texture, and organoleptic properties.

When flour and water are mixed to develop dough, wheat protein gluten forms a viscoelastic network that is capable of trapping and retaining gas bubbles. A series of aeration processes take place during baking, starting from mixing and then proving to baking. These initially incorporated bubbles divide into numerous nuclei as mixing continues and are inflated by carbon dioxide produced during fermentation.[37] As bubbles are intrinsically unstable, a film of dough confines each bubble in a manacle and protects it from coalescing. These are assumed to be the main features responsible for the typical cellular structure of the bread as well as for the softness of the final product. After going through several fundamental and complex physical processes (evaporation of water, volume expansion, gelatinization of starch, denaturation of protein, crust formation, *etc.*), gas bubbles settle into the sponge structure and stabilize the internal pressure in the bread with the atmospheric pressure and protect it from collapsing after taking it out of the oven. It is necessary to understand the different processing steps in order to gain insights into the formation of the cellular structure of bread.

4.4.1 Bubble Inclusion Steps

4.4.1.1 *Mixing*

Mixing is an important step in baking where dough structure forms by homogenization of the ingredients (such as flour, water, fats, saccharides, and yeast) by mixing and hydration of flour particles. The aeration during this step and rheological properties of dough critically influence the gas cell structure in the bread.[38] During mixing, wheat protein gluten forms a viscoelastic network with a potential for trapping and retaining gas bubbles. As mixing progresses, gas bubbles disperse into discrete cells within a semi-solid starch-protein matrix. Retention of this structure is important to produce bread with high specific volume. An estimated 10^{11} to 10^{13} gas nuclei are required to obtain a light bread with an even crumb texture.[39] Mixing causes rheological changes of dough by altering the gluten viscoelastic network. Larger loaf volume depends on the stretchability of the dough, as dough strain hardens during stretching.[40] It is assumed that with the hardening up, dough limits the growth of bubbles, thus reducing coalescence among bubbles and ultimately resulting in greater loaf volumes. Mixing speed and work input must be above a threshold value to develop the optimum gluten network to hold bubbles. As mixing progresses, there is an increasing buildup of air in the dough that peaks at a certain value of work input before decreasing. Chin and Campbell[38] defined the dough inflation system point of peak dough development (DIS-PDD). The DIS-PDD mixing time specifies the point at which the optimized gluten network is achieved for maximum bubble stability during subsequent proving and baking. Mixing at higher speeds always gives greater aeration. Higher work input increases the strain-hardening index, failure strain, and failure stress initially, followed by a decrease. Dobraszczyk and others[40] determined the degree of strain hardening using the exponential relationship of stress–strain:

$$\sigma = ke^{n\varepsilon} \tag{4.3}$$

where σ is the stress, ε is the Hencky strain, k is the coefficient, while the index n is considered a measure of the degree of strain hardening. Chin and Campbell[38] determined the failure strain and failure stress from the bubble failure point, *i.e.* when the dough bubble material fails to sustain the inflation and ruptures. They also found that as the mixing progresses, the strain-hardening index of dough increases up to a maximum, then decreases after a peak point as the dough becomes overmixed. However, the degree of strain hardening achieves its highest value at a combination of higher work inputs and faster speeds. If dough is under- or overmixed beyond its peak development, then bread of inferior quality is produced.

There is an obvious difference in dough quality when it is mixed under low pressure or vacuum instead of at atmospheric pressure. At the industry level, at first dough mixing is done under high pressure to improve oxygen availability, followed by a period of partial vacuum to have a control on cell size distribution.[41] Air bubbles are eliminated from the dough if the mixing is done under complete vacuum. Therefore the density of dough mixed under vacuum or reduced pressure is greater than that of dough mixed under atmospheric pressure.

4.4.1.2 Proving and Fermentation

Proving is the holding stage between mixing and baking, where the expanded cellular structure of the dough finally forms. Traditionally, bakers stop mixing or kneading before peak dough development, and the end point of mixing is usually identified by visual observation based on experience. However, the gluten network continues to form during proving. Additionally, shaping before fermentation induces mechanical stress, which may also contribute to gluten network formation. Air bubbles, incorporated during mixing in this network, act as nucleation sites for leavening gases to collect and thus expand the dough. The fermentation process created in the dough by the yeast occurs in two main reaction steps; the first one is amylase activity, and the second is maltose consumption by the yeast. Amylase transforms maltose into glucose, and yeast produces carbon dioxide by consuming glucose during fermentation. Nuclei formed during the mixing stage expand with the accumulation of carbon dioxide produced by fermentation and subsequently desorbed from the liquid phase. A moderate increase in fermentation temperature increases the formation of gaseous carbon dioxide due to increased fermentation rate, as well as decreasing carbon dioxide solubility in water.[42] Generally, in this stage, the bubble numbers and the dough volume gradually increase in a sigmoid-like fashion, *i.e.* gradually at the beginning of fermentation and more rapidly in the latter stages. The final result is an increase in the specific volume of the loaf with the gas constituting a significant volume. As bubbles further expand, the chance of coalescence among bubbles increases.

4.4.1.3 Baking

During baking, bubbles expand further due to the increase of vapor pressure and initially also due to the increased rate of carbon dioxide production. After a certain temperature, yeast activity, along with the metabolic reactions, terminates. Therefore, new carbon dioxide production ends and with it the associated dough rise. Starch gelatinizes, and the dough progressively losses water in the forming crust region. The thickness of the formed crust is proportional to dough porosity, *i.e.* the crust formed on porous dough is thicker than the curst formed on compact dough.[43] Eventually, gas bubbles fracture to stabilize the internal pressure in breads with the atmospheric pressure. This is why breads do not collapse when taken out of the oven.

4.4.2 Physical and Chemical Controlling Factors

4.4.2.1 Types of Flour

The production of spongy crumb structure requires the dough to be able to retain the air entrapped in the discrete gas cells during mixing and the gas formed during yeast fermentation over extended periods. Wheat flour is extensively used in bread making due to its unique ability to stretch and its capacity to retain bubbles. The elasticity of this flour plays an important role in developing larger loaves. For example, the high-protein North American wheat flour is extensively used in bread baking for its high elasticity.[44] Dough of higher elasticity prevents the coalescence of bubbles by maintaining the thickness of the dough layer more effectively, whereas, in dough with lower elasticity, bubbles grow and coalesce rapidly as the dough layer weakens more quickly.[45] Strong flour has greater void fraction, bubble size distribution, strain-hardening index, failure strain, and failure stress compared to weak dough,[46] In the case of whole wheat flour, the bran and germ tend to decrease the loaf volume of the dough and increase crumb firmness and darkening to give different flavor profiles. Additional fibers dilute gluten and distort the gluten-starch network formed during mixing.[47]

4.4.2.2 Gluten

The unique wheat protein gluten plays an important role in maintaining the gas bubble structure in bread.[39] In the cereal industry, flours with high gluten content are conventionally used for baking. A series of shearing, compressing, and stretching stages during the mixing stage after initial homogenization leads to form the gluten network. The gluten-starch matrix provides the framework for gas cell expansion. If gluten is too weak, it cannot stretch into thin films around the air bubbles produced during fermentation. The gas bubbles would then swell and burst, causing the loaf to lack volume. If gluten is too strong, then it won't stretch, so the gas bubbles can't expand, causing a very dense loaf. Also during the mixing stage, extra "shear" can impair the gluten network and thus the retaining gas bubble structure.[48]

4.4.2.3 *Water Absorbance (WA)*

The water absorbance (WA) of flour also plays a very important role in determining the loaf volume of the dough.[49] If the WA of flour is too high, it develops soft dough, which is unable to hold the cellular gas, leading to the cells merging during proofing and thus decreasing the loaf volume. A too low WA can develop hard dough as the gluten network cannot be formed without enough water absorption, which eventually leads to the incapability of holding air during fermentation, which also causes the decreased loaf volume of the bread. Therefore, the optimized water absorbance capacity of flour is very important in developing the gluten network for perfect bubble formation in the dough. WA for whole wheat flour ranges from 60.0 to 65.0 g/100 g.[49]

4.4.2.4 *Yeast/Non-yeast*

Yeast (*Saccharomyces cerevisiae*) is widely used as a leavening agent in the baking industry. It causes fermentation, which transforms a dense mass of dough to a well-risen loaf. Furthermore, yeast also helps to strengthen and develop gluten in dough and specially contributes to distinct flavors in bread. Different types of bread with variable yeast and water contents can cause distinct cellular structures in the bread.[50] With constant flour content, the percentage volume of bubbles increases to a certain value with increasing yeast amount, after which it stabilizes regardless of the increment in the quantity of yeast added. This happens as the carbon dioxide production rate is strictly related to the metabolism of the yeast with the consumption of the simple sugars present in the flour. As the flour quantity is kept constant, so is the amount of simple sugars present in the flour as well as the water content; it is therefore not surprising that the percentage pore volume expansion reaches a maximum value even though the yeast amount increases. Subsequently, in the final dough-rise step, a high amount of yeast may lead to a collapse in the structure of the cell walls and an associated loss in product quality. For non-yeasted dough, chemical leavening systems are used to raise the volume by evolving carbon dioxide, though they are unable to create the unique flavor of yeast-fermented bread.

4.4.3 Characterization of Bubbles in Bread

The aerated texture of the bread is primarily quantified by the gas bubble characteristics (such as the gas-free density, specific volume, void fraction, bubble number and distribution characteristics of dough aeration) of the dough.

4.4.3.1 Gas-free Density

Explicit measurement of these properties is not a straightforward task, requiring specialized equipment. Campbell and others[51] first introduced a dynamic dough density (DDD) apparatus, where differences in the buoyancy of dough in xylene are used to monitor dough expansion as a function of time during dough fermentation. Using this apparatus, Chin and Campbell[38] determined the gas-free dough density ρ_{gf} by extrapolating the graph of dough density *versus* mixing pressure (P) to zero absolute pressure and fitting a regression line to find the intercept:

$$\rho = \rho_{gf} - sP \tag{4.4}$$

where ρ is the dough density, and s is the slope of graph.

4.4.3.2 Specific Volume

The specific volume of dough is an important quality control parameter that gives an indication of the final degree of aeration achieved in the dough at the end of fermentation. It is also used to assess the efficacy of chemically leavening system products. Bellido and others[52] measured dynamic specific volume (DSV) in dough, where aeration was controlled over a wide range of rates using various chemical leavening systems and processing conditions. The following equation was used to measure the specific volume of the dough (Y):

$$Y = \frac{LA}{m} \tag{4.5}$$

where L is the sample thickness, A is the area of the dough as a function of time, and m is the dough mass.

4.4.3.3 Void Fraction/Gas Volume Fraction/Bubble Fraction

Void fraction is another macroproperty characterizing dough aeration. Void fraction decreases as the viscosity of the dough increases. Also, at high void fractions, gas loss takes place due to coalescence of the gas bubbles in the dough. Different gas production capacities of the various leavening systems (yeast/chemical leavening agents), as well as the flour type, play a prominent role in determining the final void fraction in the dough. For chemically leavened dough, the gas retention properties of the dough causes differences in the final void fractions. Chin and Campbell[53] determined the gas void fraction (α_p) using density of the dough at the mixing pressure:

$$\alpha_p = 1 - \frac{\rho}{\rho_{gf}} \tag{4.6}$$

Bellido and others,[46] on the other hand, determined the void fraction of the dough using the specific volume of the dough:

$$\phi = 1 - \frac{Y_{gf}}{Y} \tag{4.7}$$

where ρ is the density of the dough, ρ_{gf} is the density of the gas-free dough, Y_{gf} is the specific volume of final gas-free dough, and Y is the specific volume of the dough at a long fermentation time. Chin and Campbell[53] found that a 0.5% increase in the gas-free dough density causes a 10% increase in void fraction. During baking, the void volume fraction of bread can reach a final value between 0.8 and 0.9.

4.4.3.4 Bubble Size and Number Distribution

Gas bubbles contribute a significant proportion of bread's volume, and thus their size distribution and numbers play an important role in deciding the dough rheology, sensory properties, and final texture of the bread.[54] For example, a standard tin loaf contains numerous small bubbles, whereas baguette bread contains fewer but bigger bubbles. The precise size distribution of bubbles in the dough is much more difficult to determine due to its fragile nature. The recent development of imaging techniques enabled the *in situ* observation of bubble structure evolution during different stages of baking. Various imaging techniques, starting from flatbed scanning and conventional photography to more advanced high-resolution techniques, such as scanning electron microscopy, X-ray computed tomography, and magnetic resonance imaging (MRI), have been extensively used in studies of bread baking.[55]

4.5 Conclusions

This chapter brings to light the effect of different processing methods and conditions on the aerated structure of foods with particular reference to milk-based beverages, bread, and chocolate. Bubble sizes and their distributions play important roles in developing the structure of such foods.

The foaming properties of milk are majorly affected by milk's biochemical composition and foaming temperature, compared to the foam generation methods employed. Skim milk foams are quite stable and voluminous compared to those of whole milk.

Although aerated chocolate is preferred by consumers, there are relatively fewer studies on the role of bubbles in the development chocolate structures. Differences in bubble sizes and gas holdup in chocolate can be created by using different gases and different methods. The volumes of gas mixed, mixing methods, nozzle size, gas flow rate, and the mixing head space pressures – all these factors critically influence bubble sizes and dispersion characteristics. Carbon dioxide and nitrous oxide are capable of forming larger bubbles and lighter chocolate compared to nitrogen and argon, which result in harder and less aerated structures. β crystals in cocoa butter facilitate inclusion of small bubbles in chocolate. Correct tempering is a prerequisite to achieving the right structure in chocolates.

Bubble structure in bread is mainly affected by processing methods. The aeration during the mixing of dough, along with the viscoelastic properties of the dough, critically influence the spongy structure of bread. Additionally, fermentation during proving and baking also result in increases of the specific volume of the loaf. Flour composition and gluten content also play a determining role in characterizing the bubble-containing structure. Coalescence, disproportionation, and the drainage of water cause disintegration of the aerated structure. Microscopic imaging, coupled with image analysis, can be used to observe the structure evolution and identify the dominant destabilizing mechanisms.

References

1. T. M. Ho, T. H. A. Le, A. Yan, B. R. Bhandari and N. Bansal, *Food Res. Int.*, 2019, **116**, 379–386.
2. C. A. Jimenez-Junca, PhD Thesis, *Foaming of milk by steam injection: Relationships between foaming conditionsand foam characteristics*, School of Food Biosciences, University of Reading, Reading, UK, 2011.
3. T. Huppertz, *Int. J. Dairy Technol.*, 2010, **63**, 477–488.
4. S. Damodaran, *Protein Stabilization of Emulsions and Foams*, Blackwell Publishing Ltd, Oxford, UK, 2005.
5. S. Kamath, A. Wulandewi and H. Deeth, *Food Res. Int.*, 2008, **41**, 623–629.
6. H. C. Deeth and C. H. Fitz-Gerald, *Advanced Dairy Chemistry*, Chapman & Hall Publishers, London, UK, vol. 2, 1995, pp. 247–308.
7. H. C. Deeth, *Int. Dairy J.*, 2006, **16**, 555–562.
8. K. Borcherding, P. C. Lorenzen and W. Hoffmann, *Int. J. Dairy Technol.*, 2009, **62**, 161–169.
9. M. A. Augustin and P. T. Clarke, *Dairy Sci. Technol.*, 2008, **88**, 149–161.
10. Z. Zhang, D. G. Dalgleish and H. D. Goff, *Colloids Surf., B*, 2004, **34**, 113–121.
11. S. Kamath, T. Huppertz, A. V. Houlihan and H. C. Deeth, *Int. Dairy J.*, 2008, **18**, 994–1002.
12. J. Goh, O. Kravchuk and H. C. Deeth, *Milchwissenschaft*, 2009, **64**, 121–124.
13. A. Bals and U. Kulozik, *Int. Dairy J.*, 2003, **13**, 903–908.
14. S. Silva, DPhil Dissertation, *An investigation into milk foaming by steam injection*, University of Reading, Reading, UK, 2008.
15. L. Indrawati, Z. Wang, G. Narsimhan and J. Gonzalez, *J. Food Eng.*, 2008, **88**, 65–74.
16. I. Nicorescu, C. Loisel, A. Riaublanc, C. Vial, G. Djelveh, G. Cuvelier and J. Legrand, *Food Hydrocolloids*, 2009, **23**, 1209–1219.
17. R. D. Waniska and J. E. Kinsella, *J. Food Sci.*, 1979, **44**, 1398–1402.
18. R. J. Germick, A. S. Rehill and G. Narsimhan, *J. Food Eng.*, 1994, **23**, 555–578.
19. C. Balerin, P. Aymard, F. Ducept, S. Vaslin and G. Cuvelier, *J. Food Eng.*, 2007, **78**, 802–809.
20. R. K. Thakur, C. Vial and G. Djelveh, *J. Food Eng.*, 2005, **68**, 335–347.
21. N. Müller-Fischer and E. J. Windhab, *Colloids Surf. Physicochem. Eng. Aspects*, 2005, **263**, 353–362.

22. M. Papara, X. Zabulis and T. D. Karapantsios, *Chem. Eng. Sci.*, 2009, **64**, 1404–1415.
23. J. Haedelt, S. T. Beckett and K. Niranjan, *J. Food Sci.*, 2007, **72**, E142.
24. J. Haedelt, D. L. Pyle, S. T. Beckett and K. Niranjan, *J. Food Sci.*, 2005, **70**, E164.
25. S. T. Beckett, *The Science of Chocolate*, Royal Society of Chemistry, Cambridge, 2nd edn, 2008.
26. M. S. Jeffery, *Manuf. Confect.*, 1989, **69**, 53–56.
27. A. H. Massey, PhD Thesis, *Air inclusion mechanisms and bubble dynamics in intermediate viscosity food systems*, School of Food Biosciences, University of Reading, Reading, UK, 2002.
28. Y. H. Hui, *Bailey's Industrial oil and Fat Products*, Wiley, New York, Chicheser, 1996.
29. B. W. Minifie, *Chocolate, Cocoa, and Confectionery: Science and Technology*, Avi Pub. Co, 1980.
30. K. Asama, K. Yamada and A. Nago, *EU Pat.* 124433, 2002.
31. R. Heemskerk, *Food Market. Technol.*, 1990, **4**, 54–55.
32. O. J. Wan, *J. Am. Oil Chem. Soc.*, 1991, 16–49.
33. W. Hanselmann and E. Windhab, *J. Food Eng.*, 1998, **38**, 393–405.
34. R. Luiten, *Food Market. Technol.*, 1993, **7**, 40–41.
35. M. D. Morey, PhD Thesis, *Rheological studies on molten chocolate*, University of Birmingham, Birmingham, UK, 2000.
36. L. Trinh, T. Lowe, G. M. Campbell, P. J. Withers and P. J. Martin, *J. Food Eng.*, 2015, **150**, 9–18.
37. G. M. Campbell and E. Mougeot, *Trends Food Sci. Technol.*, 1999, **10**, 283–296.
38. N. L. Chin and G. M. Campbell, *J. Sci. Food Agric.*, 2005, **85**, 2194–2202.
39. Z. Gan, P. R. Ellis and J. D. Schofield, *J. Cereal Sci.*, 1995, **21**, 215–230.
40. B. J. Dobraszczyk, J. Smewing, M. Albertini, G. Maesmans and J. D. Schofield, *Cereal Chem.*, 2003, **80**, 218–224.
41. H. M. Elmehdi, J. H. Page and M. G. Scanlon, *Cereal Chem.*, 2007, **84**, 250–252.
42. E. Chiotellis and G. M. Campbell, *Food Bioprod. Process.*, 2003, **81**, 194–206.
43. F. Bajd and I. Serša, *Magn. Reson. Imaging*, 2011, **29**, 434–442.
44. S. Chakrabarti-Bell, S. Wang and K. H. M. Siddique, *Food Res. Int.*, 2014, **64**, 587–597.
45. S. Chakrabarti-Bell, S. Wang, M. J. Patel, R. M. Weiss and P. J. Austin, *J. Cereal Sci.*, 2013, **57**, 504–513.
46. G. G. Bellido, M. G. Scanlon, J. H. Page and B. Hallgrimsson, *Food Res. Int.*, 2006, **39**, 1058–1066.
47. V. H. Pham, M. Tomoko and M. Naofumi, *Food Res. Int.*, 2007, **40**, 273–279.
48. S. H. Peighambardoust, E. Fallah, R. J. Hamer and D. van der Goot, *J. Cereal Sci.*, 2010, **51**, 89–95.

49. G. Huang, Q. Guo, C. Wang, H. H. Ding and S. W. Cui, *LWT – Food Sci. Technol.*, 2016, **71**, 274–280.
50. V. Lampignano, J. Laverse, M. Mastromatteo and M. A. Del Nobile, *Food Res. Int.*, 2013, **50**, 369–376.
51. G. M. Campbell, R. Herrero-Sanchez, R. Payo-Rodriguez and M. L. Merchan, *Cereal Chem.*, 2001, **78**, 272–277.
52. G. G. Bellido, M. G. Scanlon and J. H. Page, *J. Cereal Sci.*, 2009, **49**, 212–218.
53. N. L. Chin, P. J. Martin and G. M. Campbell, *J. Sci. Food Agric.*, 2005, **85**, 2203–2212.
54. G. M. Campbell, C. D. Rielly, P. J. Fryer and P. A. Sadd, *Trans. Inst. Chem Eng. Lond, Part C*, 1991, **69**, 67–76.
55. M. G. Scanlon and M. C. Zghal, *Food Res. Int.*, 2001, **34**, 841–864.

CHAPTER 5

Food Structure Development in Oil and Fat Systems

R. A. NICHOLSON AND A. G. MARANGONI*

University of Guelph, Food Science Department, 50 Stone Road East, Guelph, ON N1G 2W1, Canada
*Email: amarango@uoguelph.ca

5.1 Introduction

Fats and oils are readily found in food products. From a nutritional point of view, lipids are the most calorie-dense of the macronutrients, providing 9 kcal g^{-1}. Several lipids are essential to the body's metabolic processes and must be consumed as a regular part of the diet. Fats and oils in foods act as carriers of micronutrients, helping to improve their absorption and delivery upon digestion. In addition, the consumption of lipids reduces the rate of digestion, slowing the absorption of carbohydrates.

Lipids are primarily composed of triglycerides or triacylglycerols (TAGs), which are made up of three fatty acids linked by ester bonds to a glycerol molecule. Depending on the fatty acid composition, the lipid may be a solid (fat) or a liquid (oil) at room temperature. Furthermore, the textural and sensory attributes of foods are greatly affected by fats and oils. Lipids contribute greatly to the mechanical and rheological properties along with the melting characteristics (mouthfeel) of foods. Additionally, lipids act as flavor carriers and impart desirable aromas to the food. The effects that fats and oils have on the palatability of foods are dependent on the structure of the lipids. This includes not just the amount of crystalline (solid) fat but also the

Food Chemistry, Function and Analysis No. 18
Handbook of Food Structure Development
Edited by Fotis Spyropoulos, Aris Lazidis and Ian T. Norton

Published by the Royal Society of Chemistry, www.rsc.org

size and shape of the crystals along with their arrangement within the bulk of the fat.[1]

Fats have a very complex polycrystalline structure. This structure, along with its functionality, is highly influenced by the crystallization process. TAG crystallization begins with nucleation. This occurs when the TAGs in the system are either undercooled or become supersaturated. In both cases, this is achieved by reducing the temperature of the system below the melting point of the highest melting triglyceride or lipid in the multicomponent mixture. During the nucleation stage, stable aggregates of TAG molecules are formed. Crystal growth can then occur as additional TAG molecules are added onto the nuclei.[1,2] Furthermore, this nucleation and growth process affects the nanoscale, which in turn affects the mesoscale structure of the crystal network. At the nanoscale, TAGs stack and align themselves to form layers known as lamella. The lamella stack homoepitaxially to form nanoplatelets, the primary crystals within the fat network. Nanoplatelets cluster in a variety of ways depending on their size, shape, surface properties, and external fields (heat and mass transfer limitations). These larger crystal structures will then undergo further aggregation, producing even larger structures making up the fat network.[3] The characteristics of the crystal structure at each of these length scales affect the various properties of the fat (rheological, mechanical, oil-binding capacity, *etc.*). A comprehensive understanding of fat crystal structure at all length scales and of the effects that processing conditions have on this structure is crucial in the proper tailoring of fats for food applications.

5.2 Characterizing the Structure of Fats

5.2.1 Wide-angle X-ray Scattering and Polymorphism

The unit cell is the smallest building block within a fat crystal. Furthermore, the smallest repeating unit within the unit cell is the subcell. Polymorphism refers to the packing arrangement of the fatty acid chains within the TAGs (*i.e.* the subcell) when they undergo crystallization. Three broad families of polymorphs are found in fats: α, β', and β. These correspond to hexagonal, orthorhombic perpendicular, and triclinic packing arrangements, respectively.[2] Crystal polymorphism is determined using powder X-ray diffraction (XRD) in the wide-angle region and is referred to as the wide-angle X-ray scattering (WAXS) region, in the range of 10–25° 2θ. The reflections, or Bragg peaks, observed are referred to in this field as *short spacings* because they correspond to the smaller, or "shorter", crystalline structural features within the TAG crystal. It is worth mentioning here that the subcell concept and the way TAG crystals are still described belong in the distant past. The subcell concept is a throwback of times without computers or Rietveld refinement techniques. The more appropriate way of describing polymorphism is by reporting the crystal planes responsible for the observed Bragg peaks. However, the field has not advanced to this extent yet; hence the present discussion will be based on the *status quo*. The α form displays a characteristic spacing of 4.1 Å, the β′ displays spacings at 3.8 and 4.2 Å, while the β form displays a spacing at 4.6 Å. The β form is the most stable, while α is the least stable. The melting behavior of the fat is dependent on the polymorphic form present. A more stable polymorph has a greater subcell density and both a higher melting point and enthalpy of melting.[2] In addition, polymorphism indirectly affects the size of fat crystals *via* nucleation behavior. However, these effects differ from fat to fat.[4] The macroscopic characteristics of fat products are also affected by the polymorphic form present. In shortening and margarine, the β′ polymorph is most desirable, as the smaller crystals yield a product that is smooth and plastic in nature. While in properly tempered cocoa butter, the form V (β polymorph) crystals yield a product with the gloss and snap characteristics associated with fine chocolate. Furthermore, many factors have been found to influence polymorphism, including TAG composition, shear, cooling rate, and crystallization temperature.[2]

5.2.2 Small-angle X-ray Scattering and Scherrer Analysis

Small-angle X-ray scattering (SAXS) is performed at smaller diffraction angles (1–10° 2θ). This technique is used to determine the stacking arrangement of TAGs within the crystal in the direction of the long axis of the unit cell, or (001) plane. Since these distances are much greater than the characteristic distances between ethylene groups between fatty acid chains, they are referred to as *long spacings*. While stacking into a crystal, TAGs can generally adopt two arrangements. When the fatty acids in the sn-1 and sn-2 positions are located on the same side, the TAG has a chair conformation. If the fatty acids at the sn-1 and sn-3 positions are on the same side, and the fatty acid at the sn-2 position is opposite them, then the TAG is in the tuning fork conformation. Because of the nature of these conformations, when the TAGs stack, they produce lamella with a height of either two or three fatty acids.[3] The height of the lamella and therefore whether it is 2L or 3L can be determined based on the position of the peaks observed in the long spacing region. The arranging of the TAGs to form lamella and the stacking of these lamellae-producing crystalline domains is represented in Figure 5.1.

Using the Scherrer equation, the average size of the nanoplatelets can be determined from the XRD peak width:

$$D = \frac{K \cdot \lambda}{\text{FWHM} \cdot \cos\theta} \tag{5.1}$$

where D is the nanoplatelet size in the $[hkl]$ direction; K is the Scherrer constant, which is a dimensionless magnitude for the shape of the crystallites (usually 0.9; 1 when the crystallites are spherical); λ is the X-ray wavelength (a copper source has a wavelength of 1.54 Å); FWHM is the full width at half of the maximum peak height in units of radians; and θ is the diffraction angle in radians.[5,6]

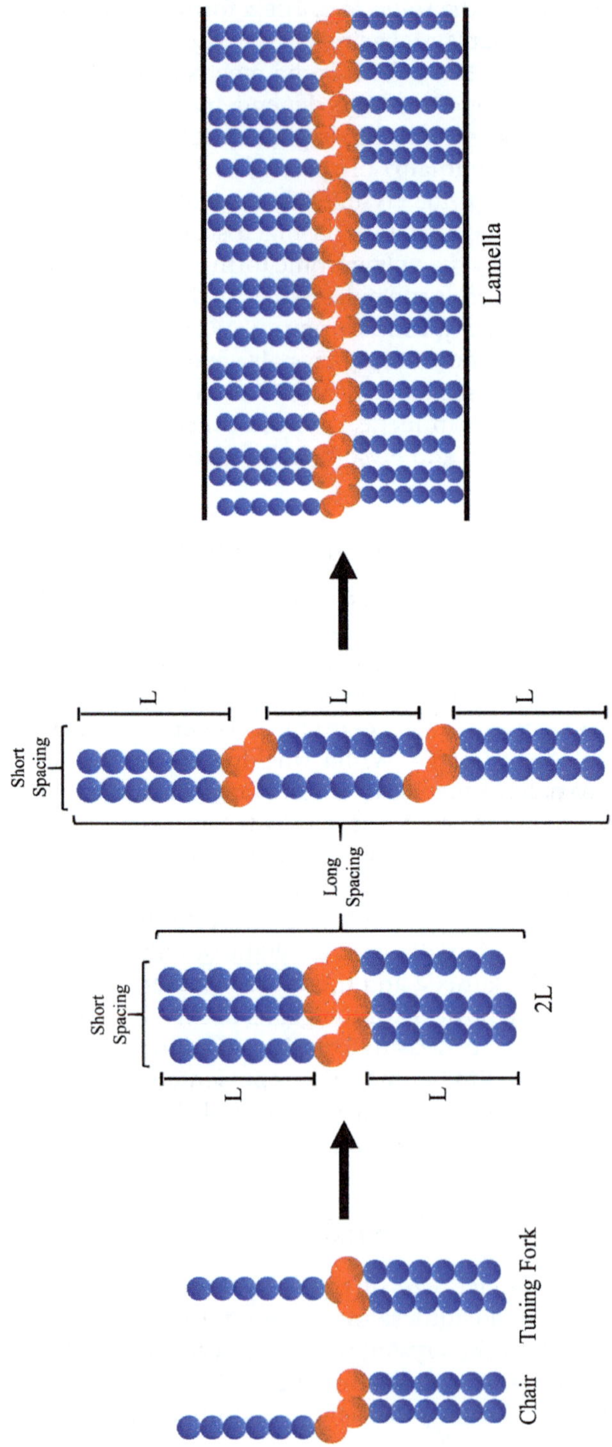

Figure 5.1 Representation of a triacylglycerol (TAG) molecule in chair and tuning fork configurations. Stacking of TAG molecules (chair configuration) in pairs and subsequently into lamella. Abbreviation: L, fatty acid chain length.

The actual dimension quantified depends on the peak chosen. Usually this analysis is carried out on the (001) reflection, thus corresponding to the size of nanoplatelet in the (001) direction, *i.e.* the thickness of the nanoplatelet. Both the length and the width of the nanoplatelet are much larger than the thickness and are outside the size range that can be estimated by the Scherrer equation. Moreover, since we still have no idea which planes are associated with a wide-angle reflection, this cannot be done. Please note that nanoplatelet size is inversely related to the peak width. As the width of the peak becomes narrower, it becomes increasingly difficult to determine whether the change in peak width is due to a change in the size of the crystallites or due to instrumental broadening.[3] As a result, the average thickness of the nanoplatelets can be estimated only up to 100 nm.

5.2.3 ^{1}H Spin Diffusion NMR

Recently, the effectiveness of using ^{1}H spin diffusion NMR to determine the nanoplatelet thickness has been demonstrated. Voda *et al.*[5] used this NMR technique to quantify the thickness of the nanoplatelets formed in different fat blends crystallized at two different cooling rates. They compared the thickness of the nanoplatelets measured through NMR to the domain size quantified with the Scherrer formula and found that the two methods were in agreement. Nanoplatelet thickness was affected by both the TAG composition and the cooling rate in much the same way as will be discussed later in the chapter. An advantage of using spin diffusion NMR to quantify nanoplatelet thickness is that this technique remains effective even in dilute systems of crystalline material, whereas there is difficulty in using traditional XRD in systems with a low amount of diffracting material.

5.2.4 Cryogenic Transmission Electron Microscopy

Transmission electron microscopy (TEM) has allowed researchers to study the nanostructure of fat systems. Viewing the fat crystals under TEM requires the experiment to be performed under cryogenic conditions (Cryo-TEM), such that the fat matrix does not melt during the investigation. Additionally, the liquid oil must be removed from the sample, or else the entrapped frozen oil will not allow for the nanoscale-sized crystals to be properly viewed. Acevedo and Marangoni[7] developed a method of extracting the nanoplatelets using cold-isobutanol as the solvent and viewing them under Cryo-TEM. This major development was built upon the work of Heertje's group.[8–11] It is important to note that the nanoplatelets must be viewed immediately after extraction because they quickly begin to form agglomerates, making it difficult to view and quantify the structures. Extracted nanoplatelets of fully hydrogenated canola oil viewed under Cryo-TEM are depicted in Figure 5.2. Using this technique, the crystallites can be visualized, and the length, width, thickness, lamellar height, and the size distribution can all be quantified. Nanoplatelets formed from fully

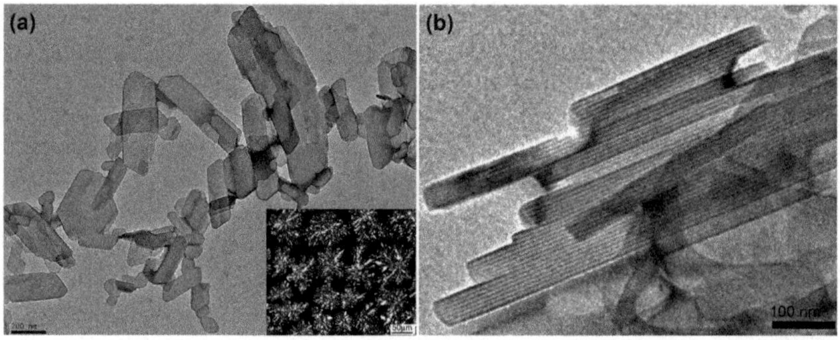

Figure 5.2 (a) Extracted nanoplatelets of a fully hydrogenated canola oil sample in the β form viewed under Cryo-TEM. Inset: Polarized light microscopy (PLM) image of the fat sample. (b) Cryo-TEM image of stacked nanoplatelets showing the multiple lamella making up the nanoplatelets. Reproduced from ref. 12 with permission from Elsevier, Copyright 2011.

hydrogenated canola oil in high oleic sunflower oil had dimensions of between 150×60×30 nm and 370×160×40 nm, depending on the concentration of fully hydrogenated canola oil and the conditions of crystallization.[7] In addition to being a useful method in itself, the development of this Cryo-TEM technique improved upon the usefulness of the XRD methods. Prior to the introduction of this Cryo-TEM method, the specific dimension of the crystallite corresponding to that calculated using the Scherrer equation was unknown. It was only after the quantification of all of the nanoplatelet dimensions through Cryo-TEM that this dimension could be correctly interpreted as the thickness.[3] Furthermore, quantification of the nanoplatelet thickness through Cryo-TEM and Scherrer analysis of XRD yield comparable results. The use of these techniques allows for a comprehensive understanding of the nanoplatelet structure.

5.2.5 Microstructure

The microstructure is typically studied using polarized light microscopy (PLM). PLM is one of the most common methods used to investigate the structure of fat crystal networks, and it has been in use for many years.[13,14] Because fat crystals are anisotropic, they are birefringent and appear bright when viewed between two cross-polarized filters, while the liquid oil is dark.[3] PLM requires minimal sample preparation, allowing for the crystals and polycrystal aggregates within the micrometer range to be viewed quickly and easily. This allows for quantification of the crystal size and for the morphology to be described. Processing conditions, such as temperature gradients and shear fields, have been shown to strongly influence the crystalline network of various fats. Herrera and Hartel[15] found that both a slower cooling rate and a higher crystallization temperature produced larger crystals in blends of high and low melting fractions of milk fat. Conversely, when cooled quickly or at a low temperature, the crystals produced were smaller with a narrower size distribution. Additionally, the researchers found that smaller crystals were produced when samples were crystallized under high levels of agitation. Martini *et al.*[16] showed that blends of high melting fraction milk fat and sunflower oil behaved in the same manner. Smaller fat crystals are formed when the fat is crystallized at a lower temperature or a faster cooling rate, as nucleation is favored over crystal growth. This relation was confirmed by Martini *et al.*,[17] as samples crystallized at lower temperatures demonstrated reduced Avrami coefficient values, correlating to a higher nucleation rate.

In addition to PLM, confocal scanning laser microscopy (CSLM) is quite useful for investigating the microstructure of fat systems. CSLM requires much less sample disruption compared to PLM because of a much greater depth resolution. Additionally, the three-dimensional structure of the fat network can be viewed from the reconstruction/stacking of a series of two-dimensional optical sections imaged. The ability to selectively stain the fat phase allows for the fat structure to be viewed in multicomponent food systems. For this reason, CSLM is used for viewing the structure of various emulsions and spreads (butter, margarine, low-fat spreads, *etc.*), as well as cheese products.[18] Ramel and Marangoni[19] used CSLM to view the structure of processed cheese products. Nile red and fast green stains were used to stain the fat and protein, respectively, to differentiate between the two components. Furthermore, CSLM can be used to examine the microstructure of systems where only fat is present as well. Rousseau *et al.*[20] viewed the structure of butterfat at 4 °C under CSLM. The sample was stained with Nile blue to differentiate between the crystalline fat and liquid oil. This stain causes the liquid oil to fluoresce yellow, while the solid fat does not fluoresce.

Recently, confocal Raman microscopy has been used to image fat crystal networks within fat dispersions[21] and fat spreads.[22] This technique allows for the microstructure of the fat network to be spatially resolved. Due to Raman spectroscopy's sensitivity to the varying degrees of unsaturation in fatty acid chains, the fat and oil phases of the network can be differentiated. The solid fat content and the distribution of the solid fat within the sample can be investigated using confocal Raman microscopy. Additionally, the size and porosity of the crystal clusters can be examined.[21] van Dalen *et al.*[22] used multivariate curve resolution to decompose the Raman spectra of emulsion systems, allowing them to observe where the solid fat, liquid sunflower oil, water, and emulsifier were concentrated in different fat spreads. The Raman concentration map of a 60% fat spread is shown in Figure 5.3. Furthermore, this is a noninvasive technique requiring no sample preparation, resulting in an unaltered fat crystal structure.

5.2.6 Ultra-small-angle X-ray Scattering

Bridging the gap between the study of the nanostructure and microstructure of fat crystal networks, the use of ultra-small-angle X-ray

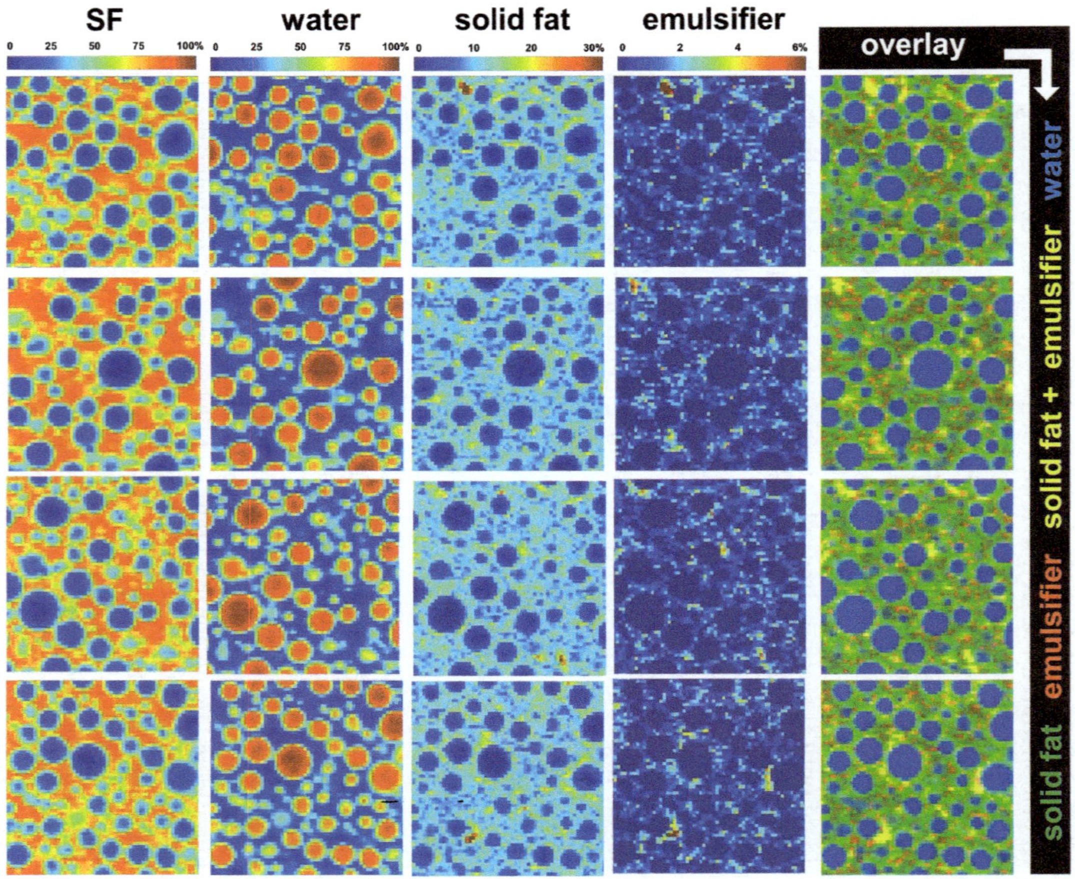

Figure 5.3 Raman concentration maps of a 60% fat spread, containing sunflower oil (SF), solid fat triacylglycerols, water, and monoacylglycerols as the emulsifier, highlighting the composition of the emulsion, at four different locations within the sample. Image size = 20×20 μm.
Adapted from ref. 22 with permission from John Wiley & Sons, Copyright © 2017 John Wiley & Sons, Ltd.

scattering (USAXS) allows for the investigation of the structural properties at the nanoscale through to the mesoscale. Performing X-ray scattering with smaller scattering vector (q) values allows for the fat crystal structure, at increasing length scales, to be examined. In the WAXS region, we explore crystal structure at a length scale of 0.2–1 nm, corresponding to the higher scattering vectors ($d = 2\pi/q$). At lower scattering vectors, the SAXS region is probed, investigating crystal structure within the 1–10 nm length. The USAXS region is examined at even smaller scattering vectors. This allows for the structural characteristics to be determined over a very wide range of length scales. USAXS can be used to study the fat crystal network from a size of 10 nm all the way to 5 μm.[23] Because of the energies required for the collimation of the X-ray beam to perform USAXS, these experiments are generally carried out at synchrotron facilities. However, some equipment manufacturers are beginning to offer lab-scale capabilities (Rigaku NANOPIX mini and SAXSLAB TR-USAXS). Nonetheless, this technique allows for the rapid and nondestructive characterization of a wide range of structural features. It is important to stress that this technique allows for the fat crystals to be studied without disrupting the structure of or physically confining the space in which the sample is allowed to crystallize, as would often be the case with various forms of microscopy.[23,24]

At the smallest order within the USAXS range of analysis is the nanoplatelets. The dimensions and surface characteristics of these structures can be deduced from the USAXS patterns obtained. The next structural level within crystalline fats is formed through the aggregation of the nanoplatelets. The third and final structure that can be deduced from USAXS is formed as a result of the clustering of the primary clusters to form larger agglomerates. Both the Unified Fit model and the Guinier–Porod model can be used to interpret the USAXS pattern in order to provide structural information corresponding to each of the three regions. In the Unified Fit model, all particles are represented as spherical, while the Guinier–Porod model recognizes the objects as being not only spheres but also asymmetrical particles.[23,25]

The Unified Fit model is usually fitted to an arbitrarily chosen number of structural levels within the USAXS pattern. Each of these structures will give a slope (P) and a radius of gyration (R_g).[25] The smallest structural region corresponds to the nanoplatelets. The absolute value of the slope for Structural Level 1 ($\sim 2 \times 10^{-3} < q < 1 \times 10^{-2}$ Å^{-1}) describes the surface characteristics of the nanoplatelets. When the value of this first slope is equal to 4, the nanoplatelets are smooth. A slope of less than 4 corresponds to rough nanoplatelets, while a slope of greater than 4 indicates that the nanoplatelets are coated in oil. The subsequent slopes describe the aggregation and formation of the higher structural levels. This will be explained later in the chapter. A summary of the interpretation of a USAXS pattern is given in Figure 5.4. Furthermore, the first radius of gyration provides information about the size of the nanoplatelets. Since they are not spherical,

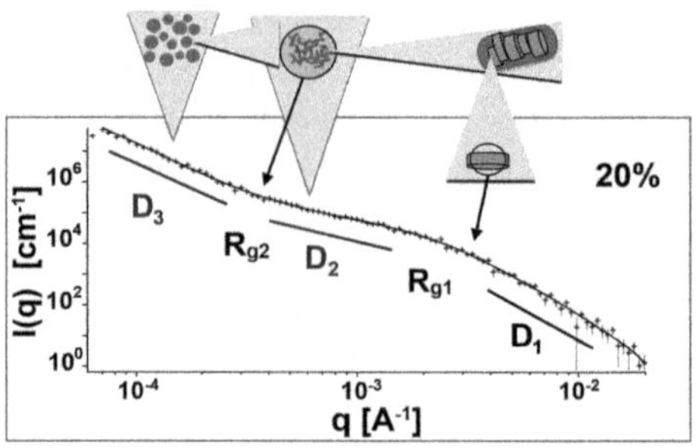

Figure 5.4 Summary of the USAXS pattern interpretation based on the unified fit and Guinier-Porod models. At the largest q values, thin disk-like nanoplatelets are the fundamental structure. The nanoplatelets aggregate into multilayered cylinders (TAGwoods), that subsequently form larger agglomerates.
Reproduced from ref. 25 with permission from AIP Publishing, Copyright 2013.

the radius of the nanoplatelets (R) can be calculated from R_g using the equation:

$$R = \sqrt{5/3}\, R_g \tag{5.2}$$

The second radius of gyration corresponds to the size of the TAGwoods or other structures formed from the aggregation of the nanoplatelets.

The Guinier–Porod model also contains radii of gyration (R_g) corresponding to the sizes of the different structures, along with a slope (P) and a parameter (s) that describes the shape of the particles.[25] Spherical particles are represented by $s = 0$ and have only one R_g. $s = 1$ represents long cylinders, which have two R_g values corresponding to the radius of the base and the length of the structure. Thin two-dimensional structures, such as very short TAGwoods, are indicated by $s = 2$. The radius of the nanoplatelets can be calculated from the equation:

$$R_c = \sqrt{2} R_{g1} \tag{5.3}$$

Using both radii, the length of the TAGwoods (L) can be calculated with the equation:

$$L = \sqrt{12(R_{g2}^2 - R_{g1}^2)} \tag{5.4}$$

5.2.7 Fractal Dimension

Fractal dimensions are used to describe the distribution of mass within an object when this distribution is disordered. Since fat crystal networks are disordered systems, their distribution cannot be described through simple integer dimensions, and it instead must be described using fractal dimensions. The relationship of the radius to the mass of the system is:

$$M \propto R^D \tag{5.5}$$

where D is the fractal dimension. The fractal dimension of a fat crystal network was traditionally determined through the use of microscopy. The box counting, particle counting, and Fourier transform methods can all be applied to determine the fractal dimension from polarized light microscopy images and more recently confocal Raman microscopy images. In addition, the fractal dimension can be determined using rheological measurements. When the logarithm of the storage modulus (G') is plotted against the logarithm of the volume fraction of fat crystals (with φ taken as the SFC/100 [SFC, solid fat content]), the slope of the resulting curve is the fractal dimension. The fractal dimension of the network is indicative of the process through which the fat crystal network was formed.[4] A fractal dimension of 1.7–1.8 indicates that the aggregation occurred through diffusion-limited cluster–cluster aggregation (DLCA), while fractal dimensions of 2.0–2.1 correspond to reaction-limited cluster–cluster aggregation (RLCA).[26] Higher fractal dimensions (2.5) are associated with 3D diffusion limited primary particle aggregation. Values of the fractal dimension can vary depending on the aggregation mechanism as well as on the number of rearrangement steps.[27]

More recently, USAXS has been used to determine the fractal dimensionality of the fat crystal network.[23–26] In addition to providing information about the surface characteristics of the nanoplatelets (Structural Level 1), the slope (P) obtained from the Unified Fit model indicates the fractal dimension of the aggregates. The q range of $\sim 4\times10^{-4} < q < 2\times10^{-3}$ Å^{-1} corresponds to the structures formed through the aggregation of the nanoplatelets (Structural Level 2). The distribution of the larger aggregates is indicated within the q range of $\sim 5\times10^{-5} < q < 4\times10^{-4}$ Å^{-1} (Structural Level 3). When $3 \leq |P| < 4$, the fractal dimension is equal to $6 - |P|$. While when $1 \leq |\mathrm{P}| < 3$, this suggests a mass fractal aggregate where the fractal dimension is equal to the value of the slope. A diffuse surface on the aggregating particles is indicated by $|P| > 4$. Furthermore, the fractal dimension determined from the slope of the q range corresponding to Structural Level 2, indicates whether the nanoplatelets aggregated *via* RLCA or DLCA. This interpretation was used to analyze 5% tristearin crystallized statically at 30 °C in triolein. It was determined that tristearin crystallized through RLCA, while the nanoplatelets in systems with 10% and 20% solids aggregated through (DLCA).[25] The fractal dimension calculated based on the slope of the lowest q range (Structural Level 3) indicates the distribution of the larger aggregates formed from the TAGwoods or other nanoplatelet structures. A fractal dimension of 3 for this structure level indicates a uniform distribution for the aggregates.

5.3 Factors Affecting the Fat Structure

5.3.1 TAG Composition

The structure of the fat network formed upon crystallization is affected by many factors. These can be extrinsic factors, such as the various processing techniques that are commonly used to achieve the desirable fat structure and functionality, or they can be intrinsic factors, such as the TAG composition of the fat as well as the minor components present.[2] Because either supersaturation or undercooling is necessary for nucleation to occur and both of these processes are inherently dependent on the TAG composition, the crystallization of a fat is therefore dependent on the TAG composition. Undercooling is achieved when the temperature of a neat TAG liquid phase is lowered to below that of the crystallization point of the crystallizing TAG species. Undercooling of TAGs with higher crystallization temperatures (*i.e.* saturated, long fatty acid chain length) is achieved at higher temperatures, and thus crystallization is easily induced. Additionally, supersaturation is achieved when the concentration of a TAG exceeds the solubility limit of the solution or melt at a given temperature. This limit can thus be exceeded by reducing the temperature of the solution further or by increasing the concentration of the particular TAG species. At higher levels of undercooling or higher degrees of supersaturation, nucleation is favored over crystal growth, and more numerous small crystals are formed. Conversely, when the degree of either undercooling or supersaturation is low, fewer, larger crystals are produced. The effects of supersaturation on the nanostructure of fats were highlighted by Acevedo and Marangoni.[28] These researchers found that higher amounts of fully hydrogenated canola oil (high melting saturated fat) in high oleic sunflower oil produced nanoplatelets with smaller dimensions.

Furthermore, the uniformity of the fatty acids within the TAGs will affect the crystallization and structure of the fat. A fat containing a greater amount of saturated fatty acids (straight chains), as well as TAGs containing fatty acids of a similar length, will crystallize into a more stable polymorphic form. The presence of kinks in the fatty acid chains, as is the case with the *cis* bonds in unsaturated fatty acids, reduces uniformity. Additionally, large differences in the fatty acid lengths within TAGs reduce uniformity further. This in turn results in the packing of the TAGs into less stable polymorphs during crystallization. The presence of a variety of TAG species in a fat system will also affect the crystal structure. For example, milk fat, which has a complex mixture of TAG species, forms crystals with a wide size distribution.[2]

Through the use of USAXS measurements, it was determined that in addition to affecting the size of the nanoplatelets, the TAG composition also affects the surface characteristics of these nanostructures. Peyronel *et al.*[24] found that tristearin in triolein, as well as tristearin in shea and triolein, formed nanoplatelets with smooth surfaces. Nanoplatelets with rough surfaces were formed by partially hydrogenated canola oil in triolein, while tristearin in cottonseed oil and triolein produced nanoplatelets with a diffuse surface. Peyronel *et al.*[25] showed that the amount of solids in the lipid system also affects the size of the linear nanoplatelet aggregates present, termed TAGwoods. Mixtures of 20% tristearin in triolein formed TAGwoods that were shorter in length than those formed from 5% tristearin in triolein.

5.3.2 Crystallization Temperature and Cooling Rate

The cooling rate and crystallization temperature used to crystallize a fat affect both the polymorphism as well as the size of the crystals. When fats are crystallized at a low temperature or a fast cooling rate, the TAGs pack into a less stable crystal polymorph (α). However, when a slower cooling rate or higher crystallization temperature is used to crystallize a fat, a more stable crystal polymorph is formed (β′, β)[2]. Furthermore, when crystallization is performed at a lower temperature or a higher cooling rate, nucleation is favored over crystal growth, and smaller crystals are formed. The effect that this has on the microstructure of the crystals was discussed earlier. More recently, it has been found that differences in the size of the crystals are particularly evident at the nanoscale. Acevedo and Marangoni[28] worked with a blend of fully hydrogenated canola oil and high oleic sunflower oil, crystallizing it at a slow (1 °C min^{-1}) and a fast rate (10 °C min^{-1}). The authors found that the size of the nanoplatelets was reduced when the system was cooled at the faster cooling rate. The reduction in size of the nanoplatelets at faster cooling rates can be seen in Figure 5.5. In the case of a blend containing a solid mass fraction of 0.4, the size of the nanoplatelets decreased from approximately 300×140×40 nm, when cooled slowly, to 150×60×40 nm, when cooled quickly. Furthermore, USAXS demonstrated that the length of the TAGwoods formed from the nanoplatelets, changed based on the cooling rate of the system. Peyronel *et al.*[25] showed

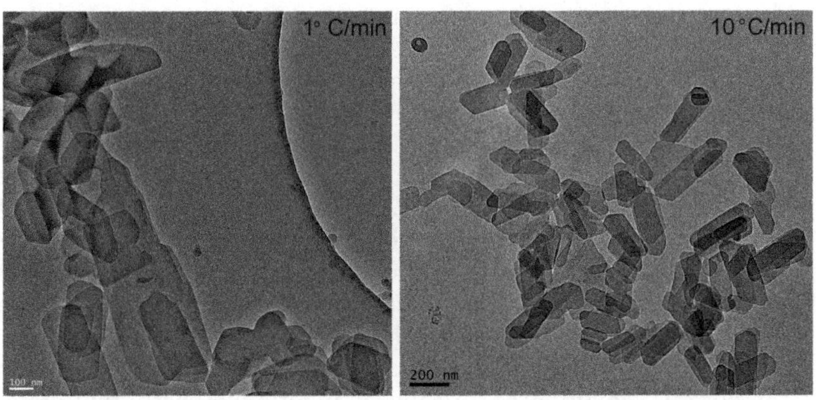

Figure 5.5 Cryo-TEM images of extracted nanoplatelets from a 1 : 1 blend of fully hydrogenated canola oil in high oleic sunflower oil that underwent crystallization at a fast (10 °C min^{-1}) and slow (1 °C min^{-1}) cooling rate.

that tristearin in triolein formed shorter TAGwoods, when the sample was crystallized at a higher cooling rate. Changing the cooling rate or the temperature of crystallization can be used to alter the crystal structure of a fat such that the desired functionality of the material is achieved.

5.3.3 Crystallization Under Shear

Crystallization under shear has been known to improve the texture and plasticity of lipids, thereby improving the functionality and sensory attributes of various food products containing high amounts of fat. Shear favors crystal nucleation, inducing the formation of numerous small crystals. Its application helps to remove the heat of crystallization and lowers the activation energy required for polymorphic phase transition to occur, promoting the transition to more stable polymorphic forms, often resulting in TAGs packing in the β arrangement. In addition, crystallization under shear helps form mixed crystals of the different TAG species present in the fat.[2,3]

Acevedo and Marangoni[28] investigated the effects of laminar shear on the nanostructure of fat crystals in blends of fully hydrogenated canola oil in high oleic soybean oil. They found that when the blends were crystallized under shear at a rate of 300 s^{-1}, the length, width, and thickness of the nanoplatelets decreased. The effects of high rates of shear on the size of the nanoplatelets, viewed under Cryo-TEM, can be seen in Figure 5.6. This effect was confirmed with cocoa butter. Maleky *et al.*[29,30] crystallized cocoa butter under laminar shear at a rate of 340 s^{-1} and observed a decrease in the size of the nanoplatelets along with a narrower size distribution, relative to cocoa butter crystallized statically. In a follow-up study by Acevedo *et al.* (2012),[31] fully hydrogenated soybean oil was crystallized in soybean oil at lower shear rates (30 and 240 s^{-1}). A large increase in both the nanoplatelet length and width was observed at the 30 s^{-1} shear rate. Increasing the shear rate to 240 s^{-1} reduced the size of the nanoplatelets; however, the length and width were still larger than that of the samples crystallized statically. Interestingly, the thickness of the nanoplatelets decreased in these systems. These results indicate the presence of a critical shear value, in the area of 300 s^{-1}, above which the crystallite size decreases with increasing shear rate. Below this critical shear value, the size of the nanoplatelets increases, as was observed in the studies mentioned. Moreover, working with a scraped surface heat exchanger that achieved a shear rate of 25 s^{-1}, Acevedo and Marangoni[32] crystallized blends of fully hydrogenated soybean oil in nonhydrogenated soybean oil. Nanoplatelets showed large increases in length; however, both the width and the thickness of the crystallites were unaffected by the turbulent shear. Furthermore, the effect of shear on the microstructure of a milk fat crystal network was previously mentioned, with high rates of agitation reducing the size of the fat crystals at the mesoscale level. This decrease in crystal size in the microstructural range can also be attributed to the damage induced by the shear in addition to its effects on nucleation.[31]

Figure 5.6 Cryo-TEM images of nanoplatelets extracted from fully hydrogenated canola oil crystallized statically and under laminar shear at a rate of 300 s^{-1}. Reproduced from ref. 28 with permission from American Chemical Society, Copyright 2010.

It has been suggested that growth in the longitudinal [100] direction, resulting in an increase in length, is preferred by nanoplatelets. Because of the anisotropic nature of the crystallites, the particles become oriented in the direction of the shear field.[33] Thus, the nanoplatelets become elongated in the longitudinal direction when crystallized under shear. In each of these previous studies, the ratio of the length to the width of the nanoplatelets increased. This was even the case when high shear rates were applied and the overall size of the crystallites decreased. Shear therefore will promote longitudinal growth of the fat crystals in addition to inducing nucleation.

5.3.4 Addition of Emulsifiers

Emulsifiers have been shown to have an effect on the crystallization of TAGs. The effect that the emulsifier has on the overall process depends on how it interacts with the TAGs and integrates itself into the crystal structure. Acevedo and Marangoni[34] found that glyceryl monostearate, polyethylene glycol sorbitan monostearate, and glyceryl monopalmitate increased the size of the nanoplatelets, while both sodium stearoyl lactylate and sorbitan monopalmitate resulted in smaller nanoplatelets in blends of fully hydrogenated soybean oil in liquid soybean oil, crystallized under shear. Additionally, soybean phosphatidylcholine was found to reduce the size of the crystals at the mesoscale range. This was attributed to its possible obstruction of crystal growth, preventing further growth of the crystals. Furthermore, monoacylglycerols are able to act as seeds for nucleation. In some cases, they are thought to disrupt the packing of TAGs onto the crystal structure following nucleation, while in other situations they may accelerate growth. This depends on the compatibility of the fatty acid of the monoacylglycerol with the TAGs of the fat. The differing properties of various emulsifiers mean that their effects on the crystal structure are highly varied.[2,3]

5.4 Relating Structure to Functional Properties

5.4.1 Oil-binding Capacity

An important characteristic of a fat crystal network is its ability to hold onto oil; this is referred to as the oil binding capacity. The oil binding capacity contributes to the functionality of fats in many ways. The liquid oil, bound within the crystalline structure of the network, affects the texture and provides the fat with plasticity. In chocolate, a high oil binding capacity is necessary to prevent bloom formation. A low oil binding capacity in either a shortening, spread, or topping would lead to oil separation and would greatly limit the shelf life of the product. The oil binding capacity is inherently related to the sensory attributes, including the mouthfeel, of any high-fat spread or confection. In a mixed food system, unbound oil will result in oil migration, leading to textural changes in the other components. The oil is retained within the fat crystal network through capillary forces generated by the network and by the solid fat acting as a barrier trapping the oil. Additionally, oil is adsorbed onto the surface of the fat crystals.[2]

The oil binding capacity is not related to the polymorphic form of the fat crystals, but it is effected by the size of the crystals.[2] Smaller crystals result in a greater surface area that is able to interact with the liquid oil and hold it within the network. The size effect at the microstructural level was demonstrated in cocoa butter.[35] A higher oil loss was seen in a system composed of larger crystals. This was attributed to the increased mobility of the oil phase through the crystalline network. Furthermore, the effects of nanoplatelet size on the oil binding capacity were tested by Acevedo *et al.* (2012).[36] In this study, the size of the nanoplatelets, in blends of fully hydrogenated soybean oil in liquid soybean oil, was altered by changing the temperature of the walls in a laminar shear cell. The authors found a strong relationship between oil loss and the size of the nanoplatelets; oil loss increased as the size of the crystals increased. This correlation was observed with the changing crystal size at both shear rates tested (30 and 240 s^{-1}). It is important to note that the level of oil loss increased with increasing rates of shear, even though the size of the nanoplatelets decreased when the shear rate was increased from 30 s^{-1} to 240 s^{-1}. This can be attributed to the high degree of structural damage that the shear is inducing on the crystal network.

The oil binding capacity of a fat crystal network is affected by the compatibility of the TAGs within the liquid phase with the TAGs of the crystalline material, which allows them to associate. This association of the liquid TAGs with the solid TAGs means that the liquid oil is able to become bound onto the surface of the crystals, thereby increasing the oil binding capacity of the material. MacDougall *et al.*[37] investigated the effect of TAG composition on the oil binding capacity of tristearin nanoplatelets. The researchers found that dielaidin-olein within the oil phase will readily associate with the tristearin crystals, adsorbing to the crystal surface, thereby improving oil binding capacity compared to a system in which the oil phase is composed of diolein-elaidic or triolein. This research shows that TAG structures with a higher degree of similarity have a greater compatibility, and therefore adsorption of the liquid TAGs to the crystal surface occurs more readily.

5.4.2 Rheological Properties

The functionality of a fat is essential to its use in various food products (shortening) or on its own as a food (chocolate, spreads, *etc.*). These include melting properties along with the mouthfeel and the spreadability of the fat. Spreadability is one of the main functional properties of a fat and is dependent on its mechanical properties. A very complex relationship exists between the various hierarchical structures of the fat crystal network and the mechanical properties that the system exhibits as a whole. The rheological properties are related to polymorphism, the size of the nanocrystals, the microstructural crystal size, and the fractal dimension of the system. While the relationship between polymorphic form and the rheological properties of fats is still uncertain, it is generally understood that the β′ polymorph produces fat products that are spreadable, and fats that are more brittle are obtained when the polymorph is in the β form. It is well recognized that larger polycrystals or crystal aggregates at the microstructural level produce softer, spreadable fats, while smaller crystals produce harder, less spreadable products, but this will also be a strong function of the amount of crystalline mass present in the system.[2] The storage modulus (G') is related to both the SFC and the fractal dimension of the fat through a power law:

$$G' \propto (\mathrm{SFC}/100)^m \tag{5.6}$$

where m is dependent on the fractal dimension and whether or not the system is within the strong-link or weak-link regime.[4]

The effect of the size of the nanoplatelets on the rheological properties of fats was investigated by Acevedo *et al.* (2012).[36] These researchers found that when the size of the nanoplatelets, in blends of fully hydrogenated soybean oil and liquid soybean oil, were altered by changing the wall temperature of a laminar shear cell during crystallization both G' and the yield stress (σ^*) were affected. Crystallization at higher wall temperatures led to an increase in the size of the nanoplatelets, and this resulted in a decrease in both G' and σ^*. The relationship between the size of the nanoplatelets and σ^* was described using the model proposed by Marangoni and Rogers (2003):[38]

$$\sigma^* \approx \frac{6\delta}{a}\left(\frac{\mathrm{SFC}}{100}\right)^{\frac{1}{d-D_f}} \tag{5.7}$$

where δ is the crystal-melt interfacial tension, a is the nanoplatelet diameter, d is the Euclidean dimension of the embedding space, and D_f is the mass fractal dimension of the crystal network. In this model, the yield stress of the fat network is inversely proportional to the size of the nanoplatelets.

This same study also demonstrated decreases in both G' and σ^* with increasing rates of shear. Although the crystal size was affected by the shear rate, the rheological properties are likely affected more so by the large amounts of structural damage inflicted by shearing the samples. This demonstrates that the effects of processing conditions on the rheological properties of fats can be positive or negative.

5.5 Conclusions

With the multitude of methods available for analyzing the various structural levels of fat crystal networks, the understanding of each of the hierarchical levels is becoming more and more comprehensive. These techniques have allowed for the structural effects of internal (composition) and external factors (cooling rate, shear) to be studied, all the way from the nanoscale through to the arrangement of the microstructure. Furthermore, this more comprehensive look at the structure of fat networks has enabled researchers to investigate the influence that a variety of structural features have on the functional properties of fats, thus allowing for the characteristics of fats to be appropriately modified to suit the needs of all sorts of applications. The knowledge of the fat structure will only continue to grow in the years to come as the use of these experimental techniques improves and expands.

Acknowledgements

The authors would like to thank the Natural Sciences and Engineering Research Council of Canada for their financial support.

References

1. A. G. Marangoni, N. C. Acevedo, F. Maleky, E. Co, F. Peyronel, G. Mazzanti, B. Quinn and D. Pink, *Soft Matter*, 2012, **8**, 1275.
2. P. R. Ramel, E. D. Co, N. C. Acevedo and A. G. Marangoni, *Prog. Lipid Res.*, 2016, **64**, 231.
3. N. C. Acevedo and A. G. Marangoni, *Annu. Rev. Food Sci. Technol.*, 2015, **6**, 71.
4. S. S. Narine and A. G. Marangoni, *Food Res. Int.*, 1999, **32**, 227.
5. A. Voda, R. den Adel, K. van Malssen and J. van Duynhoven, *Cryst. Growth Des.*, 2017, **17**, 1484.
6. P. Scherrer, *Nachr. Ges. Wiss. Goettingen, Math.-Phys. Kl.*, 1918, **2**, 98.
7. N. C. Acevedo and A. G. Marangoni, *Cryst. Growth Des.*, 2010, **10**, 3327.
8. I. Heertje, M. Leunis, W. J. M. van Zeyl and E. Berends, *Food Microstruct.*, 1987, **6**, 1.
9. I. Heertje, J. van Eendenburg, J. M. Cornelissen and A. C. Juriaanse, *Food Microstruct.*, 1988, **7**, 189.
10. I. Heertje, *Food Struct.*, 1993, **12**, 77.
11. I. Heertje and M. Leunis, *LWT–Food Sci. Technol.*, 1997, **30**, 141.

12. N. C. Acevedo, F. Peyronel and A. G. Marangoni, *Curr. Opin. Colloid Interface Sci.*, 2011, **16**, 374.
13. T. S. Awad, M. A. Rogers and A. G. Marangoni, *J. Phys. Chem. B*, 2004, **108**, 171.
14. R. Campos, S. Narine and A. G. Marangoni, *Food Res. Int.*, 2002, **35**, 971.
15. M. L. Herrera and R. W. Hartel, *J. Am. Oil Chem. Soc.*, 2000, **77**, 1197.
16. S. Martini, M. L. Herrera and R. W. Hartel, *J. Am. Oil Chem. Soc.*, 2002, **79**, 1063.
17. S. Martini, M. L. Herrera and R. W. Hartel, *J. Am. Oil Chem. Soc.*, 2002, **79**, 1055.
18. J. C. G. Blonk and H. van Aalst, *Food Res. Int.*, 1993, **26**, 297.
19. P. R. Ramel and A. G. Marangoni, *Food Struct.*, 2017, **12**, 15.
20. D. Rousseau, A. R. Hill and A. G. Marangoni, *J. Am. Oil Chem. Soc.*, 1996, **73**, 973.
21. K. J. A. Martens, G. van Dalen, P. C. M. Heussen, M. A. Voda, T. Nikolaeva and J. P. M. van Duynhoven, *J. Am. Oil Chem. Soc.*, 2018, **95**, 259.
22. G. van Dalen, E. J. J. van Velzen, P. C. M. Heussen, M. Sovago, K. F. van Malssen and J. P. M. van Duynhoven, *J. Raman Spectrosc.*, 2017, **48**, 1075.
23. F. Peyronel, J. Ilavsky, D. A. Pink and A. G. Marangoni, *Lipid Technol.*, 2014, **26**, 223.
24. F. Peyronel, B. Quinn, A. G. Marangoni and D. A. Pink, *J. Phys.: Condens. Matter*, 2014, **26**, 464110.
25. F. Peyronel, J. Ilavsky, G. Mazzanti, A. G. Marangoni and D. A. Pink, *J. Appl. Phys.*, 2013, **114**, 234902.
26. D. A. Pink, B. Quinn, F. Peyronel and A. G. Marangoni, *J. Appl. Phys.*, 2013, **114**, 234901.
27. R. Jullien, *Contemp. Phys.*, 1987, **5**, 477.
28. N. C. Acevedo and A. G. Marangoni, *Cryst. Growth Des.*, 2010b, **10**, 3334.
29. F. Maleky, A. K. Smith and A. G. Marangoni, *Cryst. Growth Des.*, 2011, **11**, 2335.
30. F. Maleky, A. K. Smith and A. G. Marangoni, *Cryst. Growth Des.*, 2011, **11**, 2429.
31. N. C. Acevedo, J. M. Block and A. G. Marangoni, *Langmuir*, 2012, **8**, 16207.
32. N. C. Acevedo and A. G. Marangoni, *Food Bioprocess Technol.*, 2013, **7**, 575.
33. G. Mazzanti, S. E. Guthrie, E. B. Sirota, A. G. Marangoni and S. H. J. Idziak, *Cryst. Growth Des.*, 2003, **3**, 721.
34. N. C. Acevedo and A. G. Marangoni, *Food Biophys.*, 2014, **9**, 368.
35. S. Marty and A. G. Marangoni, *Cryst. Growth Des.*, 2009, **9**, 4415.
36. N. C. Acevedo, J. M. Block and A. G. Marangoni, *Faraday Discuss.*, 2012, **158**, 171.
37. C. J. MacDougall, M. S. Razul, E. Papp-Szabo, F. Peyronel, C. B. Hanna, A. G. Marangoni and D. A. Pink, *Faraday Discuss.*, 2012, **158**, 425.
38. A. G. Marangoni and M. Rogers, *Appl. Phys. Lett.*, 2003, **82**, 3239.

CHAPTER 6

Food Structure Development in Chocolate

EMMA J. MCLEOD[a] AND PETER J. FRYER*[b]

[a] Mondelez International, Bournville, UK; [b] University of Birmingham, UK
*Email: P.J.Fryer@bham.ac.uk

6.1 Introduction

Consumers today demand ever higher-quality food products at very competitive prices, and this even applies to indulgent treats such as chocolate. Hence food manufacturers are focused on ways to improve the taste and texture of their products while maximizing manufacturing output efficiency and minimizing cost and energy usage. To achieve this, manufacturers need a much greater understanding of the complexity of food material (for the purposes of this chapter, chocolate), how it interacts with processing machinery, and the conditions required to create consumer-preferred products.

Alongside this, legislators and to a lesser extent consumers are demanding changes in the nutritional profiles of products. Reductions in fat and sugar, two of the bulk constituents of chocolate, are particular challenges. Over the years, strict legislation has been developed in most regions of the word to define exactly what constitutes "real" chocolate *versus* chocolate-"flavored" candy, so flexibility of formulation is limited. Looking back on the history of chocolate, it is hard to believe that what started as a bitter cocoa drink created by the Incas and Aztecs in South America has developed into the multistep manufacturing process to create the solid sweet treat we enjoy today.

A typical powder milk chocolate manufacturing process is shown in Figure 6.1; for details on all chocolate processes, Beckett *et al.*[1] is a

Food Chemistry, Function and Analysis No. 18
Handbook of Food Structure Development
Edited by Fotis Spyropoulos, Aris Lazidis and Ian T. Norton

Published by the Royal Society of Chemistry, www.rsc.org

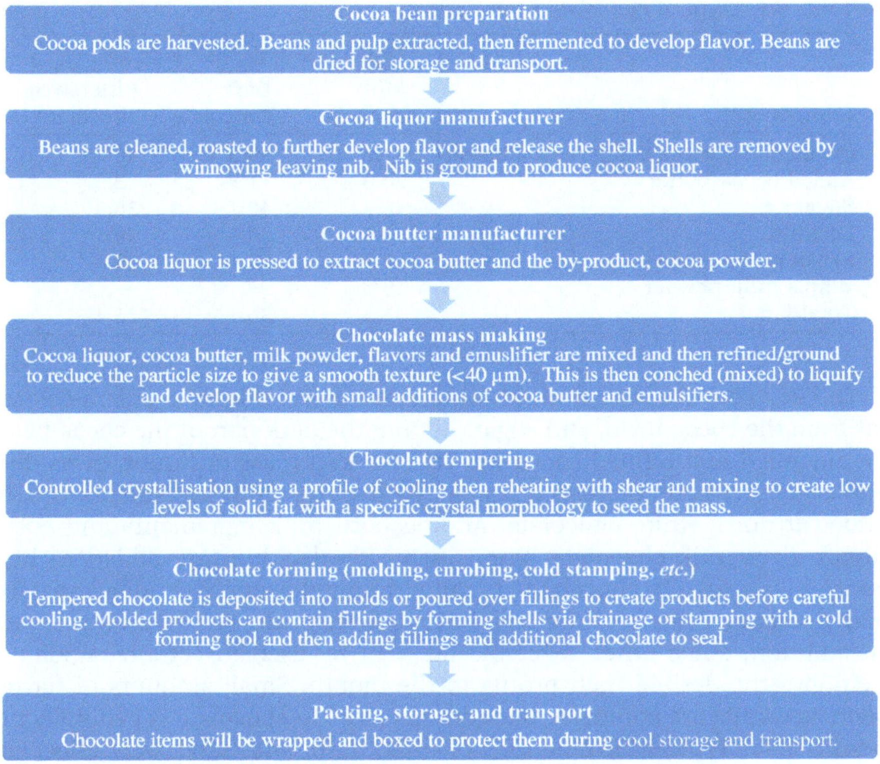

Figure 6.1 A schematic diagram of the stages of chocolate manufacture.

recommended text. The aim of this chapter is to introduce some of the product formulation and processing challenges, as well as the recent scientific and engineering research that is starting to help explore these areas. Difficulties arise in chocolate manufacture because of the complexities of the materials involved, particularly the polyphasic confectionary fats used. The challenges of fat crystallization and the cooling rates needed to create chocolate in the form desired by the consumer are covered in more detail here to demonstrate the required depth of understanding.

6.2 Developments in Chocolate Formulation

History has seen intentional adulteration of chocolate and its ingredients when raw materials are in short supply and/or has become expensive. A relatively recent example is the 2008 Chinese milk scandal, when melamine was found in milk powders, a key constituent of milk chocolate. But further back in time, other components such as starch have been used. Hence most regions of the world have legislation, such as the European Parliament and Council Directive 2000/36/EC that relates to cocoa and chocolate products across Europe, that defines chocolate and what ingredients can be used.

Table 6.1 Typical chocolate formulation.

Component	Typical (weight) percentages (%) in		
	Milk chocolate	Dark chocolate	Bittersweet chocolate
Cocoa mass (cocoa solids, cocoa butter)	11.8	39.6	60.7
Added cocoa butter	20.0	11.8	2.6
Sugar	48.7	48.1	36.3
Lecithin	0.4	0.4	0.3
Flavor compounds	0.1	0.1	0.2
Whole milk powder	19.1	—	—
Total fats	31.5	36.4	35.4

The key constituents of all chocolate are cocoa butter, a cream-colored fat from the cocoa bean, and sugar. Adding the dark part of the cocoa bean, *i.e.* liquefied and refined cocoa nib called cocoa mass or liquor, gives dark chocolate. Without the cocoa liquor (dark part), milk ingredients can be added to form white chocolate. Adding both milk ingredients and cocoa liquor gives milk chocolate. Emulsifiers, like lecithin, are added to help control viscosity. As chocolate contains minimal water, the added surface-active agents form a boundary layer between the solids and the fat.[2] In addition, some other vegetable fats can be used in certain regions to help give the desired melt profile in the mouth. Small amounts of flavors such as vanilla are common, but adding colors to chocolate is generally not permissible. Typical formulations of milk, dark, and bittersweet chocolate are given in Table 6.1.

Legislation limits the ingredients that can be used to class the product as chocolate, plus consumers want natural, low-calorie, stable products, so creativity in structuring products is required. Sugar provides structure and sweetness to the chocolate, so reducing the level of sugar needs particular inventiveness. Polyols can be used, but these may have a laxative effect, so are not ideal. An example of the creativity required is a recent patent[3] on bulk sugar replacers that uses porous amorphous particles of sugar and bulking agents such as milk powders that are permissible ingredients.

Reducing fat within the legislation only really allows for chocolate down to 25% fat, but at this level, viscosity for processing and texture for the consumer become an issue using conventional manufacturing processes. One example[4] from the patent literature proposes to first create a powder by coating one or more of the ingredients with fat and then use this in a chocolate-making process to produce a reduced fat product. Again, this is structuring standard ingredients in a different way to achieve a novel solution.

Many attempts have been made to replace high-calorie parts of the chocolate recipe with lower-impact materials. For example, Norton *et al.*[5] showed that stable water in oil (w/o) cocoa butter emulsions can be produced that are fully emulsified with small droplets (typically 3–5 μm) and with fat in the desirable polymorphic form V. The sintered crystalline shells observed at the oil/water droplet interface appear to act as Pickering particles. This both gives physical and microbial stability and acts to osmotically separate hydroscopic particles from the aqueous phase.

The process design was based on margarine manufacturing rather than on a conventional chocolate process, in which shear is provided first in a scraped surface heat exchanger and then in a pin stirrer. The work shows that the conventional route of Figure 6.1 is not the only one that can lead to tempered chocolate, suggesting that margarine technology offers a continuous one-step process for both emulsification and "tempering" with precise temperature control.

Another key challenge is the heat sensitivity of chocolate. Cocoa butter gives a suitable melt profile at body temperature, but when it is subjected to warm ambient conditions, melting, migration, and then recrystallization in an alternative polymorphic form leads to fat bloom, which gives a poor visual appearance to the chocolate. The shape and definition of the product can also be lost as the product melts. There is a significant landscape of patents in this area from all the major chocolate manufacturers. An extensive literature review[6] has been published and discusses the main strategies to produce heat-resistant chocolate; these are enhancing network microstructure, addition of an oil/fat binding polymer, and increasing the melting point of the fat phase.

All of these different formulations can give very different material properties and structure to the complex final product. In addition to formulation, the structure of the chocolate is very dependent on the size of the solid particles, which create the dense suspension discussed in the next section.

6.3 Chocolate Manufacturing and Structure Control

6.3.1 Liquid Chocolate Mass Making

During chocolate mass making, it is important to carefully control the size of the particle size to give the consumer a smooth texture and viscosity.[7] Milk powder, sugar, and cocoa solids need to be reduced in size, and this reduction can be done by various methods – *e.g.* roll refining, ball milling, air classifier milling.

Commonly, roll refining is used to reduce the particle dimensions down to the 20- to 30-μm size required to make a chocolate that is smooth in the mouth. Figure 6.2 shows a schematic of a two-roll premill and a five-roll mill, together with the particle size distributions that result. Two-stage refining makes it possible to start the refining process with a relatively coarse paste.

There are a limited number of studies of the refining process; for example, Bolenz and Mansche[8] study a range of processes and show that bimodal particle size distributions may be found. A study of multiroll milling[9] showed that the mechanical energy input could be correlated with size reduction and produced a model to predict the torque required.

Conching is a key part of the chocolate making process. Its function is to ensure that the material is converted from a dry flake to a liquid. In addition,

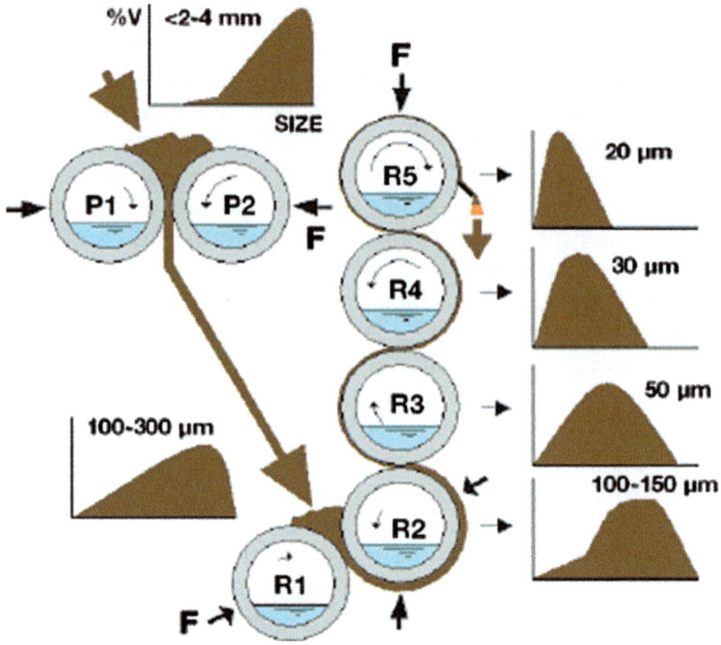

Figure 6.2 Schematic diagram of a typical multiroll chocolate refining process, with a two-roll prerefining stage (P1 and P2) and a five-roll (R1-5) refiner, with sketches of the particle size distribution generated through the process. Reproduced from ref. 9 with permission of Xabier Legarreta Basabe.

conching acts to remove undesirable flavors arising in the fermentation and roasting process, while developing a slightly cooked flavor. Temperature, time, and shear are all critical to this process. Moisture can be lost during conching, but water levels are quite low, limiting the Maillard reaction (nonenzymic browning). High temperatures can be used to develop Maillard flavors; chocolate flavor generation has been recently reviewed.[10] Viscosity control is accomplished by evenly distributing the fat phase over the particles leading to a fat continuous mass at the end of conching.

Viscosity and particle size are very closely linked. An example of controlling rheology with particle size is given in a Mars patent,[11] which describes mixing different particle size distributions to manipulate the rheological properties of a chocolate by influencing the packing of the solids.

Many authors have studied the complex rheology and viscosity characterization of chocolate, and a large number of models have been tested, such as the Casson, Bingham, and Herschel-Bulkley models. Wells[12] introduces the subject of chocolate rheology, while Gonçalves and Lannes[13] give a summary of investigations in this field. The rheology of different chocolates in terms of the particle size distribution has also been considered;[14] white chocolate, with the smallest particles, was found to possess the lowest viscosity.

6.3.2 Tempering and Forming

Following on from liquid chocolate mass making, manufacturers use and consider chocolate in three main states:

- First, there is *liquid chocolate*, which is stored warm (40–50 °C) as a melted phase and which can be characterized as a shear thinning suspension. The particles created during the refining step will influence how smooth the chocolate is perceived by the consumer but will also influence the viscosity due to the total surface area of the particles that need coating in fat to form the continuous fat phase.
- Next is *tempered chocolate*, which is used to coat and mold products. In this state, the fat in the chocolate has undergone an initial controlled crystallization. The lipid crystallization is influenced by shear. The chocolate can begin to thicken with time and has to be carefully temperature controlled (around 30 °C) to stop fat seed crystals melting completely or the chocolate setting in the equipment before it is used. Characterizing and controlling the rheology of this material is a particular challenge due to its transient nature. The rheology needs to be ideal to suit the forming method and conditions that will be used (*e.g.* flowing through a depositor and across a mold). There is very little rheological research on the complexity of this tempered state.
- The final state is the *solid chocolate*, which gets wrapped and delivered to the consumer. If it has been molded, it will need to contract enough on crystallization to allow it to fall out of the mold, and the mold surface energy will need to be low enough to allow separation from the mold surface.[15] It also has to be robust enough to withstand this demolding process and transit through the wrapping, packing, and supply chain systems.

The transitions between these states are controlled by the key process steps of tempering and cooling. To understand these processes in more detail, we need to understand the fat system in the product, particularly the polymorphic nature of the cocoa butter.

6.4 Polymorphism in Cocoa Butter

Correctly tempered and cooled chocolate will have high gloss, good snap, an appropriate melting temperature, and bloom resistance. One of the major problems in chocolate manufacture is ensuring that the fat crystallizes into the correct crystal form. Fats can exist in six different polymorphs (see Table 6.2); the requirement is to reach Form V, which has the qualities – taste, texture, and gloss – that the consumer associates with chocolate. Incorrect fat composition gives a material that is physically unattractive and possesses poor texture and melt properties in the mouth during consumption.

Table 6.2 Polymorphs of cocoa butter.

Polymorph	Conditions under which polymorphs arise	Melting point (°C)
I	Rapid cooling of melt.	17.3
II	Cooling of melt at 2 °C min^{-1}. Rapid cooling of melt followed by storage of up to an hour at 0 °C; this form is stable at 0 °C for up to an hour.	23.3
III	Solidification of melt at 5–10 °C. Transformation of form II by storage at 5–10 °C.	25.5
IV	Solidification of melt at 16–21 °C. Transformation of form III.	27.3
V	Solidification of melt. Transformation of form IV.	33.8
VI	Transformation of form V (after extended period of time at room temperature).	36.3

Once cocoa butter is mixed with the other ingredients in the chocolate recipe, it becomes difficult to predict and measure exactly what crystal form is present. Interactions between various fats in a recipe will adjust the melting points given in Table 6.2. The sugar and solids in the recipe can make techniques such as X-ray diffraction (XRD) difficult to employ, and often the polymorphic form can really be inferred only by techniques such as differential scanning calorimetry (DSC).

During the tempering stage, a series of cooling and heating steps, which select fat crystals of form β_V, are carried out. These crystals, as opposed to the β' crystals, have a melting point close to body temperature and are stable at room temperature.[16]

Alternatives to conventional processing exist. One is to preseed the material, the method of β_{VI}-seeding in which 0.05–1% (w/w) cocoa butter crystals in their most stable form VI are mixed with precooled chocolate (32–34 °C), giving a large number of small well-defined crystal nuclei.[17] Although the seed crystal nuclei are in polymorphic form VI, the surrounding chocolate solidifies and grows in the preferred form β_V.[18] From a manufacturing point of view, this requires more equipment than traditional tempering, as manufacturing lines need to handle chocolate and at the same time create the seeding mass.

Traditional chocolate processing typically involves a slow cooling (<2 °C min^{-1}) after the tempering stage.[19] Once tempered, the microstructure of the cocoa butter in the chocolate is strongly dependent on the cooling rate applied to the chocolate during processing. Svanberg *et al.*[20] studied the effect of different crystallization methods on the properties of dark chocolates and showed that different crystal densities can be developed, which in turn can affect the stability of the material.

Cold stamping is a forming process that is used to create molded chocolate shells that can be filled (*e.g.* with carmel, truffle, fondant). The FrozenCone® method uses very rapid cooling,[21] in which the shape of the chocolate is stamped out using a very cold plunger, leading to a very high cooling rate (≈100 °C min^{-1}) at the interface between the chocolate and the plunger, and these fast cooling rates do not necessarily ensure that only the β_V form of cocoa butter will be crystallized. The process raises the viscosity of the chocolate to allow it to keep its very accurate form while it is being filled but still requires the whole molded chocolate to be subsequently cooled in the more traditional, slower way.

Changing temperature histories during processing directly affects the type and extent of crystals formed. Such control might be useful in the design of novel structured products to overcome some of the challenges in reducing fat and sugar. However, the crystallization of lower-stability polymorphs is a problem for manufacturers since it can lead to an acceleration of the blooming problem (during the polymorphous transition), making the chocolate appearance unacceptable for the consumer.

The modeling of chocolate cooling has been extensively investigated.[22–25] The challenges presented for such a model are the polymorphism of cocoa butter and the lack of temperature-dependent data for the phase change kinetics of cocoa butter.

The adhesive force (between the chocolate and the mold surface) and the cohesive force (between elements of the chocolate itself) that affect both demolding and the subsequent mold cleaning process are also strongly dependent on the structure of the product. In the case of chocolate, this is related to the crystal structure obtained during processing and the interaction between the chocolate and the mold surface. Adhesive or/and cohesive failure will determine the ease of demolding stage and eventually influence the product quality characteristics.[26]

6.4.1 Chocolate Cooling Model

To model all the polymorphic changes that occur in solidifying chocolate would be very complicated due to the number of simultaneous processes occurring. A simplified model of chocolate cooling has been validated using XRD data and a range of temperature profiles.[27–29] The model was also extended to the FrozenCone® process[21] using commercial finite element software.

The model used to describe the cocoa butter phase change kinetics and polymorphism is based on the hypothesis that the polymorphs can be split into two categories of *unstable* and *stable* crystals. This is a reasonable assumption, as can be seen in the two sets of peaks seen in Figure 6.3, where stable crystals are created by increasing shear, and the temperature of melting thus changes. This hypothesis is also supported by XRD and DSC data. The unstable crystals are α and β' while the stable crystals are β crystals. *Unstable* crystals have a very strong d-spacing around 4.3 Å while *stable* crystals have very strong d-spacings at 4.6 Å.[30,31] The model considers six different reactions, each of them associated with a kinetic rate dependent on temperature (step change around the phase change temperature). To model crystallization, the heat transfer eqn (6.1) was solved,

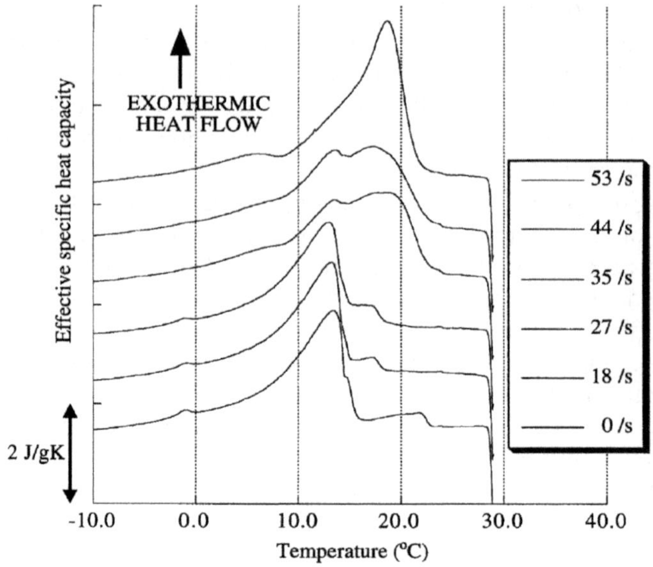

Figure 6.3 Differential scanning calorimetry of chocolate sheared at different rates; all have been held at 22 °C for 400 s and then rewarmed to 32 °C prior to cooling and shear.
Reproduced from ref. 19 with permission from John Wiley & Sons, Copyright © 1999 American Oil Chemists' Society (AOCS).

together with three differential equations for mass balance on the three phases of the system (2):

$$\rho \cdot Cp \cdot \frac{\partial T}{\partial t} = \nabla \cdot (k \cdot \nabla T) + Q_{\text{source}} \tag{6.1}$$

where ρ is the density of the chocolate, Cp is its specific heat, k its thermal conductivity, and Q_{source} is the heat source associated with the phase changes of cocoa butter. The phase changes of cocoa butter were solved according to the system of PDEs defined in eqn (6.2). The system was simplified to only two polymorphs and a liquid phase, with equations relating them:

$$\begin{aligned}
\frac{[stable]}{\partial t} &= k_{\text{ms}} \cdot [melt] \cdot [stable]^c + k_{\text{us}} \cdot [unstable] \cdot [stable]^c - k_{\text{sm}} \cdot [stable] \\
\frac{[unstable]}{\partial t} &= k_{\text{nuc}} \cdot [melt] + k_{\text{mu}} \cdot [melt] \cdot [unstable]^c - k_{\text{um}} \cdot [unstable] \\
\frac{[melt]}{\partial t} &= -k_{\text{ms}} \cdot [melt] \cdot [stable]^c + k_{\text{sm}}[stable] - k_{\text{nuc}} \cdot [melt] - k_{\text{mu}} \cdot [melt] \\
&\quad \cdot [unstable]^c + k_{\text{um}}[unstable]
\end{aligned} \tag{6.2}$$

where [*stable*] is the fraction in *stable* crystals, [*unstable*] is the fraction in *unstable* crystals, and [*melt*] is the fraction in *melt*. Q_{source} can then be calculated from the rates of phase change of the different components. Since the kinetic rates are dependent on temperature, this is a fully coupled model between phase change and heat transfer.

$$Q_{\text{source}} = \Delta H_{stable} \cdot \frac{\partial[stable]}{\partial t} + \Delta H_{unstable} \cdot \frac{\partial[unstable]}{\partial t} \tag{6.3}$$

The model has been validated against XRD and temperature data.[27] A typical comparison between the model and the experimental data obtained from XRD analysis is shown on on Figure 6.4. The agreement between the model and the experimental data is very good ($R^2 = 0.978$). The model tends to overestimate the values under 0.5; however, this could also be due to a lack of sensitivity of the technique when small amounts of crystals are present.

The temperature distribution simulated for a typical application time of the cold press ($t = 3$ s) is shown in Figure 6.5. The low thermal conductivity of chocolate ($k \cong 0.3$ W mK^{-1}), coupled to the short time of the process, leads to a heterogeneous distribution of temperature in the product at the end of the application of the cold plunger.

This leads to only a small amount of material (a thin layer just thick enough to hold the desired shape) being crystallized in the unstable form because of the rapid cooling, as shown in Figure 6.5. The maximum amount of unstable crystals is located at the interface with the cold press, reaching about 20% (in terms of weight) of the overall cocoa butter content. Very few

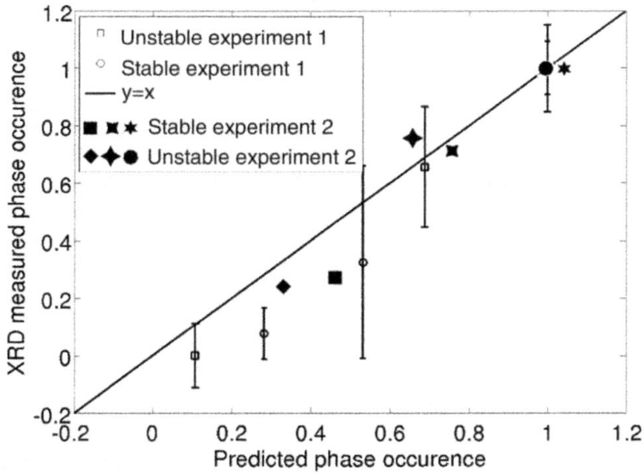

Figure 6.4 Comparison between the XRD measured and model prediction of the crystal structure of cocoa butter in chocolate at different locations in the chocolate.
Reproduced from ref. 32 with permission from Elsevier, Copyright 2011.

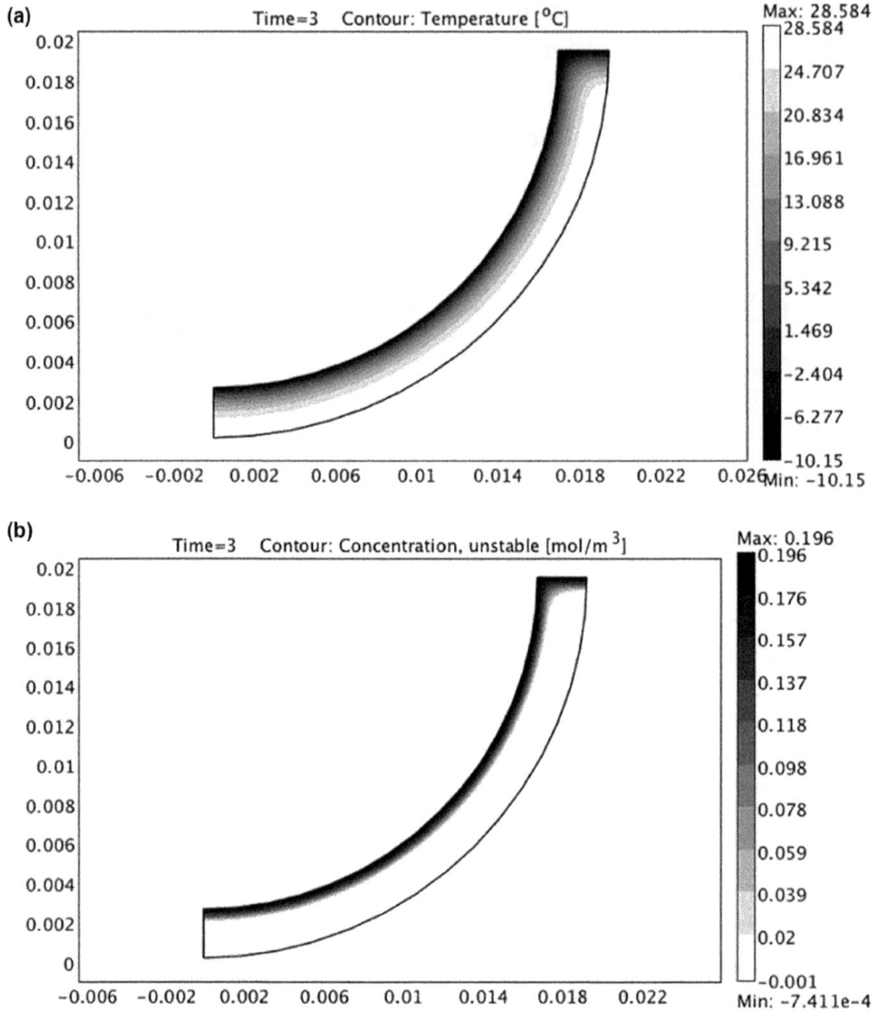

Figure 6.5 (a) Temperature distribution simulated after 3 s of application of the FrozenCone® process to chocolate; (b) unstable crystals distribution simulated after 3 s of application of the FrozenCone® process to chocolate. Reproduced from ref. 32 with permission from Elsevier, Copyright 2011.

stable crystals were formed during this process increasing from the initial 3% to 3.5% of the overall cocoa butter content.

One of the most interesting results is that the cocoa butter in chocolate is still mainly liquid (only 20% of the shell has a solid fat content over 10%). The FrozenCone® process therefore has to be followed by a series of cooling tunnels to set the cocoa butter. The further cooling in these cooling tunnels is slower, on the order of the 1–2 °C min^{-1} required in conventional processing, which allows the growth of the stable crystals seeds from the

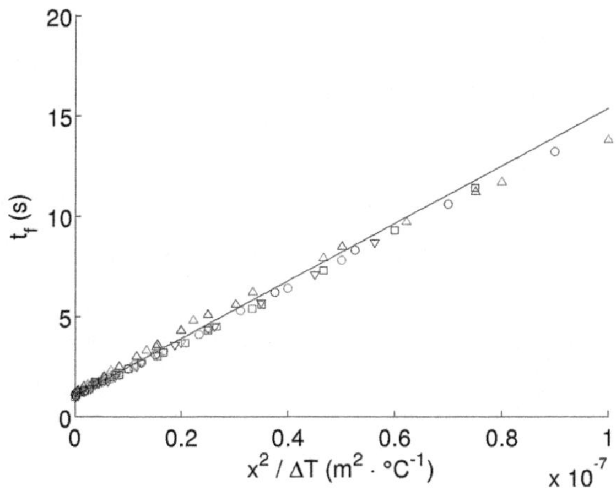

Figure 6.6 Freezing time required to crystallize the thin layer during the application of the cold press.
Reproduced from ref. 32 with permission from Elsevier, Copyright 2011.

tempering stage, explaining the success of the FrozenCone® system to produce fine chocolate in spite of the initial fast cooling rates.

From an engineering perspective, tuning process parameters to reach this crystallization target are difficult to achieve using an iterative trial-and-error approach. However, using the model to predict the level of crystallization in the shell can solve this problem. A typical approach in the calculation of freezing times of foods is to use the Plank equation:[33]

$$t_f = \frac{\rho \cdot \Delta H}{a \cdot 2 \cdot k} \cdot \frac{x^2}{\Delta T} \quad (6.4)$$

where a is a constant dependent on the geometry, x is the thickness of the geometry to freeze (here 20% of e_{shell}), ΔH is the latent heat of fusion (here 10% of $\Delta H_{unstable}$), and ΔT is the difference between the cooling media temperature (T_{FC}) and the initial temperature of chocolate (T_0).

A typical processing parameter that has to be changed from one product to another is the thickness of the shell (e). The model was used[32] to see if a modified Plank equation could be used to adjust the processing parameters (t and T_{FC}) when a different product was to be made. The results of this analysis are shown in Figure 6.6. The linear behavior as predicted by Plank's theory is valid, and this is a reasonably simple method to obtain processing time.

6.5 Mechanisms for Bloom Formation

Bloom on chocolate, due to the structure of the fat phase breaking down from the desired form, is one of the major processing problems that can arise. Bloom on chocolate is a condition in which the structure of the

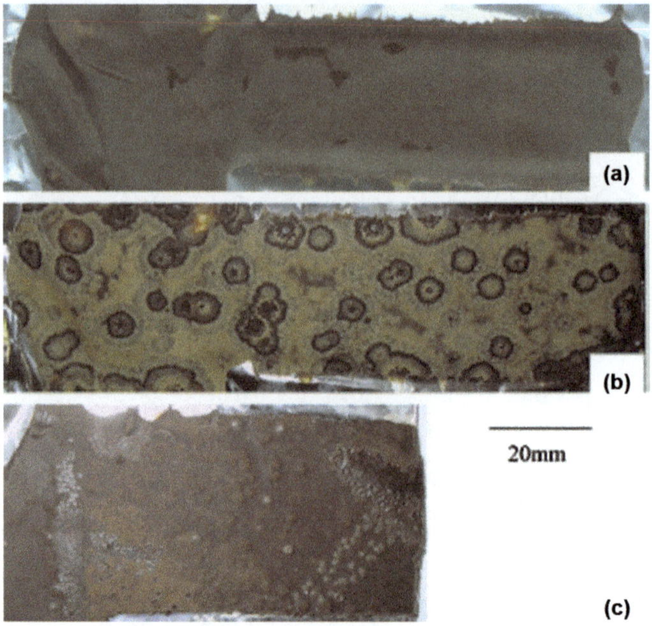

Figure 6.7 Images of 72% cocoa mass chocolate samples captured using flatbed scanner: (a) untempered chocolate immediately after solidification; (b) untempered chocolate after 7 days storage at room temperature; (c) poorly tempered chocolate after 7 days storage at room temperature. Reproduced from ref. 34 with permission from Elsevier, Copyright 2009.

chocolate changes, giving a white powdery appearance. This structure is less attractive to the consumer due to both the appearance of and the change in the properties of the chocolate because the melting point of some of the components has changed to above-mouth temperature. Figure 6.7 shows images of the surface discoloration caused by bloom.[34]

Bloom arises because form V, which is the desirable one for the consumer, is not thermodynamically stable – conventional chocolate is kinetically trapped upon cooling into an essentially unstable form. During storage or when the chocolate is reheated and then cooled, the unstable structures of the fats can break down into a bloomed form. Avoiding bloom is thus critical in manufacture and the supply chain.

At a microscale, bloom arises due to the formation of both fat and sugar crystals; this is seen clearly in the work of James and Smith,[34] who present a series of SEM (scanning electron microscope) and ESEM (environmental scanning electron microscope) images of bloom. Figure 6.8 shows the formation of fat crystals on the surface of chocolate; the top image is well-tempered material, and the lower two pictures show the morphological changes due to bloom, in which fat crystals have been forced out of the surface of the chocolate. The characteristic "corona" shape of bloom at a macroscale involves both fat and sugar crystallization.

Figure 6.8 SEM images of the surface of fresh and bloomed dark chocolate: (a) cryo-SEM image of fresh, well-tempered chocolate; (b) ESEM image of lightly bloomed, well-tempered chocolate stored at 30 °C for 30 days; (c) cryo-SEM image of surface detail of fat crystals; surface striations on fat indicate an extrusion mechanism from the surface of the chocolate. Reproduced from ref. 34 with permission from Elsevier, Copyright 2009.

A number of studies have looked at the mechanism and kinetics of bloom. The higher the temperature of storage is, the greater the bloom formation will be.[35] In another work,[36] X-ray diffraction was used to study a series of chocolate formulations cycled between 19 °C and 29 °C and showed form VI production in all cases. However, the rate of bloom was delayed if the higher melting point milk fats was used and suppressed altogether if amorphous sugar was used. It was thus proposed that the chocolate microstructure, rather than fat transitions alone, gave rise to bloom. Further evidence for this was given in a separate study,[37] which demonstrated that removing minor lipids changed the crystal form within the chocolate and then the rate and extent of bloom. Temperature cycling can also promote bloom.[38]

Jin and Hartel[39] describe a storage blooming process in which phase separation, initiated by temperature fluctuation, leads to bloom during storage *via* the (1) dissolution of the higher melting part of the fat phases, (2) migration of liquid fats to the surface, (3) recrystallization of higher melting fats on the surface into the more stable polymorph, form VI. Recrystallization is not enough to cause visual bloom; needle-shaped fat crystals must be formed on the surface to create bloom – as seen in Figure 6.8. These surface protrusions change light reflection leading to visual bloom.

Zhao *et al.*[40] used a model chocolate containing black sand to simplify the system and showed that bloom occurred more quickly in systems of larger particle size, in which fats could be more mobile. Another study,[41] however, reports that smaller particles enhanced bloom. A number of studies *e.g.*[42] have investigated how bloom can be suppressed, but these generally involve the addition of extra ingredients, which may be undesirable. In practice, bloom control is still semiempirical, and no predictive model has been successfully developed, unlike in crystallization. Fat migration from fillings contained within a chocolate shell can also impact the formation of bloom, making the science of formulation in this area even more complex.

6.6 Conclusions and Future Needs

Chocolate is a fascinating material to study due to its complex nature. Many of the mechanisms that create chocolate are still unquantified. Chocolate manufacture is a major industry, but, as in many parts of the food industry, it originates from an artisan background and hence lacks some of the fundamental science and engineering research that many other manufacturing sectors have. The lack of basic material properties and mechanistic science understanding are holding back the development of modeling and simulation tools that many other industries use routinely to develop and optimize products and processes.

There is a good mechanistic understanding of the sequence of events in processing and of how they give rise to chocolate properties. Crystallization can be quantified into a series of equations that have some predictive value; other processes, such as bloom, have not yet been understood to the level that a predictive model can be developed. While there is a place for working on model systems to gain insight and understanding, actual chocolate recipes are complex, and interactions between ingredients, particularly in the fat system, are not well understood. Work on formulation science and ingredient interactions will be key areas in building mechanistic models, speeding up new product development, and improving product performance to address the demands of regulators and consumers.

References

1. S. T. Beckett, M. S. Fowler and G. R. Ziegler, *Beckett's Industrial Chocolate Manufacture and Use*, WILEY Blackwell, Chichester, 5th edn, 2017.
2. S. T. Beckett, *Science of Chocolate*, Royal Society of Chemistry, Cambridge, 2nd edn, 2008.
3. V. D. M. Meunier, M. Dupas-Langlet, J. P. N. Mahieux, A. S. Whitehouse and L. Forny, *WO Pat.*, 093302, 2017.
4. M. De Brouwer and V. Derycke, *U. S. Pat.*, 0159091A1, 2010.
5. J. E. Norton, P. J. Fryer, J. Parkinson and P. W. Cox, *J. Food Eng.*, 2009, **95**, 172.
6. T. A. Stortz and A. G. Marangoni, *Trends Food Sci. Technol.*, 2011, **22**(5), 201.
7. G. R. Ziegler, G. Mongia and R. Hollender, *Int. J. Food Prop.*, 2001, **4**(2), 353.
8. S. Bolenz and A. Manske, *Eur. Food Res. Technol.*, 2013, **236**, 863.
9. X. Legarreta Basabe, Eng D Thesis, University of Birmingham, 2017.
10. N. J. Engeseth and M. F. Ac Pangan, *Curr. Opin. Food Sci.*, 2018, **21**, 84.
11. J. M. Kaiser and S. A. S. Purwo, *U. S. Pat.*, 6391373B1, 2002.
12. M. Wells, in *Science and Technology of Enrobed and Filled Chocolate, Confectionery and Bakery Products*, ed. G. Talbot, Woodhead, Oxford, 2009, vol. 13, pp. 255–284.
13. E. V. Gonçalves and S. C. da Silva Lannes, *Cienc. Tecnol. Aliment.*, 2010, **30**, 845.
14. V. Glicerina, F. Balestra, M. Dalla Rosa and S. Romani, *J. Food Eng.*, 2016, **169**, 165.
15. E. L. Keijbets, J. Chen, E. Dickinson and J. Vieira, *J. Food Eng.*, 2009, **92**, 217.
16. G. Chapman, E. Akehurst and W. Wright, *J. Am. Oil Chem. Soc.*, 1971, **48**, 824.
17. E. J. Windhab, *J. Therm. Anal. Calorim.*, 1999, **57**, 171.
18. Y. Zeng, P. Braun and E. J. Windhab, *Manuf. Confect.*, 2002, **82**, 71.
19. A. G. F. Stapley, H. Tewkesbury and P. J. Fryer, *J. Am. Oil Chem. Soc.*, 1999, **76**, 677.
20. L. Svanberg, L. Ahme, N. Loren and E. J. Windhab, *J. Food Eng.*, 2013, **114**, 90.
21. L. Aasted, *EP Pat.*, 1472229, 1992.
22. K. Franke, *J. Food Eng.*, 1998, **36**, 371.
23. G. Mazzanti, A. G. Marangoni and S. H. J. Idziak, *Phys. Rev.*, 2005, **71**, 1.
24. H. Tewkesbury, A. G. F. Stapley and P. J. Fryer, *Chem. Eng. Sci.*, 2000, **55**, 3123.
25. B. Tremeac, A. K. Datta, M. Hayert and A. Le-Bail, *Int. J. Refrig.*, 2007, **30**, 958.
26. E. L. Keijbets, J. Chen and J. Vieira, *J. Food Eng.*, 2010, **98**, 133.
27. B. J. D. Le Révérend, P. J. Fryer and S. Bakalis, *Soft Matter*, 2009, **5**, 891.

28. B. J. D. Le Révérend, P. J. Fryer, S. Coles and S. Bakalis, *J. Am. Oil Chem. Soc.*, 2010, **87**, 239.
29. B. J. D. Le Révérend, I. Smart, P. J. Fryer and S. Bakalis, *Chem. Eng. Sci.*, 2011, **66**, 1077.
30. H. Schenk and R. Peschar, *Radiat. Phys. Chem.*, 2004, **71**, 829.
31. R. Wille and E. Lutton, *J. Am. Oil Chem. Soc.*, 1966, **43**, 491.
32. S. Bakalis, B. J. D. Le Révérend, N. Z. Rois Anwar and P. J. Fryer, *Procedia Food Sci.*, 2011, **1**, 340.
33. M. López-Leiva and B. Hallström, *J. Food Eng.*, 2003, **58**, 267.
34. B. J. James and B. G. Smith, *LWT – Food Sci. Technol.*, 2009, **42**, 929.
35. A. Ali, J. Selamat, Y. B. Che Man and A. M. Suria, *Food Chem.*, 2001, **72**, 491.
36. J. A. Bricknell and R. W. Hartel, *J. Oil Fat Ind.*, 1998, **75**, 1609.
37. R. A. Tietz and R. W. Hartel, *J. Am. Oil Chem. Soc.*, 2000, **77**(7), 763.
38. H. Zhao and B. J. James, *J. Food Eng.*, 2019, **249**, 9.
39. J. Jin and R. W. Hartel, *J. Am. Oil Chem. Soc.*, 2015, **92**, 1473.
40. H. Zhao, G. Bingol and B. J. James, *J. Food Eng.*, 2018, **225**, 12.
41. H. Dahlenborg, A. Millqvist-Fureby and B. Bergenståhl, *J. Food Eng.*, 2015, **146**, 172.
42. M. H. M. Buscato, L. M. Hara, E. C. Bonomi, G. de Andrade Calligaris, L. P. Cardoso, R. Grimaldi and T. G. Kieckbusch, *Food Chem.*, 2018, **256**, 390.

CHAPTER 7

Food Structure Development in Cereal and Snack Products

G. DELLA VALLE,*[a,c] H. CHIRON,[a,c] A. LE-BAIL[b,c] AND L. SAULNIER[a,c]

[a] INRA, UR 1268 Biopolymers Interactions and Assemblies (BIA), BP 71627, 44316 Nantes, France; [b] Oniris, UMR CNRS 6144 GEPEA, CS 82225, 44322 Nantes, France; [c] SFR IBSM 4202, INRA-BIA, CNRS-GEPEA, BP 71627, 44316 Nantes, France
*Email: guy.della-valle@inra.fr

7.1 Introduction

Cereal foods are widely consumed in different manners throughout the world, although they may no more be the basis of the diet pyramid in Western countries. Through their content in starch and proteins, they are the most important energy source in our diet. They are already the main provider of dietary fiber in the Western diet, although whole grain products are not yet widely accepted/consumed despite their high dietary fiber and micronutrient contents. Besides, their protein fraction brings a large range of amino acids.[1] In spite of the culture dependence of food consumption, technology, envisioned as an overall chain of processes, plays a major part in providing consumers with cereal foods that possess high sensory and nutritional qualities. Processing encompasses a large range of operations from grain harvesting and milling to flour production, which results from a compromise between the necessity to avoid contamination, while taking advantage of all components from the grains' outer layers.[2] In this chapter, we focus on baking processes, which involve all operations from flour to

Food Chemistry, Function and Analysis No. 18
Handbook of Food Structure Development
Edited by Fotis Spyropoulos, Aris Lazidis and Ian T. Norton

Published by the Royal Society of Chemistry, www.rsc.org

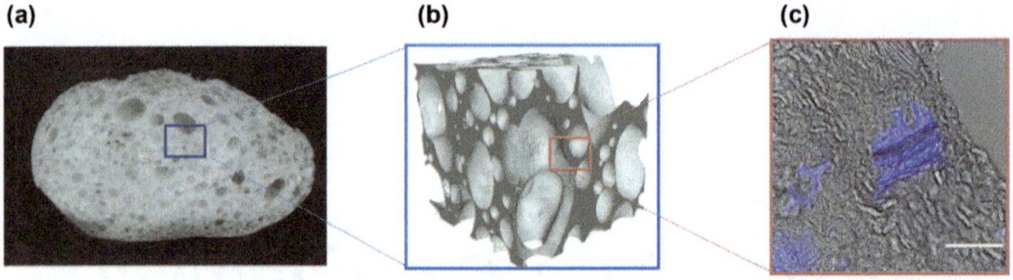

Figure 7.1 Different structural levels of an airy cereal food (crumb of current French bread): (a) crumb and crust at macroscopic level (width ≈ 10 cm); (b) 3D image of cellular structure by X-ray microtomography with wall materials in dark gray and gas cells in light gray (ESRF, F38-Grenoble, vol ≈ 1 cm^3); (c) composite microstructure imaged by light microscopy (white bar = 100 μm) of the cell wall material with flattened starch granules cemented by gluten blue stain showing presence of nonstarch polysaccharides (fibers)—and gas cell on upper right. All images from INRA.

cereal food product. These processes not only change the physicochemical properties of food components but also create the structure of the food. The knowledge of both features is needed to understand the link between composition and properties, as suggested by the concept of food matrix, which integrates the interactions with components at different levels of matter organization from the molecular to the macroscopic scale.[3,4]

Due to their aerated structure, the matrix of many cereal foods (bread, biscuits, cakes, extruded snacks, *etc.*) can be envisioned as a solid foam,[5] or cellular solid; *i.e.* because their density is much lower than the that of their components. Their different levels of matter organization may be illustrated, as in the case of bread, from the macroscopic to the microscopic scales (Figure 7.1). At the microscopic level, cereal foods may be considered as composite material, primarily made of starch, more or less amorphous, and aggregated proteins, containing more or less polysaccharides from dietary fibers coming from the outer part of the kernel (whole grain products) or from the starchy endosperm (white flour), blended with other ingredients (water, fat, sugar, and other dietary fibers). A more detailed vision of the different levels of matter organization can be found in the case of bread.[6]

Among the resulting properties, texture has an outstanding role. It can be defined as the perception of mechanical properties, either by an instrument or senses (seeing, hearing, touching) and mouthfeel. At first glance, this property depends mainly on the moisture content (MC ≈ +/−) and the density ($\rho^* \approx +/-$), which gives an approximate classification, illustrated by simple examples: extruded snack (MC ≈ −, $\rho^* \approx -$), bread (+, −), biscuit (−, +), cooked pasta (+, +). More precisely, the mechanical properties of cereal foods are functions of their density, their cellular structure, and intrinsic material properties. The latter depends on the morphology of the composite material and on the properties of each component. So, together with density, the composition, including moisture content, will govern the texture, as indicated by the few numerical values of mechanical properties

Table 7.1 Some values of physical properties for several cereal foods,[5] including biscuits,[7] and sponge cake and brioche:[8] *MC* moisture content (on total basis), T_g glass transition temperature (°C), ρ^* bulk density (kg m^{-3}), E^* Young modulus (MPa). Fat content is indicated on a total basis.

Product	MC (% bh)	T_g (°C)	ρ^* (kg m^{-3})	Modulus E* (MPa)	Maximum or rupture stress (MPa)
Extruded starch	10	82	100–300	7–140	0.3–5
Extruded maize flakes	16	106	120	5.5	0.2
Extruded wheat flakes	9	96	130–700	0.03–12.5	0.2–19
Biscuit (16% fat)	4–10.5	—	474	175–12.5	0.8–0.15
Biscuit (16–9% fat)	3–4	—	690–790	—	0.04–0.1
Baguette crumb	45	−57	200	0.004	—
Pan bread	40	−43	100–165	0.008–0.016	0.0007–0.003
Brioche (17% fat)	31	—	330	0.02	0.004
Sponge cake (6% fat)	30		210	0.005	—

reported in Table 7.1, for various cereal foods. From these values, the importance of water is clearly evidenced, but its role is ubiquitous as it continuously impacts processing and product structure at various levels of matter organization.

In the following, we will give an overview of the various components involved in cereal foods and how they are modified under processing, with emphasis on rheological properties and how these can be used in models that contribute to the better understanding of cereal foods processing. At each step of this comprehensive approach, we will strive to provide the reader with necessary information about the methods available in the literature and underline the limitations and lack of knowledge, taking bread, biscuits, and extruded snacks as the main examples. Finally, we will conclude by adopting an outlook on their potential to design cereal foods with target sensory and nutritional properties.

7.2 Products and Materials

Cereal foods generally contain less than 50% water (total basis). Besides their water content, they cover a wide range of compositions, in sugar and fat (Figure 7.2a). However, the main component is the cereal flour, and, in this section, we will focus on wheat flour components and stress the role of these biopolymers in dough and food structuring.

7.2.1 Wheat Components

Generally, the perception of wheat quality is mainly driven by the content of proteins capable of forming a suitable gluten network, which is the case for bread dough structuring. This is not a real constraint for biscuits, cakes, and snacks, which have a more composite morphology, in which gluten may not be the continuous phase (Figure 7.2b). Wheat storage proteins consist of

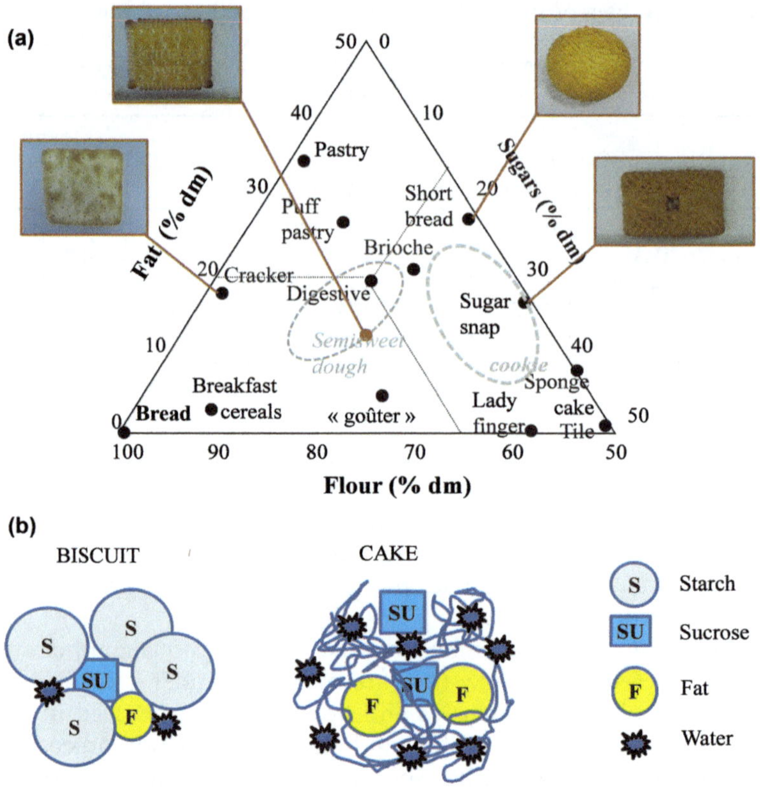

Figure 7.2 Composition and schematic view of cereal food structure: (a) ternary diagram plotting the approximate composition of several cereal foods, on dry matter basis (flour + fat + sugar =100),[9] and images of commercial biscuits (from left to right, Cream Cracker, Petit Beurre, Palet Breton, and Speculoos) at the relevant composition; (b) scheme of biscuit versus cake structure, enlightening the role of water (proteins may act on both side starch or fat + sugar, according to water content).

gliadin and glutenin, two protein classes roughly in balance in the grain and playing different roles in the viscoelastic properties of the dough. Balance between gliadin and glutenin is an important determinant of dough behavior.[7] Glutenin is also referred to as glutenin polymer because it is formed by the concatenation of several polypeptides (or subunits) linked by disulfide bonds. Their average molar mass is in the range of 10^3 to 10^6 g mole^{-1}. Flour content in very large glutenin polymers has been identified as a determinant of gluten elasticity and dough strength.[8] Their aggregation status can be followed using classical size exclusion high-performance liquid chromatography (SE-HPLC) analysis of flour proteins.[9] improved by revealing glutenin polymers rearrangement potentialities.[10] Considering that the gluten network is formed by disulfide bonds (–S–S–) between the wheat reserve proteins from the sulfhydryl groups (–S–H) that are initially present and that hydrogens atoms of the thiol functions (–S–H) are exchangeable, Magnetic resonance imaging (MRI) can be thought to use this property to image the gluten network. Recently, based on microscopy and imaging techniques, protein network analysis (PNA) has led to the characterization of the gluten network in terms of junctions' density and branching ratio, among other descriptors,[11] Likely, the fundamental understanding of gluten network would gain from the synergy between these methodological approaches, in order to improve the (supra)molecular basis of the mechanical properties of gluten and dough.[12]

Starch is by far the most abundant compound of cereals flour and plays many important roles in final product texture. However, differences in structure development during flour processing, especially wheat milling, cannot be attributed to differences in starch composition for standard wheats; *e.g.* with an amylose-to-amylopectin ratio of 25 to 75, but they can rather be attributed to the fraction of starch damaged during milling.[13,14] This is due to the granular structure of starch, which is generally maintained until heat treatment during baking. In snack extrusion, however, all the structural features of starch are modified, and processed flour may be envisioned as a continuous starch molten phase in which other components are dispersed.[15,16] These changes can be determined at all levels of matter organization (macromolecules, crystal, granule) by asymmetric forced field flow fractionation (A4F), differential scanning calorimetry (DSC), and light microscopy,[17] respectively. More accurate information on matter organization and macromolecular order in amorphous starch may be found using solid state nuclear magnetic resonance (NMR) and second harmonic generation imaging.[18]

Other components, such as cell wall components (pentosans) and lipids, also influence the processing behavior of flours. Lipids account for about 2% of the dry weight of wheat flour. About one-quarter of total lipids are embedded within starch granules and do not play a major role in processing, except during storage; in staling for instance. Nonstarch lipids are composed of nonpolar lipids, mono-, di-, and triglycerides and free fatty acids. Variation in the proportion of nonpolar-to-polar lipids can affect significantly the final quality of cereal baked products, since it has been generally shown that the nonpolar fraction depresses bread volume while the polar one enhances it. This effect is due to the role of polar lipids on the formation and stability of the lipid–protein interfaces of gas bubbles, from dough mixing to the early stages of baking, including expansion during fermentation.[19] Actually, the content and composition of lipids in flour are mainly driven by milling, which determines the proportion of germ and aleurone (nonpolar lipid-rich tissue) *versus* starchy endosperm (polar lipid-rich tissue) of wheat flour. The milling properties, as well as the mass proportions of germ, aleurone, and starchy endosperm, vary according to the genetic background and phenotype, determined by the growth conditions of wheat. That is why it is important to know the content of wheat flour lipid and their composition.

The cell walls and their polysaccharides represent 12–14% of wheat grain (4–6% of white flour). They are the main constituents of dietary fiber, having major beneficial effects on human health and significant technological effects on the end use of cereal grains.[20] The cereal grain is a complex organ constituted of several tissues. Depending on their tissue of origin, cell wall polysaccharides exhibit different functionalities (solubility, viscosity, gelling properties, water holding capacities, *etc.*) that impact food processing. The cell walls of the starchy endosperm (white flours) are thin and hydrophilic and mainly formed of arabinoxylans (AX, also known as pentosans). AX have long been reported to have a marked influence on cereal food properties despite their low amount in white flours (2–3%),[21] They are present as water-extractable (WE-AX) fractions that create a high viscosity related to specific nutritional and technological effects and as water-unextractable (WU-AX) fractions that have high water binding capacities. These components play multiple roles: By interacting with water and the gluten or amorphous starch matrices, they affect the rheological properties of the dough, the growth and stabilization of gas bubbles, and the final volume and cellular structure of the food.[22,23] The AX content in wheat flour exhibits a large extent of variation, mainly under genetic control.[24] The balance between WE-AX and WU-AX can be modified using exogenous endoxylanase, which preserves a high viscosity, beneficial for bread making and associated with the presence of high molecular weight WE-AX fraction,[25] while decreasing the amount of water linked to the insoluble matrix by solubilizing the WU-AX fraction.[25] However, such modulation is not as efficient with the bran fractions, *i.e.* the outer part of cereal grains isolated by the milling industry, which are an assembly of various tissues. In these tissues, cell walls are thick, hydrophobic, and primarily formed by cellulose and xylans, essentially water unextractable, and they also contain lignin.[23] Bran can absorb and entrap water within its porous matrix, swell, and therefore impact cereal processing, but its water-binding properties are lower than those of the cell walls originating from the starchy endosperm.[26]

7.2.2 State Diagrams

All these components undergo several state changes during processing, and, especially during baking, heat transfer contributes to the food structure development. These changes can be mapped on state diagrams[27–29] that represent the variations of temperature of state change (glass transition, denaturation, melting, *etc.*) as a function of water content, which underlines the central role of water (Figure 7.3). These variations have been determined by DSC: in the case of starch, melting temperature T_m.[30,31] At intermediate moisture content (MC $\approx$ 0.5, wb), both phenomena, starch melting and gelatinization, coexist. Due to the presence of other components (proteins, cell wall components, oligosaccharides, *etc.*), T_m values for starch in wheat flour are slightly lower than those of starch alone. Protein denaturation and especially glutenin coagulation develop between 55 °C (for MC = 0.65) and 80 °C for (MC = 0.2), inducing gluten

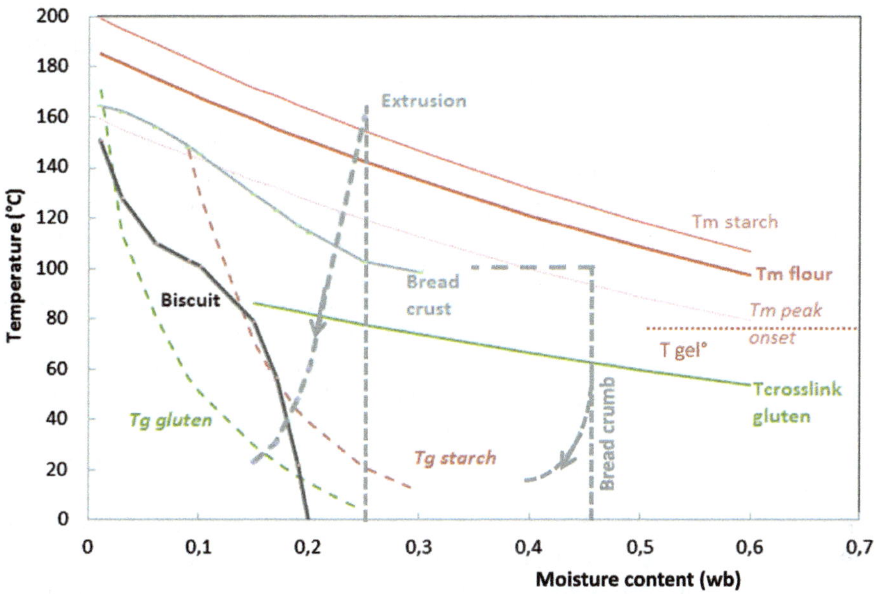

Figure 7.3 Example of state diagram for starch and gluten changes during biscuit, bread (crumb and crust), baking, and extrusion. Red and green continuous lines reflect transition temperature variations for starch and gluten, respectively (continuous: starch melting and gluten cross-linking; dotted: long glass transition; short: starch gelatinization). Gray lines refer to the hydrothermal pathways, including cooling (arrows), for each product mentioned in the graph: bread crumb[39] and crust,[40] extruded snack[41] (dotted lines), and biscuit[42] (continuous line).

cross-linking.[32,33] In the case of glass transition (T_g), DSC measurements of T_g have been ascertained by dynamic mechanical analysis (DMA) and other techniques like NMR.[34,35]

The diagrams in Figure 7.3 were completed by the processing pathways in the case of baked products and directly expanded snacks. They have also been used for practical developments of cereal foods.[36] During baking, the temperature of bread dough first reaches 100 °C, without significant loss of moisture content, at a rate inversely linked to the size of the dough. The transition from dough to crumb occurs at an intermediate temperature because of gluten cross-linking and starch granules swelling. Then, during cooling, there is a small moisture decrease in the crumb to balance the external loss, but crumb temperature is still much larger than T_g; starch remains in the rubbery domain, which explains the soft texture of bread crumb. Conversely, regarding the dough outer envelope, in the case of freestanding bread like baguette, the formation of crust is due to the concomitant increase of temperature up to about 160 °C and the decrease of moisture down to less than 0.05. This sharp drop of moisture content explains why some residual granular starch may still remain in the crust. Then after cooling and in spite of water migration from the crumb, the crust still has a low moisture content (<0.1); starch and gluten remain in the glassy state, which explains the brittle behavior of French bread crust.

The extrusion of starchy products, like cereal flour, is generally achieved at a lower moisture content (≤0.3) but at a higher temperature. The starch melting line is crossed, explaining that after extrusion, starch is amorphous, or more casually called gelatinized, although the main mechanisms are different from those involved in the presence of excess water. At the extruder die outlet, upon cooling, water vaporizes, leading to a drop of moisture content, close to its initial value in native starch. Hence, starch is in the glassy domain, and consequently directly expanded extruded snacks have a brittle behavior. Detailed diagrams have been drawn to cope with the various cases and explain the specific cellular structure of extruded snacks,[37] including the indirect expansion completed, for instance, with expansion by microwave.[38,39]

Biscuits, from sweet short dough to crackers and cookies, can display a large range of composition (see Figure 7.2), but they have generally a low moisture content and small thickness, comparatively to bread. These conditions lead to a rapid increase in temperature during baking, below the starch melting temperature. Hence, biscuits generally contain a large amount of intact starch granules, and the gluten network is weak. According to biscuit composition, the continuous phase may rather be a blend of sugar and fat,[40] leading to a glassy matrix after cooling. Conversely, in cake, starch gelatinization and melting contribute to the cohesion to the product, together with sugar and lipids providing the cohesiveness of the structure. For these three product families, some structural and processing features are recalled in Table 7.2.

Table 7.2 Overview of cereal food systems, processing operations, flow conditions, and related rheological behavior (— means not relevant or unavailable).

Product	Extruded snack	Biscuit dough	Bread dough	Bread dough
Process/operation	Flow in screw and die	Sheeting	Mixing	Proofing, fermentation
Shear rate (s^{-1})	$10–10^3$	$10^2–10^3$	$1–10^2$	—
Extensional strain rate (s^{-1})	1–10	1–10	$10^{-1}–10$	$10^{-4}–10^{-3}$
Structural model	Viscous polymeric fluid	Granule and aggregate suspension in fat/sugar emulsion	Transient gluten network entrapping starch granules	Starch/gluten matrix and liquid films separating bubbles
Rheological model	Power law (eqn (7.2) and (7.3))	—	Power law (eqn (7.2) and (7.3))	Power law (eqn (7.3))

7.3 Processing

Most cereal foods can be envisioned as solid foams, or cellular solids, due to their low density and their cellular structure. To generate this cellular structure, starting from the raw materials and ingredients, cereal processing encompasses a wide range of operations, as schematically illustrated in Figure 7.4. Two main groups of operations may be distinguished: those implementing the dough (or melt) formation and those modifying the formed dough, including cellular structure building. The case of extrusion cooking is specific since it includes all operations—except fermentation—in a single process that achieves mixing, shaping, and structure building by expansion in a characteristic time of less than 1 min.

7.3.1 Operations and Development of Cellular Structure

The bread-making process is based on the transition from a divided solid medium (the flour) to a continuous one (the dough), through hydration, mechanical, biochemical, and thermal actions (mixing, yeast fermentation, heating, *etc.*). The dough, a continuous starch/gluten matrix, becomes porous because of the expansion of bubble nuclei during fermentation, *i.e.* carbon dioxide production by yeast, thanks to the action of *Saccharomyces cerevisiae*; after baking, bubble expansion gives the crumb a cellular structure. During mixing, the first operation of bread making, ingredients are first homogenized, leading to the distribution of water in the flour components, and then the gluten network is woven, leading to the dough's extensional properties and subsequent gas cell retention and stabilization. A significant amount of air (porosity, or air volume fraction ≈ 0.1) is entrapped during mixing,[41] with further bubble growth, by fermentation, taking place during resting. Implementing partial vacuum, by modifying pressure during mixing, brings another degree of freedom to handle gas cells distribution and consequently cellular structure, as initiated by the Chorleywood bread process;[42] now widely used in the baking industry for pan bread. A succession of positive/negative pressure (relative to atmospheric pressure) mixing steps can be applied to modulate the size of the gas cells embedded in the dough; during depression, the embedded gas cells are expected to expand and then subdivide during mixing (see Figure 7.4b). This operation may be controlled using the following equation:[43]

$$V(t) = \nu/k + (V_0 - \nu/k)\exp(-k \cdot t) \tag{7.1}$$

where V is the volume of gas entrained by the dough, V_0 its initial value, ν the volumetric rate of gas entrainment, whereas k, the coefficient of disentrainment or turnover rate, increases from 10^{-2} to 10^{-1} s^{-1} according to pressure step ratio and wheat flour quality.

Cellular structure can be accurately determined by X-ray microtomography (XRT), a technique now widely applied to determine the structure of food and plant products,[44] with 3D image analysis to measure density, or

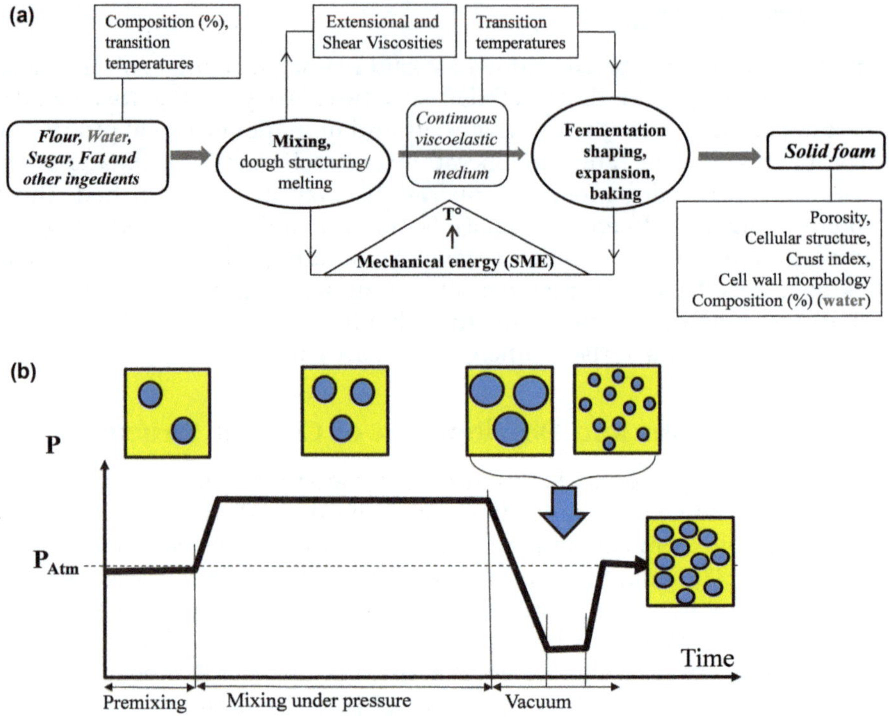

Figure 7.4 Systemic representation of cereal food process: (a) with main product states (rounded rectangles), operations (ellipses), operating variables (triangle, arrow between *SME* and *T* illustrates dissipation, especially in the case of extrusion) and product characteristics (rectangles). Note that the extrusion process integrates the two sequences of operations whereas, in the baking process, the second one involves at least three operations. Large arrows indicate the flow matter, whereas thin ones mean "influence";[48] (b) diagram of pressure–vacuum mixing process based on successive steps of dough aeration under elevated overhead pressure in mixer and pressure drop at end of mixing, to refine the cellular distribution in the dough before fermentation.[49]

porosity, and the distribution of cell size and cell wall thickness. It has been applied to fermenting wheat flour dough and its evolution just after mixing,[45] during proofing,[46,47] and baking.[48,49] These works have shown that the cellular structure of wheat flour dough is produced during fermentation when porosity increases from 0.1 to about 0.7 and that baking does not modify it significantly. The final porosity of bread dough reached at the end of fermentation may be modified by scoring its surface before baking, maintaining it close to 0.5 in the case of crispy rolls (a baguette), whereas it can reach 0.8 for pan breads and buns. During proofing, the homogeneity of cell structure decreases, and fermented dough may be envisioned as a triphasic medium with closely packed gas cells separated by liquid films, embedded in a starch-gluten network.[50] After mixing and resting, shaping can induce a porosity decrease, since gas is expelled from the dough, which will modify cellular structure. For instance, laminating increases cellular structure homogeneity[51] Consequent stress relaxation and strain recovery affect the rheological properties of the starch-gluten matrix.

The final volume increase of bread dough, dough-to-crumb transition, and gas cell coalescence occur during baking[52] by the so-called oven rise. These phenomena are affected by the air humidity during the first minutes of baking because air moisture condenses on the surface of the dough as long as the temperature stays below its dew point value. The condensed moisture plasticizes the surface of the dough, which may favor starch melting, resulting in amylose leaching and possible dextrinization; this mechanism qualifies the general trend suggested in Figure 7.3a regarding the bread crust pathway. Consequently, the higher the condensed moisture is, the more phenomena that are linked to Maillard reactions (coloration, aroma development, but also risk of acrylamide occurrence) will develop.[53] If air humidity is too low, larger pressure will develop inside the dough, resulting in tearing and unacceptable loaf deformation. Hence, oven air humidity is controlled by steam injecting, steam temperature, and humidity extraction kinetics. The precise control of air humidity during baking will affect all kinds of bakery products. About 200 g to 300 g of water per kilogram of dry air (absolute humidity) may be recommended for French baguette, for instance.[54] In addition, the decrease of the oven air humidity at the end of baking will modify the final water content of the loaf, with moderate moisture extraction for bread and higher moisture extraction rate for dry cereal products such as crackers, biscuits, *etc.*

Regarding biscuit dough, mixing plays a similar role as for bread dough. However, it is carried out less intensively, and specific flours are used with lower protein content and a higher level of starch damage. Laminated "hard dough" is the most hydrated and requires a minimal development of the gluten network in order to form a dough sheet onto which the biscuit dough pieces will be cut. An excess of gluten network development may increase checking risk, which is mitigated by the appropriate control of the degree of mixing and gluten network setting. Biscuit dough properties are mainly affected by shaping operations, such as laminating, wire cutting, or molding. Biscuit cellular structure is gained during baking thanks to gas creation by chemical leaveners (*i.e.* baking powders); these result in larger density but a fine and homogeneous cellular structure.[55] Among baking powders, ammonium carbonate is the most used for dry cereal products as it completely decomposes into ammonia and carbon dioxide, provoking expansion of the dough during baking without leaving any soapy or salty residues, as sodium carbonate may do, if not properly neutralized by an acidic coingredient. Conversely, sodium carbonate is more expansive than ammonium and is rather used for pound cake, for instance.

During extrusion, the material is transformed into a melt (Figure 7.3), a continuous starch matrix in which macromolecules may be degraded because of shear,[18,56] whereas pressure itself does not induce direct structural modifications. The cellular structure is created and largely set by expansion at the die. Here, the fast kinetics of vapor expansion at the die outlet prohibit the use of XRT to follow the cellular structure development. Indeed, expansion is a succession of mechanisms occurring within very short timescales (fractions of seconds); bubbles nucleate in the starch melt and grow before they coalesce and set, as the melt becomes glassy, or possibly collapses, leading to highly heterogeneous cellular structures.[44–46]

The comparison between the three products suggests that the quicker the expansion and setting processes are, the more heterogeneous the cellular structure will be, which can be imparted to the different timescales of the mechanisms governing cereal-based foam processing.[32]

7.3.2 Rheological Properties

Rheology is the science of deformation and flow, which aims to connect the material flow properties with its structure and its behavior during processing, by determining constitutive equations, also called rheological models. For this purpose, experiments are performed by submitting the studied product to rheological tests, which can be classified according to the strain mode implemented, the variable imposed, and its variation with time. The strain may be shear or extension, as during mixing or shaping, respectively. The material may be submitted to a given strain; hence it is necessary to measure the stress. Last, variation with time may be periodic (dynamic or oscillatory test) or steady (creep test). This approach has been considerably developed within polymer science in general. Like polymers, during processing, cereal dough is subjected to a wide range of stress and strain values, also of different type, *i.e.* not only shear but also, perhaps more importantly, extension. However, application of rheology to biopolymers and cereals is less straightforward due to the number of components, their variability, thermal sensitivity, uneven behavior, *etc.*

Dynamic measurements are difficult to convert into practical assessment for dough processing[57] because they are achieved at low strain, far from the conditions experienced during processing. However, using dynamic mechanical analysis (DMA), dough behavior during baking can be assessed by following the variations of storage modulus E' with temperature. DMA curves $E'(T)$ display similar shapes, whatever dough composition, with a minimum value (E'_{min}) obtained around 45 °C and a maximum value (E'_{max}) close to 70 °C.[39,58,59] The ratio (E'_{max} to E'_{min}) reflects the amplitude of structural changes, such as the aggregation of gluten proteins (*i.e.* network cross-linking) and, to a lesser extent, the swelling of starch granules.

Conversely, empirical measurements, like the Brabender farinograph and extensograph, the Chopin alveograph and rheofermentometer, are carried out under conditions that mimic the process. However, they are difficult to interpret in terms of structural mechanisms because they involve various uncontrolled strain modes and cannot lead to a relevant and robust determination of rheological models. Hence, in spite of their popularity, the comparison between results gained by these methods and their use for predicting bread-making performance is challenging. Results from accurate rheological measurements, performed under large strains, can be linked to process behavior. Indeed, wheat flour quality is primarily determined by the rheological properties of the dough, shear and extensional viscosities, and strain hardening.

For a given wheat flour, during mixing, shear viscosity changes may reflect the degree of cross-linking of the gluten network. Dough shear viscosity can be usually determined by conventional creep tests in rotational rheometry. Results can be represented by the Cross model,[60] which at larger shear rate values can be approximated by a power law (eqn (7.2)). During mixing, shear viscosity can be assessed by the measurements of torque and specific mechanical energy (*SME*), assuming that *SME* is linked to viscous dissipation, and since viscosity follows a power law in this range of shear rate:

$$SME/t_m = \eta \cdot \dot{\gamma}^2 = K \cdot \dot{\gamma}^{n+1} \tag{7.2}$$

where t_m is the mixing time, η is the dough shear viscosity (Pa s), $\dot{\gamma}$ is the shear rate (s^{-1}), linearly related to the rotation speed of the mixer arm, K is the consistency index (Pa s^n), and n is the flow index. By using eqn (7.2) for a single mixer, the shear viscosity of different flours can be evaluated. Conversely, using the same flour, different mixers can be characterized for their capacity to dissipate or store energy to weave the gluten network. Mixers also implement stretching (extensional or elongational strain), and stretching is thought to be more favorable to gluten network weaving, whereas intense shearing can lead to gluten network degradation.[61,62]

Elongational, or extensional, viscosity is the property that relates to dough resistance to the growth of gas bubbles during fermentation, and strain hardening reflects the gas retention and freestanding capacity of dough.[63,64] It has often been considered as an important variable for the assessment of bread-making performances.[65] The most current test for determining these properties is the uniaxial compression, or squeezing flow, under lubricated conditions (LSF), as reviewed in the case of wheat flour dough.[66] The measurement of stress σ, equibiaxial deformation ε_b in the sense of Hencky, and strain rate $\dot{\varepsilon}_b$ lead to the determination of the elongational viscosity according to a power law:

$$\eta_E = K_E \, (\dot{\varepsilon}_b)^{m-1} \exp(\mathrm{SHI}\ \varepsilon_b) \tag{7.3}$$

where K_E and m are indices of consistency and flow, respectively defined for a constant deformation ε_b, and SHI is the strain-hardening index. Although dough never reaches steady state and behaves more like a soft solid rather than as a viscous liquid, extensional behavior can be described by elongational viscosity, at constant strain, and eqn (7.3) reduces to a power law, similar to the second part of eqn (7.2). For dough of different compositions, K_E and flow index m can be negatively correlated, suggesting that increasing liquid fraction plasticizes gluten network, whereas changing the mixing conditions impacts upon gluten network cross-linking, as reflected by a positive correlation between K_E and m.[67] The correlation between the relative consistency with the ratio of storage modulus (E'_{max} to E'_{min}), as determined by DMA, suggests that the larger the elongational viscosity is, the more cross-linked is the gluten network (Figure 7.5a).

By its resistance to bubble growth and coalescence, elongational viscosity contributes to stabilizing the dough during fermentation. Therefore, its increase with the addition of fibers (Figure 7.5b), soluble or insoluble, likely contributes to the increase of the density of breads enriched with dietary fibers, although this is certainly not the only factor that explains the texture changes for these breads. However, given their significance in dough processing and sensitivity to formulation as well as structural modifications, extensional properties can be used for comparison with results from imaging experiments of dough proofing and also as input to the deterministic models of proofing mentioned thereafter.

Such measurements, at large strain, either under shear or extension, have scarcely been applied in the case of biscuit dough, except to underline the role of the fat component.[68] At low strains, within the linear viscoelastic domain ($\gamma \leq 0.2\%$), biscuit dough behaves as a weak gel ($G' > G''$), which can easily lose its structure at larger strains.[69] Conversely, the variations of loss tangent ($\tan\delta = G''/G'$) at low frequency may reflect dough structuring after rest or when modifying the recipe, *e.g.* modifying fat content or replacing fat by cellulose compounds.[70,71]

For extruded snacks, once starch is molten (see Figure 7.3), the flow of the material is controlled by its shear viscosity η, which is generally described by a power law relation like eqn (7.2), where K and n depend on temperature T, moisture content MC, and thermomechanical history (SME). To take into account the sensitivity of the material with respect to the thermomechanical history, shear viscosity can be determined using specific rheometers, inline slit die rheometers[72–74] and offline, like a capillary rheometer with pre-shearing, Rheoplast.[75] By measuring the entrance pressure effects, this device also allows the evaluation of elongational viscosity. The values of the coefficients linking K and n to T, MC, and SME can be found in literature[76] for starches; however, they are seldom available for more complex recipes. A complete characterization was obtained for a typical oat rice–based breakfast cereal showing that the effect of SME was not significant, likely because of the presence of minor components, such as lipids and sugars, that acted as internal lubricants.[77] Conversely, the addition of salt and sugar was found to increase the shear viscosity of a wheat-corn blend.[78]

7.3.3 Modeling Approaches

Given the complexity of structural changes, the wide variety of formulations and processing conditions involved in cereal food processing, there is a need to use models in order to facilitate the design of food products with targeted properties and to optimize process operation. Since the beginning of the century, likely because of the development of computing tools, several works

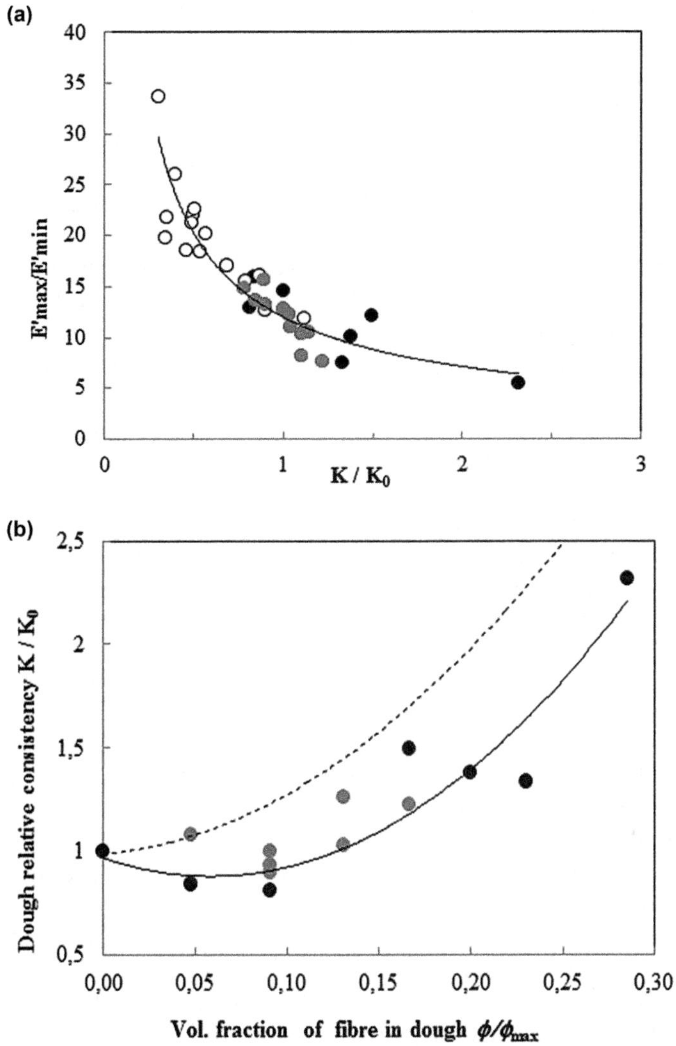

Figure 7.5 Summary of dough rheological properties variations: (a) modulus ratio (E'_{max} -to- E'_{min}, small strains) with relative consistency K/K_0 ($r^2=0.84$), and (b) relative consistency K/K_0 with relative volume fraction of fiber in dough ($r^2=0.86$); dotted line refers to Krieger–Dougherty's model[77] for suspension of spheres with a maximum packing fraction $\phi_m=0.64$. Dough prepared under various mixing conditions (○, data from ref. 78 and 56), with various amounts of insoluble fibers (●, wheat bran, data from ref. 79) and soluble fibers (●, maltodextrins, pectins, and inulins, data from ref. 80). Consistency K refers to eqn (7.2), and K_0 is its value for the initial wheat flour dough mixed under standard conditions, as indicated in corresponding references.

have been dedicated to cereal processing. However, in spite of their interest, it is still difficult to predict a target structure and even more so a desired texture, from the operating parameters and material composition. Deterministic, or physics-based, models about food processing rely on solving differential equations of flow and heat and mass transfers, thanks to sophisticated numerical approaches (*e.g.* finite element method, FEM), in order to locally describe flow and temperature fields and sometimes the evolution of a specific component.[79]

Regarding cereal processing, bread baking has drawn the most attention, and such models have considered multiphase transport in dough, which at the macroscale is regarded as a homogeneous deformable porous medium.[80,81] The main challenges rely mostly on accommodating the liquid-to-gas transition, provoked by CO_2, ethanol, and water vaporization during temperature rise, and the dough-to-crumb transition, reflected by the jump in rigidity due to starch melting and gluten aggregation.[68] The case of a freestanding loaf is even more complicated than the case of a bread in a pan, as for the latter the limits of expansion are geometrically set.

Considering mixing, the non-Newtonian viscoelastic behavior of dough has been addressed by a 2D numerical approach,[82] but the continuous changes of its free surface due to the motions of the mixing arm and bowl have not been taken into account yet. Numerical models of sheeting or laminating (*i.e.* the main shaping operations) have focused on the impact of roll speeds on roll forces, or torque, and dough recovery[83] and thickness[84] by taking into account dough shear-thinning behavior[85] but disregarding dough compressibility,[86] an assumption that, although perhaps sound for biscuit dough, could be potentially risky in the case of fermented bread dough. Finally, the creation of cellular structure could be predicted using a 3D-FEM model that takes into account bubble–bubble interactions during fermentation.[87]

Likely because of the similarity to polymer processing, FEM numerical modeling of extrusion has drawn more attention, either for simulating the influence of extrusion variables or predicting mixing abilities.[88,89] But still the complicated geometry of (twin screw) extruders and the lack of knowledge on the rheological behavior of cereal foods challenge its application.

So, in many cases, these numerical models still suffer either from insufficient knowledge about structural changes associated with material properties (in order to feed the model) or from the difficulties in accurately describing such a complex system. As a noticeable exception, a micromechanical composite model that takes into account starch–gluten debonding fully depicts the rheological behavior of wheat flour dough in shear and extension.[90] However, at present, no deterministic model, based on continuum mechanics, can describe satisfactorily all physical mechanisms involved in the structural changes of cereal food components.

For these reasons, likely, data-driven approaches have also been attempted to relate input and output process variables, like artificial neural networks applied to extrusion[91] and dough mixing,[92–94] in order to predict dough performance during mixing and baking. But, clearly, these models apply only in the experimental domain in which they have been determined.

From a practical viewpoint, at industry level, the design of cereal solid foams such as bread or biscuit, baking and extrusion are all largely based on trial-and-error approaches and also significantly rely on expert know-how. So the scientific understanding of material changes during processing should be combined with the ability to predict the final properties of the product. For this purpose, basic knowledge models (BKMs, defined as elementary models that integrate our knowledge of the mechanisms of material structural changes during processing and that are expressed in a simple mathematical form, can be used by food industry engineers to provide information on end-use properties.[95] These BKMs can be deterministic or more phenomenological based on rheological models. The approach can be first illustrated by a map, which gathers the concepts (conceptual map) defining the model of the system (product + operation) to be addressed; taking bread dough proofing, for instance (Figure 7.6a), these concepts would include the class of the BKM—or mathematical formalism, its hypotheses, its input and output variables, its implementation, *etc.* Then output variables (*e.g.* porosity kinetics) can be plotted as a graphical representation for selected values of input variables (*e.g.* specific power), in order to predict the behavior of dough during proofing in a large range of conditions while reducing the time dedicated to experimental trials.

The application of this approach to the mechanisms of solid-to-liquid (flour-to-dough) transition and liquid-to-solid foam expansion has been detailed in the case of extrusion and breadmaking.[48] It aims at linking the structural features, the rheological behavior, and the process conditions in order to predict the end use properties.

7.4 Conclusion and Main Prospects

In this chapter, we have strived to give an overview of the development of cereal foods with an emphasis on several common products (bread, biscuits, and extruded snacks). We first described the various components of wheat flour, then emphasized how their modification during processing impacts the overall structure, especially using state diagrams and systemic representation of the main operations. We have shown that the link between structure and process may be strengthened by determining accurately the rheological behavior of the dough system under the relevant thermomechanical conditions. At each step of this comprehensive approach, we have provided the reader with some methodological details about the most recent analytical developments, particularly about imaging. Although modeling is not widely used in practice, we have seen that it has contributed to improving the overall understanding of the structure-process properties relationships.

Clearly, integrating these approaches can have the potential to ultimately design and engineer cereal foods with targeted sensory and nutritional

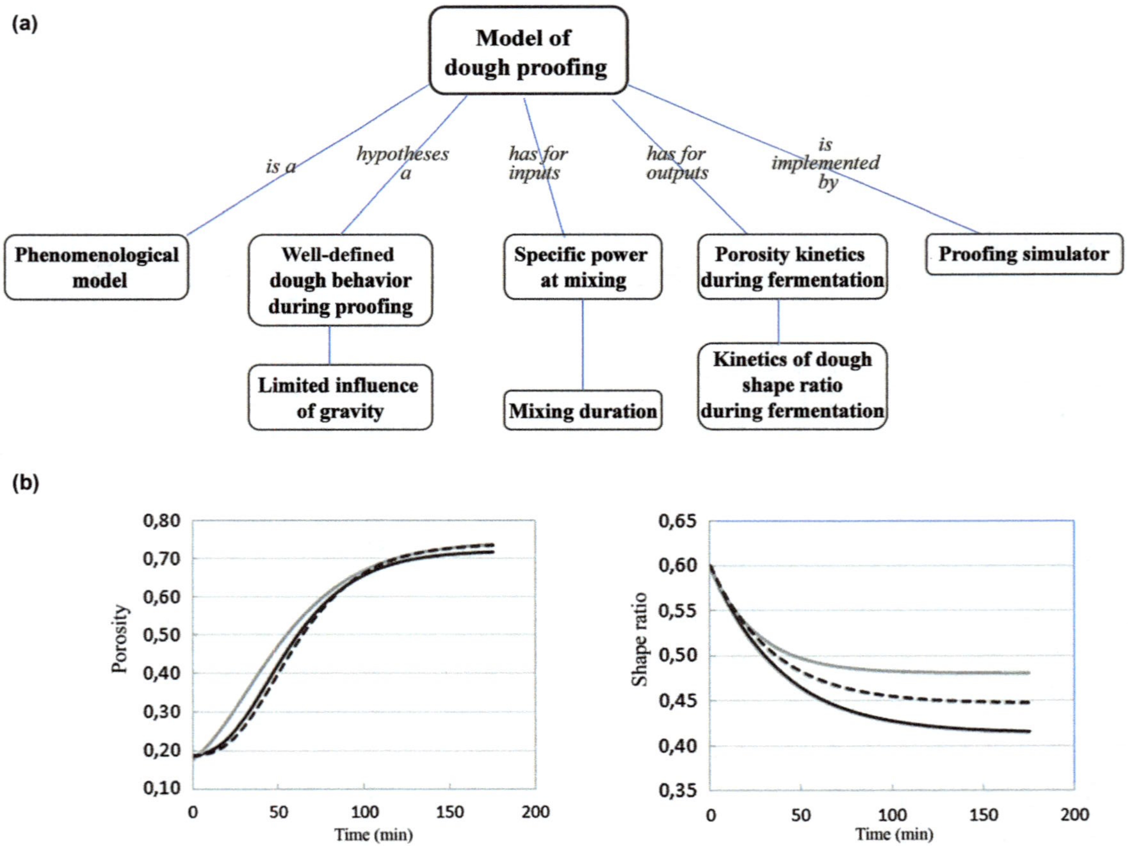

Figure 7.6 Examples of conceptual map for a basic knowledge model, here of dough proofing (a) and of results obtained for dough porosity and shape ratio kinetics for various dough mixing conditions (specific power/time: black [−,0], gray [0, +], dotted [0, −])[109] (b). In this case, the volume and type of final cellular structure (coarse, fine) of bread crumb can be inferred from these kinetics.

properties. In this regard, the growing scientific interest about food oral processing (FOP) shows how both properties may be addressed, *e.g.* by studying more accurately the texture of foods and the impact of texture, as focused in Chapters 8 and 9 of this book. Regarding raw materials, we have not dealt with cereal sources other than wheat or with other ingredients (*e.g.* sugar, fat, *etc.*), not only because these have been appropriately addressed in other monographs but also because we are convinced that the approach described here can be equally applied to a wide range of recipes and components. For instance, the enrichment of cereal and snack foods with legume proteins can be envisaged to address protein transition (from animal to plant) and to cater for the specific dietary requirements of particular populations. Tailored foods may also be tackled thanks to process innovations for culinary design, such as 3D printing, the recent development of which emphasizes the need for more accurate understanding of cereal foods structure at different levels of matter organization. Finally, a more complete exploitation of the grains (whole grain) can help to deliver a healthier diet while coping with the challenges of sustainable development.

Acknowledgements

The authors wish to thank all the colleagues involved in various EU and national multidisciplinary projects we have participated in for fruitful exchanges and helpful discussions.

References

1. K. Poutanen, N. Sozer and G. Della Valle, *J. Cereal Sci.*, 2014, **59**, 327.
2. F. Brouns, Y. Hemery, R. Price and N. M. Anson, *Crit. Rev. Food Sci. Nut.*, 2012, **52**, 553.
3. A. M. Donald, *Nature Mat.*, 2004, **3**, 579.
4. J. M. Aguilera, *J. Sci. Food Agric.*, 2006, **86**, 1147.
5. S. Guessasma, L. Chaunier, G. Della Valle and D. Lourdin, *Trends Food Sci. Technol.*, 2011, **22**, 142.
6. Z. Liu and M. G. Scanlon, *Trans. Inst. Chem. Eng.*, 2003, **81C**, 224.
7. M. Southan and F. MacRitchie, *Cereal Chem.*, 1999, **76**, 827.
8. P. Shewry, R. D'Ovidio, D. Lafiandra, J. A. Jenkins, E. N. C. Mills and F. Békés, in *Wheat Chemistry and Technology*, ed. K. Khan and P. Shewry, AACC International, St-Paul, MN, 4th edn, 2009, p. 223.
9. M.-H. Morel, P. Dehlon, J. C. Autran, J. P. Leygue and C. Bar-L'Helgouac'h, *Cereal Chem. J.*, 2000, **77**, 685.
10. M. S. Larraz Ferreira, PhD Thesis, University of Montpellier, 2011. Available at http://www.theses.fr/2011NSAM0009.
11. I. Bernklau, L. Lucas, M. Jekle and T. Becker, *Food Res. Int.*, 2016, **89**, 812.
12. P. S. Belton, *J. Cereal Sci.*, 2005, **41**, 203.
13. A. H. Bloksma, *Cereal Foods World*, 1990, **35**, 237.

14. A. Buléon, P. Colonna, V. Planchot and S. Ball, *Int. J. Biol. Mac.*, 1998, **23**, 85.
15. P. Colonna, J. Tayeb, and C. Mercier, in *Extrusion-Cooking*, ed. C. Mercier, P. Linko and J. M. Harper, AACC, St Paul, MN, 1989, p. 247.
16. M. Li, J. Hasjim, F. Xie, P. J. Halley and R. G. Gilbert, *Starch*, 2014, **66**, 595.
17. N. Logié, G. Della Valle, A. Rolland-Sabaté, N. Descamps and J. Soulestin, *Carb. Pol.*, 2018, **184**, 57.
18. V. Nessi, A. Roland-Sabaté, D. Lourdin, F. Jamme, C. Chevigny and K. Kansou, *Carb. Pol.*, 2018, **194**, 80.
19. D. Marion, L. Dubreil and P. J. Wilde, in *Interactions: The Keys to Cereal Quality*, ed. R. Hamer and R. Hoseney, AACC International, St. Paul, MN, 1998, p. 131.
20. G. B. Fincher and B. A. Stone, in *Advances in Cereal Science and Technology*, ed. Y. Pomeranz, Vol. 8, 1986, St. Paul, MN, USA.
21. L. Saulnier, F. Guillon, P.-E. Sado, A.-L. Chateigner-Boutin and X. Rouau, in *Reference Module in Chemistry, Molecular Sciences and Chemical Engineering*, Elsevier, 2013, p. 1.
22. C. M. Courtin and J. A. Delcour, *J. Cereal Sci.*, 2002, **35**, 243.
23. F. M. Robin, H. P. Schuchmann and S. Palzer, *Trends Food Sci. Technol.*, 2012, **28**, 23.
24. L. Saulnier, P.-E. Sado, G. Branlard, G. Charmet and F. Guillon, *J. Cereal Sci.*, 2007, **46**, 261.
25. M. S. Izydorczyk and C. G. Biliaderis, *Carb. Pol.*, 1995, **28**, 33.
26. M. Bonnand-Ducasse, G. Della Valle, J. Lefebvre and L. Saulnier, *J. Cereal Sci.*, 2010, **52**, 200.
27. B. Cuq, J. Abecassis and S. Guilbert, *Int. J. Food Sci. Technol.*, 2003, **38**, 759.
28. R. G. M. van der Sman and A. J. van der Goot, *Soft Matter*, 2009, **5**, 501.
29. S. Mack, M. A. Hussein and T. Becker, *Food Biophys.*, 2013, **8**, 69.
30. J. W. Donovan, *Biopolymers*, 1979, **18**, 263.
31. R. G. M. van der Sman and M. B. J. Meinders, *Soft Matter*, 2011, **7**, 429.
32. P. L. Weegels, J. A. Verhoek, A. M. G. de Groot and R. J. Hamer, *J. Cereal Sci.*, 1994, **19**, 31.
33. I. Toufeili, I. A. Lambert and J. L. Kokini, *Cereal Chem*, 2002, **79**, 138.
34. R. J. Nicholls, A. M. Appelqvist, A. P. Davies, S. J. Ingman and P. J. Lillford, *J. Cereal Sci.*, 1995, **21**, 25.
35. M. T. Kalichevsky and J. M. V. Blanshard, *Carb. Pol.*, 1992, **19**, 271.
36. L. Slade and H. Levine, *Crit. Rev. Food Sci. Nut.*, 2018, **58**, 972.
37. M. Kristiawan, L. Chaunier, G. Della Valle, A. Ndiaye and B. Vergnes, *Trends Food Sci. Technol.*, 2016, **48**, 13.
38. C. I. Moraru and J. L. Kokini, *Compr. Rev. Food Sci. Food Saf.*, 2003, **2**, 147.
39. R. G. M. Van der Sman and J. Broeze, *J. Phys. Condens. Matter*, 2014, **26**, 464103.
40. S. Chevallier, P. Colonna, A. Buleon and G. Della Valle, *J. Agric. Food Chem.*, 2000, **48**, 1322.

41. G. G. Bellido, M. G. Scanlon, J. H. Page and B. Hallgrimsson, *Food Res. Int.*, 2006, **39**, 1058.
42. S. P. Cauvain and L. S. Young, *Technology of Breadmaking.*, CRC, Woodhead Publishing Ltd, Cambridge, UK, 1998.
43. N. L. Chin, P. J. Martin and G. M. Campbell, *Food Bioprod. Process*, 2004, **82**, 261.
44. L. Schoeman, P. Williams, A. du Plessis and M. Manley, *Trends Food Sci. Technol.*, 2016, **47**, 10.
45. L. Trinh, T. Lowe, G. M. Campbell, P. J. Withers and P. J. Martin, *Chem. Eng. Sci.*, 2013, **101**, 470.
46. S. Chakrabarti-Bell, S. Wang and K. H. M. Siddique, *Food Res. Int.*, 2014, **64**, 587.
47. A. Turbin-Orger, E. Boller, L. Chaunier, H. Chiron, G. Della Valle and A.-L. Réguerre, *J. Cereal Sci.*, 2012, **56**, 676.
48. M. Whitworth and J. M. Alava, in *Bubbles in Food*, ed. G. M. Campbell, C. Webb, S. S. Pandiell and K. Niranjan, Eagan Press, St Paul, MN, USA, 1999, p. 221.
49. P. Babin, G. Della Valle, H. Chiron, P. Cloetens, J. Hoszowska, P. Pernot, A. L. Réguerre, L. Salvo and R. Dendievel, *J. Cereal. Sci.*, 2006, **43**, 393.
50. A. Turbin-Orger, P. Babin, E. Boller, L. Chaunier, H. Chiron, G. Della Valle, R. Dendievel, A. L. Réguerre and L. Salvo, *Soft Matter*, 2015, **11**, 3373.
51. A. J. Sandoval, L. Chaunier, H. Chiron, G. Della Valle and A. L. Réguerre, *J. Cereal Sci.*, 2017, **73**, 91.
52. A. Mondal and A. K. Datta, *J. Food Eng.*, 2008, **86**, 465.
53. R. Altamirano-Fortoul, A. Le-Bail, S. Chevallier and C. M. Rosell, *J. Food Eng.*, 2012, **108**, 128.
54. A. Le-Bail, T. Dessev, D. Leray, T. Lucas, S. Mariani, G. Mottollese and V. Jury, *J. Food Eng.*, 2011, **105**, 379.
55. S. Chevallier, A.-L. Réguerre, A. Le Bail and G. Della Valle, *Food Biophys.*, 2014, **9**, 219.
56. W. C. Liu, P. J. Halley and R. G. Gilbert, *Macromolecules*, 2010, **43**, 2855.
57. V. Stojceska and F. Butler, *Trends Food Sci. Technol.*, 2012, **24**, 13.
58. A. Angioloni and M. Dalla Rossa, *J. Cereal Sci.*, 2005, **41**, 327.
59. F. M. Vanin, C. Michon, G. Trystram and T. Lucas, *J. Cereal Sci.*, 2010, **51**, 277.
60. J. Lefebvre, *Rheol. Acta*, 2006, **45**, 525.
61. R. Haraszi, O. R. Larroque, B. J. Butow, K. R. Gale and F. Bekes, *J. Cereal Sci.*, 2008, **47**, 41.
62. D. Peressini, S. H. Peighambardoust, R. J. Hamer, A. Sensidoni and A. J. van der Goot, *J. Cereal Sci.*, 2008, **48**, 426.
63. T. van Vliet, A. M. Janssen, A. H. Bloksma and P. Walstra, *J. Texture Studies*, 1992, **23**, 439.
64. A. Shehzad, H. Chiron, G. Della Valle, K. Kansou, A. Ndiaye and A. L. Réguerre, *Food Res. Int.*, 2010, **43**, 1999.
65. T. van Vliet, *J. Cereal Sci.*, 2008, **48**, 1.

66. B. Launay and C. Michon, *J. Texture Stud.*, 2008, **39**, 496.
67. A. Turbin-Orger, A. Shehzad, L. Chaunier, H. Chiron and G. Della Valle, *J. Food Eng.*, 2016, **168**, 129.
68. A. Baltsavias, A. Jurgens and T. van Vliet, *J. Cereal Sci.*, 1999, **29**, 33.
69. A. S. Contamine, J. Abecassis, M. H. Morel, B. Vergnes and A. Verel, *Cereal Chem.*, 1995, **72**, 516.
70. A. Baltsavias, A. Jurgens and T. van Vliet, *J. Cereal Sci.*, 1997, **26**, 289.
71. P. Tarancon, M. J. Hernandez, A. Salvador and T. Sanz, *LWT–Food Sci. Technol.*, 2015, **62**, 640.
72. B. Vergnes, G. Della Valle and J. Tayeb, *Rheol. Acta*, 1993, **32**, 465.
73. M. Horvat, D. Ladiges and H. P. Schuchmann, *Food Bioprocess Technol.*, 2014, 7, 654.
74. F. Robin, J. Engmann, D. Tomasi, O. Breton, R. Parker, H. P. Schuchmann and S. Palzer, *Chem. Eng. Technol.*, 2010, **33**, 1672.
75. B. Vergnes and J. P. Villemaire, *Rheol. Acta*, 1987, **26**, 570.
76. F. Xie, P. J. Halley and L. Averous, *Progr. Polym. Sci.*, 2012, **37**, 595.
77. M. Núñez, G. Della Valle and A. J. Sandoval, *Food Res. Int.*, 2010, **43**, 2093.
78. K. F. Pitts, J. Favaro, P. Austin and L. Day, *J. Food Eng.*, 2014, **127**, 58.
79. A. K. Datta, *J. Food Eng.*, 2016, **176**, 9.
80. E. Purlis and V. O. Salvadori, *J. Food Eng.*, 2009, **91**, 434.
81. V. Nicolas, P. Salagnac, P. Glouannec, J. P. Ploteau, V. Jury and L. Boillereaux, *J. Food Eng.*, 2014, **130**, 23.
82. R. K. Connelly and J. L. Kokini, *J. Non-Newtonian Fluid Mech.*, 2004, **123**, 1.
83. W. F. Xiao, M. N. Charalambides and J. G. Williams, *Int. J. Food Sci. Technol.*, 2007, **42**, 699.
84. E. Mitsoulis and S. G. Hatzikiriakos, *Food Bioprod. Process*, 2009, **87**, 124.
85. J. Engmann, M. C. Peck and D. I. Wilson, *Food Bioprod. Process*, 2005, **83**, 175.
86. M. J. Patel and S. Chakrabarti-Bell, *J. Food Eng.*, 2013, **115**, 371.
87. J. Bikard, T. Coupez, G. Della Valle and B. Vergnes, *J. Food Eng.*, 2008, **85**, 259.
88. A. Ficarella, M. Milanese and D. Laforgia, *J. Food Eng.*, 2006, **73**, 103.
89. R. K. Connelly and J. L. Kokini, *J. Food Eng.*, 2007, **79**, 956.
90. M. A. P. Mohammed, M. N. Tarleton, M. N. Charalambides and J. G. Williams, *J. Rheol.*, 2013, **57**, 249.
91. G. Ganjyal, M. A. Hanna, P. Supprung, A. Noomhorm and D. Jones, *Cereal Chem.*, 2006, **83**, 223.
92. B. Grote, T. Zense and B. Hitzmann, *Food Control*, 2014, **38**, 8.
93. B. Lamrini, G. Della Valle, I. C. Trelea, N. Perrot and G. Trystram, *Food Control*, 2012, **26**, 512.
94. B. Hermannseder, A. M. Haseeb, P. Kugler and B. Hitzmann, *J. Cereal Sci.*, 2017, **76**, 64.
95. G. Della Valle, H. Chiron, L. Cicerelli, K. Kansou, K. Katina, A. Ndiaye, M. Whitworth and K. Poutanen, *Trends Food Sci. Technol.*, 2014, **36**, 5.

Part B
Food Structure Development for Performance and Functionality

CHAPTER 8

Food Structure Development for Rheological/Tribological Performance

H. M. SHEWAN AND J. R. STOKES*

School of Chemical Engineering, The University of Queensland, Brisbane 4072, Australia
*Email: jason.stokes@uq.edu.au

8.1 Introduction

Rheology is the study of deformation and flow of matter. It is widely recognized as a key parameter for the optimization of food manufacturing processes, stability during storage, and in-mouth performance, particularly during the early stages of oral processing.[1–3] *Tribology* is the study of friction, lubrication, and wear. In contrast to rheology, which is a material property, tribology is a system property, *i.e.* a function of the relative movement of surfaces and lubricant. This occurs in the mouth through the movement of oral surfaces (tongue, teeth, and palate) and is lubricated by saliva, food, and/or beverages.

This chapter focuses on methods for control of food structures to achieve the desired rheological and tribological performance of complex food products. To put this need for rheological and tribological performance into context, we include a discussion of existing structure-instrumental-sensory relationships for complex foods, particularly dairy products. An overview is included regarding the theory of established and novel rheological techniques and soft contact tribology. Food structuring is discussed to achieve

Food Chemistry, Function and Analysis No. 18
Handbook of Food Structure Development
Edited by Fotis Spyropoulos, Aris Lazidis and Ian T. Norton

Published by the Royal Society of Chemistry, www.rsc.org

three key parameters relevant to food stability and sensory perception: (1) control of viscosity, (2) tuning of the yield stress, and (3) control of fat lubrication. Rheology and tribology theory will also be introduced.

8.2 Rheology Theory and Methods

Rheology theory and methodology have been previously covered, both in general terms[4,5] and, more specifically, for foods.[1,2,6,7] We strongly recommend our 2012 book chapter titled "Oral Rheology"[7] for a detailed review of rheological methods and insights pertaining to the performance of foods and beverages during oral processing. We have also previously provided a brief account on the role of rheology in process control and stability.[3,8] Here we briefly review the theory, measurement techniques, and more specific rheology-structure relationships relevant to semisolid foods to provide a basis for understanding the relationship between food oral processing and rheology.

8.2.1 Rheology Theory

Rheology begins at two ends of a spectrum from simple liquids to elastic solids. For simple liquids, also called Newtonian fluids after Isaac Newton who first proposed the concept in 1687, the force per unit area, or shear stress (σ), required to produce motion is proportional to the shear rate ($\dot{\gamma}$):

$$\sigma = \eta\,\dot{\gamma} \tag{8.1}$$

where η is the viscosity. This linear relationship is applicable to very dilute solutions or suspensions and single-phase systems, including foods such as water, milk, and sugar syrups. This is determined using dynamic rheology measurement. For example, where torque is applied to achieve the required shear stress and the corresponding shear rate is measured or *vice versa*.

At the other end of the spectrum, an equally simple linear equation applies for perfectly elastic solid behavior. In this case, an applied stress (σ) results in an instantaneous deformation (γ) that persists at the same magnitude as long as the stress is applied. This is known as a purely elastic Hookean solid for which:

$$\sigma = G\,\gamma \tag{8.2}$$

where G is the rigidity or elastic modulus. This is measured using oscillatory rheology in which the measuring geometry is oscillated back and forth at a required frequency (ω) and with an applied stress (τ) or strain (γ).

The rheology of many foods, particularly semisolid foods, does not follow either of these linear models and can be described as non-Newtonian. In many cases, foods show both liquid-like and solid-like characteristics and can be described as viscoelastic. Their viscous behavior is also time and shear rate dependent and may be irreversible. For example, a pot-set yogurt demonstrates solid-like behavior at low shear rates; however, once shear is applied, it yields and flows like a viscous fluid; after shearing, some degree of structure may be recovered over time. Other examples of viscoelastic foods include ice cream, mayonnaise, ketchup, chocolate, and cream cheese.[9]

8.2.2 Dynamic Rheology Measurement

The main type of rheometer encountered in the food industry is the stress-controlled rotational rheometer; strain-controlled rheometers are also available but less commonly encountered. The stress-controlled rheometer can be operated in strain-controlled mode and *vice versa*. However, this is less accurate and can be difficult to control as it relies on feedback through the sample to adjust the motor torque to give the required shear rate. Stress-controlled rheometers are best for creep, yield stress measurements, and fluids that are shear thickening, with their main limitation being the achievable torque range.

Cone and plate geometry is usually used for steady shear measurement as it has a consistent shear rate ($\dot{\gamma}$, s^{-1}) across the plate diameter, given by eqn (8.3), where Ω is the angular rotation (rad s^{-1}), and θ is the cone angle.[1] Equal shear rates are particularly important for non-Newtonian fluids where behavior is shear rate dependent. Shear stress (τ, Pa) for cone and plate is given by eqn (8.4) from the applied torque (M, Nm) and plate radius (R, m).

$$\dot{\gamma} = \frac{\Omega}{\tan\,\theta} \tag{8.3}$$

$$\tau = \frac{3M}{2\pi R^3} \tag{8.4}$$

The cone and plate geometry has the disadvantage of a small gap between the tip of the cone and the bottom plate, making it unsuitable for foods containing particles as they jam in this gap, causing significant measurement errors. Parallel plates eliminate this problem and also allow measurements with variable temperatures as gap can be adjusted to compensate for thermally induced changes in the geometry.

For parallel plates, shear rate is not constant across the plate, with the fluid at the outer edge experiencing larger shear rates than at the center, given by eqn (8.5). The equation for shear stress for Newtonian fluids (eqn (8.6)) is often applied for all fluids but may cause large errors due to the variation in shear rate experienced across the plate. To determine the true stress experienced by the fluid and hence the viscosity, an appropriate model must be fitted to the data to determine the relationship between the torque (M) and the rotation rate (Ω). A variety of models are available for different fluid types, as shown in Figure 8.1. The equation used is dependent on the type of behavior displayed by the test sample and must be modeled to determine whether it is a power law fluid or fits a model, such as the Cross model. Detailed equations have been presented by Davies and Stokes.[10]

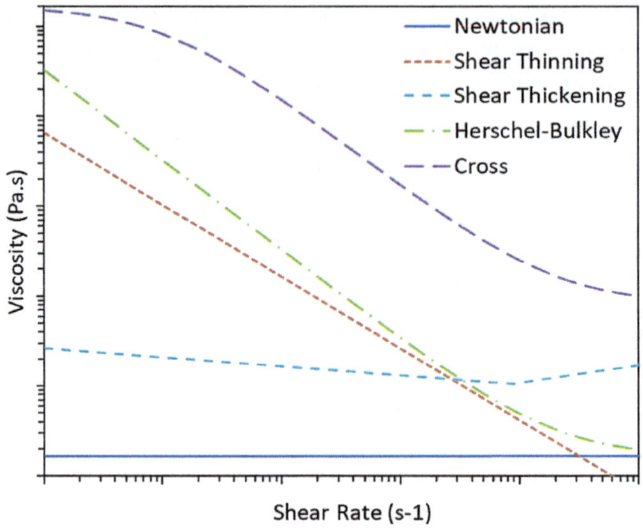

Figure 8.1 Rheological models for time-independent fluids.

The other effect of the shear rate being low in the center and high at the edge means any particles present are likely to migrate from regions of high shear to regions of low shear giving a nonuniform sample. Calculation of the Reynolds number (R_E) using eqn (8.7) allows screening for secondary flow effects, which become significant at R_E greater than 100[10]. Parallel plate geometry can also be roughened to eliminate slip, which is a measurement artifact that occurs due to depletion of the included phase, leaving only the continuous phase at the solid plate boundary when measuring complex samples, especially those containing particles.

$$\dot{\gamma} = \frac{R\Omega}{h} \tag{8.5}$$

$$\tau = \frac{2M}{\pi R^3} \tag{8.6}$$

$$R_E = \frac{\rho \dot{\gamma}_R h^2}{\eta} \tag{8.7}$$

The vane geometry, as shown in Figure 8.2, is ideal for measuring the yield stress of solid-like samples that yield and flow. It has the advantage over other geometries that there is no structure breakdown during loading of the sample into the rheometer. This has been used to measure the yield stress of samples such as mayonnaise[11] and pot-set yogurt[12] in their original containers to prevent irreversible structure breakdown during loading. The vane and cup should have a length (L) to diameter (D) ratio of between 1 and 3.5 and cup depth and diameter at least twice that of the vane to ensure that

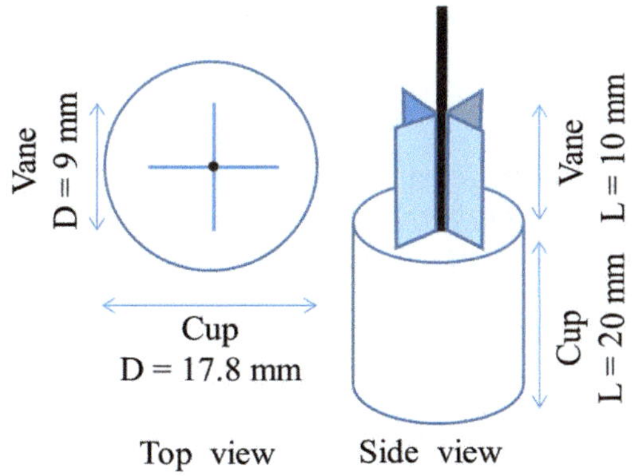

Figure 8.2 Vane and cup geometry.

edge and wall effects are minimized.[13] The shear rate can be taken as that at the edge of the vane assuming the sample is sheared in a perfect cylinder at the exact diameter of the vane. Although the actual area of shearing can be up to 5% larger, Liddel and Boger[13] found the errors were not significant, and these data were found to give the best fit to parallel plate data. Stokes and Telford[11] present an alternative, where the shear rate is taken across the gap between the cup (R_c) and the vane tip (R_v), shown in eqn (8.8) and (8.9).

$$\tau = \frac{M}{2\pi R_v^3}\left(\frac{L}{R_v} + \frac{2}{3}\right)^{-1} \tag{8.8}$$

$$\dot{\gamma} = \frac{4\Omega R_v R_c}{R_c^2 - R_v^2} \tag{8.9}$$

8.2.3 Oscillatory Rheology Including Large and Small Oscillatory Shear

Oscillatory rheology is used to probe the rheology of viscoelastic substances. Food structures including those based around emulsions, foams, and colloidal gel structures commonly fall into this category. Viscoelasticity covers the range of behavior between perfectly viscous (eqn (8.1)) and perfectly elastic (eqn (8.2)). Small amplitude oscillatory shear rheology (SAOS) is a macroscopic measurement to probe the relaxation time, which is the time taken for the constituents of a food to rearrange at the microscopic level when a stress is applied. Where a sinusoidal shear strain (γ) is applied with a frequency (ω):

$$\gamma(t) = \gamma_0 \sin(\omega t) \tag{8.10}$$

where γ_0 is the maximum shear strain. Viscous and viscoelastic samples respond to the strain input with a phase lag (δ) dependent on the type of material. In the linear viscoelastic region (LVR), where structural breakdown does not occur, oscillatory stress (σ_0) is proportional to γ_0 and the following equation applies:

$$\sigma(t) = \gamma_0 \underbrace{\left[\left(\frac{\sigma_0}{\gamma_0}\right)(\cos\delta \sin\omega t)\right]}_{\textit{in-phase}} + \underbrace{\left[\left(\frac{\sigma_0}{\gamma_0}\right)(\sin\delta \cos\omega t)\right]}_{\textit{out-of-phase}} \tag{8.11}$$

For the *in-phase*, or elastic component, of the equation $\delta = 0 \rightarrow \cos\delta = 1$ and $\sin\delta = 0$, and for the *out-of-phase*, or viscous component, of the equation $\delta = \pi/2 \rightarrow \cos\delta = 0$ and $\sin\delta = 1$.[14]

We then define two terms:

- The in-phase component, $G' = (\sigma_0/\gamma_0)\cos\delta$, is the storage modulus and describes the amount of energy stored during application of strain, *i.e.* the solid-like behavior of the suspension.
- The out-of-phase component, $G'' = (\sigma_0/\gamma_0)\sin\delta$, is the loss modulus relating to the liquid-like behavior or energy dissipated during periodic application of strain.

SAOS is relevant to small deformations, such as those that occur during the transport and storage of food. During processing and eating, foods are often subject to large and sudden deformations. A relatively new measurement technique, large amplitude oscillatory shear (LAOS), may be more relevant in this case. The rheological technique and interpretation have been developed over the last two to three decades. However, it has seen relatively little use for food materials.[15] In this method, very large strains are applied to the material without allowing time for it to relax before further stress is applied in the opposite direction. This is simple to measure but difficult to interpret as the assumption that σ_0 is proportional to γ_0, required to apply eqn (8.11), is not valid.[16] The simplest method for comparison between samples is to plot Lissajous-Bowditch curves of stress against strain (or strain rate). These provide a visual interpretation of the data, a so-called LAOS fingerprint.[16] More complex mathematical approaches to interpret the data have been proposed. Most recently Rogers *et al.*[16] demonstrated that their methodology, involving viewing the LAOS waveforms as a series of physical processes, provided an interpretation with relevant physical meaning, overcoming previous limitations of alternative analyses using Fourier transforms and Chebychev polynomials that show anomalies in their application to physical processes. LAOS has been applied to foods in recent years including gluten dough,[17,18] tomato paste,[15] and model hydrocolloid-based solutions,[19] or hydrocolloid particle suspensions.[20] Much of this work has been interpreted simply comparing Lissajous fingerprints to identify rheological differences without relating these back to product structure. However, it has been shown that LAOS can be related to structural differences in foods, such as the ability to recover structure in suspensions of tomato paste compared to mayonnaise and the differences in network strength brought about by varying the gluten content of hard and soft wheat dough.[15] Analysis of LAOS plots has also enabled separation of the elastic contribution of the gluten component and viscous contribution of the starch component in wheat doughs.[18]

8.2.4 Thin Film Rheology

Rheology in a thin film or narrow gap is a subset of several different rheological techniques broadly known as microrheometry. These rheological measurements have been proposed to be relevant to oral processing when a thin film of fluid is present between oral surfaces such as the tongue and palate.[10] In thin film conditions, the fluid may be sheared between hard or soft surfaces with a micron scale gap between them; thus, the gap is of similar size to particles included in the fluid. Under such conditions, effects such as confinement, depletion, or phase separation could occur during shearing.[10,21] Measuring at narrow gaps also makes high shear rates (above 10^3 s^{-1}) accessible, which would not be measurable at larger gaps.[10] Plate–plate geometry, used in a stress- or strain-controlled rotational rheometer, is used for thin film rheology, as plate gaps can be as low as 20 μm. Plate geometry has limitations that can be accounted for, such as gap error, which includes errors such as nonparallelism of plates and non-concentricity. Davies and Stokes[10] observed that measured viscosity decreases with decreasing commanded gap height, indicative of a gap error; therefore, results from gaps of similar magnitude to the calculated gap error should be treated with caution.[10] To calibrate and determine gap error, the gap is zeroed at a force of 4 N to account for the squeeze flow of air,[22] and gap error (ε) is calculated by measuring a 1000 cSt silicon oil at gaps (h) from 800 to 50 μm and completing nonlinear least squares regression to estimate ε.[23] Measured viscosity can then be corrected for gap error using eqn (8.12):[10]

$$\eta_{\text{actual}} = \eta_{\text{measured}} \frac{h + \varepsilon}{h} \tag{8.12}$$

Davies and Stokes[10] show measurements for a structured fluid, Carbopol 934™, using parallel plates at gaps from 500 down to 20 μm. The results, corrected for gap error, show a decrease in viscosity with decreasing gap down to 60 μm, followed by an increase in viscosity due to particle confinement at smaller gaps with an apparent low shear viscosity plateau and yield stress, which is shown to be due to slip.

Yan *et al.*[24] used an alternative technique, squeeze flow rheology, to determine the effect of confinement on structured fluids. They measured commercial food products and Carbopol™ suspensions using the squeeze flow method with profiled plates for comparison with the rheology from a rough parallel plate geometry. Their data were in close agreement with that from squeeze flow measurements by Meeten[25] for Heinz Tomato Ketchup, Tesco English Mustard, and Tesco Value Lemon Curd and from parallel plate measurements by Davies and Stokes[10] for Carbopol™ suspensions where a significant increase in apparent viscosity was seen at and below a critical gap, which occurs at approximately 5 to 10 times the average particle diameter.

8.3 Tribology Theory and Methods

Tribological studies are increasing our ability to understand the behavior of foods during the process of eating. With an increased understanding of oral tribology comes the potential to be able to select or manipulate food products to behave in a certain way and have a desirable feel within the mouth (mouthfeel). This may allow the production of food products that are either healthier or less costly yet are equally or more acceptable by consumers. In addition to this chapter, we also recommend our 2012 book chapter titled "Oral Tribology"[26] and review article[27] for a detailed account of tribological methods and insights pertaining to the performance of foods and beverages during oral processing and in relation to sensory properties.

The development of tribological techniques essentially arose because of the inability of rheology to capture the dynamics occurring during oral processing, where foods and beverages are confined to thin films between tongue and palate. In the past, many foods and especially low-fat products have been rheologically matched to their full-fat counterparts, yet sensory perception is still adversely affected. This was demonstrated for several foods by Selway *et al.*,[28] where soft foods with identical rheology exhibited vastly different tribological responses. It was highlighted that while the tribological response is a function of viscosity, other nonrheological factors also play a role, such as surface adsorption of components to substrates and phase separation in the tribological contact.

The surfaces within the mouth are rough and compliant, and therefore soft tribological studies with a compliant ball and disk surfaces are used to investigate the behavior of food during oral processing. *Ex vivo* systems such as pigs' tongues have been used to replicate surfaces of the human mouth. These surfaces have a resemblance to human oral surfaces, however are highly variable between samples and can degrade quickly.[29] For these reasons, model systems such as polydimethylsiloxane (PDMS), a silicon-based polymer, are commonly used to replicate oral surfaces. PDMS is compliant due to its elasticity, as well as being nontoxic, chemically inert, nonflammable, commercially available in large quantities, readily molded to form rough or smooth surfaces, and easily modified from hydrophobic to

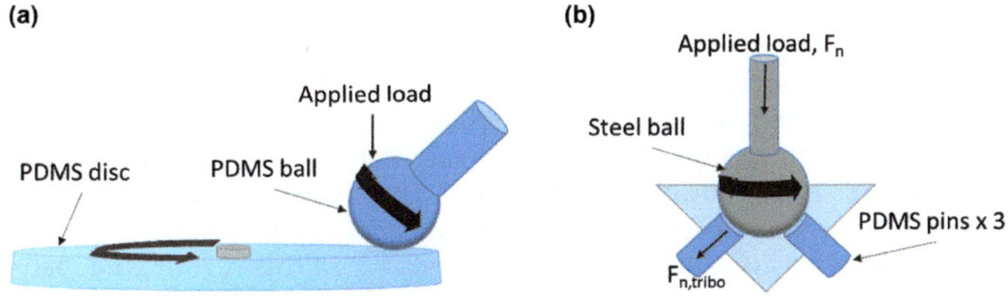

Figure 8.3 Schematic of (a) mini traction machine (MTM) and (b) Anton Paar Tribology accessory with PDMS pins.

hydrophilic. As such, it is a suitable and popular choice for soft tribological studies.

The mini traction machine (MTM) has been widely used for soft tribological studies, and much of the literature concerning oral tribology involves its use. The MTM is a multipurpose instrument for measuring friction and traction properties of lubricated and unlubricated surface contacts, under a wide range of rolling and sliding conditions, and involves a rotating ball and a rotating plate, as seen in Figure 8.3a. As such, both rolling and sliding friction are created during its operation. A number of Newtonian and non-Newtonian fluids, including hydrocolloids and emulsions, have been investigated to provide a baseline relevant to a wide range of food structuring ingredients. More recently, tribology attachments have been designed to fit standard rheometers. These attachments tend to be more cost-effective than the MTM and hold particular appeal to those who already possess rheometers. However, recent data from these attachments suggest that, to allow the reliable use of these devices and interpretation of the measurements, more work must be done to establish baseline data and determine their capabilities.[30] One example of a rheo-tribocell is that supplied by Anton Paar, pictured in Figure 8.3b, which uses a rotating ball and three stationary plates. In this configuration, only sliding friction is created. There is a potential difference in the adsorption and wear of adsorbed species between the two tribometer configurations, due to differences in the types of friction and contact area. As such, existing baseline data from the MTM cannot be used to interpret measurements from the rheo-tribocells, as tribology is a system property and hence dependent on the configuration and surfaces used for measurement.

8.3.1 Stribeck Curves

To interpret tribology, a consistent basis is required for each measurement setup. This can be achieved through the use of a Stribeck curve, which shows the friction coefficient plotted against the viscosity multiplied by entrainment speed. The baseline for this curve is generated through measurement of the traction coefficient (although usually referred to as friction coefficient) as a function of entrainment speed for Newtonian fluids of known viscosity, as shown by Bongaerts *et al.*[31] The tribological behavior can be broken into several regimes, as seen in Figure 8.4. The boundary regime occurs at low speeds. In this regime, there is negligible fluid entrainment, the load is supported by contacting asperities, and the friction factor is dependent on both surface and interfacial properties.[32] The hydrodynamic or elastohydrodynamic regime occurs at high speeds where the thickness of the film depends on viscosity and entrainment speed.[32,33] In this regime, the load is supported by the fluid, and the friction factor depends on bulk rheological properties. The mixed regime shows decreasing friction with increasing speed and occurs between the boundary and hydrodynamic regimes. In the mixed regime, the load is partially supported by asperities, and the friction factor is dependent on both surface properties and rheological properties of the entrained fluid.[32]

When working with complex test fluids, such as whole food products, it is often difficult to attribute observed results to any one component or property of the mixture. The Stribeck curve can be used to assist in this area.[31] For example, an emulsion can be collapsed onto the Stribeck curve by multiplication with the viscosity of the oil phase, water phase, or the mixed emulsion.[34] By determining which fluid viscosity collapses the Stribeck curve (or a region of the curve) onto the Mastercurve, we have an indication of which phase is entrained between ball and disk in each region.

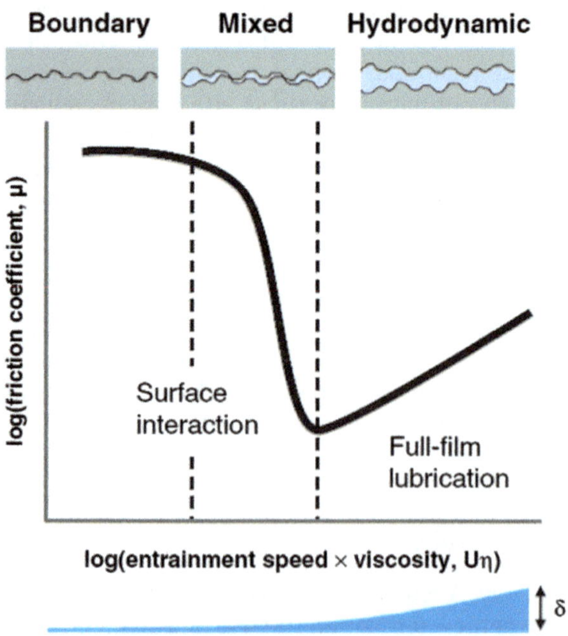

Figure 8.4 Stribeck curve showing the boundary, mixed, and hydrodynamic regime and the relationship with film thickness in each of these regimes (δ). Reproduced from ref. 28 with permission from Elsevier, Copyright 2013.

It should be noted that saliva plays an extremely important role in the lubrication of oral surfaces. It has remarkable lubrication properties and a low viscosity and is extremely elastic. Many studies do not consider the role of saliva as there are no equivalent substitutes, and thus there is a need for using human samples, which biologically vary.[27] One approach has been to adsorb expectorated saliva to PDMS substrates and measure friction dynamically in response to exposure to a beverage[35] or soft food sample,[28] where the rate of change and/or final friction value may relate to sensations such as astringency, dryness, and roughness. Another approach is to mix samples with saliva and measure their tribological response.[36]

8.4 Food Structure Design

8.4.1 Control of Viscosity

Viscosity is one of the main design variables in food and beverage design. While a majority of foods and beverages are non-Newtonian, in quality control and food development, it is common to measure viscosity only at a single shear rate (*e.g.* 50 s^{-1}) and to treat fluids as Newtonian. Sugar is a common ingredient in foods/beverages (in aqueous solution it remains Newtonian up to high concentrations), providing a pleasant taste as well as acting as a preservative. With the current focus on reducing the sugar content of foods and beverages, the loss of viscosity from removing sugar is often compensated using hydrocolloids with artificial sweeteners added to provide a sweet flavor. Hydrocolloids are also used to compensate the rheology of foods upon removal of fat from the product.

The main structure-forming capabilities of hydrocolloids in foods are thickening, gelation (discussed further in Section 8.4.2), and stabilizing emulsions or foams.[37] The most widely used hydrocolloids include polysaccharides from plant sources (starch, pectin, cellulose, guar gum, and locust bean gum), from marine plants (agar, carrageenan), from microbial sources (xanthan, gellan), and from proteins from animal sources (gelatine, whey, and egg). The formation, structure, rheology, and tribology of these systems have been covered in detail in two separate publications by Stokes.[6,26] The reader is also referred to Chapters 1 and 2 of this book, which are specifically concerned with the role of hydrocolloids in the development of food structure. Here we provide further discussion of the relationship between rheology and tribology for a range of hydrocolloids for food structure design. The rheological behavior of a dilute dispersion (polymers or particles) is usually described in terms of its phase volume (ϕ), which is simply a function of particle number and density for hard particles. This is due to the disruption in the flow streamlines during flow at low shear rates; according to Einstein's theory (shown by eqn (8.13), at concentrations tending to zero ($c \to 0$), viscosity of a suspension (η_0) is increased above that of the solvent viscosity (η_s) by an amount proportional to the (phase) volume of a particle (or polymer). This equation is also used as the basis to obtaining the intrinsic viscosity $[\eta]$, described by eqn (8.14). However, for soft particles or polymer suspensions, ϕ cannot be simply based on the dry weight of the polymer due to the solvent also being incorporated into the particle.[38] An apparent hydrodynamic phase volume ϕ_0 can be determined for soft particles and polymers using rheology of dilute suspensions. The intrinsic viscosity can then be substituted into Einstein's equation, rearranged as shown in eqn (8.15), to determine ϕ_0.

$$\eta = \eta_s\,(1 + 2.5\phi_0) \tag{8.13}$$

$$[\eta] = \lim_{c \to 0}\left(\frac{\eta_0 - \eta_s}{c\eta_s}\right) \tag{8.14}$$

$$\phi_0 = [\eta]\; c/2.5 \tag{8.15}$$

One challenge in correcting viscosity using hydrocolloid systems is that many of these species are non-Newtonian even at low concentrations, and hence it is challenging to match the viscosity of a Newtonian solution across a range of shear rates. An example of a range of hydrocolloids with matched viscosity at 100 s^{-1} is shown in Figure 8.5a in comparison to a Newtonian corn syrup. Often the shear rates of 50 to 100 s^{-1} are used as a point of comparison, as this shear rate appears to be comparable to initial in-mouth thickness perception.[39] Other in-mouth sensory attributes related to longer time in mouth and during swallow, where high shear rates are more dominant, show less dependence on viscosity at 50 s^{-1}.[40] The friction coefficient for a PDMS soft-contact (MTM) lubricated by these hydrocolloids varies greatly despite a similar viscosity at 100 s^{-1}. As shown in Figure 8.5b, this can partially be accounted for by multiplying the entrainment speed by an effective viscosity, *i.e.* the viscosity at high shear rates (*ca.* 10 000 s^{-1}). However, the variation in the friction coefficient in the mixed and boundary regimes is due to another factor that arises from adsorption of the polymer to the substrates. Stokes *et al.*[41] used quartz crystal microbalance (QCM) measurements to elucidate the thickness of the adsorbed hydrocolloid film on smooth PDMS and demonstrated that there is a relationship between the mass of adsorbed film and friction coefficient.[41] This has been shown to be true for commercially available low-fat (0.1%) and high-fat (4.7%) yogurt, in which fat was replaced by a range of hydrocolloid stabilizers, and bulk viscosity is shown to be almost identical.[28] However, their friction coefficient in the boundary regime was 0.06 for the high-fat yogurt and almost an order of magnitude higher, at 0.3, for the low-fat version.[28] Hence by designing food structures to control rheology, tribology will also be impacted as a function of both viscosity and adsorbed film thickness.

An alternative to hydrocolloids as polymeric gels is to use inclusions, as shown in Figure 8.6, at low phase volume to increase the viscosity. These may be colloidal- or noncolloidal-scale particles, droplets, vesicles, micelles, or globular proteins. Hydrocolloids may be used in the formation of these

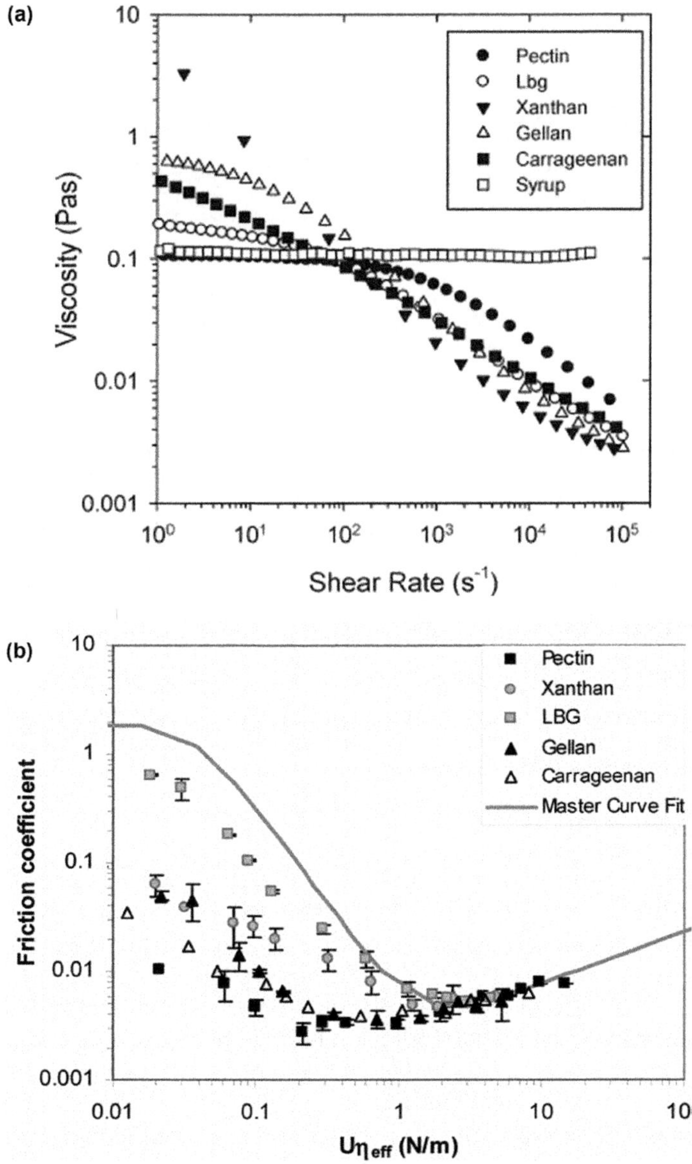

Figure 8.5 (a) Viscosity against shear rate for polysaccharide solutions (1.27% pectin, 0.48% locust bean gum (LBG), 1% xanthan gum, 0.54% gellan, and 0.91 κ/ι carrageenan), and 97% corn syrup at 25 °C; and (b) friction coefficient as a function of the product of entrainment speed and an effective viscosity of polysaccharide solutions ($U\eta_{eff}$) between smooth hydrophobic PDMS tribopairs. The line is the Newtonian Master Curve for smooth PDMS surfaces. The effective viscosity (η_{eff}) is determined for each of the hydrocolloids by collapsing the data in the hydrodynamic regime to that of the Master Curve.
Reproduced from ref. 41 with permission from American Chemical Society, Copyright 2011.

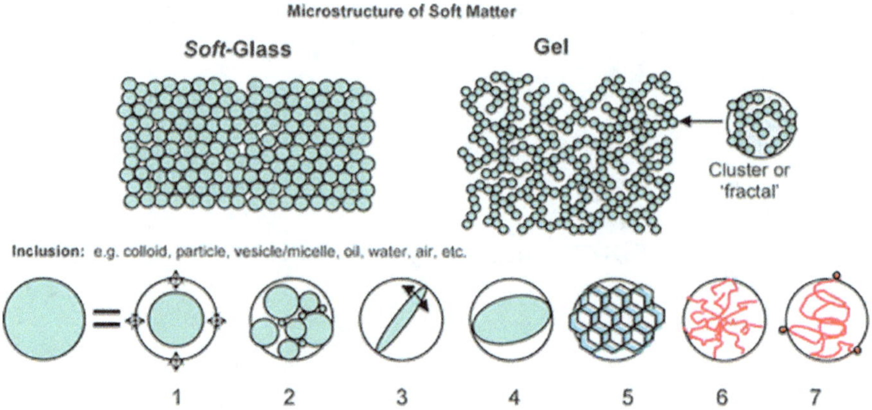

Figure 8.6 Schematic representation of microstructures of semisolid food systems represented in the form of a soft glass, in which the high-volume-fraction included phase is dynamically arrested, or gel, in which the low-volume-fraction included phase forms an interconnected network: (1) inclusion with interparticle interactions (attractive/repulsive) and excluded volume effects; (2) polydisperse size of inclusion and/or aggregates; (3) anisotropic inclusions (*e.g.* clay platelet, rod-like vesicles, plant cells); (4) deformable inclusion (*e.g.* microgel, hydrated core, polymer-coated particle, liquid droplet, air bubble [wet foam]); (5) inclusion is made up of lamaller bilayers (sheets, gel phase), foam/cellular structures, *etc.*; (6) inclusions consists of polymer, peptides, worm-like micelles, star polymers, *etc.*; (7) inclusion contains a polymer with potential to form cross-links (*e.g.* polymer gel).
Reproduced from ref. 8 with permission from the Royal Society of Chemistry.

inclusions, in the form of microgels or through phase separation of biopolymer mixtures. Eqn (8.13) to (8.15) can be used to determine phase volume for these soft particle suspensions. However, due to the large number of uncertainty-generated by errors in dilution to very low particle concentrations, it is preferable to extend the concentration range over which ϕ_0 can be determined. Batchelor's eqn (8.16) builds on Einstein's theory but is applicable at higher concentrations as it is designed to account for two particle interactions. The constant in front of the second-order term (C_2) can have a range of values from 4.2 to 6.2, depending on the underlying assumptions and approach to the derivation.[42,43] Tan *et al.*[44] used Batchelor's equation to determine the apparent phase volume of soft particles (ϕ_0), up to ϕ_0 of 0.3, substituting ϕ_0 for $k \cdot c$, where c is the mass concentration of the dispersion in g ml^{-1} and k (specific volume in ml g^{-1}) is constant and the only adjustable parameter.

$$\eta = \eta_s(1 + 2.5\phi_0 + C_2{\phi_0}^2) \tag{8.16}$$

At higher phase volumes, multibody particle interactions must be taken into account.[45] The model developed by Maron and Pierce[46] and made popular by Quemada[47] (MPQ model, eqn (8.17)) does this by including the maximum packing fraction (ϕ_m). The Krieger–Dougherty model is an alternative higher-order model with an exponent of $[\eta]\phi_m$ in place of -2; this model has the advantage that it collapses to Einstein's equation at low phase volumes. Both of these models are empirical and are fitted to experimental data in order to determine ϕ_m.

We have recently shown that the MPQ model can be used without free fitting by replacing ϕ_m with the random close packing fraction (ϕ_{rcp} determined from the particle size distribution of the suspension,[48] according to the method of Farr and Groot[49]). This modified MPQ model, shown in eqn (8.18), provides a theoretical framework by which the behavior of food products can be interpreted to determine the influence of particle structure, size distribution, and interactions on the rheology of food suspensions.[50] Plotting this data in the linear form demonstrates the point at which rheology deviates from the modified MPQ model due to particle interactions or changes in particle volume from that at ϕ_0 or due to measurement artifacts such as slip or particle migration, as shown in Figure 8.7.

$$\eta/\eta_s = (1 + \phi_0/\phi_m)^{-2} \tag{8.17}$$

$$\eta/\eta_s = (1 + \phi_0/\phi_{rcp})^{-2} \tag{8.18}$$

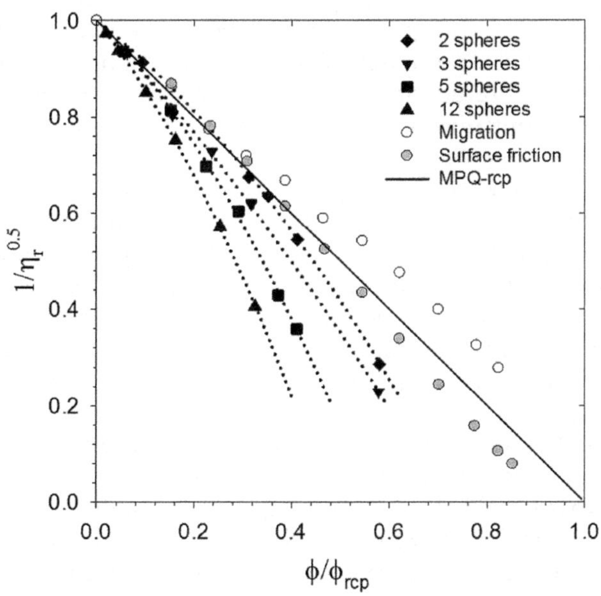

Figure 8.7 The linearized form of the MPQ model against ϕ/ϕ_{rcp} for suspensions that deviate from the MPQ model for hard spheres including aggregates made of different numbers of glass spheres;[75] migration of spherical polystyrene; and suspensions in which friction between particles in close contact occurred (supported by simulations).[76] Dotted lines are a guide for the eye.[48]

Reproduced from ref. 48 with permission from Elsevier, Copyright 2015.

At low phase volumes ($\phi_0 < 0.2$), suspensions of noninteracting spheres are Newtonian or weakly shear thinning. In terms of their flow behavior, noninteracting soft and hard, colloidal and noncolloidal particles in suspension behave like hard spheres at low phase volume, following Einstein's model, but deviate from that behavior with increasing phase volume.[38,51] In the range up to $\phi_0 \approx 0.2$, it is relatively simple to increase the viscosity of the solvent by approximately 1.5 times (estimate based on Einstein's equation) and maintain a desirable Newtonian viscosity. Between $\phi_0 = 0.2$ and ϕ_{rcp}, viscosity becomes increasingly shear thinning. Charged entities or those with irregular shapes, especially rod-like shapes that orient themselves in the shear flow direction, will rapidly become highly shear thinning and will develop a yield stress at $\phi_0 < \phi_{rcp}$.

8.4.2 Tuning of the Yield Stress

Many viscoelastic foods with a yield stress can be considered as either a gel or a glass. A "soft glass" occurs when the high-volume fraction included phase is dynamically arrested, which differs from a "gel," in which the low volume fraction included phase forms an interconnected network. Examples of soft food systems that can be classified as gels or glasses include ketchup—a system of tomato cells in a continuous aqueous phase, which behaves as a soft glass—and molten chocolate—aqueous sugar droplets dispersed in a fat phase, which behave as a percolated gel network due to attractive interactions among the hydrophilic sugar entities in a hydrophobic continuous phase. These are shown schematically in Figure 8.6.

At high phase volumes ($>\phi_{rcp}$) and zero or low-shear conditions, soft particle suspensions display solid-like rheological properties. At these high phase volumes, the number and strength of interparticle interactions and suspension microstructure strongly influence suspension rheology.[8,52,53] In this regime, interparticle interactions and microstructure define suspension behavior. Semi-empirical and theoretical models have been proposed for understanding the rheology, particularly the viscoelastic behavior (G' and G'' [Pa]) of these suspensions.

For systems that follow the soft-glassy structure, Evans & Lips[54] developed a model for the suspension modulus (G) and fitted data for Sephadex, a chemically cross-linked dextran microgel. This model was developed for suspension phase volumes where particle deformation has begun, meaning that interparticle forces are dominated by elastic deformation of contacting particles. The Evans and Lips[54] model:

$$G = \left[\left(\phi_r^{2/3} - \phi_r^{1/3} \right)^{1/2} - \left(\frac{8}{3} \right) \left(\phi_r - \phi_r^{2/3} \right)^{3/2} \right] \left(\frac{\phi_m n G_p}{5\pi(1-\nu)} \right) \tag{8.19}$$

where ϕ_r describes the packing fraction relative to the maximum packing fraction, and n is the number of nearest neighbors. This model was found to fit for Sephadex microgels in the range $0.7 < \phi_r < 1$ and agarose microgels;[55] however, it was found not to be a good fit for agar microgels possibly due to the slight charge on the agar.[56] The particle modulus (G_p) can be assumed to be the same as the modulus of a bulk gel, or, for particles that are susceptible to osmotic deswelling, the osmotic pressure required to deswell the particle can be determined.

For systems that are better approximated by the gel network structure, a percolation model such as the simplified form of the van der Poel model[57] is more appropriate:

$$G' = \beta\ (\phi - \phi_c)^{\alpha} \tag{18.20}$$

where ϕ_c is the percolation threshold, or the minimum phase volume at which the polymers or particles form a space spanning network; β is a free-fitting parameter; and exponent α is usually estimated as a value of 2, as the number of elastically effective stretched strands required to form a gel network[42] (this applies regardless of its deformation). This model describes the network formation due to an increased number of polymers, weakly attractive particles, or particle aggregates, creating structures that span the system and are able to transmit stress.[14,58]

8.4.3 Control of Fat Lubrication

Fat lubrication is the most widely studied area of tribological control in food products, and it has been suggested that with increasing fat content, friction continues to decrease. For example, recent work shows a strong correlation between fat percentage and friction measurements on a discovery hybrid rheometer, using ring-on-plate triborheometry (TA Instrument, USA) on a rough plastic surface of 3M Transpore Surgical Tape 1527-2 (3MHealth Care, USA).[59] However, this is not a fundamental or mechanistic relationship that could be applied across any system as it is dependent on measurement surface, speed, and configuration[60] and on fat droplet morphology, droplet size, narrower size distribution, and smaller effective phase volume.[61] Tribological studies on emulsion lubrication using MTM or a similar device highlight how a single phase (oil or water) can 'plate out' in the tribological contact that depends on the wetting properties of the substrates, drop size, viscosity ratio, and surfactant coverage.[34,62,63]

It is not simply the presence or absence of fat but also the influence of food structure and physical and chemical interaction with other ingredients, such as hydrocolloids and particles, that can affect tribology.[64] For example, Liu *et al.*[65] found that gelled emulsions with a dispersed phase of bound oil droplets exhibit lower friction in the boundary regime than unbound oil droplets, as shown in Figure 8.8. The authors suggest this may be due to the release of fat droplets to coat the tribopair in the unbound case. It is hypothesized that fat lubrication is dominated by the oil–surface contact area; this is known as the droplet spreading mechanism. Systems containing fat/oil are commonly presented as emulsions that generally contain natural

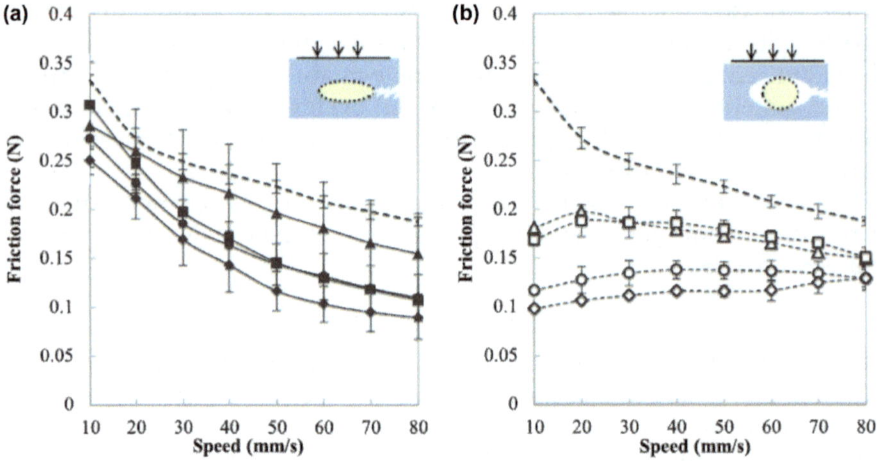

Figure 8.8 Friction force as a function of speed for gelatine gels (dash line with no symbols) and emulsion-filled gelatine gels with (A) bound droplets (see schematic inset) and (B) unbound droplets (see schematic inset) under compression between surfaces in the optical tribometer configuration. Droplets also include solid fat at 5% (▲), 10% (■), 15% (●), and 20% (◆).[65] Adapted from ref. 65 with permission from Elsevier, Copyright 2015.

or added emulsifiers to stabilize droplets. These emulsifiers may be only a small weight percentage of the overall food. However, they can dominate the tribological response due to their surface adsorption ability.[64]

8.5 Using Rheology and Tribology as Tools for Food Structure Development in Relation to Sensory Perception

Rheology, tribology, and structural information are increasingly being used in combination in predicting food sensory perception. This is a good starting point as an indication of the way in which changes in product structure may be reflected in instrumental and sensory measures. However, much of this work is empirical in nature and does not tend to build a mechanistic understanding of the relationships between sensory outcomes, instrumental measures, and the structural parameters of the food.

Dairy systems are one example where this approach has been utilized, and recent work suggests that by mapping a combination of measurements, including rheology, tribology, and particle size, the mouthfeel of dairy products can be predicted empirically.[66–68] It was found that *creaminess* prediction for yogurt was dependent on rheology at a shear rate of 100 s^{-1} and on particle size and tribology at a friction coefficient at 1 mm s^{-1}, according to a mathematical relationship.[68] The authors hypothesize that sensory perception is related to gaps between oral surfaces and that viscosity/tribology or structural properties are dependent on the stage of oral processing. However, this relationship is valid only for this type of yogurt within the viscosity range studied and cannot be related to product microstructure. The study also overlooks the fact that lubrication occurs at all stages of oral processing and is not limited to the point of swallow. A similar study was carried out by Nguyen *et al.*[69] to investigate the influence of four commonly used hydrocolloids on the rheology, tribology, and sensory perception of yogurts. They found no correlation between rheology and sensory perception of their yogurts. They did find some correlations between sensory descriptors of *lumpiness* and *smoothness* and the transition from boundary to mixed and mixed to hydrodynamic regime. Again, this could not be related to the mechanism of sensory perception of lubrication, particularly due to the uncertainty as to which component of the yogurt was entrained into the gap between tribosurfaces.[69] Laiho *et al.*[70] demonstrated that the proportion of the serum phase had the greatest influence on the tribology of yogurts with varying whey-to-casein ratios. They also showed that tribology was significantly correlated to sensory attributes including *smoothness* and *creaminess*. Particle size distribution was also a good predictor of sensory properties, as increased whey protein content resulted in a firmer, coarser gel structure. Mathmann *et al.*[66] suggested a fluid mechanics approach to correlating viscosity to the mouthfeel of yogurt. Based on viscosity measurements and their mouth model, they found correlations between viscosity and a range of mouthfeel attributes using neural networks and fuzzy logic. It is not clear how this could be applied to the design of real systems, and again this model is only applicable to yogurt. Even within a single yogurt system, time and shear rate history of the sample will influence product structure (aggregate size), as shown by Javanmard *et al.*,[12] which is likely to influence sensory perception. This highlights the need to treat instrumental and sensory samples identically prior to measurement if any comparison is to be carried out and using the basis, such as a structural parameter for comparison and elucidation of mechanisms of sensory perception.[12]

Another key factor is that tribology, rheology, and sensory elements should not be looked at in isolation from food structure. This is supported by the work of Liu *et al.*,[65] previously discussed, which demonstrated better sensory perception of fatty mouthfeel from unbound oil droplets than that achieved by bound oil droplets in a gelled emulsion, in agreement with measured friction. They hypothesize that this is related to fat spreading on the oral surface. These authors also demonstrate the influence of particles as fat droplet replacers.[71,72] They suggest that microparticulated whey protein (MPWP) particles exhibit similarly low friction to that of oil droplets in water. However, they hypothesize that the lubrication mechanisms are different – with a rolling or "ball bearing" mechanism occurring for MPWP particles compared to a fat spreading mechanism for oil droplets.[72] This difference can be perceived sensorially in terms of *creaminess* and *fattiness*, with oil droplets scoring highly on both *creaminess* and *fattiness* attributes, whereas small MPWP particles contributed to *creaminess* but not *fattiness* scores, as shown in Figure 8.9.[71]

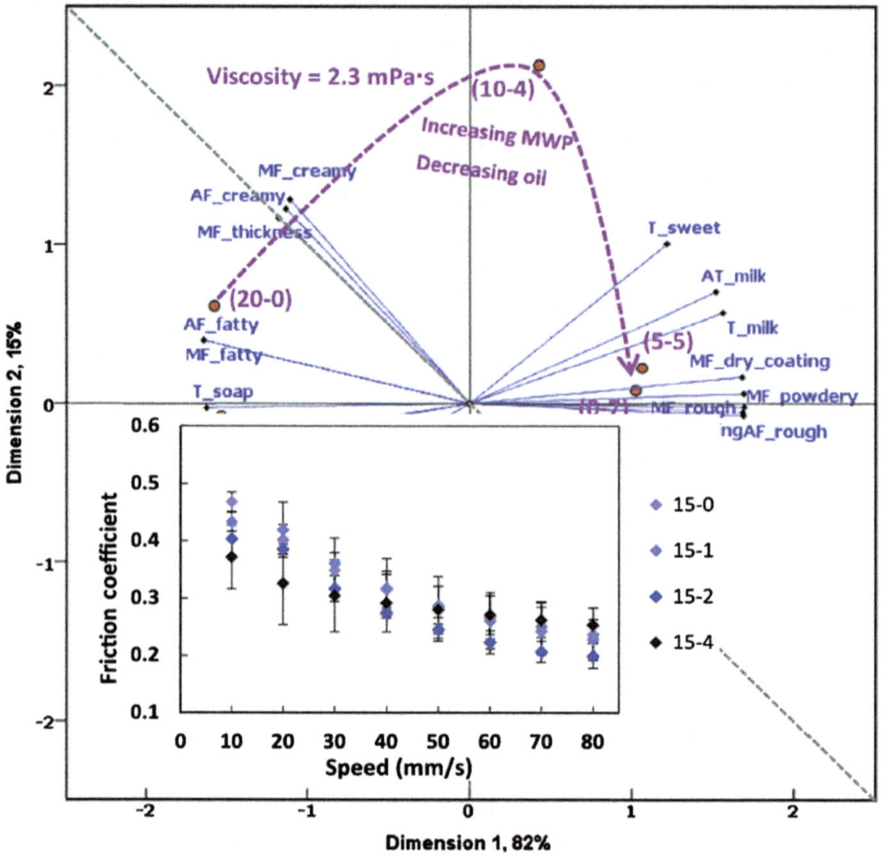

Figure 8.9 Principle component analysis plot showing physical properties of emulsions of varying oil content and microparticulated whey protein (MWP) content shown in the diagram in brackets as (Oil-MWP) for samples of matched viscosity at 2.3 m Pa s. This shows that samples with decreasing oil content from 20% down to 10%, while simultaneously increasing MWP content from 0% to 4%, maintain their creamy texture. Inset shows friction as a function of speed for the same set of samples showing that friction decreases with increasing MWP content from 0% to 4% in a 15% oil emulsion.
Adapted from ref. 71 with permission from Elsevier, Copyright 2016.

It is critical that food structure, assessed instrumentally, is relevant to the stage of food oral processing of interest. For example, evaluation of potato chips has shown no difference in the fracture properties of the whole chip. However, using an *in vitro* comminution process and mixing with physiological buffer solution highlighted rheological differences between the simulated bolus for the different type of chip.[58] A similar effect has been shown using two chocolates of identical composition and viscosity, which were evaluated sensorially with findings suggesting that one was more *mouthcoating* than the other. Upon analysis of the orally processed samples, it was observed that the more mouthcoating of the two samples was more flocculated than the less mouthcoating one.[73] Tribology of the melted chocolate samples was also carried out, and differences in the lubrication profile were shown.[73] However, no mechanism was suggested for these differences in tribology, and this measurement was not completed for the mixed chocolate and saliva samples. Recent work by He *et al.*[74] on expectorated chocolate samples shows correlations between sensory mouthcoating and friction in the mixed regime of the Stribeck curve. They also show correlations between the hydrodynamic friction and thickness perception of the chocolate samples. This paper also takes the next step and begins to elucidate mechanisms for the observed tribology of milk chocolate samples by probing emulsion droplet structure in the expectorated chocolates using confocal microscopy and Raman spectroscopy. They hypothesize that the mouthfeel is dependent on whether the droplet is stabilized by milk proteins alone or by both cocoa and milk proteins, with the latter giving a lower friction coefficient in the mixed regime and a lower mouthcoating score.

8.6 Conclusion and Future Prospects

Rheology and tribology are both useful tools in developing food structures. These tools should be used together as alteration to one will affect the other. Rheology-structure relationships are well established and regularly used in food design that controls product stability and format. The field of tribology for oral processing studies is still developing, but a lot is now known on how many different components and types of foods and beverages behave within tribological contacts. The use of visualization techniques, such as confocal and Raman microscopy, and structure characterization techniques, such as particle size distribution, are also critical in developing structure-property-oral process relationships for foods.

An area still in development is making fundamental links between rheology, tribology, and sensory perception that include texture and mouthfeel. While some publications suggest a link between rheology and tribology with such textural percepts as smoothness, creaminess, and afterfeel characteristics (for example), these are yet to be generalized and typically can apply only to set systems. In addition, they are often not linked back to the underlying structure of the food/beverage. Further work is required in this area with reference to fundamental studies using both simple model systems, as well as complex food products, to elucidate the influence of structural components and interfacial interactions between food components and oral surfaces, especially in the presence of saliva.

Acknowledgements

We acknowledge support from the Australian Research Council project LP160100239.

References

1. J. F. Steffe, *Rheological Methods in Food Process Engineering*, Freeman Press, East Lansing, MI, 2nd edn, 1996.
2. J. R. Stokes, M. W. Boehm and S. K. Baier, *Curr. Opin. Colloid Interface Sci.*, 2013, **18**, 349–359.
3. N. Selway and J. R. Stokes, *Annu. Rev. Food Sci. Technol.*, 2014, **5**(5), 373–393.
4. H. A. Barnes, J. F. Hutton and K. Walters, *An Introduction to Rheology*, Elsevier, Amsterdam, 1989.
5. C. W. Macosko, *RHEOLOGY Principles, Measurements, and Applications*, Wiley-VCH, New York, 1994.
6. J. R. Stokes, in *Food Materials Science and Engineering*, ed. B. Bhandari and Y. Roos, Wiley-Blackwell, UK, 2012, ch. 6.
7. J. R. Stokes, in *Food Oral Processing*, ed. J. S. Chen and L. Engelen, Wiley-Blackwell, UK, 2012, ch. 11.
8. J. R. Stokes and W. J. Frith, *Soft Matter*, 2008, **4**, 1133–1140.
9. J. R. Stokes, in *Food Oral Processing: Fundamentals of Eating and Sensory Perception*, ed. J. S. Chen and L. Engelen, Wiley-Blackwell, UK, 2012, pp. 227.
10. G. A. Davies and J. R. Stokes, *J. Non-Newtonian Fluid Mech.*, 2008, **148**, 73–87.
11. J. R. Stokes and J. H. Telford, *J. Non-Newtonian Fluid Mech.*, 2004, **124**, 137–146.
12. M. Javanmard, E. Wong, T. Howes and J. R. Stokes, *J. Food Eng.*, 2018, **222**, 250–257.
13. P. V. Liddel and D. V. Boger, *J. Non-Newtonian Fluid Mech.*, 1996, **63**, 235–261.
14. T. Van Vliet, *Rheology and Fracture Mechanics of Foods*, CRC Press, Boca Raton, 2014.
15. O. C. Duvarci, G. Yazar and J. L. Kokini, *J. Food Eng.*, 2017, **208**, 77–88.
16. S. A. Rogers, B. M. Erwin, D. Vlassopoulos and M. Cloitre, *J. Rheol.*, 2011, **55**, 435–458.
17. G. Yazar, O. C. Duvarci, S. Tavman and J. L. Kokini, *J. Cereal Sci.*, 2017, **77**, 201–210.
18. G. Yazar, O. C. Duvarci, S. Tavman and J. L. Kokini, *J. Food Eng.*, 2016, **190**, 195–204.
19. D. Szopinski and G. A. Luinstra, *Carbohydr. Polym.*, 2016, **153**, 312–319.
20. H. M. Shewan, J. R. Stokes and M. Cloitre, *Soft Matter*, 2017, **13**, 2099–2106.
21. C. Clasen and G. H. McKinley, *J. Non-Newtonian Fluid Mech.*, 2004, **124**, 1–10.
22. G. A. Davies and J. R. Stokes, *J. Rheol.*, 2005, **49**, 919–922.
23. O. Kravchuk and J. R. Stokes, *J. Rheol.*, 2013, **57**, 365–375.
24. Y. Yan, Z. Zhang, D. Cheneler, J. R. Stokes and M. J. Adams, *Rheol. Acta*, 2010, **49**, 255–266.

25. G. H. Meeten, *Rheol. Acta*, 2007, **46**, 803–813.
26. J. R. Stokes, in *Food Oral Processing*, ed. J. S. Chen and L. Engelen, Wiley-Blackwell, UK, 2012, ch. 12, pp. 265–287.
27. C. Pradal and J. Stokes, *Curr. Opin. Food Sci.*, 2016, **9**, 34–41.
28. N. Selway and J. R. Stokes, *Food Res. Int.*, 2013, **54**, 423–431.
29. J. M. Kim, F. Wolf and S. K. Baier, *Tribol. Int.*, 2015, **89**, 46–53.
30. H. M. Shewan, C. Pradal and J. R. Stokes, submitted, 2019.
31. J. H. H. Bongaerts, K. Fourtouni and J. R. Stokes, Soft-tribology: Lubrication in a Compliant PDMS-PDMS Contact, *Tribol. Int.*, 2007, **40**(10–12), 1531–1542.
32. B. J. Hamrock, S. R. Schmid and B. O. Jacobson, *Fundamentals of Fluid Film Lubrication*, Marcel Dekker, New York, Basel, 2nd edn, 2004.
33. P. M. Lugt and G. E. Morales-Espejel, *Tribol. Transact.*, 2011, **54**, 470–496.
34. J. de Vicente, H. A. Spikes and J. R. Stokes, *J. Tribol.*, 2006, **128**, 795–800.
35. D. Rossetti, J. H. H. Bongaerts, E. Wantling, J. R. Stokes and A. M. Williamson, *Food Hydrocolloids*, 2009, **23**, 1984–1992.
36. N. Brossard, H. F. Cai, F. Osorio, E. Bordeu and J. S. Chen, *J. Texture Stud.*, 2016, **47**, 392–402.
37. R. Lapasin and S. Pricl, *Rheology of Industrial Polysaccharides, Theory and Applications*, Blackie Academic and Professional, Glasgow, 1995.
38. J. R. Stokes, in *Microgel Suspensions – Fundamentals and Applications*, ed. A. Fernandez-Neives, H. Wyss, J. Mattson and D.A. Weitz, Wiley-VCH Verlag, Weinheim, 2011.
39. J. L. Kokini, *J. Food Eng.*, 1987, **6**, 51–81.
40. Q. He, J. Hort and B. Wolf, *Food Hydrocolloids*, 2016, **61**, 221–232.
41. J. R. Stokes, L. Macakova, A. Chojnicka-Paszun, C. G. de Kruif and H. H. J. de Jongh, *Langmuir*, 2011, **27**, 3474–3484.
42. G. K. Batchelor, *J. Fluid Mech.*, 1977, **83**, 97–117.
43. R. C. Ball and P. Richmond, *Phys. Chem. Liq.*, 1980, **9**, 99–116.
44. B. H. Tan, K. C. Tam, Y. C. Lam and C. B. Tan, *Polymer*, 2004, **45**, 5515–5523.
45. *Colloid Science: Principles, Methods and Applications*, ed. T. Cosgrove, Blackwell Pub. Professional, Ames, Iowa, 2005.
46. S. H. Maron and P. E. Pierce, *J. Colloid Sci.*, 1956, **11**, 80–95.
47. D. Quemada, *Rheol. Acta*, 1977, **16**, 82–94.
48. H. M. Shewan and J. R. Stokes, *J. Non-Newtonian Fluid Mech.*, 2015, **222**, 72–81.
49. R. S. Farr and R. D. Groot, *J. Chem. Phys.*, 2009, 131.
50. H. M. Shewan and J. R. Stokes, *J. Colloid Interface Sci.*, 2015, **442**, 75–81.
51. M. S. Wolfe and C. Scopazzi, *J. Colloid Interface Sci.*, 1989, **133**, 265–277.
52. R. Hughes, in *Colloid Science: Principles, Methods and Applications*, ed. T. Cosgrove, Blackwell Pub. Professional, Ames, IA, 2005, ch. 11, p. 287.
53. S. P. Meeker, R. T. Bonnecaze and M. Cloitre, *Phys. Rev. Lett.*, 2004, **92**, 198302–198304.

54. I. D. Evans and A. Lips, *J. Chem. Soc., Faraday Trans.*, 1990, **86**, 3413–3417.
55. H. Shewan, PhD, The University of Queensland, 2015, DOI: 10.14264/uql.2015.533.
56. S. Adams, W. J. Frith and J. R. Stokes, *J. Rheol.*, 2004, **48**, 1195–1213.
57. C. van der Poel, *Rheol. Acta*, 1958, **1**, 198–205.
58. M. W. Boehm, S. K. Baier and J. R. Stokes, *Food Res. Int.*, 2013, **54**, 544–551.
59. P. T. M. Nguyen, B. Bhandari and S. Prakash, *J. Food Eng.*, 2016, **168**, 27–34.
60. A. Chojnicka-Paszun, H. H. J. de Jongh and C. G. de Kruif, *Int. Dairy J.*, 2012, **26**, 15–22.
61. J. Chen, *Crit. Rev. Food Sci. Nutr.*, 2007, **47**, 583–598.
62. M. Douaire, T. Stephenson and I. T. Norton, *J. Food Eng.*, 2014, **139**, 24–30.
63. J. R. Stokes, in *Aqueous Lubrication*, ed. N. D. Spencer, World Scientific, 2014, pp. 73–101.
64. H. S. Joyner, C. W. Pernell and C. R. Daubert, *Tribol. Lett.*, 2014, **56**, 143–160.
65. K. Liu, M. Stieger, E. van der Linden and F. V. De Velde, *Food Hydrocolloids*, 2015, **44**, 244–259.
66. K. Mathmann, W. Kowalczyk and A. Delgado, *J. Texture Stud.*, 2009, **40**, 16–35.
67. M. A. Rao, in *Rheology of Fluid, Semisolid, and Solid Foods*, ed. M. A. Rao, Springer, US, New York, 2014, ch. 7.
68. A. Sonne, M. Busch-Stockfisch, J. Weiss and J. Hinrichs, *LWT–Food Sci. Technol.*, 2014, **59**, 342–347.
69. P. T. M. Nguyen, O. Kravchuk, B. Bhandari and S. Prakash, *Food Hydrocolloids*, 2017, **72**, 90–104.
70. S. Laiho, R. P. W. Williams, A. Poelman, I. Appelqvist and A. Logan, *Food Hydrocolloids*, 2017, **67**, 166–177.
71. K. Liu, M. Stieger, E. van der Linden and F. van de Velde, *Food Hydrocolloids*, 2016, **60**, 186–198.
72. K. Liu, Y. Tian, M. Stieger, E. van der Linden and F. van de Velde, *Food Hydrocolloids*, 2016, **52**, 403–414.
73. A. M. Carvalho-da-Silva, I. Van Damme, W. Taylor, J. Hort and B. Wolf, *Food Funct.*, 2013, **4**, 461–469.
74. Q. He, F. Bramante, A. Davies, C. Elleman, K. Fourtouni and B. Wolf, *Food Funct.*, 2018, **9**, 3532–3546.
75. T. B. Lewis and L. E. Nielsen, *Trans. Soc. Rheol.*, 1968, **12**, 421–443.
76. F. Boyer, O. Pouliquen and E. Guazzelli, *J. Fluid Mech.*, 2011, **686**, 5–25.

CHAPTER 9

Developing Food Structure for Mechanical Performance

I. K. MOHAMMED, C. G. SKAMNIOTIS AND M. N. CHARALAMBIDES*

Imperial College London, Department of Mechanical Engineering, London SW7 2A7, UK
*Email: m.charalambides@imperial.ac.uk

9.1 Introduction

The structure–property link has been an active topic of research for several decades in materials science and engineering. Numerous studies are reported in the literature driven by the common need to develop know-how enabling the design of structures at the various scales such that the desired bulk properties are obtained. This chapter will concentrate on mechanical properties, while there is also parallel research on the effect of structure on other properties such as thermal and electrical. It will also focus on food materials, *i.e.* food proteins and polysaccharides (hydrocolloids) with other common ingredients being salts, sugars, nutrients, and emulsifier surfactants;[1] however, it should be pointed out that experimental and modeling advances in other materials fields (*e.g.* synthetic polymers and their composites) have direct relevance to food biopolymers as well, even though the latter are significantly more complex.[2]

Many foods undergo large-scale industrial processing during manufacturing. Common processes such as extrusion, baking, and cutting of foods[3–10] often lead to intricate and complex structures that will subsequently impact on the properties of the product downstream in the

Food Chemistry, Function and Analysis No. 18
Handbook of Food Structure Development
Edited by Fotis Spyropoulos, Aris Lazidis and Ian T. Norton

Published by the Royal Society of Chemistry, www.rsc.org

manufacturing chain (*e.g.* during packaging), as well as its behavior while it is being consumed by humans or animals, *i.e.* during the oral and gastric processes. In the latter processes, breakdown occurs due to the mechanical action (*e.g.* mastication, peristalsis of the gut), as well as due to hydration and mixing with enzymes in the oral cavity[11,12] and gastrointestinal tract.[13] Food structure plays a key role on both of these causes of breakdown. During oral processing, the behavior of the food will largely affect consumer perception and textural sensory profiles, while during the gastric processes food deformation and breakdown may be linked to gastric emptying and satiety as well as nutrient bioavailability. Therefore, it is evident that the optimization of a food product, whether it is for industrial or human (body) processing, heavily relies on the development of design tools that can elucidate this crucial structure–property relationship.

As foods are highly complex biomaterials, several structural and material parameters are often at play simultaneously. Therefore, product development through experimentation alone is often a prohibitively costly process. In addition, it is often not possible to isolate parameters to study their individual effect on the tracked property as any alteration has an impact on several other parameters simultaneously. For example, an extruded snack product will respond to external loads during industrial processes (*e.g.* cutting) or the oral process (*e.g.* first bite) largely according to its cellular structural architecture as well as the material properties of the solid cell walls. It would be extremely difficult to devise experimental protocols to change the architecture but not the solid cell wall and *vice versa*. Therefore, simulations and models are crucial in enabling the food engineer to design new and optimized products as they readily enable such parametric studies. Having said that, simulations and models need several parameters as inputs, which can be determined only through careful, independent experiments. In addition, experimental validation is always a crucial step in the development of a model. Therefore, both experimental and modeling studies are needed. Such studies are reviewed in the following subsections, which for convenience and clarity divide foods into three distinct classes:

- particulate composites and gels, with dispersed emulsion droplets, fat crystals, starch granules and/or protein aggregates[1]
- cellular solid foams with gas bubbles as the dispersed phase
- short fiber–reinforced foods

It is acknowledged that several foods often fall into more than one of these categories. In this case, the main findings from each category still apply; however, the effects cannot be simply additive, and potential phase– and reinforcement-type interactions should also be considered.

9.2 Particulate Composites, Colloids, and Gels

This subsection will focus on materials that can be described as particulate composite materials, colloids, or gels. It is split into two subsections, the first summarizing work on "real" food products (*e.g.* dough) and the second on gels as model systems.

9.2.1 Structure: Property Studies on Real Food Products

A food material falling in this category is bread dough. The rheology of this material has been heavily studied due to bread being a staple of the human diet for millennia. The processing of dough (*e.g.* mixing, sheeting), its baking performance, and the quality of the final product are all highly affected by the mechanical/rheological response of the dough to the imposed external forces during these industrial processes. The microstructure of the dough itself has a large influence on mechanical behavior, as summarized in a review by Jekle and Becker.[14] A review on the use of hydrocolloids in wheat bread making, as used for improving the sensory quality of whole grain breads and composite breads, is given by Ferrero.[15] Even though a large number of dough rheological studies are reported, here we concentrate only on those that treat dough as a multiphase composite.

Tanner *et al.*[16] regarded dough as a suspension of rigid starch particles of two distinct geometries (oblate spheroids and spherical) in a viscoelastic matrix. They developed an analytical suspension model in order to investigate and predict the change of the rheological behavior of the system in the linear regime as a function of starch volume fraction. The model was able to explain the rapid softening of the bread dough with increasing strain using the notion of stress-intensification factors in the matrix close to the particles, as was initially suggested for filled rubbers.

Mohammed *et al.*[17] also treated bread dough as a two-phase particulate composite, consisting of starch particles dispersed in a continuous gluten protein matrix with a volume fraction of about 45% (see Figure 9.1b). They separated the composite into its two constituents by washing with water. Therefore, they were able to perform a comprehensive series of tests on the composite materials as well as on their two constituents. The data were fitted with constitutive laws, a nonlinear viscoelastic model based on rubber elasticity coupled with a Prony series was calibrated for the gluten phase, whereas a viscoplastic model was found to be suitable for the starch phase. It is worth noting here that, for the same strain rate, the starch particles appeared to be stiffer than the gluten matrix. Therefore, through composite analysis, it was expected that the behavior of the composite dough should also appear stiffer than that of the gluten matrix alone. However, it was found that although this was indeed the case for smaller strains (see Figure 9.2 for test data in tension), the composite dough stress fell to lower levels for strains larger than approximately 0.7. The authors hypothesized that this was due to the interface of the filler and matrix being weak, enabling damage through debonding to take place at larger tensile strains, which therefore lowered the stress for the dough. They tested this hypothesis by developing a numerical micromechanical model for the dough composite. First, the microstructure of the dough composite was obtained through cryo-SEM (scanning electron microscope) (Figure 9.1a). This was then

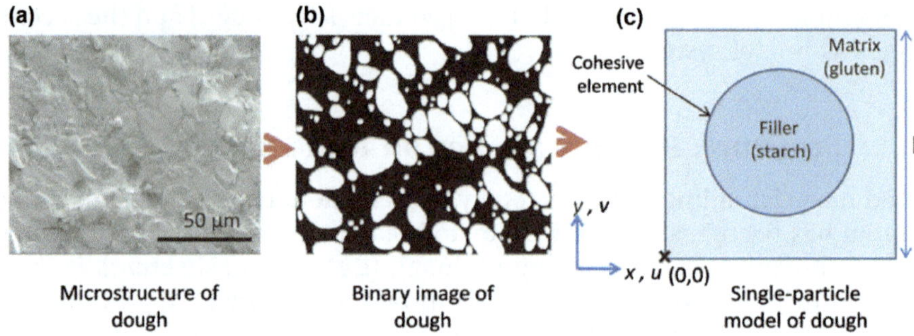

Figure 9.1 Microstructure studies of bread dough: (a) cryo-SEM image of dough; (b) binary image of the cryo-SEM image of dough; (c) a single-particle model of dough.
Reproduced from ref. 17 with permission from The Society of Rheology, Copyright 2013.

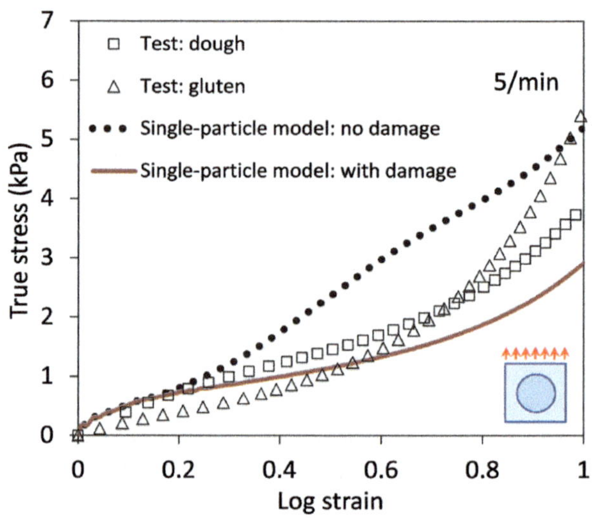

Figure 9.2 Dough and gluten tensile test data at 5 min^{-1} and calibration of the single-particle model under tensile load. Two model curves are also shown for comparison with the experimental data, one that allowed damage in the form of debonding along the starch/gluten interface and one with a perfect interface bonding assumption.
Reproduced from ref. 17 with permission from The Society of Rheology, Copyright 2013.

digitized (Figure 9.1b) such that it could be fed into a 2D numerical finite element (FE) analysis model, as shown in Figure 9.3. A simpler, single-particle model, as shown in Figure 9.1c, was also used to enable a comparison of the results to the real microstructure model. A cohesive zone model (CZM), which is now a very common way of modeling interface behavior in fracture mechanics,[18] was employed (see Figure 9.4).

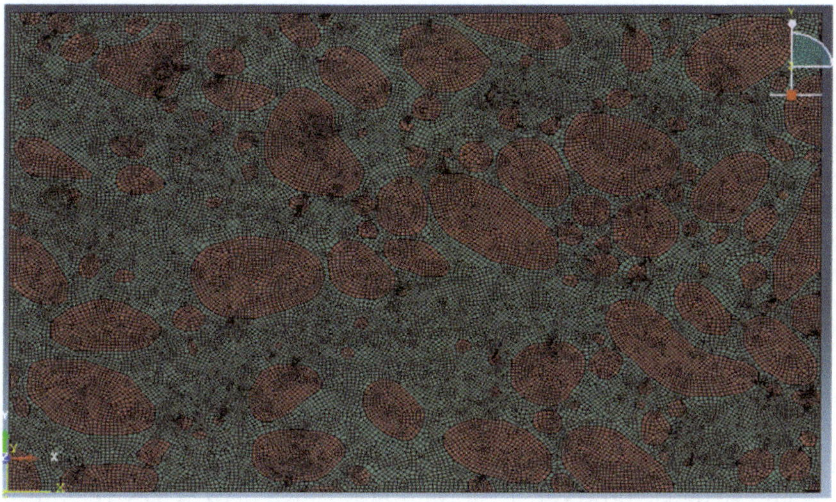

Figure 9.3 A finite element model mesh based on the digitized image shown in Figure 9.1b.
Reproduced from ref. 17 with permission from The Society of Rheology, Copyright 2013.

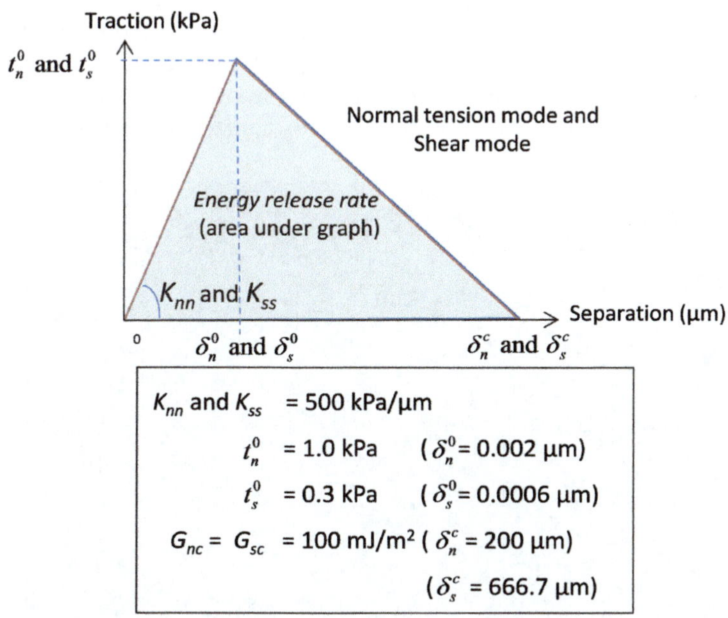

Figure 9.4 Cohesive contact parameters used in the micromechanics model of bread dough under normal and shear loading.
Reproduced from ref. 17 with permission from The Society of Rheology, Copyright 2013.

The interaction between the two phases in both the opening (normal) as well as the sliding (shear) directions was therefore modelled through a bilinear traction *versus* separation behavior. In the first linear region, the behavior is linear elastic with no damage taking place at the interface between the filler and matrix. After the peak stress is reached, a damage variable D is introduced that takes the value of zero when traction is equal to t_n^0. This damage variable subsequently increases, the traction decreases, and D reaches its maximum value of 1 when the separation is equal to δ_n^c. At this point, complete decohesion takes place. The area under the bilinear CZM law is equal to the energy release rate of the interface G, which is a true material property characterizing the fracture toughness. It defines how much energy is needed to create a new fracture surface of unit area; low values of G denote brittle materials, whereas a high G implies a tough response. The results of the micromechanical models were compared with the experimental data corresponding to the composite, and the correct trends were predicted. It was acknowledged, however, that calibration of the CZM law (Figure 9.4) needs to be further examined as the parameters themselves could indeed be functions of strain rate and loading conditions.

Cheese is another example of a food for which several structure–function studies have been performed. Cheese can be treated as a particulate composite with fat particles dispersed in the continuous protein network (see Figure 9.5). A driver for this research activity is to replace some of the fat with low-calorie alternatives in order to lower the energy content associated with cheese. A review article, including this topic, by Dickinson[19] stresses the temperature dependence and phase change of the fat filler over a range of 10 °C to 30 °C, which has a direct influence on the modulus of the fat particles and hence on the modulus of the composite cheese material.[20]

Figure 9.5 Scanning electron microscope image of parmesan cheese.

It is expected that at low temperatures, therefore, the fat particles lead to a strong modulus enhancement, whereas at high temperatures the liquid fat could lead to a weakening effect. A study, aiming to replace milk fat with canola oil in an effort to reduce *trans* and saturated fats in processed cheese, showed that adding oat fiber particles compensated for the changes in textural properties when 50% of the fat was replaced with canola oil.[21] The linear viscoelastic properties of cheese with various fat contents and water/protein ratios were measured by Faber *et al.*;[22] a fractional Maxwell model was used to fit the data and derive analytical expressions for estimating *firmness*, *springiness*, and *rubberiness* textural attributes for the various cheese structures.

Another example of a micromechanical model developed for foods is the one developed for chocolate.[23] Chocolate was modeled as a suspension of approximately 70% weight fraction of particles consisting of sucrose and cocoa particles dispersed in a continuous cocoa fat phase. A simulation was performed to determine the stress concentration developing around the particles upon cooling of the chocolate from 32 °C to 10 °C. The microstructure was taken from X-ray microtomography images that were fed into a finite element analysis simulation model. The stress distributions were used to predict where cracks in the chocolate would form; these are important as it is hypothesized that these small cracks act as channels for lipid migration to the surfaces of the chocolate, eventually leading to the undesirable blooming effect, *i.e.* the whitening of the external surfaces of the chocolate.

9.2.2 Structure: Property Studies for Gels and Filled Gels

Protein gels have also received significant attention in the food rheology literature due to their versatility in modifying food texture but also due to their potential to entrap targeted molecules for delivery to physiological targets.[1,24,25] In addition, further interest has arisen because of the possibility of extracting plant-derived proteins. This has become a worldwide concern due to projected water shortages and their direct impact on the availability of animal-derived proteins, as well as the fact that one-third of the world's arable land is used to grow crops to feed animals.[26] Alternative sources of protein are therefore urgently needed. Structuring plant proteins into desirable textures, though, is still an unresolved issue, and literature has focused on that field.

Yang *et al.*[24] considered deriving protein from canola seeds as a by-product of oil production. Under different pH and temperature processing conditions, contrasting gel structures were obtained ranging from aggregated particulate to porous dense-wall network structures. These led to markedly different mechanical properties. Foegeding[27] also found processing conditions to change the structure of protein gels, in particular observing two general types designated as fine-stranded and particulate gels. The different structure had a direct impact on the food breakdown pattern during mastication. Specifically, fine-stranded whey protein gels are firmer, release less water, and fracture into larger particles with more irregular shapes, as compared to particulate gels. Munialo *et al.*[28] studied the effect of the microstructure of pea protein gels on the mechanical response. Various network structures were generated by varying pH and protein concentration. Dynamic rheological measurements as well as large strain uniaxial compression tests were undertaken. The microstructures were quantified through analyzing the inhomogeneity of pixel intensity in images obtained from the scanning electron microscope and the confocal laser scanning microscope. The images were taken at various magnifications to enable characteristic length scales to be analyzed in terms of the pair correlation function. It was found that the rheological and fracture properties correlated with the structural changes. In another related study,[29] fine and coarse stranded protein gels were used to study water removal upon gel deformation, as this has a direct impact on sensory perception. Water removal is both a function of gel structure and gel stiffness; in fine-stranded gels with small characteristic length scales, the stiffness of the structures was found to be less important in water removal, in contrast to coarse gels where the stiffness of the network was an important parameter, even though gel coarseness was still found to be the dominant factor.

Emulsion-filled protein gels have also been considered by several authors as a way of structuring foods for optimizing texture. These are soft solid-like materials consisting of a protein gel containing dispersed oil-in-water emulsion droplets.[19,30] In this field of study, reference is often made to active and inactive fillers; the former denote fillers that are mechanically connected to the gel matrix and hence may contribute to the gel strength and modulus. Inactive fillers are those that are not well connected to the matrix and therefore behave as small holes in the matrix; these fillers always lead to a reduced modulus and strength of the soft solid composite. Dickinson[19] presented a review of the effect on solid composites from such particulate reinforcement, including large-deformation rheology and fracture properties. Polymeric gels are compared to aggregated particle gels where it is seen that adding a high volume fraction of emulsion droplets into the protein network leads to a much decreased linear viscoelastic strain region and fracture strains. It was concluded that biopolymer-based emulsion gels are suitable model systems for investigating the structural parameters affecting the breakdown of food polymers during oral and gastric processes. This was based on the fact the various real foods, such as cheese and reformulated meats, have already been considered as multiphase composite systems[20] with reasonable success regarding the understanding of the effect of the structure on the rheological behavior. Another example of this is the study by Rosa *et al.*,[30] who used a rheological model of concentrated dispersions, developed by van der Poel,[31] to compare to experimental shear storage modulus data derived from various protein and oil concentrations of emulsion-filled gels, reporting a good agreement. Biopolymer gels were seen as excellent model systems for understanding the relationships between food structure and texture by Foegeding[27] as well. The emphasis here was the link between microstructure and sensory texture. Gels formed by whey protein isolate–containing emulsified lipids were found to possess increased *firmness* and *chewiness*.[32]

Geremias-Andrade *et al.*[25] studied emulsion-filled gels, although in their study solid lipid microparticles (SLM) were used instead of an oil-in-water emulsion. A literature search of the term "SLMs" results in numerous drug delivery applications; however, these systems are now studied for possible use in foods as they may be used for the incorporation of bioactives such as natural antioxidants that are beneficial to human health. Such lipid-based microencapsulation is thought to enhance the bioavailability of such antioxidants while also protecting them from light degradation. Geremias *et al.*[25] found that increasing the weight fraction of the SLMs beyond a critical value led to significant rise in mechanical properties, such as Young's modulus, fracture stress and strain, all measured from uniaxial compression tests. It was suggested that at lower concentrations of SLM, the porous structure of the gel matrix may have allowed mobility of the SLMs under compression, which hindered significant interactions between the matrix and the dispersed phase. At higher weight fractions, the mobility of the SLMs may have been restricted (called the *crowding* effect), and therefore the composite gel was stiffer as a result of the stronger interaction between the phases. Surprisingly, however, the rheological measurements showed the opposite effect; *i.e.* a higher SLM concentration led to lower storage and loss moduli. The amount of surfactant added to stabilize the interface of the gel-SLM was also highlighted as an important parameter affecting the behavior of these complex soft solids.

The important point concerning the role of water in biopolymeric composites has also been emphasized in several studies.[2,3] In contrast to synthetic polymers, composite foods often employ a large amount of water in their formulation which is managed through adding hydrodynamically active (water-binding) biopolymeric ingredients.[2] So not only is the composite constitutive response influenced by the mechanical properties of the constituent phases—the interface, volume fraction, dispersion and shape of the fillers—but also by the competition for moisture between the matrix and fillers. Water binding and distribution in a composite will also affect the properties of the matrix, filler, and interface as well as the swelling of the filler. Therefore, complex interrelationships may exist, making the prediction and design of the composite food's response a real challenge. An example of this difficulty is when biopolymeric ingredients are added in an effort to replace fat, whose water dynamics may differ.

In an effort to avoid such complicating issues, model food systems using glass microspheres dispersed in gelatine and protein gels, were studied in order to investigate the complex response of filled gels.[33–36] The shape and size of the particles were varied; the modulus enhancement was found to be insensitive to particle size as expected, whereas shape led to a marked effect on the failure load in agreement with data published on synthetic polymers.[37] The interface effect was studied through producing hydrophilic and hydrophobic filler surfaces resulting in matrix fractures in the former and in debonding along the interface in the latter, as evidenced by scanning electron micrographs.

Finally, the rate-dependent fracture properties of gelatine gels were studied by Forte *et al.*[41] The rate dependence in the fracture of gels has been observed by several studies;[34,38,39] however, it was found previously that the deformation properties (the stress-strain curve before fracture point) were largely independent of strain rate[40] (see Figure 9.6). The reader is directed to the references cited in Forte *et al.*[41] for more information on various theories put forward to explain this rate dependency. In order to determine fracture properties, Forte *et al.*[41] performed wire cutting tests, and a fracture mechanics-based analysis allowed the energy release rate G of the gelatine to be determined as a function of strain rate. The rate dependence of the fracture toughness was stipulated to be a direct consequence of the pressure-driven flow of water within the porous gel structure. A simple analytical model was developed combining the pressure-driven flow of water through the pores and the stress-driven growth of the fracture process zone. The power law dependence of G on strain rate at larger rates with a power index of 0.5 was in very good agreement with experimental data (see Figure 9.7). Physically, it was suggested that at low rates, fluid flow occurs readily and does not contribute to the load-carrying ability of the gel. At such low rates, G is constant, as it would be for an elastic solid (see Figure 9.7). At greater rates, the contribution of the fluid flow to the fracture resistance is

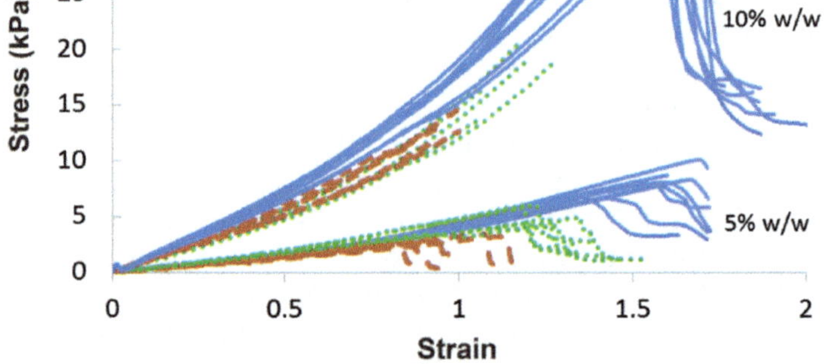

Figure 9.6 Compression response for the 5% w/w and 10% w/w gelatine gel at strain rates of 0.25, 2.5, and 25 min^{-1}. The fracture points correspond to the maximum or last points on each curve.
Reproduced from ref. 41 [https://doi.org/10.1016/j.foodhyd.2014.12.028] under the terms of a CC BY 4.0 license [https://creativecommons.org/licenses/by/4.0/].

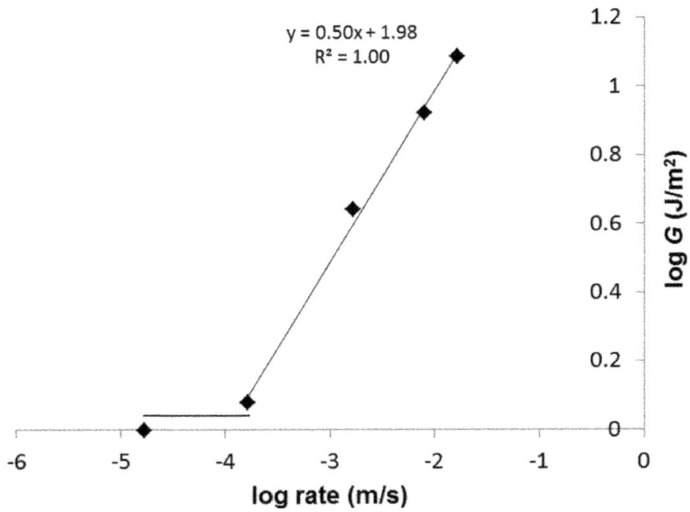

Figure 9.7 Logarithmic plot of *G versus* cutting rate for 10% w/w gelatine gel. Data obtained from extrapolations of wire cutting test data performed at speeds of 1–1000 mm min^{-1} using wire diameters varying from 0.05 mm to 0.5 mm. Note that the cutting rate is in units of m s^{-1} before taking its logarithm.
Reproduced from ref. 41 [https://doi.org/10.1016/j.foodhyd.2014.12.028] under the terms of a CC BY 4.0 license [https://creativecommons.org/licenses/by/4.0/].

significant, and therefore the measured apparent *G* rises. A similar rate effect was noted in the fracture stress determined through uniaxial compression (see Figure 9.6). A numerical simulation that treated the gel as a poroelastic material enabled the stress in the solid network and pore pressure contributions to be decoupled. A simple critical maximum stress criterion in the solid network was used, which captured the dependence of fracture strain on the rate while the initial moduli indeed remained constant and rate independent.

9.3 Cellular Structures

This subsection on cellular structures will focus on brittle confectionery products that have been manufactured *via* traditional baking and extrusion techniques. The products were characterized in terms of their mechanical response and microstructural makeup and were then modeled numerically using 3D image data.

9.3.1 Manufacturing Methods of Edible Foam Food Products

Edible solid foams are a class of cellular food products that are porous, consisting of air within the solid food material. These foods are made by dispersing gas into a biopolymeric matrix using aeration or by blowing agents released during a sudden change of pressure and/or temperature.[3,42] During the dispersion of the gas phase or blowing agent into the matrix, the polymers are in the rubbery state; however, the final form is a solid foam at a gelled or glassy state. Foams with a glassy polymer matrix are brittle, such as expanded potato snacks, wafers, cereal, and keropok,[43–45] while foams with a gelled polymer matrix are ductile, such as bread and sponge cake.[46–48] These foams can be produced *via* extrusion, baking, frying in oil, or popping in hot air or microwave. In this section, brittle confectionery wafer products produced by baking[49,50] and extrusion[51] will be used as a case study.

Oven-baked wafers are used in several popular confectionery products and have been manufactured and marketed successfully for decades.[52] The wafer batter is poured into a heated metal mold, and the steam generated has difficulty in escaping from the batter, resulting in the formation of bubbles. Near the surface of the mold, a dense layer with tiny bubbles is formed, due to compression of the larger bubbles in the middle of the wafer.[53] For the wafers presented in this study, liquid batter was spread and then baked between oven plates with engravings known as *reedings*, which result in the geometric shape of the wafers.

An alternative method of producing similar lightweight and crispy products is the extrusion process, which has many advantages over the conventional baking process since it is a continuous unit operation with the flexibility of online process adjustments for achieving the desired product characteristics.[51] The extruded products are made by passing the mixture through a twin-screw extruder at high pressure and temperature where the semicooked material exits the die and then solidifies. During both the baking and extrusion processes, most of the moisture evaporates, resulting in a porous cellular structure that is lightweight and crisp. Due to the nature of the two manufacturing processes, baking within hot plates produces flat sheets, whereas extrusion leads to cylindrical tubes.

9.3.2 Mechanical Testing of Baked and Extruded Foam Foods

Crispiness and *crunchiness* are sensory traits of edible brittle foams,[54,55] which are experienced during biting as the foam fractures, accompanied by sound emissions. Both the mechanical and acoustic sensations are important for crispiness and crunchiness. *Crunchiness* is associated with the food fracturing and releasing acoustic emissions after the first bite, although not completely fracturing after the first bite. A crispy foam fractures easily and has a low *hardness*; crispiness and hardness have a negative correlation.[3] The mechanical response of brittle food foams obtained *via* bending, cutting, penetration, and compression tests can be used to predict the sensory trait of *crispiness* by correlating with acoustic emissions.[56]

Uniaxial compression tests are the simplest mechanical tests to perform because there is no need to grip the specimens, and they are also relevant to mastication. The compressive load-displacement curves for both the extruded tube and baked wafer were characteristic of a brittle foam, as shown in Figure 9.8. They display an initial linear elastic region followed by a jagged plateau and then a densification region as shown in Figure 9.8a, even though, due to the shape of the extruded tube, the sample would break into pieces before densification of the foam was reached (Figure 9.8b). Young's modulus and fracture stress were calculated from the experimental curves and were found to be independent of the testing speed. However, both of these material properties were an order of magnitude higher for the

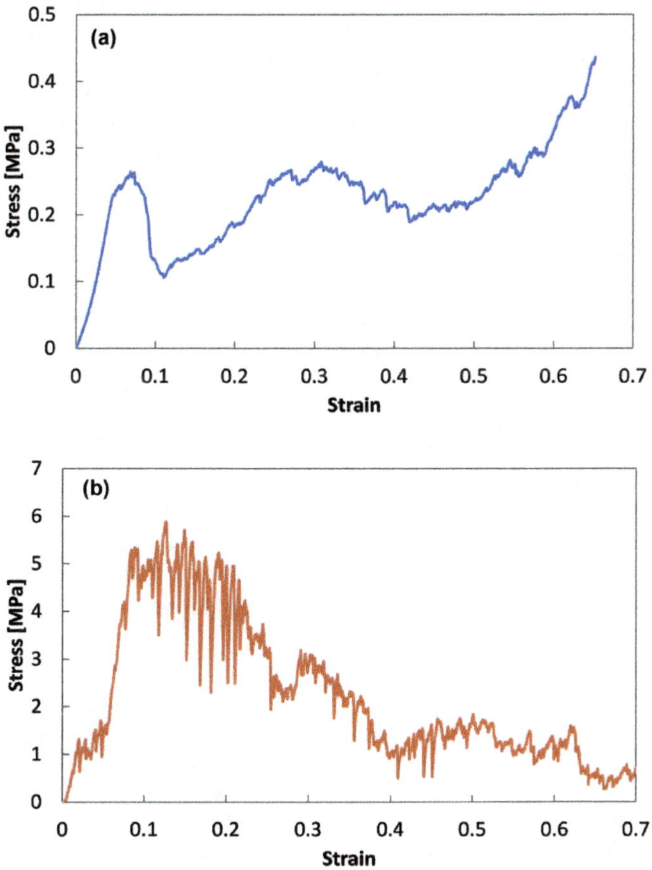

Figure 9.8 A typical stress-strain curve: (a) a baked wafer; (b) an extruded tube, under compression. For the tube, compression is parallel to the extrusion direction.
(a) Reproduced from ref. 50 with permission from Elsevier, Copyright 2014. (b) Reproduced from ref. 51 [https://doi.org/10.1016/j.jfoodeng.2018.05.039] under the terms of a CC BY 4.0 license [https://creativecommons.org/licenses/by/4.0/].

extruded tube than the baked wafer, which was in line with the higher forces required for fracture during the compression tests.[51] The load-displacement response of the extruded products displayed a large amount of jaggedness throughout the test, which correlated to more frequent fracture events. In comparison, the baked wafer had one major fracture event, signifying the initial collapse within the core,[50] followed by minor fractures of the remaining cell walls until densification.

9.3.3 Microstructural Characterization of Foam Foods Using X-ray Micro Tomography

In the classic Gibson and Ashby model for an ideal foam with uniform cell size distribution and a cubic unit cell, both the foam elastic modulus and tensile strength are related to the porosity of the foam.[57] An open cell foam structure consists only of struts, and the faces of the cell do not contain any material, whereas the opposite is true for a closed cell foam. Edible brittle foams are often a mixture of closed and open cell structures. Therefore, in order to characterize such foams, the actual morphology needs to be obtained.

X-ray microtomography (XMT) is a versatile imaging technique that has numerous applications in diverse fields such as geosciences, material science, and biology.[58] Over the last decade, 3D imaging methods have improved and have become more accessible for food microstructure analysis.[59] XMT has been used on a wide range of foods[60,61] with different aims including finding internal disorders, microstructural characterization and quantification, infestation detection, pore size distribution, and evaluating a specific ingredient or component.[58]

The internal microstructures of the baked wafer and similar extruded products were observed using scanning electron microscopy (SEM) and XMT imaging techniques.[50] The cellular nature of the baked and extruded products was visually confirmed with both techniques, as shown in Figure 9.9.

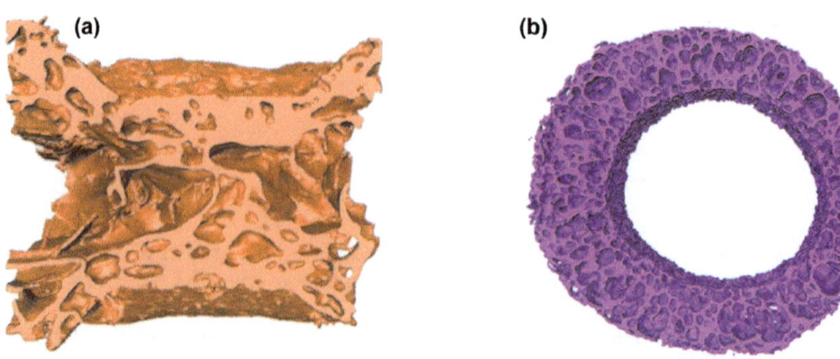

Figure 9.9 3D reconstructions: (a) baked wafer; (b) extruded tube using XMT image data.

The baked wafer exhibited distinct regions similar to a sandwich structure with two skins on either side of a core. The skins possessed many small pores, while the core of the wafer had fewer pores but was much larger in size. The pores located within the skins were all closed-cell and were separated from one another by very thick walls, which gave the skins their dense appearance. The cell wall thickness within the core was much thinner than in the skins. Conversely, the extruded tubes had more homogeneous cellular structures with a nonuniform thickness around the annulus. Image analysis was performed on the XMT data of the baked and extruded products to characterize their microstructure in terms of pore size and cell wall thickness distributions as well as porosity. Although the global porosities of the baked wafer and extruded tube were nearly identical at around 80%,[51] the microstructural features were different, which was a direct consequence of the manufacturing processes.

9.3.4 Finite Element Modeling of the Fracture of a Brittle Baked Wafer

For a complete understanding of cellular materials, the structure should be studied at two different scales: the microstructure of the constitutive material and the cellular architecture of the foam itself.[62] The cellular microstructure describes the size and morphology of the arrangement of the solid and gaseous phases in a cellular material. The constitutive material governs the global behavior because it modifies the properties of the solid part of the structure.

The 3D image data obtained from XMT scans can also be used to model the microstructure of materials, perform numerical simulations, and predict mechanical properties.[62] This has been applied to food science for the mechanical modeling of cereal products in order to predict texture.[63] XMT reconstructions and FE simulations have been combined to model the deformation of brittle food products.[50,64] The study performed by Mohammed *et al.*[50] demonstrated that an FE model can be used to inversely determine the solid modulus and fracture stress of the cell wall material of the foam. Figure 9.10 shows the fracture of the confectionery wafer at different stages during the compression simulation.

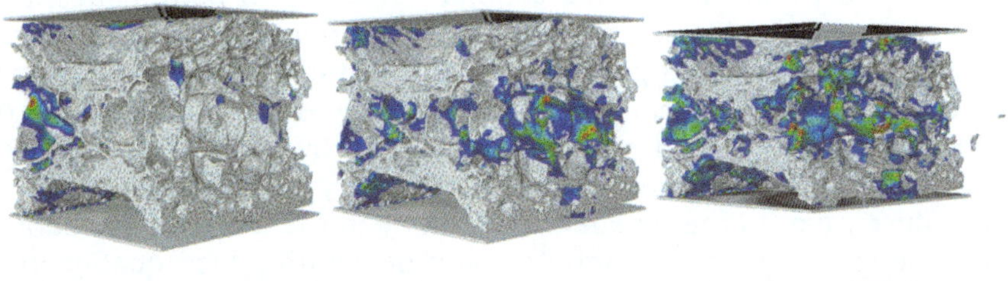

Figure 9.10 A finite element simulation of the damaged wafer at different stages during the compression.
Reproduced from ref. 50 with permission from Elsevier, Copyright 2014.

9.4 Short Fiber Composites

9.4.1 Adding Fibers in Foods

Short fibers can modify overall material stiffness, while also being particularly effective in enhancing toughness through various mechanisms: *e.g.* fiber breakage and fiber-matrix de-bonding.[65] Owing to the increasing need for renewable, biodegradable composites that address both mechanical performance criteria and environmental concerns,[66,67] starch reinforced by cellulose fiber systems displays particular interest.[66,68,69] Starch shows a high solubility and biocompatibility profile, as well as unique thermo-mechanical properties, while cellulose fibers are known to simultaneously increase stiffness and enhance resistance to moisture and microorganism attacks.[67] These attributes have been widely used to manufacture biomedical and packaging materials.[3,67]

Nevertheless, as shown by an extensive review by Van der Sman,[3] the mechanics of short fiber food composites are less explored, even though the effect of fibers on the expansion properties of extruded foods is widely studied.[70] This may be due to the complexity induced by the typical rate-dependent, nonlinear constitutive behavior of foods.[71,72] The potentially random orientation and nonstandard shape of fibers in extrusion-cooked foods complicate further the applicability of analytical/numerical predictive models for fiber reinforcement mechanisms.[73] Yet fibers are critical components in foods not only due to their potential contribution to texture[74] but also due to their nutritional value.[3,75] They are found to reduce the risk of coronary heart disease, improve colon health, while also assisting with the dietary management of obesity.[73] Similar to other food additives,[76] fibers can be also used as inexpensive, low-calorie bulking agents, reducing the need for sugar, fat, and flour in order to increase water and oil reservation.[75,77]

Incorporation of fiber has been studied for baked products (*e.g.* bread crumb manufacturing[78]) and has also been applied to extruded foods. More importantly, the extent of reinforcement currently rests on performing physical trials in order to ensure that no undesirable textural changes occur with increasing the volume fraction of the fibrous phase.[75] In this regard, models of fiber-food matrix interactions would aid in optimizing various reinforcement parameters, *e.g.* fiber volume fraction, diameter, length, and stiffness, without physical prototyping. This would allow for adjusting texture with respect to different consumer group needs systematically, not only by changing the extrusion parameters but also by adjusting fiber-specific characteristics. Such challenges are continuously set especially in the pet food industry due to the remarkable diversity in the masticatory characteristics among pet breeds,[71] combined with demanding food quality criteria that need to be simultaneously met, *i.e.* nutritional completeness, nutritional balance, safety, digestibility, and, of course, palatability. Conclusively, there is a timely need for understanding and modeling the micro- and macromechanics of fiber-reinforced foods.[3]

9.4.2 Mechanical Characterization of Fiber Food Composites

The link between food composite structure and mechanical performance was investigated by Chanvrier *et al.*[74] Through compression tests on extruded starchy foams, it was shown that the maximum stress at rupture increases either by adding oat bran concentrate fibers or wheat bran fibers. These results were further correlated to foam structure characteristics *via* the application of the Gibson-Ashby theory.[57] Similar starchy foam composites were studied by Robin *et al.*[79] at varying extrusion conditions and wheat bran fiber concentrations. Via three-point bending tests, it was found that the impact of increasing fiber concentration on the rupture stress and elastic modulus properties depends on the cell wall thickness and fiber size. On the other hand, Xiang *et al.*[77] linked the microstructural properties of microcrystalline cellulose, extracted from oil palm fibers, to their corresponding rheological and chemical properties, while more recently Omar *et al.*[80] investigated the biodegradation of tensile properties of oil palm fibers.

Skamniotis *et al.*[73] measured the effect of cellulose fibers and minerals on the stress-strain and fracture toughness properties of cereal starch pet food extrudates, *via* the tests illustrated in Figure 9.11. Similar to fibers, minerals are also essential nutrients that contribute to various body activities in pets; *i.e.* formation of bone and cartilage, transport of oxygen into the bloodstream, as well as hormone production.[73] The study involved four compositions (w/w): pure starch, starch reinforced by 2.5% fibers, starch with added minerals at 5%, as well as starch with both added fibers (2.5%) and minerals (5%). These were compared with respect to their tensile stress-strain response at different strain rates $\dot{\varepsilon}$ (see Figure 9.12), as well as their toughness properties based on both the essential work of fracture (EWF) (see Figure 9.13a) and orthogonal cutting methodologies (Figure 9.13b), extensively described in Skamniotis *et al.*[81] and Williams *et al.*,[82] respectively.

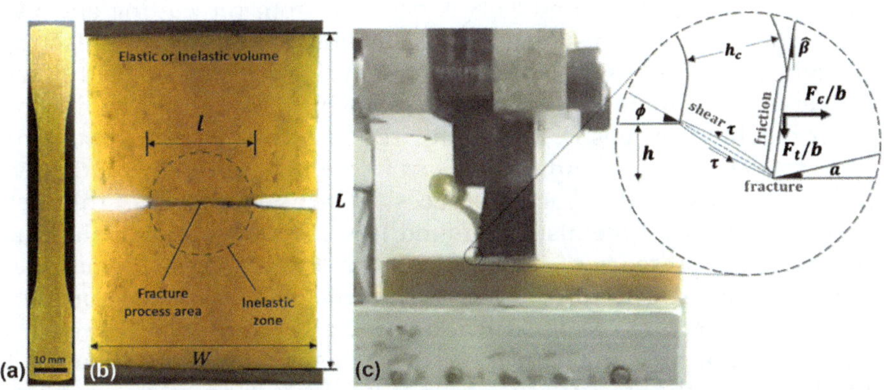

Figure 9.11 Experimental methodology: (a) dumbbell uniaxial tensile specimen; (b) DENT specimen used in EWF (l = 14 mm); (c) chip formation during orthogonal cutting—schematic of the analysis performed at the cutting tip. Reproduced from ref. 73 with permission from Elsevier, Copyright 2018.

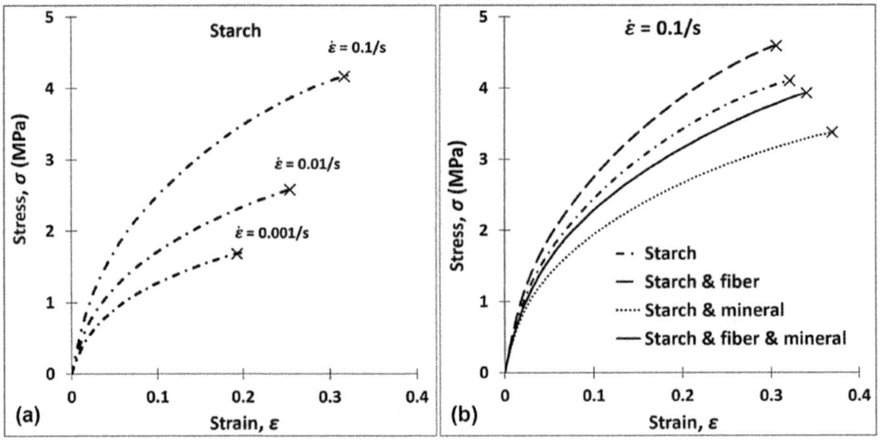

Figure 9.12 Uniaxial tensile stress-strain response until ultimate failure: (a) effect of strain rate for the "starch" composition; (b) effect of composition for a single rate of 0.1 s^{-1}.
Reproduced from ref. 73 with permission from Elsevier, Copyright 2018.

As shown in Figure 9.12a for pure starch, a strongly rate-dependent and nonlinear stress-strain response was found, while Figure 9.12b illustrates that adding fibers increased the overall stiffness but decreased the tensile failure strain ε_f (last data point x of each curve in Figure 9.12b), as opposed to minerals, which caused the opposite effect. On the other hand, Figures 9.13a and b indicate the effect of composition on fracture toughness, based on the EWF and cutting data, respectively. In both these graphs, the intercept of the lines gives estimates of the energy release rate G. Based on the difference between the extrapolation intercepts of the "starch & fiber & mineral" and "starch & mineral" recipes, a fiber-toughening effect of 8% and 21% is deduced from the cutting and EWF data, respectively. This discrepancy was attributed to a potential anisotropy in toughness, as well as the absence of fiber bridging effects in cutting due to fibers being broken/severed by the moving blade. The significantly higher toughness values in EWF compared to cutting were also explained by the viscous energy dissipation effects in the remote volume of the EWF specimens, outside the fracture process zone (see Figure 9.11b). Noteworthily, although minerals were found less effective in toughening than the fibers, the presence of both minerals and fibers gave the maximum toughness values. The former was attributed to the fiber pullout mechanism, promoted by the presence of minerals, being more effective in toughening than the fiber breakage mechanism. This was supported by SEM images of the fracture specimen faces between the starch & fiber & mineral (Figure 9.14a) and the starch & fiber (Figure 9.14b) compositions; Figure 9.14(a) clearly indicates larger fiber pullout lengths compared to Figure 9.14b and almost no evidence of fiber breakage.

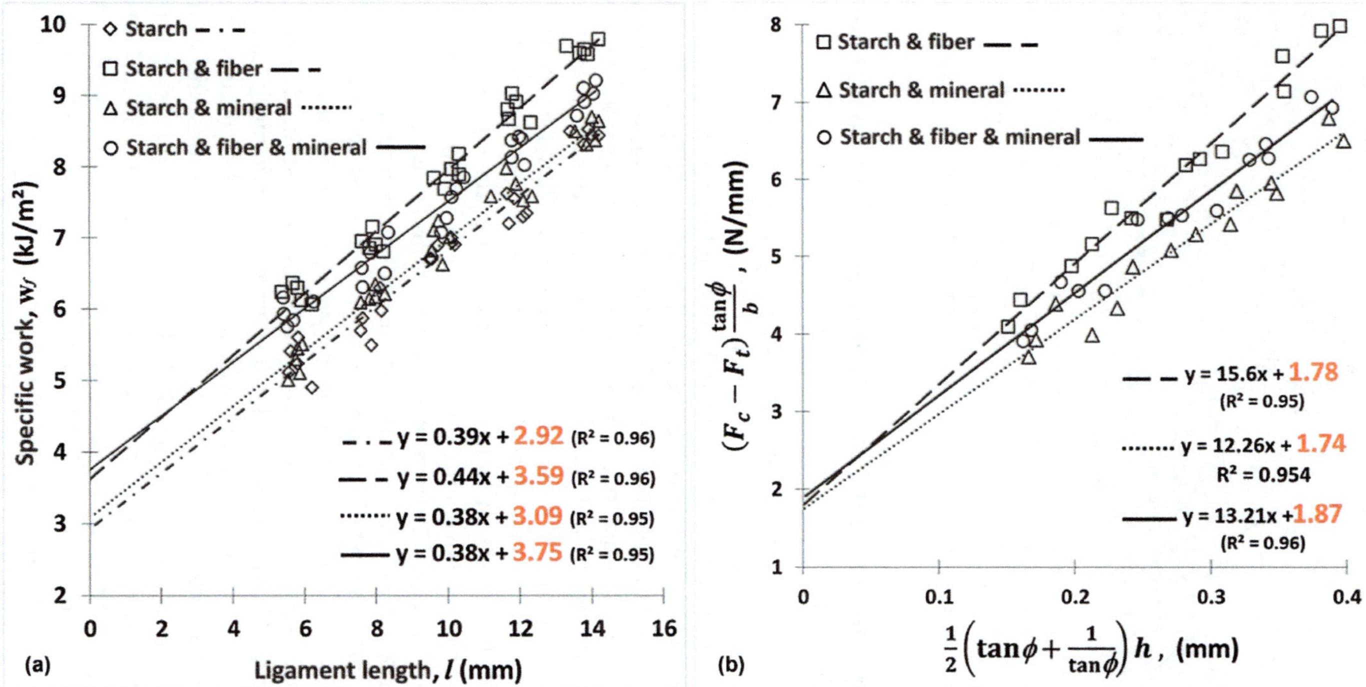

Figure 9.13 Extrapolation results obtained between four recipes based on: (a) EWF method; (b) cutting method. Intercepts of trend lines give values of G for each recipe.
Reproduced from ref. 73 with permission from Elsevier, Copyright 2018.

Figure 9.14 SEM observation of fiber pullout and fiber breakage mechanisms on fractured specimen faces for: (a) starch & fiber & mineral recipe—average pullout length 0.21 ± 0.08 mm; (b) starch & fiber recipe—average pullout length 0.15 ± 0.05 mm.
Reproduced from ref. 73 with permission from Elsevier, Copyright 2018.

9.4.3 Using Mechanical Properties to Simulate Food Oral Breakdown

Beyond understanding and quantifying the effect of fibers and other additives in foods, mechanical characterization also provides key input material parameters for numerical simulations of food oral breakdown. Such models are urgently needed in order to replace the time-consuming, costly sensory screening and challenges imposed in physically replicating the masticatory characteristics across consumers. However, despite the available literature in food characterization and material modeling, predicting realistic food oral breakdown remains a longstanding challenge.[83] Harrison *et al.*[84] predicted food breakdown in humans through a cutting-edge smoothed particle hydrodynamics (SPH) 3D model, which takes into account the complex interactions between saliva and the anatomical structures of the mouth (teeth, tongue, cheeks). Toward predicting aroma release, an extended model by Harrison *et al.*[85] also simulated the spillout of liquid caramel from an enclosing chocolate shell as the latter was compressed by the teeth. In both studies, the frictional conditions between interacting geometries were simplified, while the constitutive material laws studied included a Newtonian viscous fluid model, an elastic–perfectly plastic model, as well as a linear elastic-brittle model, coupled with a simple brittle fracture criterion.

Recently, Skamniotis *et al.*[86,94] developed an FE 3D oral breakdown model that incorporates rate dependency and the critical toughness parameter essential for fracture in the food constitutive response.[95] The model involved a soft and tough food item of a starch-cellulose fiber recipe, similar to the ones used by Skamniotis *et al.*[73,95] (described in Section 9.4.2), as well as the digital mandible and maxilla molar pet teeth geometries shown in Figure 9.15a. The numerical model's predictions were found to agree well with the total mandible force-displacement response (see Figure 9.15c),

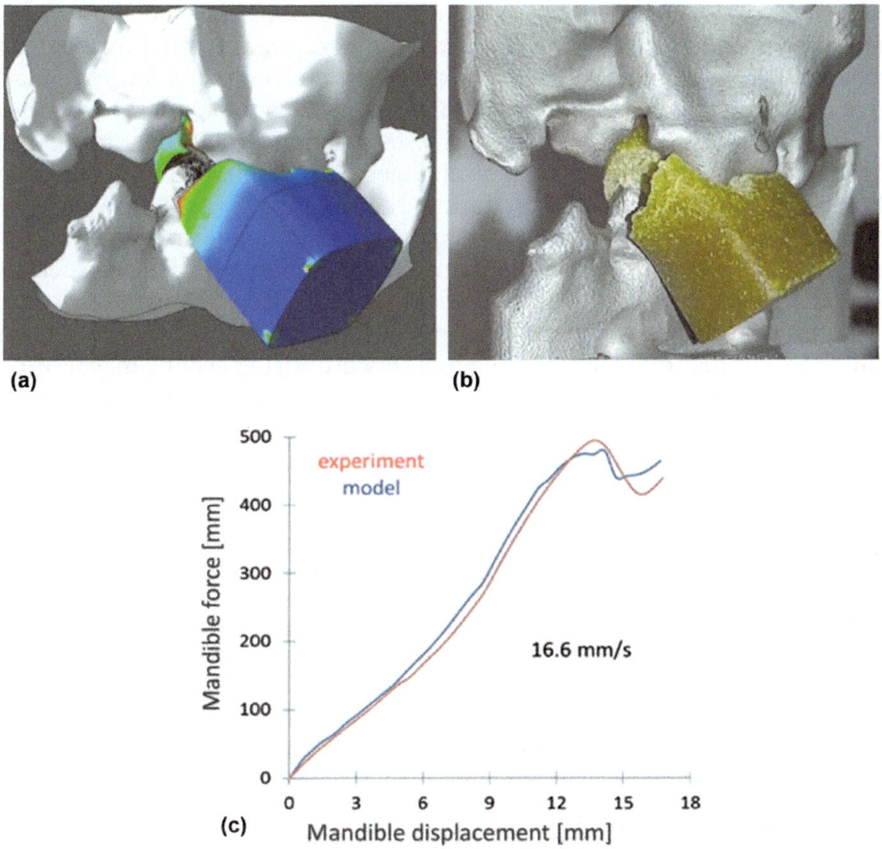

Figure 9.15 Experimental validation of the finite element food oral breakdown model : (a) model at maximum bite displacement; (b) experimental-physical replicate at maximum bite displacement; (c) experimental-model force-displacement agreement at mandible displacement rate of 16.6 mm s^{-1}.
Reproduced from ref. 94 with permission from AIP Publishing, Copyright 2017.

measured experimentally from the model's physical replicate displayed in Figure 9.15b. This has raised the potential for such a computational approach to be used toward establishing links between short fiber reinforcement parameters and food performance during mastication.

9.5 Conclusion and Future Trends

From the literature reviewed in the preceding sections, it is apparent that the link between the structure of foods and their continuum of response remains a very active research field. As biopolymers and their composites are extremely complex materials due to their multiscale microstructural features, most of the published literature has so far focused on experimental studies. This work was enabled through advances in experimental equipment, in particular imaging such as X-ray tomography and other microscopy techniques. It is now possible to perform *in situ* experiments on the small scale (*e.g.* inside a scanning electron microscope[50] or rheometer[87]) such that the bulk response is linked to the microstructural evolution. During the cooking processes, time-dependent changes in the microstructure can be observed at interrupted stages during the cooking.[88] As hardware and software continues to advance, 4D scanning techniques (time-resolved 3D imaging) will allow microstructural changes in dynamic processes such as ice cream and bread production to be captured in real time.[89,90]

At the same time, computational models with their power to perform multiscale simulations are emerging as a very promising avenue in exploring the missing link between the structure and continuum response, though they do require significant expertise and training to ensure that accurate and meaningful results are obtained.[95] For example, recently, molecular dynamics simulations were reported in investigations of the small strain rheological response of a particulate gel, using a so-called optimally windowed chirp as a boundary condition in order to speed up the computation.[91] The results were found to be consistent with the viscoelastic response as calculated by a fractional Kelvin-Voight constitutive model. Numerical models that can simulate damage and fracture in soft solid foods have also been developed.[50,86] In such work, the stiffness of severely loaded elements is degraded using a damage factor D, similar to the one used in modeling interfacial debonding[17] (see Figure 9.5). The issues here are concerned with the mesh dependency of such predictions of the damage and fracture processes, and perhaps these can be treated in the future through using nonlocal or gradient-based plasticity models.[92,95] Another important challenge concerns the presence of the water often found in soft foods. This complicates micromechanical studies in two ways. First, if a food product is a composite of two distinct phases or materials, the competition for water leads to unknown water absorption kinetics. This hinders the possibility of testing the two phases individually such that the material properties can be used as input in numerical simulations of the composite food. Second, if water (or another fluid) is present as well as solid phases in the food product, more advanced models are needed in order to consider solid–fluid interactions more accurately. Finally, another challenge remains in performing experimental studies for calibrating interface models (such as the cohesive zone Model) at the small scale, though techniques such as atomic force microscopy look promising,[93] as well as digital image correlation as reported in synthetic polymer composite studies.[37]

Developing food structure for mechanical performance is an exciting and rapidly evolving research field. It has attracted food scientists and technologists as well as engineers, mathematicians, and nutritionists both in academia and industry. The complexity of the science behind food structure and mechanical performance needs close collaboration between all these research communities. Collaborative and multidisciplinary research, empowered by advances in experimental equipment and computational power, will lead to substantial solutions and advancement of the state of the art, with potential benefits for human health as well as the environment.

Acknowledgements

The authors would like to acknowledge the Biotechnology and Biological Sciences Research Council (Grant BB/P023851/1) for funding the post-doctoral position of Dr Christos Skamniotis. The image shown in Figure 9.5 was enabled through a cryo-SEM purchased through an equipment grant by the Engineering and Physical Sciences Research Council (EP/C520629/1).

References

1. E. Dickinson, *Curr. Opin. Colloid Interface Sci.*, 2007, **4**, 155–157.
2. C. M. Lee, *J. Food Sci.*, 2002, **67**, 902–909.
3. R. G. van der Sman, *Adv. Colloid Interface Sci.*, 2016, **231**, 23–35.
4. M. A. Mohammed, L. Wanigasooriya, S. Chakrabarti-Bell and M. N. Charalambides, *J. Rheol.*, 2017, **61**, 49–65.
5. M. Kristiawan, L. Chaunier, G. Della Valle, A. Ndiaye and B. Vergnes, *Trends Food Sci. Technol.*, 2016, **48**, 13–26.
6. W. Xiao, M. N. Charalambides and J. Gordon Williams, *Int. J. Food Sci. Technol.*, 2007, **42**, 699–707.
7. E. Mitsoulis and S. G. Hatzikiriakos, *Food Bioprod. Process.*, 2009, **87**, 124–138.
8. E. B. Muliawan and S. G. Hatzikiriakos, *Int. Dairy J.*, 2008, **18**, 615–623.
9. E. Vandenberghe, M. N. Charalambides, I. K. Mohammed, B. De Ketelaere, J. De Baerdemaeker and J. Claes, *J. Food Eng.*, 2017, **208**, 1–10.
10. S. M. Goh, M. N. Charalambides and J. G. Williams, *Eng. Fract. Mech.*, 2005, **72**, 931–946.
11. B. Le Révérend, F. Saucy, M. Moser and C. Loret, *Physiol. Behav.*, 2016, **165**, 136–145.
12. P. J. Lillford, *J. Texture Stud.*, 2011, **42**, 130–136.
13. K. C. Drechsler and M. J. Ferrua, *Food Res. Int.*, 2016, **88**, 181–190.
14. M. Jekle and T. Becker, *Crit. Rev. Food Sci. Nutr.*, 2015, **55**, 369–382.
15. C. Ferrero, *Food Hydrocolloids*, 2017, **68**, 15–22.
16. R. I. Tanner, S. Uthayakumaran, F. Qi and S. Dai, *J. Cereal Sci.*, 2011, **54**, 224–228.
17. M. A. Mohammed, E. Tarleton, M. N. Charalambides and J. G. Williams, *J. Rheol.*, 2013, **57**, 249–272.
18. P. P. Camanho, C. G. Davila and M. F. De Moura, *J. Compos. Mater.*, 2003, **37**, 1415–1438.
19. E. Dickinson, *Food Hydrocolloids*, 2012, **28**, 224–241.
20. R. E. Carter, *Rheology of Food, Pharmaceutical and Biological Materials with General Rheology*, Elsevier Applied Science, 1990.
21. P. R. Ramel and A. G. Marangoni, *Food Res. Int.*, 2018, **107**, 110–118.

22. T. J. Faber, A. Jaishankar and G. H. McKinley, *Food Hydrocolloids*, 2017, **62**, 325–339.
23. S. K. Reinke, F. Wilde, S. Kozhar, F. Beckmann, J. Vieira, S. Heinrich and S. Palzer, *J. Food Eng.*, 2016, **174**, 37–46.
24. C. Yang, Y. Wang, T. Vasanthan and L. Chen, *Food Hydrocolloids*, 2014, **40**, 225–236.
25. I. M. Geremias-Andrade, N. P. Souki, I. C. Moraes and S. C. Pinho, *LWT–Food Sci. Technol.*, 2017, **86**, 166–173.
26. Food shortages could force world into vegetarianism, warn scientists. Available at https://billtotten.wordpress.com/2012/08/ (accessed April 2019).
27. E. A. Foegeding, *Food Biophys.*, 2006, **1**, 41–50.
28. C. D. Munialo, E. van der Linden, K. Ako and H. H. de Jongh, *Food Hydrocolloids*, 2015, **49**, 104–117.
29. V. Urbonaite, S. van der Kaaij, H. H. de Jongh, E. Scholten, K. Ako, E. van der Linden and L. Pouvreau, *Food Hydrocolloids*, 2016, **56**, 334–343.
30. P. Rosa, G. Sala, T. O. Van Vliet and F. Van De Velde, *J. Texture Stud.*, 2006, **37**, 516–537.
31. Van der Poel C, *Rheol. Acta*, 1958, **1**, 198–205.
32. E. A. Gwartney, D. K. Larick and E. A. Foegeding, *J. Food Sci.*, 2004, **69**, S333–S339.
33. R. K. Richardson, G. Robinson, S. B. Ross-Murphy and S. Todd, *Polym. Bull.*, 1981, **4**, 541–546.
34. S. B. Ross-Murphy and S. Todd, *Polymer*, 1983, **24**, 481–486.
35. K. R. Langley, M. L. Green and B. E. Brooker, in *Food Polymers, Gels and Colloids*, ed. E. Dickinson, Royal Society of Chemistry, 1991, Special Publication Number 82, p. 383.
36. K. R. Langley and M. L. Green, *J. Texture Stud.*, 1989, **20**, 191–207.
37. R. Zhang, J. Y. Li-Mayer and M. N. Charalambides, *Int. J. Fract.*, 2018, **211**, 125–148.
38. T. van Vliet, H. Luyten and P. Walstra, in *Food Polymers, Gels and Colloids*, 1991, vol. 82, pp. 392–403.
39. T. van Vliet, H. Luyten and P. Walstra, in *Food Colloids and Polymers, Stability and Mechanical Properties*, ed. E. Dickinson and P. Walstra, Woodhead Publishing, 1993, vol. 113, p. 175.
40. C. Gamonpilas, M. N. Charalambides and J. G. Williams, *J. Mater. Sci.*, 2009, **44**, 4976–4986.
41. A. E. Forte, F. D'Amico, M. N. Charalambides, D. Dini and J. G. Williams, *Food Hydrocolloids*, 2015, **46**, 180–190.
42. R. G. Van der Sman and J. Broeze, *J. Food Eng.*, 2013, **114**, 413–425.
43. I. S. Dogan, *Int. J. Food Sci. Technol.*, 2006, **41**, 569–576.
44. Y. Tsukakoshi, S. Naito and N. Ishida, *Food Res. Int.*, 2008, **41**, 909–917.
45. T. K. Taewee, *Int. Food Res. J.*, 2011, **18**, 855–866.
46. T. Van Dyck, P. Verboven, E. Herremans, T. Defraeye, L. Van Campenhout, M. Wevers, J. Claes and B. Nicolai, *J. Food Eng.*, 2014, **123**, 67–77.
47. J. Gao, Y. Wang, Z. Dong and W. Zhou, *Int. J. Food Sci. Technol.*, 2018, **53**, 858–8872.

48. E. Wilderjans, A. Luyts, K. Brijs and J. A. Delcour, *Trends Food Sci. Technol.*, 2013, **30**, 6–15.
49. I. K. Mohammed, M. N. Charalambides, J. G. Williams and J. Rasburn, *J. Mater. Sci.*, 2013, **48**, 2462–2478.
50. I. K. Mohammed, M. N. Charalambides, J. G. Williams and J. Rasburn, *Innovative Food Sci. Emerging Technol.*, 2014, **24**, 48–60.
51. S. Butt, I. K. Mohammed, V. Raghavan, J. Osborne, H. Powell and M. N. Charalambides, *J. Food Eng.*, 2018, **238**, 112–121.
52. R. Sundara, *New Food Magazine*, 2012, **15**, 56–60.
53. G. M. Glenn, W. J. Orts and G. A. Nobes, *Ind. Crops Prod.*, 2001, **14**, 201–212.
54. G. Roudaut, C. Dacremont, B. V. Pàmies, B. Colas and M. Le Meste, *Trends Food Sci. Technol.*, 2002, **13**, 217–227.
55. H. Luyten, J. J. Plijter and T. Van Vliet, *J. Texture Stud.*, 2004, **35**, 445–492.
56. E. Çarşanba, K. Duerrschmid and G. Schleining, *J. Food Eng.*, 2018, **229**, 93–101.
57. L. J. Gibson and M. F. Ashby, *Cellular Solids, Structure and Properties*, Cambridge University Press, 1999.
58. L. Schoeman, P. Williams, A. du Plessis and M. Manley, *Trends Food Sci. Technol.*, 2016, **47**, 10–24.
59. S. Devahastin, *Food Microstructure and Its Relationship with Quality and Stability*, Woodhead Publishing, 2017.
60. S. A. Alam and N. Sozer, Nondestructive Imaging of Cellular Solid Foods, in *Imaging Technologies and Data Processing for Food Engineers*, Springer, Cham, 2016, pp. 169–196.
61. Z. Wang, E. Herremans, S. Janssen, D. Cantre, P. Verboven and B. Nicolaï, *Annu. Rev. Food Sci. Technol.*, 2018, **9**, 323–343.
62. E. Maire, A. Fazekas, L. Salvo, R. Dendievel, S. Youssef, P. Cloetens and J. M. Letang, *Compos. Sci. Technol.*, 2003, **63**, 2431–2443.
63. S. Guessasma, L. Chaunier, G. Della Valle and D. Lourdin, *Trends Food Sci. Technol.*, 2011, **22**, 142–153.
64. R. Kadowaki, H. Kimura and N. Inou, *J. Texture Stud.*, 2016, **47**, 3–13.
65. Q. Qin and J. Ye, *Toughening Mechanisms in Composite Materials*, Elsevier, 2015.
66. A. Taguet, M. N. Bureau, M. A. Huneault and B. D. Favis, *Carbohydr. Polym.*, 2014, **114**, 222–229.
67. Y. Wan, H. Luo, F. He, H. Liang, Y. Huang and X. Li, *Compos. Sci. Technol.*, 2009, **69**, 1212–1217.
68. F. Zhao, D. Yao, R. Guo, L. Deng, A. Dong and J. Zhang, *Nanomaterials*, 2015, **5**, 2054–2130.
69. N. H. Zakaria, N. Muhammad and M. M. Abdullah, *IOP Conf. Ser.: Mater. Sci. Eng.*, 2017, **209**, 012087.
70. F. Robin, C. Dubois, N. Pineau, H. P. Schuchmann and S. Palzer, *J. Food Eng.*, 2011, **107**, 80–89.
71. C. Skamniotis, M. Elliott and M. N. Charalambides, *Phys. Fluids*, 2017, **29**, 121610.

72. A. Fuongfuchat, N. Seetapan, T. Makmoon, W. Pongjaruwat, P. Methacanon and C. Gamonpilas, *J. Food Eng.*, 2012, **109**, 571–578.
73. C. Skamniotis, Y. Patel, M. Elliott and M. N. Charalambides, *Food Hydrocolloids*, 2018, **84**, 515–528.
74. H. Chanvrier, F. Desbois, F. Perotti, C. Salzmann, S. Chassagne and J.-C. Gumy, *et al.*, *Carbohydr. Polym.*, 2013, **98**, 842–853.
75. M. Elleuch, D. Bedigian, O. Roiseux, S. Besbes, C. Blecker and H. Attia, *Food Chem.*, 2011, **124**, 411–421.
76. C. Gamonpilas, W. Pongjaruvat, P. Methacanon, N. Seetapan, A. Fuongfuchat and A. Klaikherd, *J. Food Eng.*, 2013, **114**, 262–268.
77. L. Y. Xiang, M. A. P. Mohammed and A. S. Baharuddin, *Carbohydr. Polym.*, 2016, **148**, 11–20.
78. A. Sangnark and A. Noomhorm, *Food Res. Int.*, 2004, **37**, 66–74.
79. F. Robin, C. Dubois, D. Curti, H. P. Schuchmann and S. Palzer, *Food Res. Int.*, 2011, **44**, 2880–2888.
80. F. N. Omar, H. S. Hafid, A. S. Baharuddin, M. A. Mohammed and J. Abdullah, *Planta*, 2017, **246**, 567–577.
81. C. Skamniotis, M. Kamaludin, M. Elliott and M. N. Charalambides, *Polymer*, 2017, **117**, 167–182.
82. J. Williams and Y. Patel, Fundamentals of cutting, *Interface Focus*, 2016, **6**, 20150108.
83. X. Wang and J. Chen, *Curr. Opin. Colloid Interface Sci.*, 2017, **28**, 22–30.
84. S. M. Harrison and P. W. Cleary, *Eur. Food Res. Technol.*, 2014, **238**, 185–215.
85. S. M. Harrison, P. W. Cleary, G. Eyres, M. D. Sinnott and L. Lundin, *Food Funct.*, 2014, **5**, 2792–2805.
86. C. Skamniotis, Y. Patel, M. N. Charalambides and M. Elliott, *Interface Focus*, 2016, **6**, 20160005.
87. J. Läuger and P. Heyer, *Annu. Trans. Nord. Rheol. Soc.*, 2011, 19.
88. H. M. Chong, I. K. Mohammed, B. Linter, R. Allen and M. N. Charalambides, *LWT – Food Sci. Technol.*, 2017, **86**, 148–158.
89. P. Rockett, S. Karagadde, E. Guo, J. Bent, J. Hazekamp, M. Kingsley, J. Vila-Comamala and P. D. Lee, *IOP Conf. Ser.: Mater. Sci. Eng.*, 2015, **84**, 012076.
90. P. Babin, G. Della Valle, H. Chiron, P. Cloetens, J. Hoszowska, P. Pernot, A. L. Réguerre, L. Salvo and R. Dendievel, *J. Cereal Sci.*, 2006, **43**, 393–397.
91. M. Bouzid, B. Keshavarz, M. Geri, T. Divoux, E. Del Gado and G. H. McKinley, *J. Rheol.*, 2018, **62**, 1037–1050.
92. T. Svedberg and K. Runesson, *Int. J. Plast.*, 1997, **13**, 669–696.
93. S. Cárdenas-Pérez, J. J. Chanona-Pérez, J. V. Méndez-Méndez, G. Calderón-Domínguez, R. López-Santiago and I. Arzate-Vázquez, *Innovative Food Sci. Emerging Technol.*, 2016, **34**, 234–242.
94. C. G. Skamniotis, M. N. Charalambides, M. Elliott, in *AIP Conference Proceedings 2017 Oct 16*, AIP Publishing, 2017, vol. 1896, p. 150002.
95. C. G. Skamniotis, M. Elliott and M. N. Charalambides, *Int. J. Plast.*, 2019, **114**, 40–62.

CHAPTER 10

Design Structures for Optimal Sensory Performance

MARINE DEVEZEAUX DE LAVERGNE* AND
BENJAMIN LE RÉVÉREND

Perception Physiology, Institute of Material Sciences, Nestlé Research, Lausanne, Switzerland
*Email: Marine.DevezeauxdeLavergne@rd.nestle.com

10.1 Introduction—What Is Sensory Performance?

Let us start this chapter by trying to explain what sensory performance is. Certainly, for most of us we would start by defining it as a food or beverage that tastes "good" or maybe "great." Examples in the classical literature are ubiquitous in the description of food taste and texture that are experienced by their protagonists for their sensory performance. Hans Castorp's exquisite meals in Mann's *Magic Mountain*,[1] Pantagruel's *festins* in Rabelais's *Quart Livre*,[2] and certainly the famous madeleine in Proust's *Swann's Way*[3] are written testimonies of the human sensory experience that derives from the integration of sensory signals by the human brain. We would then argue and use as basis for this chapter that *sensory performance* is how well the integration of these sensory signals compares with expectations shaped by previous experiences as well as cultural and societal standards.

Thus, sensory performance is not a matter of mechanical, chemical, or microstructural properties, which would be the ensemble of the physical properties of the food bolus Φ. It is rather a matter of sensory signalling from the peripheral nervous system for touch, taste, smell, audition, and vision

Food Chemistry, Function and Analysis No. 18
Handbook of Food Structure Development
Edited by Fotis Spyropoulos, Aris Lazidis and Ian T. Norton

Published by the Royal Society of Chemistry, www.rsc.org

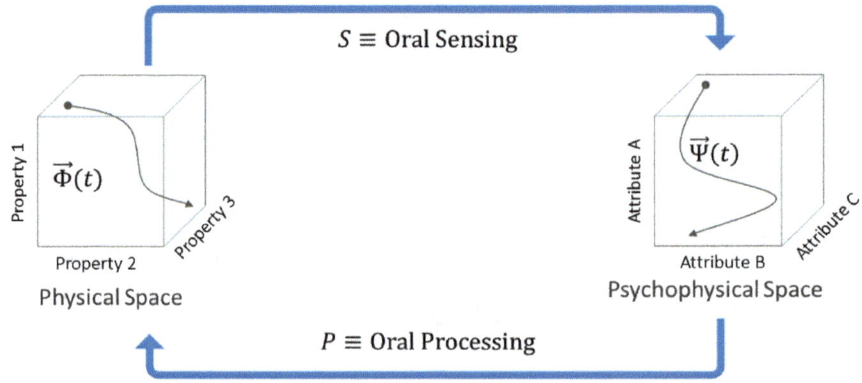

Figure 10.1 Simple scheme linking the physical and psychophysical spaces through oral sensing and processing.

sensations. These are the ensemble of sensory properties of the food Ψ, sum of electrical signals encoded after interaction of the food structures of/at different length scales with the relevant sensory physiological features. Let us call this operator of food *sensing* S, and thus $\Psi = S(\Phi)$. In the same way that tools are adapted depending on whether one wants to break a wall or a walnut, living creatures use sensory properties as insight for their eating behaviour both in the pre-oral[4] and oral[5,6] phases of feeding. The operator of food *processing* in mouth P, defining changes in physical properties of the food bolus, such as particle size, moisture content, and more generally $\mathrm{d}\Phi/\mathrm{d}t$, can be expressed as $\mathrm{d}\Phi/\mathrm{d}t = P(\Psi)$. Upon iteration of this sensing–processing cycle, both Ψ and Φ will vary over time to respectively describe either a sensory or a physical path (see Figure 10.1).

Describing analytically, numerically, or experimentally theses paths enables the targeting of an optimal sensory route by reverse engineering specific food structures at an initial condition $\Phi(t = 0)$ for the food oral process and to inform food engineers and scientists to build manufacturing processes and formulations leading to this $\Phi(t = 0)$ state, hence leading to the design of food structures for optimal sensory performance.[6,7]

10.2 How Do Humans Sense and Process Foods in the Oral Cavity?

10.2.1 Sensory Physiology (S)

At the root of our perception machinery is the peripheral sensory system, which is essentially the biological setup used to probe food materials at different scales. First, let us exclude vision from this chapter. Although certainly the first sense that we use to probe food materials, since it guides consumer choice towards specific products upon presentation of a large offering, it is also a sense that sensory analysis if often aiming to reduce, in order to remove bias and expectations of visual aspects on sensory perception. All the other senses, taste, touch, audition, olfaction, and trigeminal sensations are related to the *in-nose* and *in-mouth* perception of food material properties.

10.2.2 Olfaction

Olfaction through the orthonasal pathway (prior to ingestion) is the first chemical sensing pathway by which the food "chemicals" interact with its consumer. Volatile molecules, with a molecular weight typically below 300 Da,[8] partition from the food to the headspace above them according to Henry's law (1), linking the partial pressure in the gas phase p_i, the molar concentration in the liquid phase ξ_i^L, the volatility constant K_i, and travel by convection to the nasal cavity of consumers.

$$p_i = \xi_i^L \cdot K_i \tag{10.1}$$

In a similar fashion, after ingestion, oral processing, and swallowing, the pharyngeal mucosa is coated with the food bolus, and, upon breathing, volatile molecules are transported to the nasal cavity by convection.[9–11] This is the retronasal pathway.

Whether it is orthonasally or retronasally, the nasal cavity epithelium is lined with olfactory neurons, which express dedicated G-protein–coupled receptors (GPCRs) that have been identified in the early 1990s by Buck and Axel (1991).[12] About 900 different genes have been identified coding in this multigene family, of which about 600 are pseudogenes, not leading to the expression of a functional GPCR, leaving about 300–400 genes coding for functional olfactory receptors.[12] Each olfactory neuron is believed to project information *via* its axon in a dedicated glomerulus of the olfactory bulb of the brain. This model means that there is no data multiplexing before the olfactory bulb, behaving as a *labelled line* model. Although this seems like a very large number of functional units to encode volatiles' identity towards sensory perception, humans can recognize over 10 000 aroma compounds,[13] and a single cup of coffee alone contains over 500 volatiles.[14]

How can the human olfactory system recognize more molecules than it has receptors? It was eventually discovered that the sensory system uses a matricial code to recognize volatiles.[15,16] Essentially, each olfactory receptor can dock a number of ligands, whilst each ligand can bind to multiple receptors. This leads to a theoretical number of combinations equal to $\sum_{p=1}^{n} \binom{n}{p} = 2^n - 1 > 10^{n-4}$. If $n = 300$, this leads to a very large number of combinations, probably leading to little limitation of the olfactory capability to the available chemical space.

10.2.3 Taste

Research in the area of olfactory reception has greatly enlightened other senses, in particular taste, primarily in the way taste encoding and information trafficking has been modelled. Taste receptors and their corresponding genes were discovered in the early 2000s and can be divided into different families of proteins. A first family is G-protein–coupled receptors, such as the T2Rs family (~30) coding for bitter sensing,[17] as well as the dimers T1R2-T1R3 binding sugars[18,19] and other sweet ligands, and T1R1-T1R3 binding Na-glutamate, amino acids and other umami ligands.[19] Monosaccharides are also proposed to be perceived *via* the sodium glucose cotransporter 1 (SGLT1) and the sulfonylurea receptor 1 (SUR1), expressed in the sweet sensing cells in conjunction with T1R2-T1R3.[20] The two others taste qualities, salty and sour, are believed to involve ion channels conducting cations. The proposed candidates are, respectively, the epithelial sodium channel (ENaC) for saltiness[21] and the pH-sensitive polycystic kidney disease 2-Like 1 (PKD2L1) or, more recently, *via* a proton channel for sourness.[22] The taste cells expressing those receptors are commonly grouped in taste buds containing 50–100 cells,[23] and those taste buds are embedded in fungiform, foliate, and circumvallate papillae, which are visible and spread over the surface of the tongue.[24] Since all taste buds contain all types of cells, the whole area of the tongue is sensitive to all taste, contrary to the, unfortunately still commonly spread, belief that different zones on the tongue are responsible for sensing different tastes (Figure 10.2).[19]

The current model for the taste sensory system is also based on a *labelled line*. Each of these receptors or channels is expressed in dedicated cells, type 2 for GPCRs and types 1 and 3 for ion channels–based sensing, and upon depolarization of those cells, information travels to the brain *via* the facial (VII), glossopharyngeal (IX), and vagus (X) cranial nerve. The front of the tongue, rich in fungiform and foliate papillae, is mostly innervated by a part of the facial nerve called chorda tympani, whilst the back of the tongue, containing essentially foliate and circumvallate papillae, is innervated by the glossopharyngeal nerve.

10.2.4 Trigeminal and Texture

The trigeminal nerve conducts both touch and what are known as trigeminal sensations such as cooling, heating, tingling, and burning sensations. Trigeminal sensations are mostly coded *via* the large family of ions channels called Transient Receptor Potential channels, whose conductance to cations changes upon exposure to temperatures but also by specific chemical compounds (see Figure 10.3). For example, capsaicin triggers the activation of TRPV1, which is also activated by temperatures above 43 °C.[26] In the same way, menthol activates TRPM8, which is activated by temperatures below 25 °C.[27] The exact distribution of these channels on the surface of the tongue is not yet precisely reported, but current progress on microscopy and labelling will certainly uncover their location.[28]

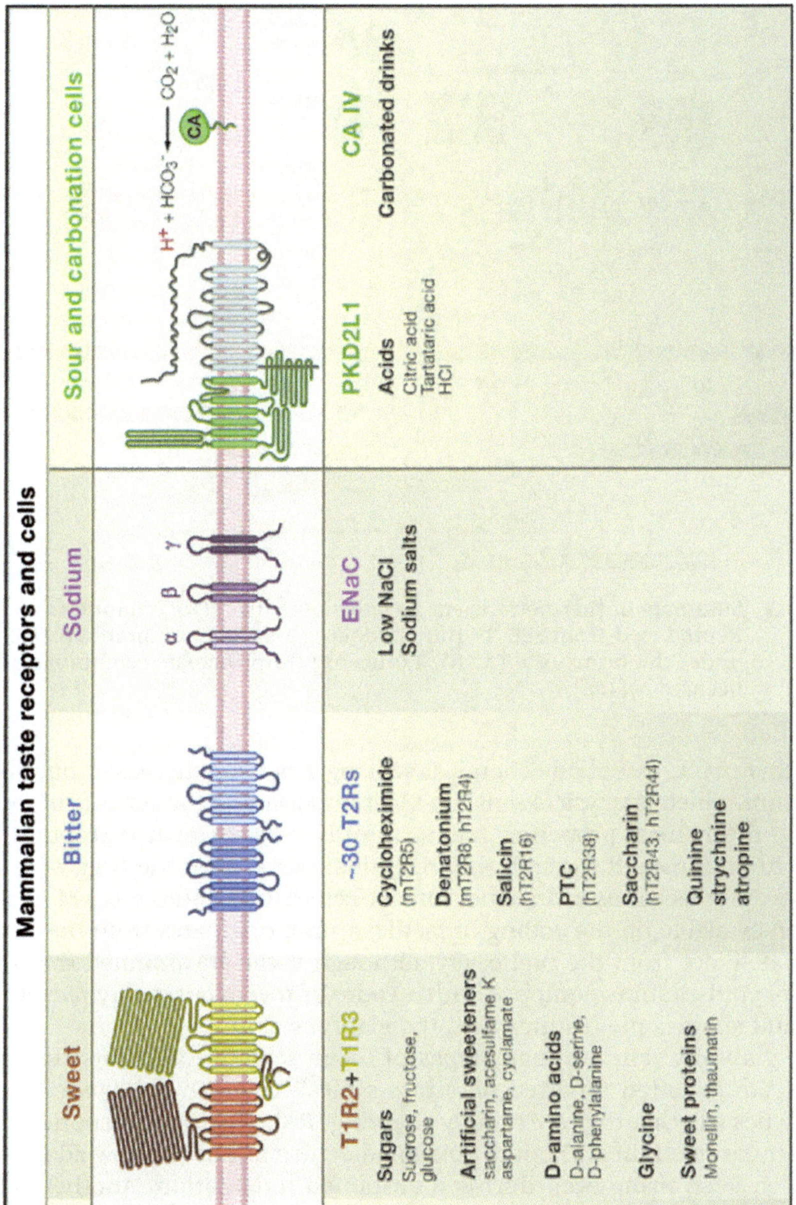

Figure 10.2 Summary of receptors for umami, sweet, bitter, and sour tastes. Schematic representation of taste receptors (and candidate receptors) mediating four of the five basic taste modalities.
Reproduced from ref. 25 with permission from Elsevier, Copyright 2009.

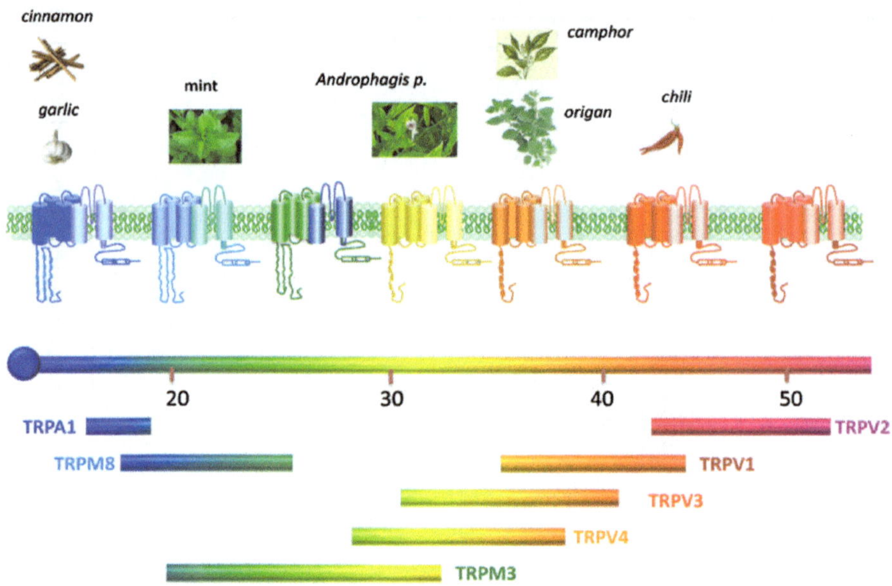

Figure 10.3 Summary of thermotransient receptor potential (TRP) channels. Reproduced from ref. 27 [https://doi.org/10.3390/membranes4030525] under the terms of a CC BY 4.0 license [https://creativecommons.org/licenses/by/4.0/].

Until this point, essentially chemical sensing has been discussed but also temperature, which is physical sensing. Most important to the sensing of food materials' mechanical properties (*i.e.* texture) is touch. Touch perception is believed to be conveyed through the trigeminal nerve, after the triggering of one or several mechanosensitive neurons or cell-neurite complexes. Most information available on the coding of tactile stimuli originates from the skin, unfortunately not from the oral cavity, although it can be assumed that the oral cavity epithelium is equipped with a cadre of mechanosensory receptors that should share some features with the glabrous skin.

In the glabrous skin are many types of touch-sensitive afferents that respond to strain (often reported on stress scales). One can subdivide these cellular types whether they are rapidly adapting (RA), those that respond only upon changes in strain and not to sustained deformations, or slow adapting (SA), which keep firing even during a sustained indentation. Another subdivision can be made depending on how far from the surface the cells are located. Meissner and Pacinian corpuscules are both SA and are, respectively, in the upper and deeper layers of the dermis. Similarly, the Merkel and Ruffini endings are both RA and, respectively, in the upper and deeper layers of the dermis. In addition to these dedicated structures, C-fibre free nerve endings also confer additional tactile sensing to the skin.[29]

Exploration of the human tongue by microneurography of single afferents in the lingual nerve revealed the presence of both SA and RA afferents firing with a threshold as low as tens of microNewtons.[30] Yet this report does not allow us to understand either the spatial organization or the coding of the food mechanical properties into texture. Since the tongue has a complex villous structure due to papillae, it has been hypothesized that filiform papillae could be at the centre of the coding of fluids' viscosity,[31] due to their deformation upon contact with food and beverages. The recent discovery of afferents expressing the stress-gated ion channel Piezo2[32,33] in the murine filiform papillae comforts this hypothesis.[34] Since the coding of texture perception is in its infancy, this will certainly be an exciting area for sensory and physical scientists to watch in the near future, as well as for food engineers.

To better understand the coding of texture in the oral cavity, one may consider the additional complexity coming from the fact that mechanosensory cells are embedded in non-sensory tissue about 10 cell diameters away from the skin surface.[29] It is therefore the oral mucosa deformation at the contact of the food material that will propagate to the mechanosensory cells, so that information on the mechanical properties of that food can be encoded as electrical signals. One can predict that the material properties that will be encoded by these cells will be vastly modulated by the tissue properties surrounding sensory cells.[35] For example, skin elasticity and hydration status have a large impact on tactile sensitivity.[36]

This is not exclusive to mechanical properties; the very topography of the surrounding tissue can have a vast influence on sensory transduction. In the skin, fingerprint structures have been found to strongly filter signals perceived by Pacinian corpuscles during typical surface scanning and probing by the human finger.[37,38] Similarly, the presence of filiform papillae at the surface of the tongue was shown to theoretically enhance the strain experienced by oral tissues, supporting their contribution to touch transduction in a direct manner.[31] This is of particular relevance since filiform papillae have often been linked to the texture perception function[39] but with no clear scientific data supporting this statement.

10.2.5 Food Oral Processing Physiology (*P*)

In addition to the perception of molecules through receptors, sensory performance also depends on food oral processing. Flavour molecules first release from the food structure and dissipate in air (retronasal aroma) or in saliva, to reach olfactory, or taste and trigeminal receptors, respectively. Moreover, perception of texture through mechanoreceptors requires the manipulation of food structures in mouth. The end goal of these complex processes is to swallow safely a broken down and lubricated bolus. In this section, oral manipulation and the role of saliva addition are discussed.

The "Philosophy of the Breakdown Path" by Hutchings and Lillford (1988)[5] suggests a different oral processing pathway depending on the initial lubrication and structure of the food. For instance, liquids require minimum oral manipulation to be transported to the back of the throat and swallowed. Soft solids require tongue–palate compression and can be comminuted by shear forces,[40–42] whereas hard solid foods require mastication by chewing. Thus, oral manipulation is required for all foods, but residence time in the mouth varies greatly. The main organs involved in food manipulation are the tongue, the soft tissues of the cheeks, the palate, the teeth, and the masticatory muscles (masseter, temporalis, pterygoid, and digastric). The teeth represent the occlusal area for food comminution, thus the number, position, and size of teeth have an impact on particle breakdown. Measurements of the development of the masticatory system in children have been studied, showing that palatal width and jawbone size increase drastically over the first year of life, then steadily until puberty, and a small increase can be observed from puberty until 18 years of age. This results in an inconsistent chewing sequence in toddlers.[43] The appearance of the 20 primary teeth happens between as early as 6 months of age until 3 years of age, while the 32 permanent teeth appear between 7 years of age until as late as 21 years of age (wisdom teeth). In addition to dentition state, the volume of the masticatory muscles defines bite force. For instance, elderly consumers suffer from a loss in muscle volume, and their bite force has been reported to decline.[44] Finally, the neuromuscular control of chewing allows for high chewing speeds and forces, which can exceed 1000 N,[45] while preventing teeth breakage and adapting chewing to food properties. Chewing can be monitored using a variety of techniques, *e.g.* electromyography (EMG) and kinetic jaw movement (KJM). Such techniques showed wide differences in mastication between foodstuff and individuals, often food hardness being the determinant parameter with higher chewing forces applied to break harder foods.[46]

During comminution by the teeth, as well as transport to the back of the throat and swallowing, the tongue plays a major role. Tongue strength has been correlated with the ability to break foods by tongue–palate compression.[47] Measuring tongue movement non-invasively is a challenge; nevertheless, it can be observed *in vivo* using ultrasound, articulography, videofluography, or X-ray techniques. These movements will determine the shear applied to the food in the mouth and thus the perception of food viscosity, as most foods elicit a non-Newtonian behaviour. The range of viscosities applied to food in mouth is predicted to be comprised between 10 and 1000 s^{-1};[48,49] however, lower shear rates are supposed to be more relevant for evaluation of thin products.[50] Indeed, oral movements often depend on food physical properties, such as viscosity, or taste, such as sweetness; however, panellists also modify oral movements when asked to rate a specific sensorial attribute.[51]

In addition to breakdown by mechanical forces, the incorporation of saliva into the foodstuff has a significant impact on the structural level of the food bolus formation as well as on a variety of reactions at the molecular level, as shown in Figure 10.4.[52] Saliva is produced by three major salivary glands (submandibular, sublingual and parotid) and several minor glands; thus, whole-mouth saliva is considered a mixed saliva. Saliva is mainly composed of water and only about 1 to 2 $mg\,mL^{-1}$ of salivary proteins, as well as electrolytes (sodium, potassium, calcium, magnesium, bicarbonate, and

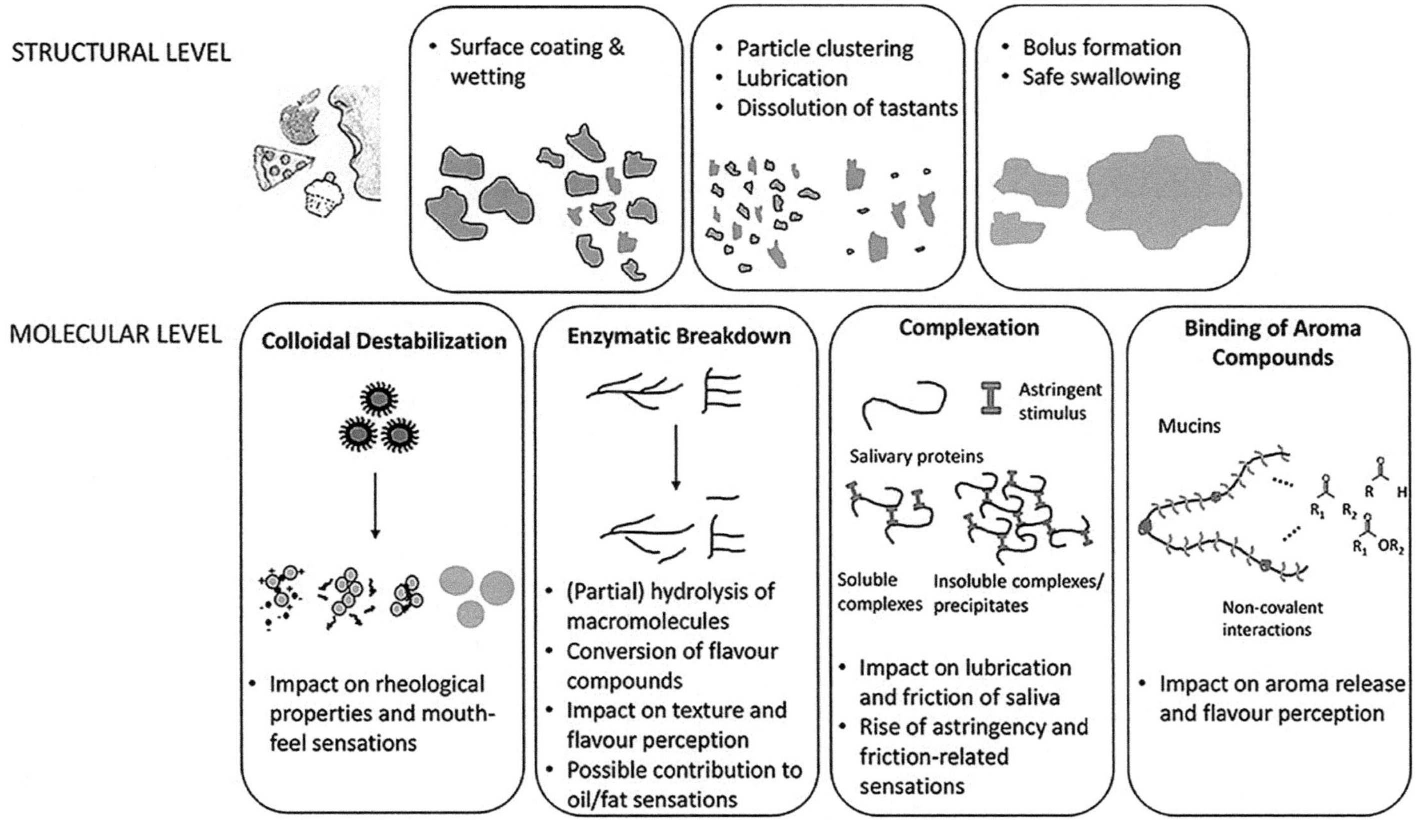

Figure 10.4 Saliva implications in food oral processing at the structural and molecular levels, as thoroughly described by Mosca and Chen (2017).
Reproduced from ref. 52 with permission from Elsevier, Copyright 2017.

phosphates) and enzymes (amylases, proteases, and lipases). This small amount of salivary proteins (especially the presence of mucins) is sufficient to generate the viscoelastic properties of saliva, which is an efficient lubricant.[53] Saliva's composition and flow vary with age, gender, health status and within individuals (circadian rhythm, stimulation, diet). Also, saliva can be stimulated by mechanical (*e.g.* chewing), taste (*e.g.* acid), and olfactory signals. Such stimulation increases salivary flow, especially from the parotid, and modifies saliva composition. For instance, stimulated saliva contains more proteins and bicarbonate and has been described to be less viscous than whole-mouth saliva,[54]

The amount of saliva added to the bolus depends foremost on the properties of the food. Often, saliva addition and mastication duration (or number of chews) are correlated.[55,56] A minimum amount of moisture has been hypothesized to be required before safely swallowing the food bolus, as was shown for wheat flakes bolus.[56] However, Motoi *et al.* (2013)[57] studied saliva addition to Dutch cakes and breads varying in initial moisture and showed that moisture at swallow increased with initial food moisture, even though the mastication time and the saliva added were less. This suggests that a target consistency rather than a target moisture content would trigger swallowing. In addition, fat-containing foods require less saliva addition before swallow, since fat contributes to bolus lubrication.[58]

Besides its rheological properties, saliva is a highly reactive bio-fluid that will modify the food matrix by enzymatic breakdown, colloidal destabilization, and complexation/binding of proteins and small molecules (*e.g.* tannins, aroma compounds).[52] For instance, amylases contribute to the digestion of starches and modify the saltiness perception and textural properties of starch-thickened foods and rice, respectively.[59,60] Another example is precipitation of tannins with proline rich proteins (PRPs), forming complexes that are insoluble in the stomach.[61,62] This precipitation is believed to cause the sensation of astringency in tannin-rich foods such as tea, wine, and some plant-based foods. Thus, the interaction of saliva with food should not be overlooked when linking food structure to sensory perception.

In conclusion, oral processing has a significant impact on the sensory performance of foods. However, several structural properties of foods can be engineered to influence oral residence time, bolus formation, and perception.

10.3 How to Reverse Engineer Sensory Performance?

In the oral cavity, the food properties dictate the type and intensity of interactions with the oral mucosa and teeth, both physically and chemically. Obtaining a detailed understanding of sensory perception at different length scales, beyond those of cells and receptors and their interplay with food properties, offers the opportunity to reverse engineer stimuli, micro-, meso-, and macrostructures and ultimately formulations and processes for optimal sensory performance.

10.3.1 Quantify Sensory Perceptions and Product Performance

As described in previous sections, sensory perception depends on numerous factors inducing inter- and intra-individual variations.[63] However, it is necessary to measure the sensory properties of products, with as little variation as possible, in order to understand and retro engineer sensory performance. A toolkit of established sensory methodologies exists and is continuously advanced through new research. Analytic tests (difference testing and descriptive analysis) focus on objective sensory characterization, while hedonic tests (preference or acceptance) evaluate the liking of consumers. Jointly, these can indicate the sensory performance target.

Sensory testing requires following so-called standard sensory practices regarding panellists (*i.e.* recruitment, training), samples (*i.e.* size, shape, temperature), test set-up (*i.e.* scales, attributes, coding, randomization, palate cleansing) and test area/environment (*i.e.* light, humidity, noise). Such standards have been described in detail elsewhere[64] and are not described in this chapter. Together with good sensory practice, the use of sturdy statistical analysis is key to avoid false negatives or false positives and to assess the variability of the measure. This section focuses on sensory test types and their contribution to a successful food structure design.

Analytic sensory testing started with early psychophysics in the 19th century. E. H. Weber introduced the concept that a stimulus variation required for a Just Noticeable sensorial Difference (JND) is proportional to the magnitude of the stimuli by a constant factor called the Weber fraction (Weber's law). G. T. Fechner complemented this law by a logarithmic relationship stating that a sensation intensity is a function of the logarithm (log) of the physical stimulus multiplied by the Weber faction (Fechner law, see Figure 10.5A). Later, more complex log–log relationships[65,66] were also introduced, but the Weber–Fechner laws remain a reference. Such laws

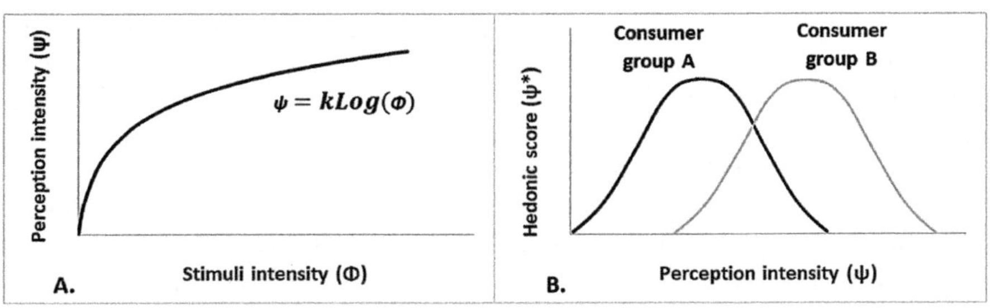

Figure 10.5 (A) Weber–Fechner law linking the intensity of a stimuli (property in the physical space: *e.g.* sucrose content, viscosity) to the intensity of a perception (attribute in the psychophysical space: *e.g.* sweetness, thickness) with a log relationship were k is the Weber fraction. (B) Wundt hedonic curve linking the hedonic score (intensity of liking of a product) to the intensity of a perception, indicating an optimal intensity depending on the consumer group selected.

implying a simple stimulus–response relationship have been criticized for not including the subjective experience of an assessor, which introduces biases and context effects. Nevertheless, the methodologies developed at that time set the basis for analytical sensory science as we know it today. Analytical tests allow answering several questions regarding a sample set, *e.g.* whether there is any difference between samples (difference testing) or how the samples differ (descriptive testing).

Difference testing allows, in most cases, concluding only whether a difference does exist between two products. A historical example is the triangle test, developed in the 1940s.[67] In this test, three products are presented to the panellist, two similar and one different, and the panellist must identify the different sample. Aside from the triangle test, many other methodologies are also used, namely, paired-comparison, duo-trio, sorting tests (tetrad, two out of five, Harris–Kalmus), dual standard, A-not-A, and ABX. All these tests ask panellists to recognize a control or to select/group samples based on differences. The set-up of each test imposes a different statistical analysis and number of repetitions. Often, the choice of test is based on the availability of products and panellists (usually 25–40), and on product-dependent fatigue from testing many samples simultaneously. Most methods do not specify the sensorial dimension of difference, except for the *n*-AFC test. When a difference exists between samples and it is not clear which dimension is concerned, descriptive sensory analysis provides further information.

Descriptive analysis started with the development of a Flavor Profile method in the late 1940s, involving extensive training of a panel, group discussion, and consensus. In the 1960s, the Texture Profile Analysis method was developed, similar to the Flavor Profile, with a focus on texture attributes.[68,69] Finally, the Quantitative Descriptive Analysis (QDA) method was developed by the Stanford Research Institute[70] and is still considered one of the gold standards of descriptive sensory analysis to date. QDA focuses on the independent judgement of each attribute using a scale and a statistical analysis of collected data. Moreover, QDA offers a broad description of a product, as the attributes are developed with the panel during training. This also implies that the results are context dependent. To avoid this contextual effect, the spectrum method involves extensive training with intensity references.[71] Often, industry uses a combination of descriptive methods, depending on the problem, product, and resources.

Descriptive analysis requires the use of sensory attributes that are relevant to the product tested. Often, appearance is rated first, followed by smell, taste, and texture, and then aftertaste and after-feel. Several glossaries and sensory wheels have been published,[72,73] but some attributes are still generated by QDA panels or through the use of consumer-friendly methods (*e.g.* Free Choice Profiling). As a guideline for texture attributes, Szczesniak[74] suggests an order of appearance for attributes rating throughout mastication, as part of attributes definition. Traditionally, descriptive sensory analysis is associated with the use of scaling to rate the intensity of attributes. Recently, the use of CATA (Check All That Apply) has proven interesting for descriptive sensory analyses by naïve panels. This method consists in selecting attributes that are relevant while tasting a product, from a predefined list.[75–77]

Since eating is a dynamic process involving the oral processing of foods, dynamic methods have been developed since the 1950s to include this time axis in sensory analysis. Time intensity focuses on a continuous intensity rating of one to two attributes,[78] and so does progressive profiling but at defined time points.[79] Since 2009, Temporal Dominance of Sensations (TDS) was developed. This dynamic method focuses on the selection of the most *dominant* attribute, often described as the sensation that catches the most attention, over the course of mastication and/or after-feel.[80] An advantage of TDS is the rating of up to 10 attributes in parallel with little requirement for training. It has been proven to correlate with QDA data while still being complementary.[81] Temporal Check-All-That-Apply (TCATA) is a temporal method based on CATA over consumption time, which allows selecting several attributes at a time, differently to TDS.

The second type of sensory assessment is hedonic testing involving consumers, which indicates the preference or liking of a product. Liking usually depends on several sensorial attributes, and each attribute's liking follows a Wundt hedonic curve, with an optimal intensity (see Figure 10.5B). Even though consumer data should be acquired through questionnaires and market analysis,[64] the sensory performance of a given product can be assessed only when consumers actually evaluate it. When assessing hedonic response, additional parameters, such as appropriateness (*e.g.* way of serving, time of the day, *etc.*), have to be taken into account. Mostly, blind testing with minimal product information (*e.g.* no branding) is used when assessing a sensorial change in a product. However, context and food-extrinsic cues, such as product branding and claims, are important to evaluate the response of consumers to a selected product positioning, which could improve product acceptability.[82]

The first type of hedonic tests is comparison testing. These tests compare two or more products with the selection of the most/least favourite one(s) or require these to be ranked based on preference. A second category of tests is acceptance testing, which involve a liking scale. The 9-point hedonic scale is a widely utilized method, but several scales comprising lines, labels, pictograms, and so on, are used. Standing out from others, the "just about right" scales combine hedonic and descriptive data, as they include liking of specific attributes and intensities (*e.g.* much too sweet). The combination of hedonic and descriptive data allows plotting a preference map, which is employed for the visualization of the liked products based on descriptive differences using multivariate statistics. For instance, Perrot *et al.*[83] found three clusters of coffee mix consumers, with different preferences for descriptive sensory attributes using multivariate analysis, as seen in Figure 10.6.

This toolkit of methods is not offering a list from which to pick a test and obtain an equivalent result. The choice of the type of sensory analysis is widely dependent on the objective of the test. For instance, to confirm

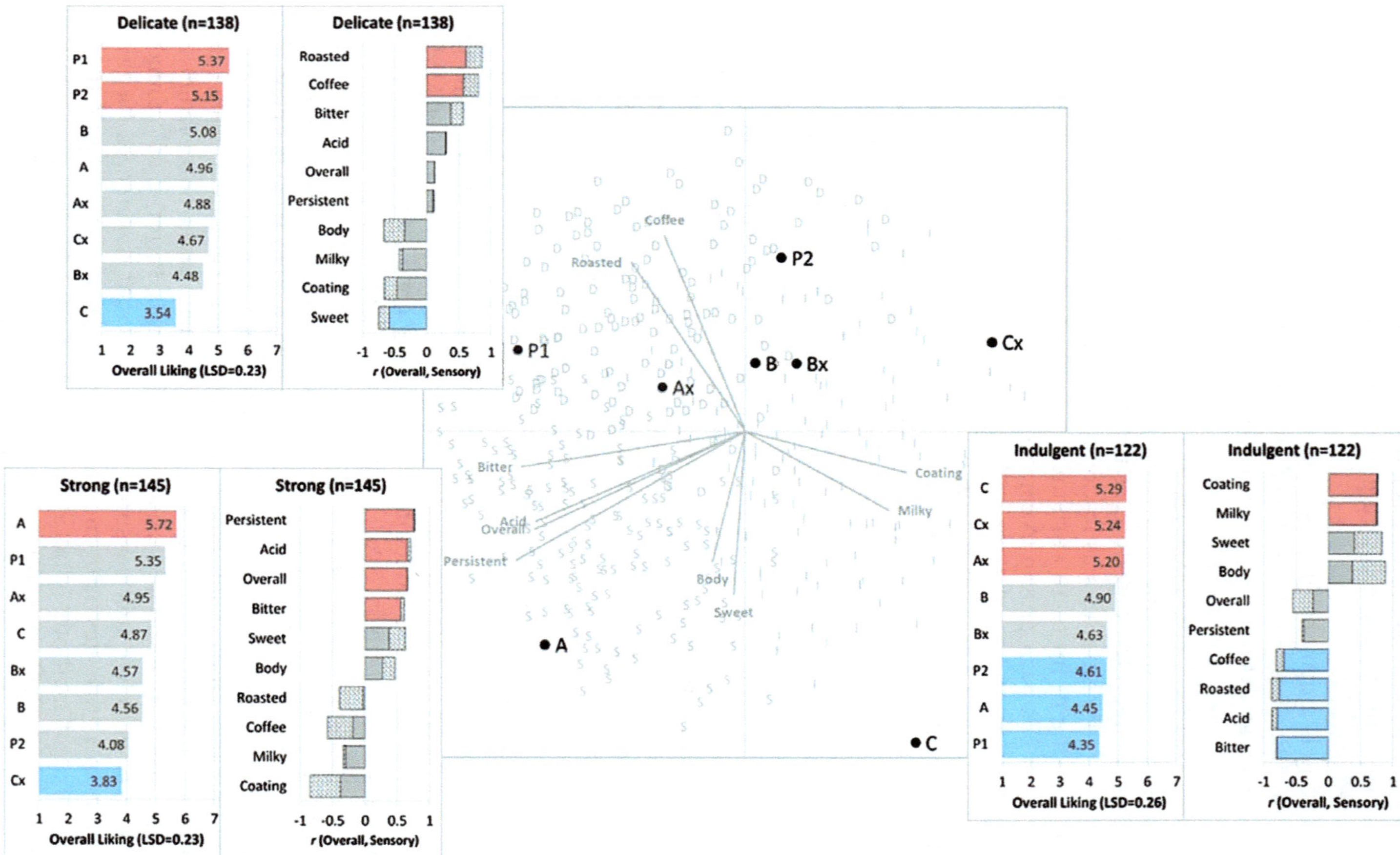

Figure 10.6 Internal Preference Mapping of coffee mixes with consumers coded according to *k*-means clustering (3 identified clusters: delicate, strong, and indulgent) and sensory attributes projected as supplementary variables. Mean overall liking and correlations (1st order and incremental for 2nd order) between overall liking and sensory attributes for all 3 clusters are given. Reproduced from ref. 83 with permission from Elsevier, Copyright 2018.

the perfect matching of a new product with the profile of an existing one, a difference test is appropriate. However, only a consumer test can confirm the necessity of a perfect match for consumer acceptance. Thus, analytical and hedonic tests both define sensory performance.

10.4 Foods as Time-dependent Stimuli for the Sensory System

10.4.1 Foods as Time-dependent Soft Matter Systems Under Mechanical Stresses

During mastication, solid foods are broken down in several small particles until the formation of a bolus that is safe to swallow. As already described, during this process, broken particles are stuck together thanks to the viscoelastic properties of saliva and should form a cohesive bolus. Of course, one can describe the mastication process on the human end, *e.g.* using biomechanics techniques to describe either the muscle activity,[84–86] jaw kinematics,[86,87] and even single-event bite force,[88–91] that can be linked with food size and its fracture strength to create diagrams determining the suitability of foods for specific populations.[92]

Conversely and most importantly in the context of this book, the transformations experienced by the food bolus during mastication can be directly described by food material science. Mastication reduces particle size according to both the ability of the chewer and the bolus material properties. Lucas and Luke[93,94] have described this process using milling engineering models based on two functions: (1) the selection function $S(x)$, the chance that a particle of size x is broken during a chewing cycle and the breakage function $B(y/x)$, and (2) the size distribution of subparticles of size y, following the selection of particles of size x. By iterating this series over many chewing cycles, one can fully describe the chewing process and predict bolus particle size distribution over time and, partially, its trigger for swallowing.[95] But is particle size the only swallowing trigger? Certainly not, since aiming to swallow a spoon of dry, fine powder would definitely be a challenge. Prinz and Lucas[96] proposed that for a safe swallow to occur, the cohesion force within the bolus must overcome the adhesion forces of individual particles. Quantitatively, the adhesion force scales with x, whilst the cohesion force scales with x^2 and inversely with the inter-particle distance within the bolus. Either of these can be calculated by Monte Carlo simulations,[96] so certainly qualitatively scaling with $x^2 \cdot \phi^{-1/3}$, ϕ being the particles' volume fraction in the bolus and therefore decreasing as saliva gets incorporated into the bolus. As the bolus breaks down, the net (cohesion less adhesion) force initially increases due to the particle size reduction and eventually decreases due to the large quantity of saliva being added. It is possible to hypothesize that chewing development involves not only motor control[43] but also sensing bolus rheology[56,95,97,98] to swallow at the global maximum of the bolus net force for maximal safety.

Although the changes that solid foods are subject to are very important with respect to bolus structure, liquid foods are also subject to mechanical stresses, and their resistance to these stresses is what essentially constitutes texture perception. The typical resistance to mechanical deformation is given by Young's modulus in elastic solids (at low deformation), viscosity in viscous liquids and storage and loss moduli in viscoelastic materials bridging the gap between the first two classes of materials.[99] This clearly highlights the relevance of rheology to study food texture.[100] In addition to the bulk properties that are typically measured by rheological techniques, an additional branch of mechanical measurements has been introduced to the food scientist's toolbox over the past 20 years.

Tribology[101] has been used to study the flow of matter in confined spaces, such as the narrow gap between tongue and palate, during the oral exploration of foods. This technique is interesting for many reasons. First, it offers the study of the flow of matter in geometries that can be soft themselves,[102–104] much like those in the oral cavity or even during the contact of *ex vivo* material, to attempt to mimic the human tongue surface with, for example, pig tongue.[105,106] In addition, tribology indicates the friction coefficient of the material between these surfaces[42] for a given entrainment speed. Given the physiology of mechanosensation previously described, it is likely that humans encode stress rather than viscosity as a sensory input. It is not impossible that the human tongue measures both normal and shear stresses using different cells at different locations and reports some kind of friction coefficient to the central nervous system. However, the method also has its limitations, given that the contact between the surfaces and the food material cannot mimic those of the human tongue; in terms of surface chemistry, wetting, mechanical properties, and topography, one can interrogate the quality of the biomimetic representation of the oral cavity only by a tribometer.[105] Also, the boundary conditions in a tribometer are less well known than those in a rheometer, *e.g.* gap, shear, and elongational stress, sometimes making data interpretation difficult and subject to conceptual models.[103]

The stresses that liquid foods experience in the oral cavity, whether approximated by rheometry or tribometry, have an impact on material properties due to disruption of polymer networks[49,176] or droplet coalescence.[154] It is therefore interesting to study such liquids that are very sensitive to the oral fluid mechanical environment as their transformation is a clear mark of the oral processing mechanical fingerprint.

10.4.2 Foods as Time-dependent Soft Matter Systems Under Biochemical Stresses

In addition to mechanical stresses in the oral environment, biochemical reactions also take place. One well known and important biotransformation in the oral cavity is the hydrolysis of starch by α-amylase. This, of course, has a major impact on solid cereal-based matrices where the presence of saliva can be linked to destructuring kinetics,[107] mastication patterns,[108] and even aroma release kinetics.[109] Of course, given that saliva is, in the case of bread and other cereal-based foods, the source of moisture, *i.e.* the fluid giving cohesion and the source of enzymatic breakdown, it is difficult in this case to disentangle the specific role of salivary amylase on food oral processing.

However, starch being also used as a thickener in many liquid foods, the specific action of salivary amylase in these systems is simpler to study. The enzymatic action of saliva is then often linked with a reduction in bolus viscosity, being halved within approximately 2–3 minutes.[110] This reduction of viscosity is directly linked with the textural perception of custards such as increasing melting mouthfeel or decreasing thickness.[111] Reversely, the addition of amylase inhibitor acarbose can be linked to the increase of thickness perception in starch-thickened custards in a dose-dependent fashion.[112]

10.4.3 Foods as Time-dependent Soft Matter Systems Under Thermal Stresses

In addition to mechanical and biochemical transformations occurring in the oral cavity, food materials undergoing a thermal phase change at the mouth temperature (35 °C) also drastically change structure transiently during their oral residence time. The two main examples that can be cited for this are chocolate and ice cream.

Focusing on the former as an example, chocolate melts in the oral cavity because cocoa butter, its main fat component, melts around 33 °C,[113] when well crystallized, into the sensorially desired polymorph.[114] Because of this melting, chocolate has a unique *food oral processing* characteristic, since it does not need to but can be chewed. Carvalho-da-Silva and coauthors (2011)[115] have shown that different population clusters were processing chocolate in different ways and could be classified in three categories: fast chewers, thorough chewers, and suckers. These clusters could be clearly differentiated thanks to their muscle activity, jaw kinematics, and oral residence time of the product. The *fast chewers* exhibit the shortest oral residence time (16 s), certainly due to faster mechanical breakdown of the material, enhancing heat transfer and phase change in small particles. At the opposite end of the scale, *suckers* relied almost exclusively on heat transfer to melt chocolate into a swallowable bolus, leading to an average oral residence time of 42 s. This clearly highlights the benefits of studying the oral breakdown of food to design food structures for different consumer targets.

10.5 Application of the Understanding of Food Sensing to Design Complex Structures Modulating Chemosensory Perception

In this section, the reader is exposed to various attempts to reverse engineer food structures based on the understanding of the processes

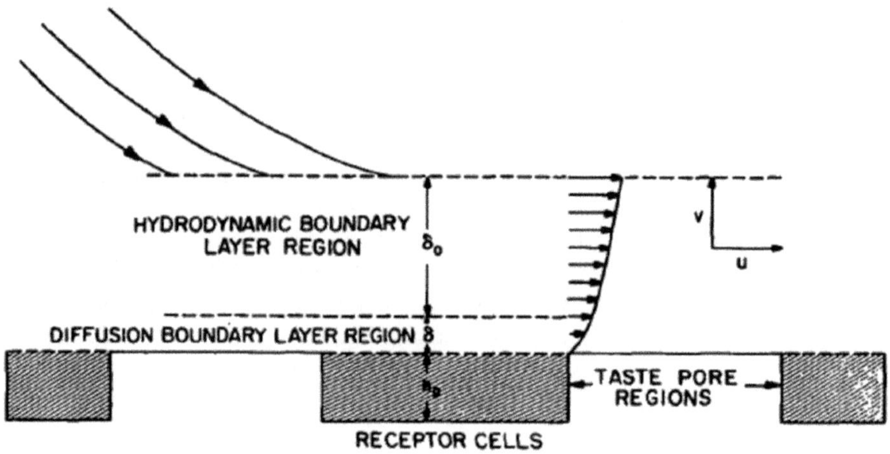

Figure 10.7 Schematic representation of taste stimulus transport from a flowing source to the receptor cells within the taste pore.
Reproduced from ref. 116 with permission from Oxford University Press.

occurring in the oral cavity leading to the optimization of chemosensory perception. Most of the examples discussed in this section have a common underlying principle, facilitating mass transport of taste and aroma-active molecules; this can be done in several ways, all based on a sound understanding of the transient interplay between the food structure and the oral mucosa.

Understanding taste sensing and flow in the oral cavity, it can be derived that taste-active molecules are convected to the boundary layer close to the surface of the tongue and then diffuse through a thin salivary layer that covers that oral surface.[116–119] By solving the equations of mass transfer, it appears that mass transfer is governed in part by the Reynolds number that is representative of the flow in the oral cavity and is thus modulated by both the velocity of the fluid and its viscosity. The neurophysiological and psychophysics community has thoroughly investigated the effect of oral velocity on taste perception[116] and has confirmed that, in line with theoretical predictions (see Figure 10.7), increasing the velocity of delivery increases taste perception. A more applicable approach for food material scientists to modulate taste perception and enable the reduction of taste-active compounds, such as sucrose and sodium, is to modulate fluid viscosity. Although the link between the addition of thickeners and taste perception has been known for decades,[120,121] the direct relationship between viscosity (in particular low shear viscosity) and taste perception, rather than the potential chemical interaction between the thickener and the tastants, has been evidenced only recently.[122,123] A reduction in the content of thickeners or in the application of thickeners with specific microrheological properties offers potential for sodium and sucrose reduction.[59]

In addition to modulating the bulk properties of fluids such as their viscosity, in the past decade, scientists have attempted to control the initial distribution of taste-active molecules within a product in order to tune their sensory properties by optimizing the contact time between the sensory active part of the product and the surface of the tongue. Successful reports of applying this approach in food systems are so far almost exclusive to solid foods. In solids, the heterogeneity can be kept for a certain amount of time during storage, and the polarity of the product can be identified from common consumer habits in order to introduce the tasting portion of the food to be in contact with the tongue.[124–126] Products exhibiting a topical application of the tastants are inherently heterogeneous, and one can also accelerate mass transfer in these cases by reducing the crystal size of taste-active crystals[127] or enhancing their intrinsic dissolution kinetics,[128,129] both of these examples definitely falling in material sciences either at the macromolecular or at the molecular scale. In liquid foods, since tastants are fully dissolved, enabling such heterogeneity is very difficult since both diffusion and convection are likely to equilibrate tastants' concentrations over the products' shelf life and storage. Yet research has shown that it was possible to tune sensory perception of liquids if one could deliver taste with a specific pattern,[130–132] although this seems highly dependent on the pattern delivered, and the results are sometimes not conclusive and only dependent on taste actives' content.[133] Most importantly, the results collected so far have been achieved only by using gustometers[134] in laboratory settings. How fluids can be structured and fight thermodynamic equilibrium? Clear theories enabling reverse engineering based on the kinetics of sensory perception have not been yet drawn, yet some examples based on double emulsions structuring have proven successful in enhancing taste perception,[135] based on controlled destructuring of the emulsion in contact with the oral cavity.

The control of in-mouth delivery of chemically active substances is not exclusive to taste. Since aroma release is activated by aroma molecules partitioning within the headspace in the oral cavity, aroma release is directly impacted by the viscosity of the food matrix.[136] Based on this, Lenfant and colleagues[137] hypothesized that the melting rate of chocolate should have a significant impact on its aroma release properties. Since melting can be optimized by adequate heat transfer between the heat source (the mouth) and the heat sink (the chocolate piece), the authors decided to use computational modelling techniques to develop a shape that would be a close fit to the human palate shape. This enhancement of the contact area between the oral cavity and the product enabled a reduction of the melting time and an increase in aroma being released, measured both by nose-space PTR-MS (proton transfer reaction–mass spectrometry) and by sensory analysis. This approach is a clear example of reverse engineering of in-mouth conditions for enhanced sensory perceptions, showing that smart food structuring can also be achieved at macroscopic scales and inspire a specific design.[138]

10.6 Application of the Understanding of Food Sensing to Design Complex Structures Modulating Mechanical Perception

This section focuses on the retro engineering of texture, which does not depend on the availability and sensing of a single molecule, contrary to chemosensing. Texture is a multimodal sensory perception resulting from a combination of food properties, simultaneously to oral manipulation.[40] proposed that the sensory space of custards was governed by a *melting-to-thick* dimension, related to viscosity, and a *creamy-to-rough* dimension, related to fat and particle perception. Retro engineering of viscosity, particles, and fat sensing have been extensively studied by modifying the continuous phase (*e.g.* hydrocolloids) or using phase-separated networks (emulsions, foams, *etc.*), and some of the main studies in this field are discussed here.

Thickness has been shown to relate to the viscosity of fluids with a log–log function of 0.3, which implies a much lower sensitivity than for chemical sensing (usually 0.6).[139,140] Just noticeable differences in thickness perception of maltodextrin solutions have been calculated at a Weber fraction of 0.26, which is higher than for taste (sucrose) but lower than for some olfactory stimuli (*e.g.* strawberry and caramel flavour molecules), indicating that humans could be more sensitive to in-mouth thickness than to aroma.[141] Even though human sensitivity to thickness and rheological or tribological settings for correlation to perception are still under discussion, several approaches in structuring fluids have proven successful for an improved thickness perception. The use of hydrocolloids, *e.g.* carboxymethylcellulose, dextrose, and guar gum, shows an increase in thickness perception, but shear-thinning hydrocolloids are perceived less thick if matched in viscosity at low shear rates,[139] shear rates at about 10^5 s^{-1} being more relevant for the evaluation of such fluids.[122] Measuring extensional viscosity was also suggested to improve the prediction of in-mouth thickness perception for solutions thickened with a mixture of xanthan and dextran.[50] Thickness is not completely uncoupled from creaminess, contrarily to what de Wijk *et al.*[40] suggested.[142] Indeed, increasing thickness of emulsions with thickeners increased creaminess perception.[143,144] However, emulsions thickened by xanthan and maltodextrins at iso-viscous concentrations were not perceived equally creamy,[145] suggesting that other properties of the hydrocolloids must be considered, including their lubricative attributes.[146]

Another food property that has a known impact on lubrication is the presence of particles. It has been shown that humans are able to perceive particles as little as 2 μm.[58] Particle sensing depends on the size,[147] shape,[148] mechanical properties, concentration, and dispersion media[149] of these particles. Simplistically, big, irregular, hard, concentrated particles, dispersed in low viscosity media, are most likely to be perceived as rough. Thus, limiting the roughness of particles in foods can be done by particle size reduction,[58] and the desired size depends on particles and matrix properties.

Moreover, small non-fat particles can be used as fat replacers to mimic the presence of oil droplets in an emulsion. An example of particle-based fat replacer is microparticulated whey proteins (MWP). In model emulsions and soft solids, MWP particles of about 1 μm were perceived creamy, possibly due to a ball-bearing mechanism, while MWP particles of about 10 μm were perceived rough,[150] confirming the concept that a maximum particle size should not be exceeded.

As shown by its modulation through the viscosity or particles, creaminess is a complex multimodal texture attribute. However, this attribute is primarily linked to the perception of fat in a food system.[151] Controlling creaminess is therefore linked to fat content and fat structuring in a formulation, even if it depends on other parameters including non-texture-related food properties (*e.g.* flavour). Increasing fat content increases creaminess perception;[143] however, the perception of increased fat content reaches a limit.[152] Moreover, it is desired to lower fat content in foods for formulating healthier food products by modulating other factors. Droplet size was not found to directly correlate with the perception of creaminess or even thickness in low-viscosity emulsions.[144,153] However, globule size is linked to the behaviour of emulsion droplets in the mouth and their ability to aggregate, coalesce,[154] and adhere to oral surfaces.[155] Thus, bigger particles are more likely to be perceived as creamy, but due to in-mouth behaviours that are also related to the properties of the emulsifiers used. Emulsions stabilized by modified starches are destabilized by amylases and coalesce in mouth, resulting in a creamy perception.[154] Charged emulsion droplets (*e.g.* stabilized by lactoferrins) are destabilized by salivary ions or salivary mucins, which results in salt-induced aggregation or bridging flocculation, respectively. It was hypothesized that such aggregation would increase creaminess perception, but irreversible flocs from bringing-flocculation were perceived as astringent, supposedly due to the loss of the lubrication of saliva or the generation of large gritty structures above 10 μm.[156,157] Contrarily, neutral or negatively charged emulsions lead to depletion flocculation in the presence of salivary mucin, which resulted in increased creaminess perception.[156] Similarly to in-mouth aggregation, aggregation of oil droplets in the food matrix increases the product viscosity and creaminess perception. In a protein-stabilized system, such aggregation can be reached by processing parameters (*e.g.* one- stage homogenization) and the destabilization by pH and ionic strength modulation,[158] or by the addition of particles or hydrocolloids that interact with oil droplets.[159]

Double emulsions allow maintaining mouthfeel by replacing part of the fat in oil droplets with smaller droplets of aqueous phase (water-in-oil-in-water: w/o/w). As long as the system is stable, such a replacement has no rheological and sensorial impact, since sensorial properties are governed by the properties of the outer droplets.[160] Even if the outer droplets of these emulsions tend to be bigger than regular o/w emulsions, double emulsions can be used in a wide variety of foods.[161] Another fat replacement strategy is the use of air bubbles. It is, for instance, known that many small air bubbles increase chocolate mousse creaminess.[147] Air bubbles are more challenging to stabilize than oil droplets in low-viscous foods, but this can be achieved through several strategies, *e.g.* Pickering stabilization, stabilization by a thick layer of denatured proteins, and stabilization by hydrophobins.[162] Such microbubble systems showed similar physical and sensorial properties as oil systems in high-concentration emulsions but not in dilute systems or in 1 : 1 replacement; thus a combination of oil droplets and microbubble seems the most promising approach.[162,163]

In solid foods, fat perception is impacted by its distribution and the properties of the food matrix. Two types of emulsion gels have been described: emulsion-filled gels, in which droplets are evenly distributed and that are dominated by the properties of the continuous matrix, and protein-stabilized emulsion gels, which are a type of particulate gel dominated by the properties of the network of aggregated emulsion droplets.[164] In emulsion-filled gels, it has been shown that matrix properties and droplets binding to the gelled matrix have an impact on fat perception.[152] Fracture strain of gelled emulsions mainly determined late chew-down texture attributes, with a low fracture strain being associated with grainy perception, while a high fracture strain was related to creamy perception.[165] Emulsion droplets containing solid fat and unbound droplets to the gel matrix also resulted in higher creaminess perception.[166] A last strategy reported in emulsion-filled gels is heterogeneous distribution of fat, which was proven to increase fat perception at a concentration of fat of 15%.[167]

All these examples show that the structuring of foods allows us to control and predict texture attributes (*e.g.* creaminess) beyond characterizing the food itself but also by considering interactions occurring during food oral processing (Figure 10.8).

10.7 Conclusions and Future Avenues for Research in Food Perception

The avenues for future research in this area are far from being fully explored. Certainly the expansion of this research field to the systematic reverse engineering of in-mouth perception offers avenues for industry to tackle specific problems such as sugar, sodium, and saturated fats reduction by tuning taste and texture perception. Of course, several experiments demonstrating the effects of in-mouth phenomena on chemosensory perception have been demonstrated, but a gap lies at least within the textural perception area. Indeed, the mechanosensory molecular targets, ultrastructure of the mechanosensing cellular units, and specific deformation patterns of the tongue during oral tactile exploration of foods are still not known. Therefore current reverse engineering approaches are based on linking food structures with mechanical properties measured either by using proxies for tactile sensing, such as rheology and tribology,[101] or by mimicking particular microstructures that are known to have specific sensory properties, such as

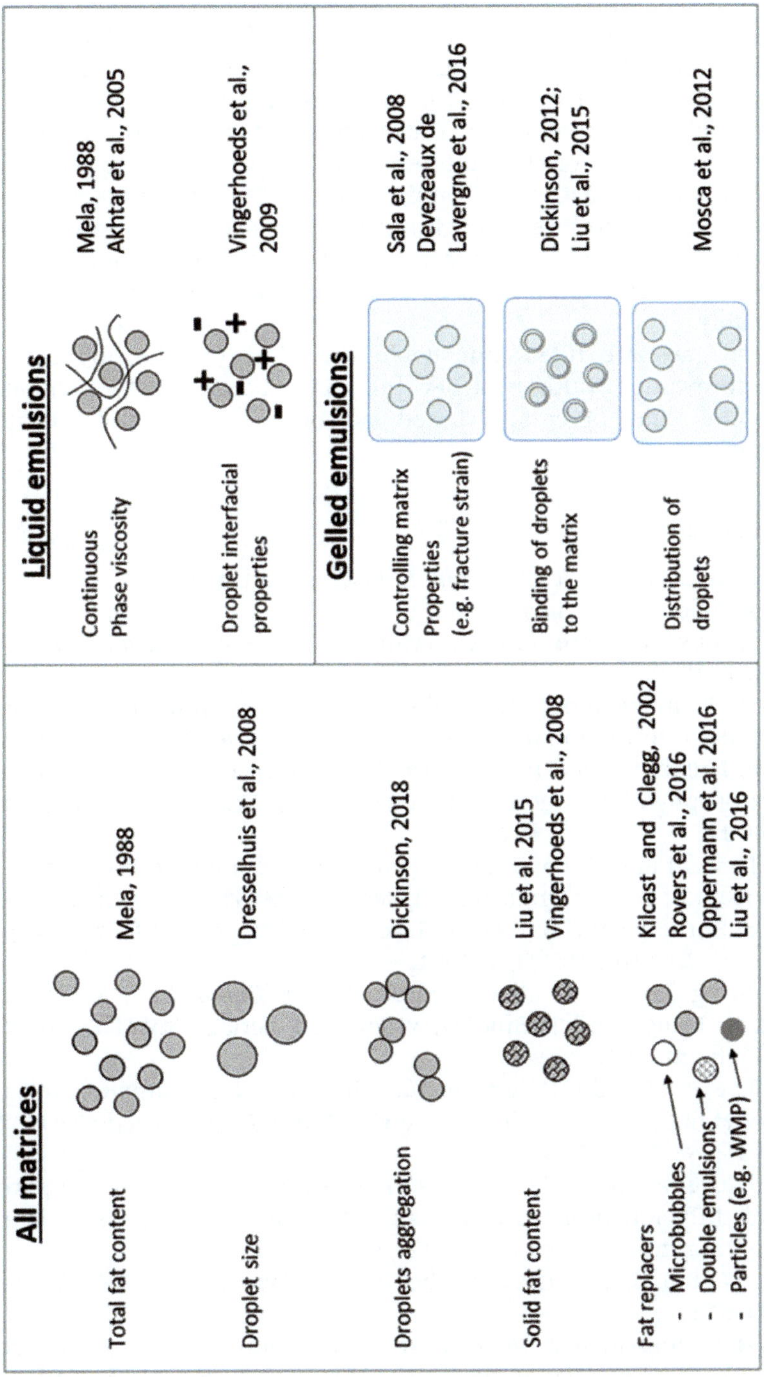

Figure 10.8 Summary of strategies to modulate fat-related sensory perception by structuring emulsions as described in this section.

microgels in lieu of emulsion droplets.[103] Defining the direct mechanisms of texture sensing will not only be a capital discovery within the life science area but will also enable us to direct *food material science* in developing the right supramolecular structures for enhanced textural sensations.

In addition to contributing to a better understanding of the behaviour of food materials in the oral cavity, the extension of models to the rest of the digestive tract is also of clear interest to industry.[168] Better qualifying the micromechanics of gastric[169,170] and intestinal[171,172] transport phenomena will also enable us to reverse engineer structures that have specific breakdown kinetics under gastrointestinal conditions for, amongst others, the targeted release of actives[173,174] or specific nutritional availability/functionality.[175]

If these gaps are filled, one could then imagine the application of global optimization routines of new product structures and accompanying formulations and processes, based on both sensory and nutritional performance.

References

1. T. Mann, *The Magic Mountain*, S. Fischer Verlag, 1924.
2. F. Rabelais, Quart-Livre, 1552.
3. M. Proust, Du coté de chez Swann, in *A la recherche du temps perdu*, vol. 1, 1913.
4. N. J. Dominy and J. D. Yeakel, *et al.*, How chimpanzees integrate sensory information to select figs, *Interface Focus*, 2016, **6**(3), 20160001.
5. J. B. Hutchings and P. J. Lillford, The perception of food texture – the philosophy of the food breakdown path, *J. Texture Stud.*, 1988, **19**(2), 103–115.
6. F. Lenfant and C. Loret, *et al.*, Perception of oral food breakdown. The concept of sensory trajectory, *Appetite*, 2009, **52**(3), 659–667.
7. A. S. Burbidge and B. J. D. Le Révérend, Biophysics of food perception, *J. Phys. D*, 2016, **49**(11), 114001.
8. S. Frings and J. Bradley, in *Transduction Channels in Sensory Cells*, ed. S. Frings and J. Bradley, Wiley-VCH Verlag GmbH & Co. KGaA, Weinheim, FRG, 2004.
9. C. Lauverjat and I. Déléris, *et al.*, Salt and aroma compound release in model cheeses in relation to their mobility, *J. Agric. Food Chem.*, 2009, **57**(21), 9878–9887.
10. C. Lauverjat and C. de Loubens, *et al.*, Rapid determination of partition and diffusion properties for salt and aroma compounds in complex food matrices, *J. Food Eng.*, 2009, **93**, 407–415.
11. D. M. Small, Flavor is in the brain, *Physiol. Behav.*, 2012, **107**(4), 540–552.
12. L. Buck and R. Axel, A novel multigene family may encode odorant receptors: a molecular basis for odor recognition, *Cell*, 1991, **65**(1), 175–187.

13. S. M. Boyle, S. McInally and A. Ray, Expanding the olfactory code by in silico decoding of odor-receptor chemical space, *eLife*, 2013, **2**, e01120.
14. S. K. Dart and H. E. Nursten, Volatile components, in *Coffee: Volume 1: Chemistry*, ed. R. J. Clarke and R. Macrae, Springer, 1985, pp. 223–265.
15. B. Malnic and J. Hirono, *et al.*, Combinatorial receptor codes for odors, *Cell*, 1999, **96**(5), 713–723.
16. L. B. Vosshall, A. M. Wong and R. Axel, An Olfactory Sensory Map in the Fly Brain, Cell, *Cell Press*, 2000, **102**(2), 147–159.
17. E. L. Beckett and C. Martin, *et al.*, Bitter taste genetics – the relationship to tasting, liking, consumption and health, *Food Funct.*, 2014, **5**(12), 3040–3054.
18. S. Damak and M. Rong, *et al.*, Detection of sweet and umami taste in the absence of taste receptor T1r3, *Science*, 2003, **301**(5634), 850–853.
19. J. Chandrashekar and M. A. Hoon, *et al.*, The receptors and cells for mammalian taste, *Nature*, 2006, **444**(7117), 288–294.
20. S. K. Sukumaran and K. K. Yee, *et al.*, Taste cell-expressed α-glucosidase enzymes contribute to gustatory responses to disaccharides, *Proc. Natl. Acad. Sci. U. S. A.*, 2016, **113**(21), 6035–6040.
21. J. Chandrashekar and C. Kuhn, *et al.*, The cells and peripheral representation of sodium taste in mice, *Nature*, 2010, **464**(7286), 297–301.
22. Y.-H. Tu and A. J. Cooper, *et al.*, An evolutionarily conserved gene family encodes proton-selective ion channels, *Science*, 2018, **359**(6379), 1047–1050.
23. N. Chaudhari and S. D. Roper, The cell biology of taste, *J. Cell Biol.*, 2010, **190**(3), 285–296.
24. G. K. Essick and A. Chopra, *et al.*, Lingual tactile acuity, taste perception, and the density and diameter of fungiform papillae in female subjects, *Physiol. Behav.*, 2003, **80**(2–3), 289–302.
25. D. A. Yarmolinsky, C. S. Zuker and N. J. P. Ryba, Common sense about taste: from mammals to insects, *Cell*, 2009, **139**(2), 234–244.
26. J. Grandl, S. E. Kim, V. Uzzell, B. Bursulaya, M. Petrus, M. Bandell and A. Patapoutian, *Nat. Neurosci.*, 2010, **13**, 708–714.
27. C. Ferrandiz-Huertas and S. Mathivanan, *et al.*, Trafficking of ThermoTRP Channels, *Membranes*, 2014, **4**(3), 525–564.
28. D. Alvarez-Berdugo and L. Rofes, *et al.*, Localization and expression of TRPV1 and TRPA1 in the human oropharynx and larynx, *Neurogastroenterol. Motil.*, 2016, **28**(1), 91–100.
29. E. A. Lumpkin, K. L. Marshall and A. M. Nelson, The cell biology of touch, *J. Cell Biol.*, 2010, **191**(2), 237–248.
30. M. Trulsson and G. K. Essick, Sensations evoked by microstimulation of single mechanoreceptive afferents innervating the human face and mouth, *J. Neurophysiol.*, 2010, **103**(4), 1741–1747.
31. E. Lauga, C. J. Pipe and B. Le Révérend, Sensing in the Mouth: A Model for Filiform Papillae as Strain Amplifiers, *Front. Phys.*, 2016, **4**, 35.

32. S. S. Ranade and S.-H. Woo, *et al.*, Piezo2 is the major transducer of mechanical forces for touch sensation in mice, *Nature*, 2014, **516**(7529), 121–125.
33. B. Coste and J. Mathur, *et al.*, Piezo1 and Piezo2 are essential components of distinct mechanically activated cation channels, *Science*, 2010, **330**(6000), 55–60.
34. Y. Moayedi, L. F. Duenas-Bianchi and E. A. Lumpkin, Somatosensory innervation of the oral mucosa of adult and aging mice, *Sci. Rep.*, 2018, **8**(1), 9975.
35. D. R. Lesniak and K. L. Marshall, *et al.*, Computation identifies structural features that govern neuronal firing properties in slowly adapting touch receptors, *eLife*, 2014, **3**, e01488.
36. L. Skedung and C. El Rawadi, *et al.*, Mechanisms of tactile sensory deterioration amongst the elderly, *Sci. Rep.*, 2018, **8**(1), 5303.
37. R. Candelier, A. Prevost and G. DebrÕgeas, The role of exploratory conditions in bio-inspired tactile sensing of single topogical features, *Sensors*, 2011, **11**(8), 7934–7953.
38. E. Wandersman and R. Candelier, *et al.*, Texture-Induced Modulations of Friction Force: The Fingerprint Effect, *Phys. Rev. Lett.*, 2011, **107**(16), 164301.
39. O. Sato and T. Maeda, *et al.*, Filiform papillae as a sensory apparatus in the tongue: An immunohistochemical study of nervous elements by use of neurofilamcnt protein (NFP) and S-100 protein antibodies, *Cell Tissue Res.*, 1988, **252**(2), 231–238.
40. R. de Wijk, L. Engelen and J. F. Prinz, The role of intra-oral manipulation in the perception of sensory attributes, *Appetite*, 2003, **40**(1), 1–7.
41. G. A. van Aken, M. H. Vingerhoeds and E. H. A. de Hoog, Food colloids under oral conditions, *Curr. Opin. Colloid Interface Sci.*, 2007, **12**(4–5), 251–262.
42. B. J. D. Le Révérend and F. Spyropoulos, *et al.*, Colloidal aspects of eating, *Curr. Opin. Colloid Interface Sci.*, 2010, **15**, 84–89.
43. B. J. D. Le Révérend, L. R. Edelson and C. Loret, Anatomical, functional, physiological and behavioural aspects of the development of mastication in early childhood, *Br. J. Nutr.*, 2014, **111**(03), 403–414.
44. M. Palinkas and M. S. P. Nassar, *et al.*, Age and gender influence on maximal bite force and masticatory muscles thickness, *Arch. Oral Biol.*, 2010, **55**(10), 797–802.
45. V. F. Ferrario and C. Sforza, *et al.*, Maximal bite forces in healthy young adults as predicted by surface electromyography, *J. Dent.*, 2004, **32**(6), 451–457.
46. J. Chen and J. R. Stokes, Rheology and tribology: Two distinctive regimes of food texture sensation, *Trends Food Sci. Technol.*, 2012, **25**, 4–12.
47. W. A. Alsanei and J. Chen, Studies of the Oral Capabilities in Relation to Bolus Manipulations and the Ease of Initiating Bolus Flow, *J. Texture Stud.*, 2014, **45**(1), 1–12.

48. F. Shama and P. Sherman, Identification of stimuli controlling the sensory evaluation of viscosity II. Oral Methods, *J. Texture Stud.*, 1973, **4**(1), 111–118.
49. F. Shama, C. Parkinson and P. Sherman, Identification of stimuli controlling the sensory evaluation of viscosity I. Non-Oral Methods, *J. Texture Stud.*, 1973, **4**(1), 102–110.
50. Q. He, J. Hort and B. Wolf, Predicting sensory perceptions of thickened solutions based on rheological analysis, *Food Hydrocolloids*, 2016, **61**, 221–232.
51. R. A. De Wijk, A. M. Janssen and J. F. Prinz, Oral movements and the perception of semi-solid foods, *Physiol. Behav.*, 2011, **104**(3), 423–428.
52. A. C. Mosca and J. Chen, Food-saliva interactions: Mechanisms and implications, *Trends Food Sci. Technol.*, 2017, **66**, 125–134.
53. J. Bongaerts, D. Rossetti and J. Stokes, The Lubricating Properties of Human Whole Saliva, *Tribol. Lett.*, 2007, **27**(3), 277–287.
54. S. Gittings and N. Turnbull, *et al.*, Characterisation of human saliva as a platform for oral dissolution medium development, *Eur. J. Pharm. Biopharm.*, 2015, **91**, 16–24.
55. L. J. Pereira, M. B. D. Gaviao and A. Van Der Bilt, Influence of oral characteristics and food products on masticatory function, *Acta Odontol. Scand.*, 2006, **64**(4), 193.
56. C. Loret and M. Walter, *et al.*, Physical and related sensory properties of a swallowable bolus, *Physiol. Behav.*, 2011, **104**(5), 855–864.
57. L. Motoi and M. P. Morgenstern, *et al.*, Bolus Moisture Content of Solid Foods during Mastication, *J. Texture Stud.*, 2013, **44**(6), 468–479.
58. L. Engelen and R. A. de Wijk, *et al.*, Relating particles and texture perception, *Physiol. Behav.*, 2005, **86**(1–2), 111–117.
59. A. L. Ferry and J. Hort, *et al.*, Viscosity and flavour perception: Why is starch different from hydrocolloids?, *Food Hydrocolloids*, 2006, **20**, 855–862.
60. G. M. Bornhorst, H. Hivert and R. P. Singh, Rice bolus texture changes due to α-amylase, *LWT–Food Sci. Technol.*, 2014, **55**(1), 27–33.
61. Y. Lu and A. Bennick, Interaction of tannin with human salivary proline-rich proteins, *Arch. Oral Biol.*, 1998, **43**(9), 717–728.
62. A. E. Hagerman, M. E. Rice and N. T. Ritchard, Mechanisms of Protein Precipitation for Two Tannins, Pentagalloyl Glucose and Epicatechin 16 (4 → 8) Catechin (Procyanidin), *J. Agric. Food Chem.*, 2002, **46**(7), 2590–2595.
63. C. Dinnella and A. Recchia, *et al.*, Saliva characteristics and individual sensitivity to phenolic astringent stimuli, *Chem. Senses*, 2009, **34**(4), 295–304.
64. H. T. Lawless and H. Heymann, *Sensory Evaluation of Food*, Springer, New York, 2010, Food Science Text Series.
65. S. S. Stevens, The surprising simplicity of sensory metrics, *Am. Psychol.*, 1962, **17**(1), 29–39.
66. S. S. Stevens (1969) Sensory scales of taste intensity, *Percept. Psychophys.*, 6(5), 302–308.

67. D. R. Peryam and V. M. Swartz, Measurement of sensory differences, *Food Technol.*, 1950, **5**, 207–210.
68. M. A. Brandt, E. Z. Skinner and J. A. Coleman, Texture Profile Method, *J. Food Sci.*, 1963, **28**(4), 404–409.
69. A. Szczesniak, General food texture profile revisited – a ten year perspective, *J. Texture Stud.*, 1975, **6**(1), 5–17.
70. H. Stone, J. Sidel, *et al.*, Sensory Evaluation by Quantitative Descriptive Analysis, in *Descriptive Sensory Analysis in Practice*, Food & Nutrition Press, Inc., Trumbull, CT, USA, 2008, pp. 23–34.
71. G. V. Civille, B. T. Carr and M. Meilgaard, *Sensory Evaluation Techniques*, CRC Press, 5th edn, 2015.
72. R. Gawel, A. Oberholster and I. L. Francis, A "Mouth-feel Wheel": terminology for communicating the mouth-feel characteristics of red wine, *Aust. J. Grape Wine Res.*, 2000, **6**(3), 203–207.
73. L. J. R. Lawless, A. Hottenstein and J. Ellingsworth, The McCormick spice wheel: A systematic and visual approach to sensory lexicon development, *J. Sens. Stud.*, 2012, **27**(1), 37–47.
74. A. S. Szczesniak, Texture is a sensory property, *Food Qual. Prefer.*, 2002, **13**, 215–225.
75. S. Adams and S. Singleton, *et al.*, *In-vivo* visualisation of mouth-material interactions by video rate endoscopy, *Food Hydrocolloids*, 2007, **21**(5–6), 986–995.
76. G. Ares and C. Barreiro, *et al.*, Application of a check-all-that-apply question to the development of chocolate milk desserts, *J. Sens. Stud.*, 2010, **45**(1), 67–86.
77. G. Ares and R. Deliza, *et al.*, Comparison of two sensory profiling techniques based on consumer perception, *Food Qual. Prefer.*, 2010, **21**(4), 417–426.
78. M. Cliff and H. Heymann, Time-Intensity evaluation of oral burn, *J. Sens. Stud.*, 1993, **8**(3), 201–211.
79. F. R. Jack, J. R. Piggott and A. Paterson, Analysis of Textural Changes in Hard Cheese during Mastication by Progressive Profiling, *J. Food Sci.*, 1994, **59**(3), 539–543.
80. N. Pineau and P. Schlich, *et al.*, Temporal Dominance of Sensations: Construction of the TDS curves and comparison with time-intensity, *Food Qual. Prefer.*, 2009, **20**(6), 450–455.
81. M. Ng and J. B. Lawlor, *et al.*, Using quantitative descriptive analysis and temporal dominance of sensations analysis as complementary methods for profiling commercial blackcurrant squashes, *Food Qual. Prefer.*, 2012, **25**(2), 121–134.
82. B. Piqueras-Fiszman and C. Spence, Sensory expectations based on product-extrinsic food cues: An interdisciplinary review of the empirical evidence and theoretical accounts, *Food Qual. Prefer.*, 2015, **40**(A), 165–179.
83. M. Perrot and N. Pineau, *et al.*, Use of multi-market preference mapping to design efficient product portfolio, *Food Qual. Prefer.*, 2018, **64**, 238–244.

84. L. Mioche, Mastication and food texture perception: Variation with age, *J. Texture Stud.*, 2004, **35**(2), 145–158.
85. E. M. Wilson and J. R. Green, The development of jaw motion for mastication, *Early Hum. Dev.*, 2009, **85**(5), 303–311.
86. B. Le Révérend and F. Saucy, *et al.*, Adaptation of mastication mechanics and eating behaviour to small differences in food texture, *Physiol. Behav.*, 2016, **165**, 136–145.
87. M.-A. Peyron and L. Mioche, *et al.*, Masticatory jaw movement recordings: A new method to investigate food texture, *Food Qual. Prefer.*, 1999, **7**, 229–237.
88. L. Mioche and M. A. Peyron, Bite force displayed during assessment of hardness in various texture contexts, *Arch. Oral Biol.*, 1995, **40**(5), 415–423.
89. T. Kamegai and T. Tatsuki, *et al.*, A determination of bite force in northern Japanese children, *Eur. J. Orthod.*, 2005, **27**(1), 53–57.
90. J. Weickenmeier and M. Jabareen, *et al.*, Experimental and Numerical Characterization of the Mechanical Masseter Muscle Response During Biting, *J. Biomech. Eng.*, 2017, **139**(12), 121007.
91. M. Simione and C. Loret, *et al.*, Differing structural properties of foods affect the development of mandibular control and muscle coordination in infants and young children, *Physiol. Behav.*, 2018, **186**, 62–72.
92. B. Le Révérend and C. Hartmann, Numerical modeling of human mastication, a simplistic view to design foods adapted to mastication abilities, *Physiol. Behav.*, 2013, **124**, 61–64.
93. P. W. Lucas and D. A. Luke, Computer simulation of the breakdown of carrot particles during human mastication, *Arch. Oral Biol.*, 1983, **28**(9), 821–826.
94. P. W. Lucas and D. A. Luke, Methods for analysing the breakdown of food in human mastication, *Arch. Oral Biol.*, 1983b, **28**(9), 813–819.
95. M.-A. Peyron and I. Gierczynski, *et al.*, Role of physical bolus properties as sensory inputs in the trigger of swallowing, *PLoS One*, 2011, **6**(6), e21167.
96. J. F. Prinz and P. W. Lucas, An optimization model for mastication and swallowing in mammals, *Proc. Royal Soc. B*, 1997, **264**, 1715–1721.
97. P. Clavé, M. De Kraa, *et al.*, (2006) The effect of bolus viscosity on swallowing function in neurogenic dysphagia, *Aliment. Pharmacol. Ther.*, 24(9), 1385–1394.
98. A. S. Burbidge and J. A. Y. Cichero, *et al.*, "A day in the life of the fluid bolus": An introduction to fluid mechanics of the oropharyngeal phase of swallowing with particular focus on dysphagia, *Appl. Rheol.*, 2016, **26**, 64525.
99. J. F. Steffe, *Rheological Methods in Food Process Engineering*, Freeman Pr., 1992.
100. H. S. Joyner (Melito), Explaining food texture through rheology, *Curr. Opin. Food Sci.*, 2018, **21**, 7–14.
101. J. R. Stokes, G. A. Davies, *et al.*, From Rheology to Tribology: Multiscale Dynamics of Biofluids, Food Emulsions and Soft Matter, in The XV International Congress on Rheology: The Society of Rheology 80th

Annual Meeting, ed. A. Co and G. L. Leal, *AIP Conference Proceedings*, 2008, pp. 1171–1173.
102. J. H. H. Bongaerts, K. Fourtouni and J. R. Stokes, Soft-tribology: Lubrication in a compliant PDMS-PDMS contact, *Tribol. Int.*, 2007, **40**(10–12), 1531–1542.
103. A. Gabriele, F. Spyropoulos and I. T. Norton, A conceptual model for fluid gel lubrication, *Soft Matter*, 2009, **6**, 4205–4213.
104. A. Gabriele, F. Spyropoulos and I. T. Norton, Kinetic study of fluid gel formation and viscoelastic response with kappa-carrageenan, *Food Hydrocolloids*, 2009, **23**(8), 2054–2061.
105. E. H. A. de Hoog and J. F. Prinz, *et al.*, Lubrication of Oral Surfaces by Food Emulsions: the Importance of Surface Characteristics, *J. Food Sci.*, 2006, **71**(7), E337–E341.
106. D. M. D. M. Dresselhuis and M. A. C. M. A. C. Stuart, *et al.*, Fat retention at the tongue and the role of saliva: Adhesion and spreading of 'protein-poor' versus 'protein-rich' emulsions, *J. Colloid Interface Sci.*, 2008, **321**(1), 21–29.
107. G. M. Bornhorst and R. P. Singh, Kinetics of *in Vitro* Bread Bolus Digestion with Varying Oral and Gastric Digestion Parameters, *Food Biophys.*, 2013, **8**(1), 50–59.
108. C. Tournier and M. Grass, *et al.*, The impact of mastication, salivation and food bolus formation on salt release during bread consumption, *Food Funct.*, 2014, **5**(11), 2969–2980.
109. P. Poinot and G. Arvisenet, *et al.*, Use of an Artificial Mouth to Study Bread Aroma, *Food Res. Int.*, 2009, **42**(5–6), 717–726.
110. B. Hanson and B. Cox, *et al.*, Effects of saliva on starch-thickened drinks with acidic and neutral pH, *Dysphagia*, 2012, **27**(3), 427–435.
111. R. A. de Wijk and J. F. Prinz, *et al.*, The role of α-amylase in the perception of oral texture and flavour in custards, *Physiol. Behav.*, 2004, **83**(1), 81–91.
112. C. I. Heinzerling, G. Smit and E. Dransfield, Modelling oral conditions and thickness perception of a starch product, *Int. Dairy J.*, 2008, **18**(8), 867–873.
113. R. Wille and E. Lutton, Polymorphism of cocoa butter, *J. Am. Oil Chem. Soc.*, 1966, **43**(8), 491–496.
114. B. Le Révérend, S. Bakalis and P. J. Fryer, Structured chocolate products, in *Food Materials Science: Principles and Practice*, Springer, 2008, pp. 525–546.
115. A. M. Carvalho-da-Silva and I. Van Damme, *et al.*, Characterisation of chocolate eating behaviour, *Physiol. Behav.*, 2011, **104**(5), 929–933.
116. J. A. Desimone and G. L. Heck, An analysis of the effects of stimulus transport and membrane charge on the salt, acid and water-response of mammals, *Chem. Senses*, 1980, **5**(4), 295–316.
117. J. F. Pfeiffer, R. B. Boulton and A. C. Noble, Modeling the sweetness response using time-intensity data, *Food Qual. Prefer.*, 2000, **11**, 129–138.

118. B. Le Révérend, I. T. Norton and S. Bakalis, Modelling the human response to saltiness, *Food Funct.*, 2013, **4**(6), 880–888.
119. J. A. Desimone, The dependence of the phasic response of the taste nerves on stimulus flow rate arises in the diffusion boundary layer region at the lingual surface: A convective-diffusion analysis, *Chem. Eng. Sci.*, 2017, **168**, 289–308.
120. H. R. Moskowitz, B. Drake and C. Akesson, Psychological measured of texture, *J. Texture Stud.*, 1972, **3**(2), 135–145.
121. C. M. Christensen, Effects of solution viscosity on perceived saltiness and sweetness, *Percept. Psychophys.*, 1980, **28**(4), 347–353.
122. A.-L. Koliandris and C. Morris, *et al.*, Correlation between saltiness perception and shear flow behaviour for viscous solutions, *Food Hydrocolloids*, 2010, **24**(8), 792–799.
123. B. Aubert, A. Lima and B. Le Révérend, Biophysical basis of taste modulation by viscous solutions in humans, *Food Hydrocolloids*, 2016, **60**, 494–499.
124. A. C. Mosca and F. van de Velde, *et al.*, Enhancement of sweetness intensity in gels by inhomogeneous distribution of sucrose, *Food Qual. Prefer.*, 2010, **21**(7), 837–842.
125. A. C. Mosca, J. H. F. Bult and M. Stieger, Effect of spatial distribution of tastants on taste intensity, fluctuation of taste intensity and consumer preference of (semi-)solid food products, *Food Qual. Prefer.*, 2013, **28**(1), 182–187.
126. L. Forny, J. Li, *et al.*, Reduced sodium bakery product, *WO Pat.*, WO2016058996A1, 2014.
127. M. Quilaqueo, L. Duizer and J. M. Aguilera, The morphology of salt crystals affects the perception of saltiness, *Food Res Int.*, 2015, **76**, 675–681.
128. H. Oertling, T. Alzieu, *et al.*, Fast-dissolving, co-crystalline forms of sodium chloride, *WO Pat.*, WO2017016907, 2015.
129. V. Meunier, M. Dupas-Langlet, *et al.*, Bulk sugar replacer, *WO Pat.*, WO2017093302A1, 2016.
130. K. M. M. Burseg and C. Brattinga, *et al.*, Sweet taste enhancement through pulsatile stimulation depends on pulsation period not on conscious pulse perception, *Physiol. Behav.*, 2010, **100**(4), 327–331.
131. K. M. M. Burseg, S. Camacho and J. H. Franciscus, Effects of pulsation rate and viscosity on pulsation-induced taste enhancement: New insights into texture – Taste interactions, *J. Agric. Food Chem.*, 2011, **59**(10), 5548–5553.
132. K. M. M. Burseg, S. M. Camacho and J. H. F. Bult, Taste enhancement by pulsatile stimulation is receptor based but independent of receptor type, *Chemosens. Percept.*, 2012, **5**(2), 179–187.
133. C. Morris and C. Labarre, *et al.*, Effect of pulsed delivery and bouillon base on saltiness and bitterness perceptions of salt delivery profiles partially substituted with KCl, *Food Qual. Prefer.*, 2010, **21**(5), 489–494.

134. J. Hort and T. A. Hollowood, Controlled Continuous Flow Delivery System for Investigating Taste – Aroma Interactions, *J. Agric. Food Chem.*, 2004, **52**(15), 4834–4843.
135. M. Lad, L. Hewson and B. Wolf, Enhancing saltiness in emulsion based foods, *Flavour*, 2012, **1**(1), 13.
136. D. J. Cook and T. A. Hollowood, *et al.*, Perception of taste intensity in solutions of random-coil polysaccharides above and below c*, *Food Qual. Prefer.*, 2002, **13**, 473–480.
137. F. Lenfant and C. Hartmann, *et al.*, Impact of the shape on sensory properties of individual dark chocolate pieces, *LWT–Food Sci. Technol.*, 2013, **51**(2), 545–552.
138. C. Hartmann and F. Lovell, Chocolate tablet, *US Pat.*, USD613028S1, 2008.
139. C. M. Christensen, Oral perception of solution viscosity, *J. Texture Stud.*, 1979, **10**(2), 153–164.
140. C. M. Christensen and L. M. Casper, Oral and Nonoral Perception of Solution Viscosity, *J. Food Sci.*, 1987, **52**(2), 445–447.
141. S. Camacho and M. Dop, *et al.*, Just Noticeable Differences and Weber Fraction of Oral Thickness Perception of Model Beverages, *J. Food Sci.*, 2015, **80**(7), S1583–S1588.
142. J. L. Kokini and E. L. Cussler, Predicting the Texture of Liquid and Melting Semi-Solid Foods, *J. Food Sci.*, 1983, **48**(4), 1221–1225.
143. D. J. Mela, Sensory assessment of fat content in fluid dairy products, *Appetite*, 1988, **10**(1), 37–44.
144. M. Akhtar and J. Stenzel, *et al.*, Factors affecting the perception of creaminess of oil-in-water emulsions, *Food Hydrocolloids*, 2005, **19**(3), 521–526.
145. M. Akhtar, B. S. Murray and E. Dickinson, Perception of creaminess of model oil-in-water dairy emulsions: Influence of the shear-thinning nature of a viscosity-controlling hydrocolloid, *Food Hydrocolloids*, 2006, **20**(6), 839–847.
146. M. E. Malone, I. A. M. Appelqvist and I. T. Norton, Oral behaviour of food hydrocolloids and emulsions. Part 2. Taste and aroma release, *Food Hydrocolloids*, 2003, **17**(6), 775–784.
147. D. Kilcast and S. Clegg, Sensory perception of creaminess and its relationship with food structure, *Food Qual. Prefer.*, 2002, **13**, 609–623.
148. P. Tyle, Effect of size, shape and hardness of particles in suspension on oral texture and palatability, *Acta Psychol.*, 1993, **84**(1), 111–118.
149. E. Imai, K. Hatae and A. Shimada, Oral perception of grittiness: effect of particle size and concentration of the dispersed particles and the dispersion medium, *J. Texture Stud.*, 1995, **26**(5), 561–576.
150. K. Liu and Y. Tian, *et al.*, Evidence for ball-bearing mechanism of microparticulated whey protein as fat replacer in liquid and semi-solid multi-component model foods, *Food Hydrocolloids*, 2016, **52**, 403–414.
151. R. de Wijk and M. E. J. Terpstra, *et al.*, Perceived creaminess of semi-solid foods, *Trends Food Sci. Technol.*, 2006, **17**(8), 412–422.

152. G. Sala and F. van de Velde, *et al.*, Oil droplet release from emulsion-filled gels in relation to sensory perception, *Food Hydrocolloids*, 2007, **21**(5–6), 977–985.
153. M. H. Vingerhoeds and R. A. de Wijk, *et al.*, How emulsion composition and structure affect sensory perception of low-viscosity model emulsions, *Food Hydrocolloids*, 2008, **22**(4), 631–646.
154. D. M. Dresselhuis and E. H. A. de Hoog, *et al.*, The occurrence of in-mouth coalescence of emulsion droplets in relation to perception of fat, *Food Hydrocolloids*, 2008, **22**(6), 1170–1183.
155. T. van Vliet and G. A. van Aken, *et al.*, Colloidal aspects of texture perception, *Adv. Colloid Interface Sci.*, 2009, **150**(1), 27–40.
156. M. H. Vingerhoeds and E. Silletti, *et al.*, Relating the effect of saliva-induced emulsion flocculation on rheological properties and retention on the tongue surface with sensory perception, *Food Hydrocolloids*, 2009, **23**(3), 773–785.
157. B. Vardhanabhuti and P. W. Cox, *et al.*, Lubricating properties of human whole saliva as affected by β-lactoglobulin, *Food Hydrocolloids*, 2011, **25**(6), 1499–1506.
158. E. Dickinson, Hydrocolloids as emulsifiers and emulsion stabilizers, *Food Hydrocolloids*, 2009, **23**(6), 1473–1482.
159. E. Dickinson, Stabilising emulsion-based colloidal structures with mixed food ingredients, *J. Sci. Food Agric.*, 2013, **93**(4), 710–721.
160. A. K. L. Oppermann and B. Piqueras-Fiszman, *et al.*, Descriptive sensory profiling of double emulsions with gelled and non-gelled inner water phase, *Food Res. Int.*, 2016, **85**, 215–223.
161. G. Muschiolik and E. Dickinson, Double Emulsions Relevant to Food Systems: Preparation, Stability, and Applications, *Compr. Rev. Food Sci. Food Saf.*, 2017, **16**(3), 532–555.
162. E. Dickinson, On the road to understanding and control of creaminess perception in food colloids, *Food Hydrocolloids*, 2018, **77**, 372–385.
163. T. A. M. Rovers and G. Sala, *et al.*, Potential of Microbubbles as Fat Replacer: Effect on Rheological, Tribological and Sensorial Properties of Model Food Systems, *J. Texture Stud.*, 2016, **47**(3), 220–230.
164. E. Dickinson, Emulsion gels: The structuring of soft solids with protein-stabilized oil droplets, *Food Hydrocolloids*, 2012, **28**(1), 224–241.
165. M. Devezeaux de Lavergne and M. Stieger, *et al.*, Uncoupling the Impact of Fracture Properties and Composition on Sensory Perception of Emulsion-Filled Gels, *J. Texture Stud.*, 2015, **47**(2), 92–111.
166. K. Liu and M. Stieger, *et al.*, Fat droplet characteristics affect rheological, tribological and sensory properties of food gels, *Food Hydrocolloids*, 2015, **44**, 244–259.
167. A. C. Mosca and J. A. Rocha, *et al.*, Inhomogeneous distribution of fat enhances the perception of fat-related sensory attributes in gelled foods, *Food Hydrocolloids*, 2012, **27**(2), 448–455.

168. D. Dupont and S. Le Feunteun, *et al.*, Structuring food to control its disintegration in the gastrointestinal tract and optimize nutrient bioavailability, *Innovative Food Sci. Emerging Technol.*, 2018, **46**, 83–90.
169. A. Tharakan and I. T. Norton, *et al.*, Mass transfer and nutrient absorption in a simulated model of small intestine, *J. Food Sci.*, 2010, **75**(6), 339–346.
170. O. Gouseti and M. R. Jaime-Fonseca, *et al.*, Hydrocolloids in human digestion: Dynamic in-vitro assessment of the effect of food formulation on mass transfer, *Food Hydrocolloids*, 2014, **42**, 378–385.
171. C. de Loubens and R. G. Lentle, *et al.*, Fluid mechanical consequences of pendular activity, segmentation and pyloric outflow in the proximal duodenum of the rat and the guinea pig, *J. R. Soc., Interface*, 2013, **10**(83), 20130027.
172. Y. F. Lim and C. de Loubens, *et al.*, Flow and mixing by small intestine villi, *Food Funct*, 2015, **6**(6), 1787–1795.
173. S. J. Hur, E. A. Decker and D. J. McClements, Influence of initial emulsifier type on microstructural changes occurring in emulsified lipids during in vitro digestion, *Food Chem.*, 2009, **114**(1), 253–262.
174. P. Rayment and P. Wright, *et al.*, Investigation of alginate beads for gastro-intestinal functionality, Part 1: In vitro characterisation, *Food Hydrocolloids*, 2009, **23**(3), 816–822.
175. I. Norton, P. Fryer and S. Moore, Product/Process integration in food manufacture: Engineering sustained health, *AIChE J.*, 2006, **52**(5), 1632–1640.
176. A. N. Cutler, E. R. Morris and L. J. Taylor, Oral perception of viscosity in fluid foods and model systems, *J. Texture Stud.*, 1983, **14**(4), 377–395.

CHAPTER 11

The Development of Food Structures for the Encapsulation and Delivery of Bioactive Compounds

NICOLAAS JAN ZUIDAM*[a] AND KRASSIMIR P. VELIKOV[a,b,c]

[a] Unilever R&D Vlaardingen, Olivier van Noortlaan 120, 3133 AT Vlaardingen, The Netherlands; [b] Institute of Physics, University of Amsterdam, Science Park 904, 1098 XH Amsterdam, The Netherlands; [c] Soft Condensed Matter Group, Debye Institute for NanoMaterials Science, Department of Physics and Astronomy, Utrecht University, Princetonplein 5, 3584 CC, Utrecht, The Netherlands
*Email: klaas-jan.zuidam@unilever.com

11.1 Introduction

Bioactive compounds can be generally considered as compounds that interact on a molecular or supramolecular level with our biosensors (*e.g.* aroma molecules, tastants) and/or participate in food digestion (*e.g.* enzymes, probiotics) or metabolism (*e.g.* micronutrients, nutraceuticals). They are added to food products to enhance their nutritional, health, or sensory benefits. Quite often the formulation and incorporation of bioactive molecules can be problematic, depending on the characteristics of the bioactive and food product (see Table 11.1).

Food Chemistry, Function and Analysis No. 18
Handbook of Food Structure Development
Edited by Fotis Spyropoulos, Aris Lazidis and Ian T. Norton

Published by the Royal Society of Chemistry, www.rsc.org

Table 11.1 Potential issues associated with commonly used bioactive compounds.

Issue	Dimension of potential solutions	Typical examples[a]
Product incompatibility	Compounds need to be dissolved, dispersed, or mixed	Phytosterols, phytostanols, flavour
Physicochemical stability	Need to avoid (redox) reactions or keep oxygen and/or light away	Fish oil, carotenoids, iron, flavour
Gastrointestinal stability/ bioavailability	Retain compound to control release in the gastrointestinal tract	Peptides, proteins, probiotics, iron
Taste/smell	Need for molecules to be retained, so they are not detected	Caffeine, polyphenols, potassium, oligopeptides
Colour/ discoloration	Need for photons to be scattered or absorbed in desired wavelength range	Carotenoids, vitamin A, iron

[a]Some bioactive compounds can have more than one challenge.

Several aspects need to be considered when bioactive compounds are formulated and delivered *via* a food product. Firstly, the *physicochemical properties* of the functional ingredient define its solubility in water or/and in commonly used edible oils. Stability with regard to oxidation and light degradation, as well as specific interactions with other components in the food product, should also be taken into account. The next important aspect is the type of *release profile* needed. That could be a burst release immediately upon use (*e.g.* upon cooking) or at a targeted location (*e.g.* in mouth, in gastrointestinal tract). Finally, one should consider the *product format* in which the delivery system is incorporated (*i.e.* powder, oil or water continuous emulsions). Here it is important that the bioactive compound is properly incorporated within the food product (*e.g.* no sedimentation, creaming, segregation) and stay intact in the product upon storage and during use.

The use of delivery systems or encapsulation,[1–5] a general process of isolating the bioactive molecule from the food matrix into particles, can be one of the possible technical solutions. However, it is certainly not the only and often not the preferred one as encapsulation is costly and complex. Proper design of the food structure, formulation, processing, and/or packaging may also optimise the performance of bioactive compounds.[1,6] To stress the latter approach, we will firstly show some examples of food products without encapsulation as a vehicle for bioactive compounds. Then we discuss several encapsulates and delivery systems for use in food products.

11.2 Food Products as a Vehicle for Bioactive Compounds

Food and beverages can be ideal vehicles for delivery of bioactive compounds. Quite often the latter can be added without problems, so without the use of a delivery system:

- Hydrophilic water-soluble bioactive compounds can be added to the aqueous parts of food or beverage products. Examples of such

compounds are water-soluble flavours, water-soluble vitamins (*e.g.* B, C), polyphenols (*e.g.* catechin), alkaloids (*e.g.* caffeine), and some mineral salts (*e.g.* ferrous sulphate, calcium chloride).
- Lipophilic oil-soluble actives can be added to the oil part of a food or beverage. Examples of such compounds are lipophilic flavours, fish oil, oil-soluble vitamins (*e.g.* A, D, E, K), and nutraceuticals (*e.g.* phytosterol esters, carotenoids, lycopene).
- Powders of bioactive compounds can be simply mixed in food products that are in powder form. To avoid segregation, the size and density of the powder particles should be in the same range as that of the food product.

Challenges may appear when ingredients are not compatible with one another. For example, some lipophilic flavour components cannot be added to a 100% aqueous beverage product, iron may oxidise fat, probiotics may not survive in a product upon storage, or the addition of polyphenols may give an unwanted colour to the product. Encapsulation may in general provide a solution in all these cases. However, one should consider the inherent challenges related to the use of delivery systems as well. These include additional ingredient and processing costs, increased complexity of production and/or supply chain, and stability challenges of encapsulates themselves (*e.g.* premature release). Therefore, it is recommendable to critically review the formulation challenges related to the corresponding bioactives before considering the use of a delivery system. Here we show a few examples to illustrate how the holistic approach in product structure design can be used to enable the stable delivery of bioactives.

11.2.1 Iron-fortified Bouillon Cube

The addition of iron salt as an essential micronutrient to a food product may cause oxidation of other food ingredients, thereby providing off-colour and/or off-flavour formation. Encapsulation may then be an option; however, this is quite costly compared to the cost of the iron salt itself.[7] A commonly used approach is to use a less water-soluble iron salt, *e.g.* ferric pyrophosphate instead of ferrous sulphate. Ionic iron catalyses the oxidation more than iron particles. Another option to further reduce the reactivity of the iron particle is to add to the food product a phosphate salt, other than iron phosphate.[8] For example, the addition of sodium pyrophosphate to bouillon cubes with both ferric pyrophosphate had significantly less off-colour formation upon storage.

11.2.2 Food Products as Regulators of Digestion

Food products can also carry bioactive ingredients that may help with or control the overall digestion and function of our gastrointestinal tract. Such examples can be the delivery of prebiotics and probiotics, digestive enzymes,[9,10] or enzyme inhibitors.[11] Prebiotics are food ingredients (like dietary fibres) that induce the growth or activity of beneficial microorganisms. Probiotics are living microorganisms that improve the indigenous microflora. Among the beneficial bacteria are *Bifidobacterium* and *Lactobacilli*. To ensure that probiotics stay alive, one may dry and encapsulate them.[12] However, another good option to ensure probiotics survive storage is to provide them with enough feed. Quite often the available probiotic products are dairy-based ones, as they provide good conditions in which the probiotics can survive and even proliferate upon storage. Examples of such dairy-based products include yogurt, cheese, and frozen desserts. Some of these products may even protect the probiotics upon consumption in the gastrointestinal tract due to their matrix structure and high-fat content (especially in cheese[13]). However, most probiotics cannot withstand the low pH in the stomach, and then encapsulation may provide protection.[12] On some occasions, food products can be used to deliver enzymes to promote digestion of gluten[14] or to deliver enzyme inhibitors to slow down lipid and carbs digestion and adsorption for weigh management purposes.[15,16]

11.2.3 Food as an Aid to Extract Bioactive Ingredients from Diet

Most bioactive compounds are added to the food product to deliver a health effect. In contrast, one may also use a food product to improve the bioavailability of bioactive compounds from other products in the dish; this is an important role of the food, which is often not well recognised. The use of some food products can facilitate the extraction and bioaccessibility of bioactive molecules such as vitamins and minerals from the rest of our diet. For example, the presence of soybean oil in dressings may enhance the bioavailability of carotenoids and fat-soluble vitamins from salad vegetables.[17] An almost linear dose–response was observed between the oil level and plasma content of women, although the effect varied widely among individuals as some showed a negligible response. Such an enhancement was also found when adding mayonnaise to a basic dressing.[18] One believes that the addition of oil (from the food product) increases the formation of dietary mix micelles, which facilitate the absorption of fat-soluble (non-)nutrients in the intestines.

A diet rich in vitamin C is well known to enhance the bioavailability of non-haem iron,[19] as vitamin C decreases the formation of less soluble and unabsorbable iron compounds and reduces ferric to ferrous iron, which is a requirement for uptake.

11.3 Delivery Systems and Encapsulation Structures for Bioactive Compounds

The use of delivery systems and encapsulation structures is another common approach for enabling the production of stable food products with purposely formulated bioactive compounds. Encapsulation can be defined as a process to entrap a bioactive compound in another substance, thereby producing particles in the range of a few nanometres to a few millimetres.[1] However, a delivery system can also be composed of (almost) all bioactive material, as we will illustrate in this chapter with homogeneous core systems. Nevertheless, delivery systems/encapsulates are often used as synonyms. In this section, we will discuss the basic structures and provide some examples to illustrate these further.

11.3.1 Basic Structures

Some very basic structures of encapsulates can be seen in Table 11.2. The simplest form is a spherical homogeneous carrier structure for the bioactive, which can be liquid (*e.g.* emulsion) or structured (*e.g.* gel, hard fat). The bioactive can be molecularly dissolved (*i.e.* inside a liquid carrier) or dispersed (*i.e.* inside a solid carrier).

When the bioactive compound is heterogeneously distributed (*i.e.* as interacting or non-interacting droplets or particles) throughout a homogenised carrier, we generally refer to it as a matrix type of carrier. Such carriers can also

Table 11.2 Basic structures of encapsulates.[a]

Basic carrier structure	Examples of encapsulates	Remarks
Homogeneous type 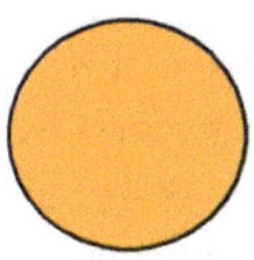	Emulsions, spray-dried/extruded particles, microgels, lipid particles, associated complexes	The bioactive is distributed at a molecular level all over the carrier.
Matrix type 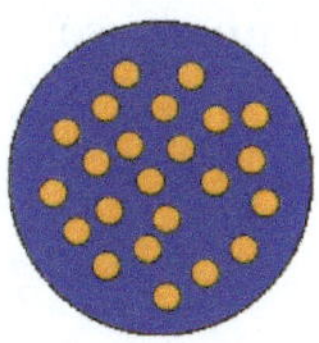	Spray-dried emulsions, prilling (spray-cooling or spray-congealing), complex coacervates with inclusions, microgels with inclusion of droplets or particles, multiple emulsions	Particulates with the bioactive are heterogeneously distributed inside the matrix.
Core–shell/multilayer type 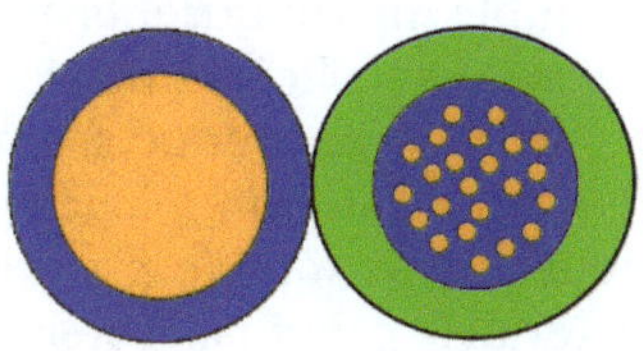	Fluid bed–coated particles, complex coacervates, co-extrudes, layer-by-layer coated particles, microgels, or emulsions, colloidosomes	Single shell or multiple shells are encapsulating molecularly or particularly distributed bioactive.

[a]Only spherical shapes are shown here, but they can also be cylindrical, oval, or irregular shaped.

be realised with both liquid and solid matrixes. The stability of these systems is very much determined by the ability of the dispersed phase to be retained within the matrix. Depending on the nature of the bioactive, compounds can diffuse from both homogeneous or heterogeneous cores and then enter into the bulk phase of the product. First- or multiple-order release kinetics are common for such types of encapsulates, as the release of the bioactive compound is dependent on the relatively long pathway through the carrier material(s) as well as on the concentration difference between the bioactive compound in the encapsulates and that in the bulk. Of course, if small particles are used, the pathway is small, and one would then notice a burst release (all-or-nothing), triggered by a change in conditions.

Both homogeneous- and matrix-type carries may have one or more layers (shells) around them. Such a structure is referred to as a core-shell or reservoir type. The release from these (saturated) cores depends on their shell properties. Typically, a zero-order release of the bioactive compound from a core-shell structure is observed; *i.e.* the release is linear with time as the rate-limiting step is the pathway through the relatively thin shell. Alternatively, core-shell type structures may break upon the application of pressure, resulting in a faster or even burst release.

There are many different processes to make delivery systems and encapsulates. The interested reader is referred to good reviews[1–4] and the references therein. In a recent review, we have also provided guidelines for the selection of delivery systems and encapsulation for application within different type of food products.[5]

Here we would like to discuss several examples of these structures in more detail. Firstly, we will discuss some types with homogeneous cores and then matrix types with heterogeneous cores. Finally, we will show some examples of encapsulates with a core-shell morphology. Please note that we are not providing a full overview of all encapsulation structures possible; the interested reader is referred for this to the reviews previously mentioned or to reviews focusing on different encapsulation approaches for certain classes of bioactive molecules (*e.g.* polyphenols[20]).

11.3.2 Homogeneous Core Systems

11.3.2.1 Micronutrient and Nutraceutical-based Colloidal Particles

Colloidal particles of many materials have been shown to be suitable for delivery of functional ingredients. Many water-insoluble micronutrients and nutraceuticals in particular can benefit from the use of colloidal delivery systems because they enable their introduction in low-fat products and beverages. The colloidal particles in this case are made from pure bioactive and are dispersed/stabilised using proper colloidal interactions (*e.g.* steric repulsion). This is in contrast to using a matrix material (*e.g.* lipids, proteins, carbohydrates) as a delivery system, where the bioactive molecules are imbedded in the homogeneous core as foreign molecules.

For example, in the case of mineral colloidal particles, the metal ion is an integral part of the structure.[21–24] Such (mineral) particles are often used to deliver the essential mineral in a form that is less reactive than the corresponding soluble form of these metal ions. In some cases, unique properties of the metal salt are used, as in the case of ferric pyrophosphate, the only very lightly coloured insoluble iron salt, which is commonly used for iron fortification[25] (see also Section 11.2.1). It is now well established that if a low-soluble salt is used in colloidal form, the rate of dissolution can be eliminated as a rate-limiting step in bioavailability, while the low solubility assures low reactivity.[26] The concept was also extended to multi-mineral particles,[27,28] as shown in Figure 11.1.

Colloidal particles have been prepared from almost any nutraceutical including carotenoids,[29–31] lycopene,[32,33] phytosterols,[34] and steroid glycosides (see Figure 11.2).[35] The synthesis method can be different depending on the solubility and melting properties of the particles. Approaches such as melt emulsification, wet milling, emulsion-assisted solvent removal, and anti-solvent precipitation are often utilised.

If a bioactive is prone to chemical stability (*e.g.* oxidation or light degradation), colloidal particles are not well suited as a delivery system as they have a higher specific surface area in comparison to larger particles.

11.3.2.2 Emulsion-based Delivery Systems

Many food products contain liquid oils, which are often emulsified, and the resulting emulsions are well suited for delivery of any oil-soluble bioactive compounds.[36] If the products are low in oil content, custom-designed lipid-based emulsions can be used for this purpose. Such examples include beverages where oil-soluble flavours have to be delivered while keeping stability and appearance of the product acceptable.[37] In such applications and in other liquid products in particular, colloidal (nanoscale) emulsions can provide particular advantages as they have very good kinetic stability, can enable translucent products, and can facilitate good bioaccessibility of the delivered bioactive compound.[31,38–40] If the bioactive compounds are not sufficiently soluble in common triglycerides, one may use specialty oils. For example, steroid glycosides having more amphiphilic molecules are not well soluble in long chain triglycerides (LCT). However, they are well soluble in medium chain triglycerides (MCT) and in diacylglycerides- (DAGs-) rich oils.[41] Steroid glycosides in nanoscale emulsions using DAG ions have excellent stability against coalescence and creaming, thanks to their negative charges. In contrast to steroid glycosides, carotenoids are more soluble in LCT and less in MCT; therefore, their bioaccessibility was higher in the LCT (*e.g.* corn oil) than in the MCT.[40] Another approach for the delivery of compounds with limited solubility is undercooled emulsions. In this approach, a supersaturated solution of the bioactive compound is formed at high temperature and then quenched. The nanoscale confinement can suppress the crystallisation leading to the formation of undercooled

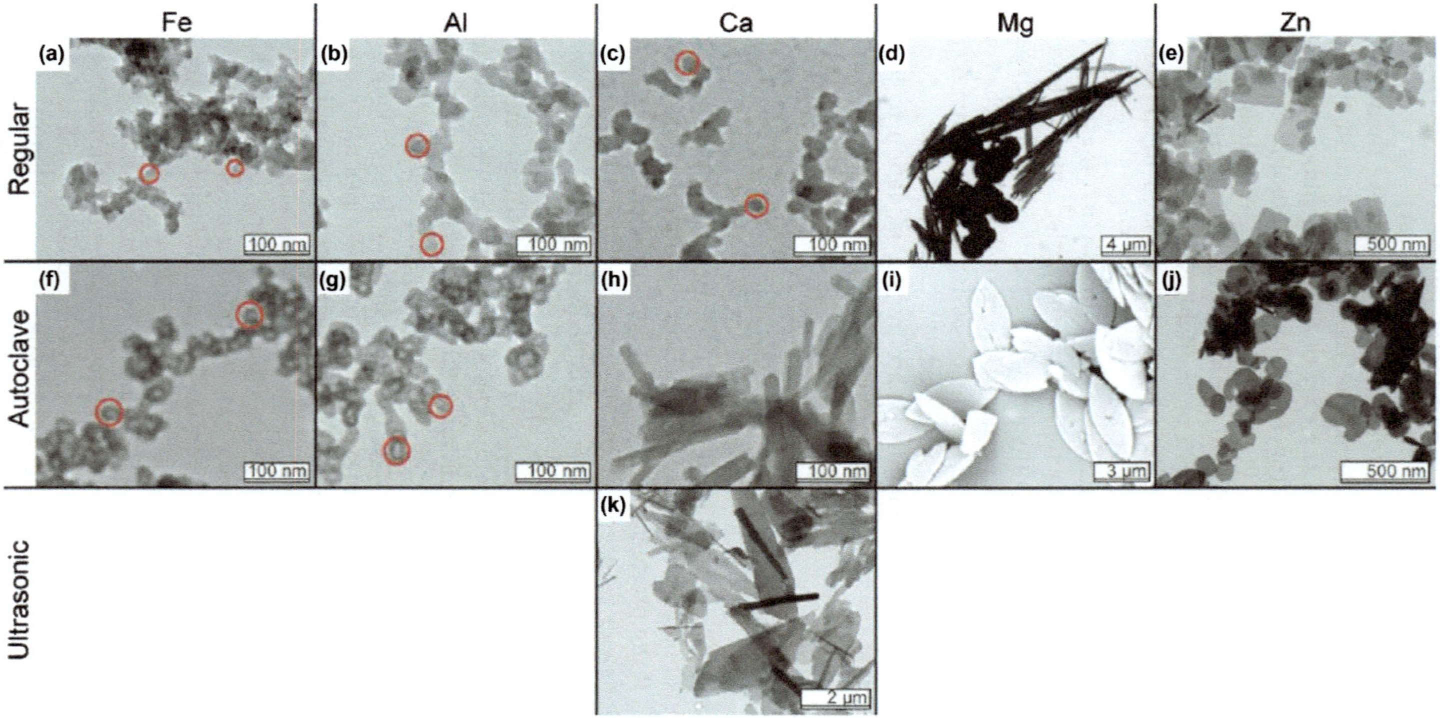

Figure 11.1 Transition electron microscopy (TEM) image matrix summarising the influence of preparation method and metal ion on colloidal mix mineral pyrophosphate particles. Rows indicate the preparation method; columns indicate the metal ion used. The red circles refer to primary particles in the amorphous aggregates (image [i] is an SEM image). Adapted from ref. 28 with permission from the Royal Society of Chemistry.

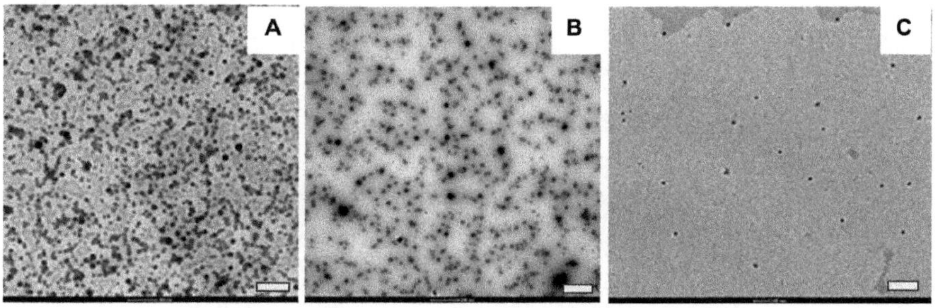

Figure 11.2 TEM micrographs of colloidal particles comprising steroid glycoside extract stabilised with (A) octenyl succinate (OSA) starch, (B) gum arabic, and (C) composite phytosterol-steroid glycoside particles. These nanoscale particles were made by a facile anti-solvent precipitation method; steroid glycoside was dissolved in ethanol and then precipitated by addition to water under high shear mixing, followed by ethanol removal under reduced pressure. Scale bars = 500 nm. Adapted from ref. 35 with permission from the Royal Society of Chemistry.

emulsions (Figure 11.3). The use of emulsion droplets as undercooled delivery systems kept phytosterol in a soluble state at a concentration several times higher than its solubility in a bulk oil phase.[42]

There are also developments in the field of Pickering emulsions for encapsulation and delivery of bioactive compounds.[43–45] Such emulsions are based on the stabilisation by particles at the oil–water interface, thereby forming a rigid but porous shell structure. Examples include emulsions with curcumin stabilised by colloidal silica particles,[43,44] modified starch,[44] lipophilic proteins (zein, denatured soy),[44] lipids,[44] and chitosan-tripolyphosphate.[45] The main potential advantage of Pickering emulsions could be their higher physical stability, compared to emulsions with standard, low-molecular-weight emulsifiers.

11.3.2.3 Colloidal Lipid and Protein-based Particles for Delivery

The use of a solid matrix made from edible material in colloidal particles can bring additional benefits by restricting diffusion of the bioactive and/or controlling its crystallisation behaviour. Solid lipid particles have extensively been studied for encapsulation for various lipophilic bioactive compounds.[31,46,47] Such particles have been used to create amorphous structures that contain imbedded nutraceuticals. An example is again the phytosterol-lipid composite particles (Figure 11.4).

The use of water insoluble proteins, such as prolamins (*e.g.* zein), is particularly suitable for encapsulation as these are fully edible and do not require chemical modification to form well defined particles.[48,49] Colloidal particles may help the formulation of water-insoluble, phenolic compounds, such as curcumin and quercetin. It was shown that both bioactives can be

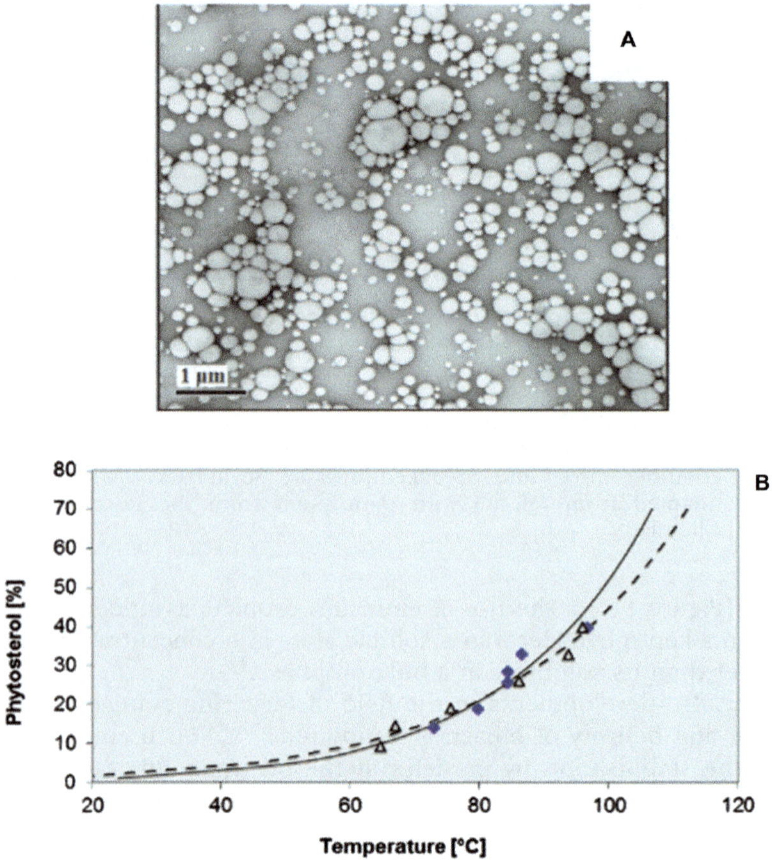

Figure 11.3 (A) Cryo-TEM micrograph of negative stained, colloidal emulsion containing 10 wt% of phytosterol (PS) in the lipid phase (myritol; melting point < −5 °C). (B) PS solubility in myritol (solid rhomb) and in trilaurin (open triangles), indicating that the equilibrium solubility of PS in the oil at ambient temperature is less than 2 wt%.
Reproduced from ref. 42 with permission from the Royal Society of Chemistry.

entrapped within amorphous particles with a good colloidal stability.[50,51] In the case of quercetin, which has the tendency to form large needle-like crystals, the crystallisation morphology was completely changed, and only spherical particles were formed (Figure 11.5).

Other edible, but non-digestible, materials such as shellac have also been shown to be suitable for formulation of colloidal delivery systems for other polyphenols.[52]

While formulations of water-insoluble polyphenols have been successfully made, the use of colloidal particles to deliver water-soluble polyphenols is much more challenging. To entrap a water-soluble polyphenol in a protein particle, a strong affinity between the two molecules is required. Such strong

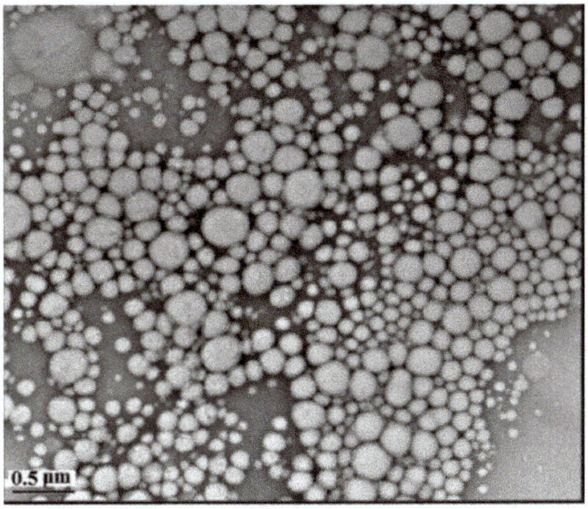

Figure 11.4 TEM negative staining of colloidal dispersion containing 10 wt% of PS in trilaurin. Trilaurin has a melting point at 46.5 °C, so it is a solid lipid at ambient temperature.
Reproduced from ref. 42 with permission from the Royal Society of Chemistry.

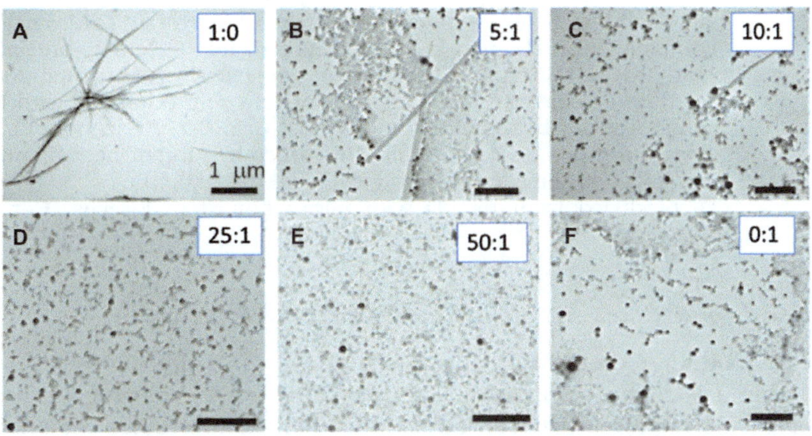

Figure 11.5 Composite colloidal particles with different zein-to-quercetin (w/w) ratios, obtained by anti-solvent precipitation.
Reproduced from ref. 53 with permission from the Royal Society of Chemistry.

affinity, however, was not found when zein-based colloids were used to encapsulate epigallocatechin gallate (EGCG), a polyphenol found in green tea.[54] Therefore, no effect of EGCG loading on particle morphology was found (see Figure 11.6). Most likely, EGCG can interact with sodium caseinate only at interfaces and scarcely interacts with zein.

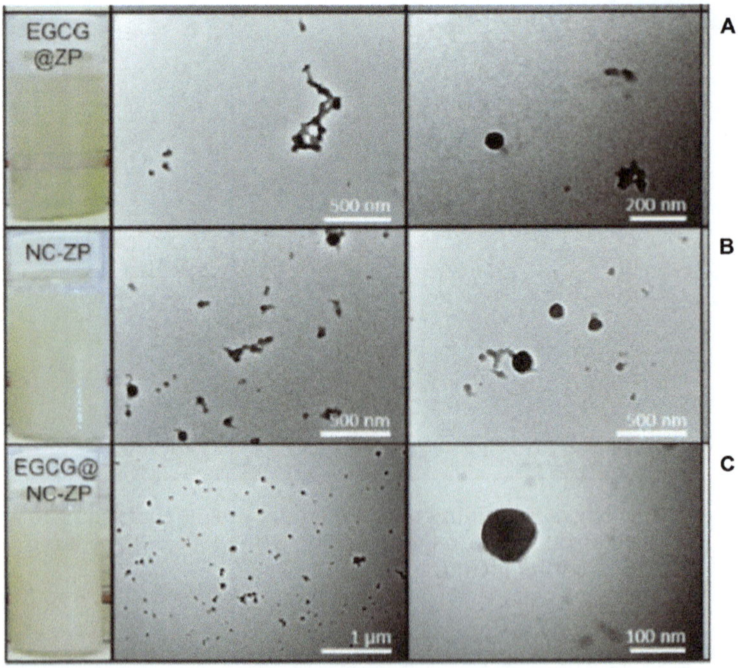

Figure 11.6 TEM images of zein-based colloidal particles without and with EGCG made by anti-solvent precipitation. A set of three TEM images at different magnification has been reported for (A) zein co-precipitated with EGCG (EGCG@ZP), (B) for composite particles of zein stabilised by sodium caseinate (NC-ZP), and (C) for zein co-precipitated with EGCG and stabilised by sodium caseinate (EGCG@NC-ZP). Each set of images is grouped in a single line and is identified by a legend accompanied by an image of the corresponding colloidal dispersion.
Reproduced from ref. 54 with permission from Elsevier, Copyright 2017.

11.3.2.4 Particles Based on Associative Complexes

Several food ingredients can strongly interact with one another to form insoluble associated complexes. This area has been exploring the use of associative complexes between oppositely charged molecules for encapsulation by complex coacervation[55] (see also Section 11.3.4.2). The field of using association between small molecules and polymers or between two small molecules to form colloidal deliver systems is still emerging. When these molecular interactions lead to precipitation, this is generally considered a problem. However, when purposefully utilised, these systems can be used to design novel biobased particles[56–58] for delivery or stabilisation of foams and emulsions.[59] An example of such an approach is demonstrated for water-soluble tea polyphenol EGCG and a water-soluble cellulose derivative; *i.e.* methyl cellulose (Figure 11.7). Upon interaction in solutions, the two molecules form spherical, hydrogel-like colloidal particles with tunable

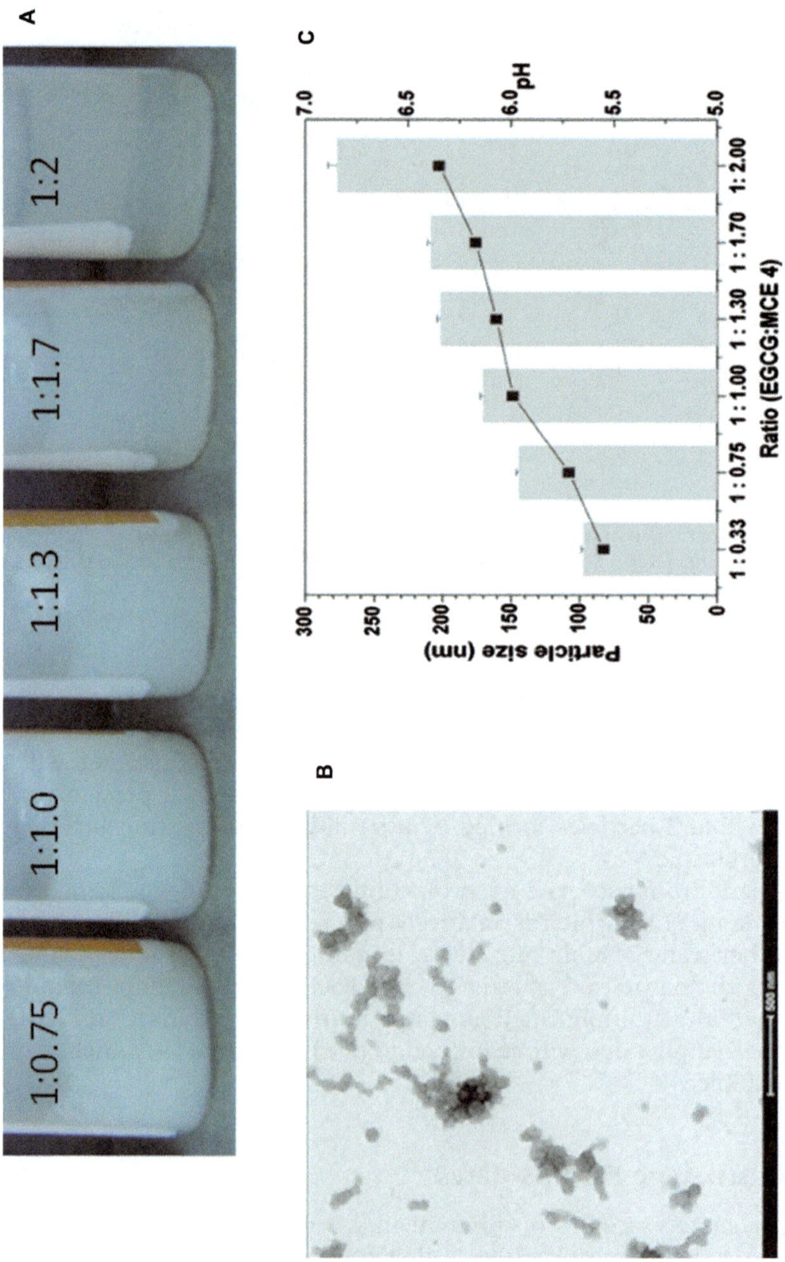

Figure 11.7 (A) Colloidal complexes of EGCG with methyl cellulose (MCE) at different polyphenol–polymer weight ratios. The complexes were made by addition of polymer solution to an EGCG solution under vigorous stirring. (B) TEM images of colloidal complexes of EGCG with MCE formed at a 1 : 1 weight ratio. (C) Particle size (mean ± SD) and pH of colloidal dispersion prepared at various w/w ratios of EGCG/MCE.
Adapted from ref. 56 with permission from the Royal Society of Chemistry.

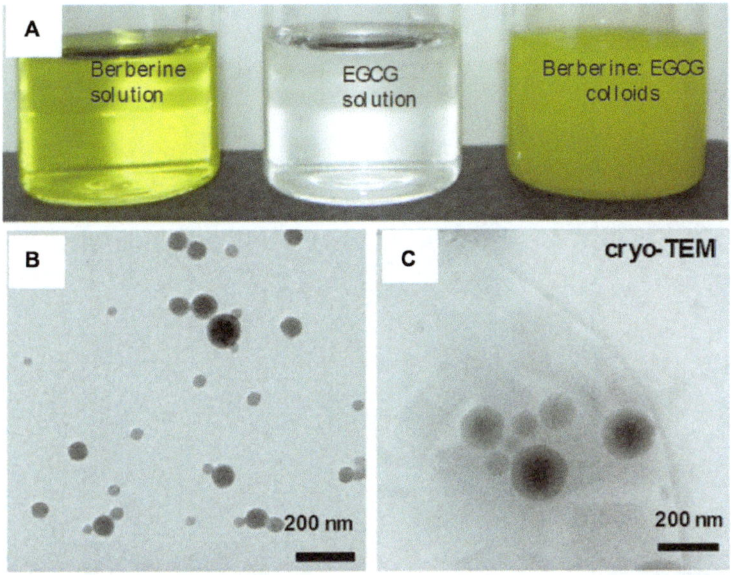

Figure 11.8 (A) Berberine and EGCG form clear solutions separately, but colloids are formed when these compounds are added together under vigorous stirring. (B, C) Cryo-TEM images of colloidal particles prepared using Berberine: EGCG at a molar ratio of 1 : 1.
Reproduced from ref. 57 with permission from the Royal Society of Chemistry.

sizes. In this way the water solubility of nutraceuticals and phytochemicals can be controlled, and adverse effects related to taste can be influenced.[56,60]

The concept of associative complexation was further demonstrated with EGCG and tannic acid with alkaloid berberine.[57] Figure 11.8 presents an example of colloidal particles formed by association precipitation between EGCG and berberine.

Particles made from associated complexes have a microgel-type structure (Figures 11.7 and 11.8), which is relatively porous and not suitable for retention of other water-soluble bioactives, in addition to the one that is already forming the particles. The strength of associative interaction depends on environmental conditions such as temperature, pH, and ionic strength. Therefore, their application will be limited to product formats in which they can remain stable.

11.3.3 Matrix-type Encapsulates

Here, we give some examples of different matrix-type delivery systems and encapsulates. Depending on the physical state of the matrix and dispersed phase, we distinguish liquid-in-liquid (*e.g.* multiple emulsions), liquid-in-solid (*e.g.* spray-dried emulsions in polymers or entrapped droplets in gel particles), and solid-in-solid (*e.g.*, solid particle, spray-dried with polymers,

solid particles spray-cooled with lipids, solid particles in extrudates or gels particles) structures.

11.3.3.1 Liquid-in-liquid: Multiple Emulsions

Multiple emulsions, especially water-in-oil-in-water (w/o/w) ones, have always been a very appealing structuring approach for the encapsulation and delivery of water-soluble bioactive molecules during food digestion.[61] Despite the great progress made in the fabrication and characterisation of multiple emulsions,[36,62,63] these systems remain difficult to utilise due to inherent stability problems related to water diffusion between the internal and external aqueous phases, the possibility for coalescence of the internal phase droplets, and the shear sensitivity of the large droplets.

11.3.3.2 Liquid in Solid Dispersions: Spray-dried and Melt Extrudated Flavours

About 20–25% of all flavour molecules sold are encapsulated, and 80–90% of these are spray-dried in the presence of carrier materials like maltodextrin, modified starch, gum, proteins, *etc.* This results in small, porous powder particles (see Figure 11.9), which release their flavour fast upon hydration and have only limited shelf-life as it does not protect against oxidation (especially at relative high moisture conditions like open air). One option to delay flavour release and to protect better against oxidation is the use of so-called melt extrudates, which are made using twin extrusion and have relatively less surface area. Such melt extrudates are often used in teabags (see Figure 11.9) and may be given a tea-like colour to make them look like

Figure 11.9 Flavour encapsulates: (A): scanning electron micrograph (SEM) of a typical spray-dried flavour; (B): melt extrudates – the white particles – within tea leaves obtained from a teabag, giving an excellent shelf life stability to flavours as the material is in its glassy state. In addition, due to their relatively large particle size (0.3–5 mm), they will not fall out of the teabags.

tea leaves. Commonly, such melt extrudates are made of starch, maltodextrins, modified starches, sugars, cellulose ethers, protein, emulsifiers, lipids, and/or gums.[1]

11.3.3.3 *Solid-in-solid: Theobromine in Lipid Particles*

Another example of matrix type encapsulates is shown in Figure 11.10. This encapsulate contains 20% of theobromine, a methylxanthine (like caffeine or theophylline), and is covered by fat and the emulsifier Dimodan HP by spray-cooling a molten dispersion in a cold chamber.[64] Theobromine is a bitter compound, and the fat coating should mask its bitterness. The encapsulate was stable in a dry environment or when frozen in an ice cream product. Upon addition to water or upon consumption, the theobromine may slowly solubilise through pores of the coating. Here a fat was used with a melting point of 42 °C, which is well above mouth temperature (especially when eating ice cream). Particles with a fat with a lower melting point were greasier and not so free-flowing at ambient temperatures, while fat with a higher melting point of 58 °C did not release much of the theobromine under *in vitro* gastrointestinal conditions.[64] Interestingly, the solubilisation of theobromine was low when no emulsifier was used in the fat-coating material with a melting point of 42 °C. The theobromine solubilisation from the lipid particles could be enhanced by the addition of hydrophobic

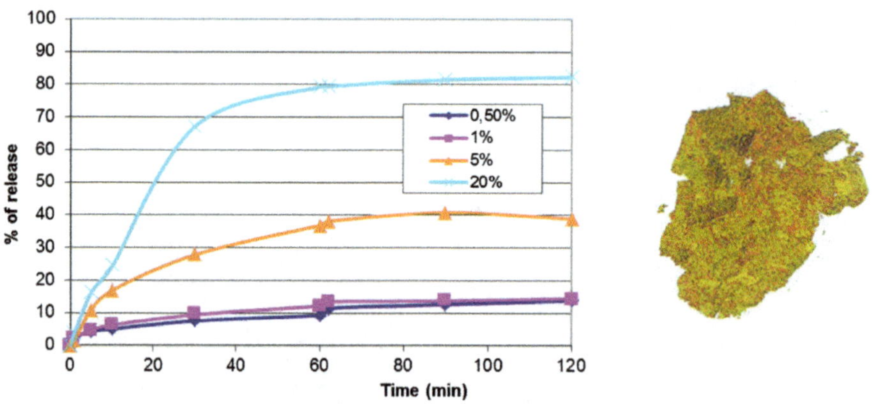

Figure 11.10 Influence of the emulsifier Dimodan HP on theobromine release from fat particles under *in vitro* gastrointestinal tract conditions.[64] The particles were composed of 20 wt% theobromine and 80 wt% of inES10 fat (melting point of 42 °C) and Dimodan HP. The particles were made by spray cooling and are 50–100 μm in size. A typical 3D XRT image of theobromine (red dots) coated by fat particle and Dimodan HP (yellow) is shown on the right. The encapsulates were firstly submitted to 1 h of gastric conditions at 37 °C, pH 2 and 0.5 g L^{-1} of pepsin, followed by 1 h of small intestine conditions at 37 °C, pH 7.5, 0.25 wt% of bile extracts and 0.08 wt% of pancreatin. No release was found at 25 °C.

emulsifiers like Dimodan HP, Datem, Span 60, or Citrem but not by hydrophilic emulsifiers like Tween 80 or whey protein.[64] Figure 11.10 shows that the more Dimodan HP was used in the fat coating, the faster the release of theobromine was under *in vitro* gastrointestinal tract conditions at 37 °C. Interestingly, no release was found at 25 °C in the presence of emulsifier in fat. This latter may indicate that no or little release of theobromine will occur in the mouth when eating an ice cream. When 0.5% of Dimodan in the fat layer was used and the encapsulates were mixed within a vanilla ice cream, the bitterness of theobromine was indeed not tasted by a sensory panel.[64] In contrast, non-encapsulated theobromine in vanilla ice cream was perceived to be bitter.

11.3.3.4 Colloidal Particle Systems

As already discussed in Section 11.3.2, the use of colloidal delivery systems may provide unique benefits. However, the design of matrix-type colloidal delivery systems is more challenging. In order to keep the structure in the colloidal range (*i.e.* benefiting bioaccessibility and stability against gravity), one may firstly make nanoscale particles from the bioactive and then incorporate these in a matrix. Although such structures are known in the general colloidal systems approach, very few food-grade matrix-type colloidal delivery systems have been demonstrated. One example is the use of the water-insoluble protein zein to entrap mineral-containing particles comprising iron (Figure 11.11).[65]

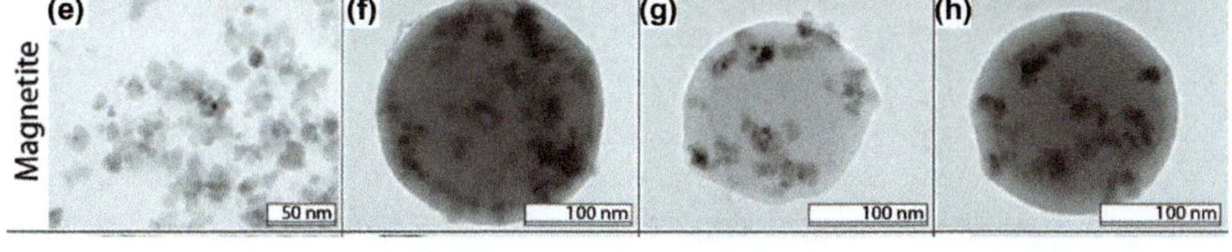

Figure 11.11 An image matrix of TEM micrographs of all prepared morphologies and compositions. Columns represent the different morphologies (left to right): bare mineral nanoparticles (NP), matrix-type structures (NP-Z or Z-NP, made by addition of zein in 80% of ethanol to NP in water or NP to zein particles in water, respectively), and a coated matrix-type structure (Z-NP-Z). The different nanoparticle systems are represented in the rows (top to bottom): iron pyrophosphate (FePP) and magnetite. Adapted from ref. 65 with permission from Elsevier, Copyright 2015.

11.3.4 Core-shell-type Encapsulates

Core shell encapsulates are particles that have a less permeable outer layer. Such an outer layer is often applied later on top of a homogeneous core or matrix particles in order to delay the release of their contents (see also Section 11.3.1). Here we will discuss a few examples; *i.e.* coating of a melt extrudate to delay flavour release, complex coacervates for fish oil delivery, and some core-shell colloidal particles.

11.3.4.1 Coated Melt Extrudates for Flavours

Only a few coatings are food grade, heat resistant, and relatively impermeable for small molecules like flavour. These include shellac and cellulose derivatives like hydroxypropyl methylcellulose (HPMC) and ethylcellulose. An ethylcellulose coating was used to delay flavour release from a melt extrudate in hot or boiling water.[66] Without a coating, such a melt extrudate solubilises quickly and releases its contents (see also Section 11.3.3.2), in contrast to coated ones (see Figure 11.12).

Interestingly, the flavour molecules did not diffuse through the coating itself but through just a few holes per particle that were created during exposure to the hot water (see Figure 11.12B). This explains why no large

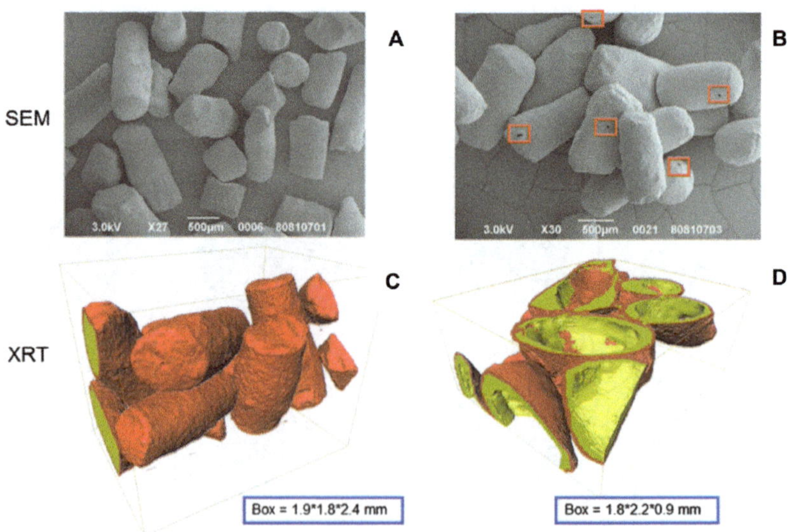

Figure 11.12 SEM (top) and X-ray topography (XRT, bottom) images of coated melt extrudes[66] before (left) and after (right) application in hot water. Maximal 15 wt% of flavour was first entrapped by melt extrudation, and then coated with 15 wt% of ethylcellulose. Particle size is about 600 μm×200 μm, and the coating was about 38 μm thick. The XRT images are 3D models obtained after tomographical reconstruction. The rectangles in the top right picture indicate presence of holes.

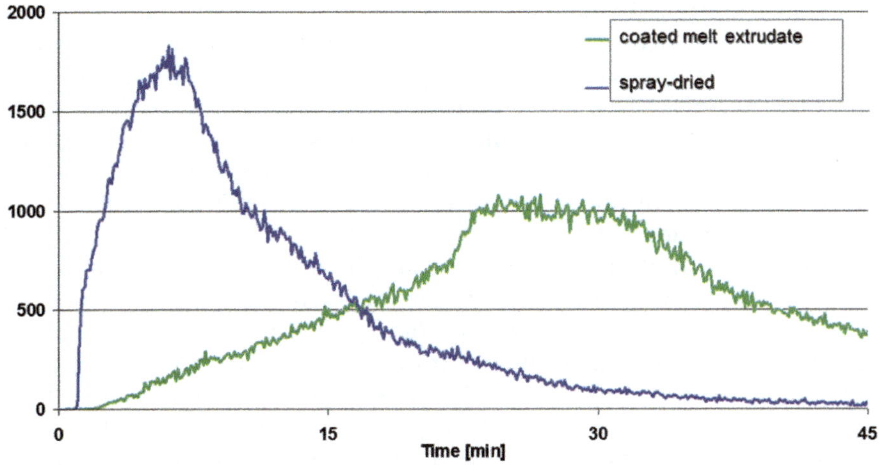

Figure 11.13 Release of methyl allyl trisulfide as part of a garlic flavour encapsulated in either spray-dried form or a melt extrudate, from a model broth at 80 °C. The dose of garlic flavour was the same in both cases. The values were measured as absolute intensity of m/z 153, using PTR-MS (Proton Transfer Reaction – Mass Spectrometry). Similar release patterns could be obtained with other flavours (*e.g.* basil). Adapted from ref. 66.

differences in release between different flavour components were observed, thereby keeping the right flavour balance. One would expect to observe such a difference if the coating stays intact due to differences in diffusion rates of the different flavour components (mainly depending on their lipophilicity). XRT images clearly show the coating on the outside, as well as their empty cores after release (see Figure 11.12B). A typical flavour release is shown in Figure 11.13. The flavour release from this coated melt extrudate was slower than the one obtained with spray-dried garlic. The latter dissolved immediately and gave a "burst" release. The more coating that was used, the slower the release from the melt extrudate became. Interestingly, further delay could also be obtained by changing the core material; when this material was less water soluble and could swell, the delay was prolonged.[66] Such a delayed release might be useful for especially long incubations (*e.g.* au bain marie), while keeping a good flavour balance, as the discharge of flavours is through tiny holes, and therefore no large differences between the release of individual flavour components are expected.

11.3.4.2 Complex Coacervates of Fish Oil

Coacervation is a process in which one or more hydrocolloids in an aqueous solution separate into a hydrocolloid-rich phase and a hydrocolloid-lean phase.[1,55,67] The hydrocolloid-rich phase can precipitate on the interface of emulsion droplets, thereby forming a shell around them. The process is called complex coacervation when two or more hydrocolloids are involved. Commonly, gelatin or whey protein is used, and the other hydrocolloid is of oppositely negative charge, like gum arabic, sodium polyphosphate, or carboxy methyl cellulose. For example, an oil-in-water emulsion with gelatin and gum arabic at a 1 : 1 (w/w) ratio and at a 2–4 wt% of each hydrocolloid can be entrapped within a coacervate phase by lowering the pH from about neutral to <5 under agitation at >35 °C, followed by cooling and optionally cross-linking with glutaraldehyde or transglutaminase.[1,55,67] Agitation during the coacervation process is important as the morphology of the complex coacervates is also dependent on the Reynolds number.[67] Mononucleated-type structures were formed under turbulent conditions when using gelatin and gum arabic as the hydrocolloids, while the production of polynucleated systems was encouraged at lower Reynolds numbers (lower shear). The smaller the Reynolds number is, the higher the contact time is of individual capsules and the higher the chance of aggregation. Figure 11.14 shows typical morphologies of either mononucleated or polynucleated complex coacervates.

Please note that for fish oil, the coacervate shell around the droplets is indeed a shell. However, this gel-like coating is quite porous for flavour molecules entrapped in the oil droplet within the complex coacervates, and thus these species will easily migrate across it until equilibrium with the outer side is reached.[69] So for fish oil, complex coacervates are a core–shell-type of encapsulate, but for flavour, complex coacervates are encapsulates of a matrix type. The same is true when using biopolymer-based microspheres,[1] like calcium alginate ones made by *e.g.* dropping

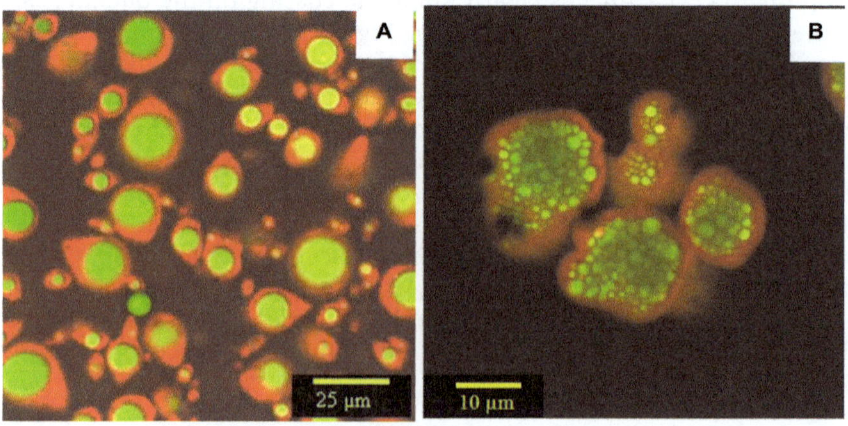

Figure 11.14 Confocal Laser Scanning Microscopy (CLSM) images of (A) mononucleated and (B) polynucleated complex coacervates using Nile blue as a fluorophore. The green light depicts the oil droplets and red signals the protein.
Reprinted from ref. 68 with permission from Springer Nature, Copyright 2010.

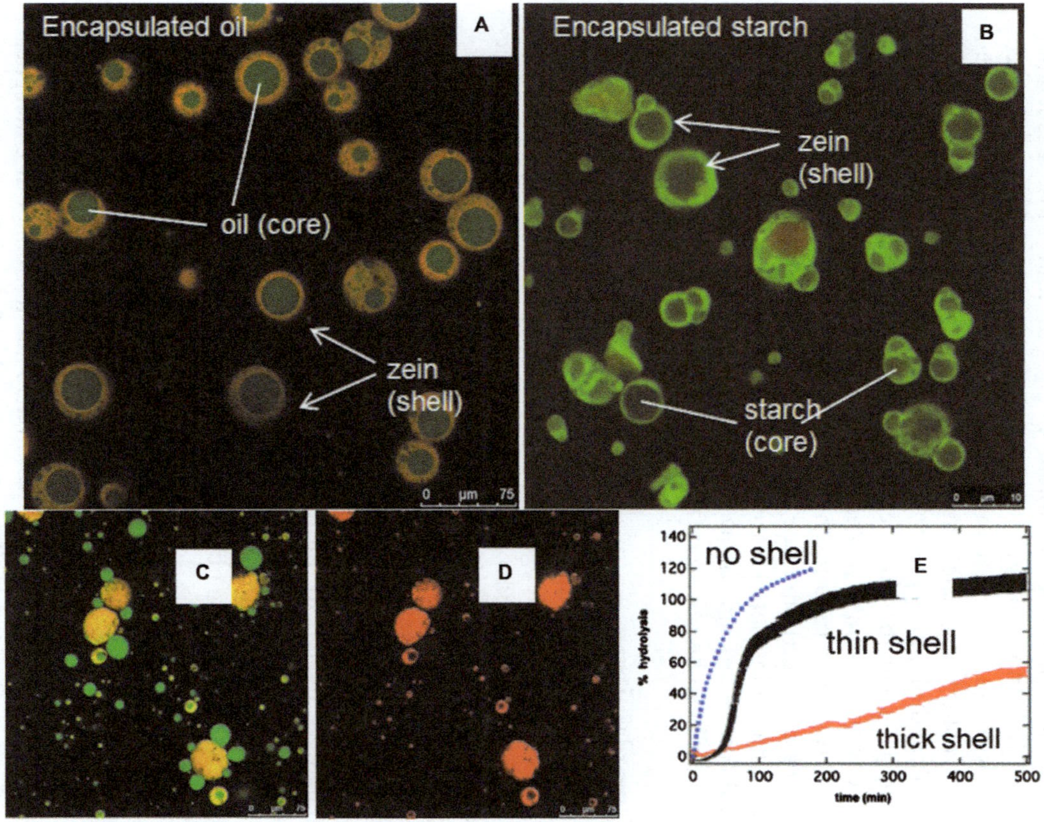

Figure 11.15 CLSM images of (A) oil droplets coated with zein and (B) starch particles coated with zein shell (Courtesy of P. Voudouris). These structures were made *via* anti-solvent precipitation by adding a zein solution in 80% of ethanol to an aqueous solution of oil droplets or starch particles under vigorous stirring. Upon compression, (C) release of oil was observed from the oil-zein core–shell, while (D) the zein shell remains largely intact as indicated by red structures (Courtesy of P. Voudouris). The rate of lipid digestion was strongly influenced by the presence of zein shell and by its thickness (see E). Figure 11.15E adapted from ref. 70 with permission from John Wiley & Sons, Copyright © 2014 WILEY-VCH Verlag GmbH & Co. KGaA, Weinheim.

sodium alginate solutions and bioactive compound (fish oil or flavour entrapped in oil droplets) into a calcium chloride bath.

11.3.4.3 Core– (Multi-)Shell Colloidal Delivery Systems

Core–shell colloidal systems are very attractive multifunctional systems combining the advantages of colloidal behaviour and those associated with the core–shell morphology. The core–shell morphology can be used in several ways to improve functionality ranging from influencing optical properties to controlling the rate of digestion. The combination of water-insoluble, edible, and preferably digestible material for the shell of core–shell particles can be quite challenging. From the class of food ingredients, the water-insoluble protein zein is a very good candidate for shell material, as it does not require chemical modification for the delivery or controlled release of nutrients. Due to the slow rate of digestion of zein, the rate of digestion of zein-coated starch particles and oil droplets (as shown in Figure 11.15) can be strongly delayed and depends on the thickness of the shell.[70] Oil could also be released from the oil droplets by compression.

11.4 Conclusions

Bioactive compounds are often added to food products to enhance their consumer value. The challenges that arise from the incorporation of bioactive compounds in food products should be addressed in a holistic manner considering both the food product as a vehicle and the use of specially designed delivery systems and encapsulates. Various approaches based on encapsulation or delivery systems can be used to address these issues. Both the internal structure and size of these delivery systems can play an important role in formulating the bioactive ingredient in order to address specific challenges related to the bioactive compound. The final approach on formulation may depend on several other factors such as costs, consumer acceptance (including Kosher and Halal considerations), regulatory aspects, freedom of use and intellectual property right status, supplier availability, *etc.*

Acknowledgements

We thank Panayiotis Voudouris for providing the micrographs of zein-coated starch particles and oil droplets. We would also like to thank our colleagues and the management of the Unilever R&D labs in UK and The Netherlands for supporting this work.

References

1. *Encapsulation Technologies for Active Food Ingredients and Food Processing*, ed. N. J. Zuidam and V. A. Nedović, Springer, New York, 2010.

2. *Encapsulation Technologies and Delivery Systems for Food Ingredients and Nutraceuticals*, ed. N. Garti and D. J. McClements, Woodhead Publishing, Cambridge, UK, 2012.
3. V. Đorđević, B. Balanč, A. Belščak-Cvitanović, S. Lević, K. Trifković, A. Kalušević, I. Kostić, D. Komes, B. Bugarski and V. Nedović, *Food Eng. Rev.*, 2015, **7**(4), 452.
4. *Encapsulation and Controlled Release Technologies in Food Systems*, ed. J. Lakkis, Blackwell Publishing, Oxford, 2nd edn, 2016.
5. N. J. Zuidam and V. P. Velikov, *Curr. Opin. Food Sci.*, 2018, **21**, 15.
6. J. Ubbink and J. Krüger, *Trends Food Sci. Technol.*, 2006, **17**, 244.
7. N. J. Zuidam, in *Encapsulation Technologies and Delivery Systems for Food Ingredients and Nutraceuticals*, ed. N. Garti and D. J. McClements, Woodhead Publishing, Cambridge, UK, 2012, ch. 17, pp. 505–540.
8. F. J. H. Janssen and K. P. Velikov, *EP Pat.* 2774497B1, 2015.
9. M. Roxas, *Altern. Med. Rev.*, 2008, **13**(4), 307.
10. G. Ianiro, S. Pecere, V. Giorgio, A. Gasbarrini and G. Cammarota, *Curr. Drug Metab.*, 2016, **17**(2), 187.
11. S. A. Tucci, E. J. Boyland and J. C. G. Halford, *Diabetes, Metab. Syndr. Obes.: Targets Ther.*, 2010, **3**, 125.
12. V. Manojlović, V. A. Nedović, K. Kailasapathy and N. J. Zuidam, in *Encapsulation Technologies for Active Food Ingredients and Food Processing*, ed. N. J. Zuidam and V. A. Nedović, Springer, New York, 2010, ch. 10, pp. 269–302.
13. C. G. Venderola, N. Bailo and J. A. Reinheimer, *Int. Dairy J.*, 2002, **12**, 579.
14. C. Mitea, R. Havenaar, J. W. Drijfhout, L. Edens, L. Dekking and F. Koning, *Gut*, 2008, **57**, 25.
15. S. Kumar, S. Narwal, V. Kumar and O. Prakash, *Pharmacogn. Rev.*, 2011, **5**(9), 19.
16. G. Williamson, *Mol. Nutr. Food Res.*, 2013, **57**, 48.
17. W. S. White, Y. Zhou, A. Crane, P. Dixon, F. Quadt and L. M. Flendrig, *Am. J. Clin. Nutr.*, 2017, **106**(4), 1041.
18. F. de Oliveira Martins, R. van den Berg, A.-R. Hoogenraad, E. Schuring, C. Verhoeven, A. Giese Laverdy Neto, R. Calvalcante Maranhao and P. Verhoef, *Int. J. Nutri. Food Sci.*, 2015, **4**(6), 644.
19. *Guidelines on Food Fortification with Micronutrients*, ed. L. Allen, B. de Benoist, O. Dary and R. Hurrell, 2018. Available at http://www.who.int/nutrition/publications/guide_food_fortification_micronutrients.pdf (accessed August 2018).
20. Z. Fang and B. Bhandari, *Trends Food Sci. Technol.*, 2010, **21**(10), 510.
21. L. M. Flendrig, C. E. Marshman, K. P. Velikov, I. Winter, *WO Pat.* 2007/062723A1, 2007.
22. C. E. Marshman and K. P. Velikov, *WO Pat.* 2007/009536, 2007.
23. C. E. Marshman and K. P. Velikov, *EP Pat.* 1792543, 2007.
24. C. E. Marshman and K. P. Velikov, *EP Pat.* 1792544, 2007.
25. L. Rossi, K. P. Velikov and A. P. Philipse, *Food Chem.*, 2014, **151**, 243.

26. F. M. Hilty, M. Arnold, M. Hilbe, A. Teleki, J. T. N. Knijnenburg, F. Ehrensperger, R. F. Hurrell, S. E. Pratsinis, W. Langhans and M. B. Zimmermann, *Nat. Nanotechnol.*, 2010, **5**(5), 374.
27. M. B. Zimmermann and F. M. Hilty, *Nanoscale*, 2011, **3**, 2390.
28. Y. M. van Leeuwen, K. P. Velikov and W. K. Kegel, *RSC Adv.*, 2012, **2**, 2534.
29. N. Anarjan, C. P. Tan, I. A. Nehdi and T. C. Ling, *Food Chem.*, 2012, **135**(3), 1303.
30. H. S. Ribeiro, B.-S. Chu, S. Ichikawa and M. Nakajima, *Food Hydrocolloids*, 2008, **22**(1), 12.
31. H. S. Ribeiro, H. P. Schuchmann, R. Engel, E. Walz and K. Briviba, in *Encapsulation Technologies for Active Food Ingredients and Food Processing*, ed. N. J. Zuidam and V. A. Nedović, Springer, 2010, ch. 8, pp. 211–252.
32. B. Mert, *J. Food Eng.*, 2012, **109**(3), 579.
33. Y. N. Shariffa, T. B. Tan, F. Abas, H. Mirhosseini and C. P. Tan, *Food Bioprod. Process.*, 2016, **98**, 210.
34. L. Rossi, J. W. M. S. ten Hoorn, S. M. Melnikov and K. P. Velikov, *Soft Matter*, 2010, **6**, 928.
35. K. P. Velikov, M. van Ruijven, A. K. Popp, A. R. Patel, L. M. Flendrig and S. M. Melnikov, *Food Funct.*, 2018, **9**(1), 485.
36. I. A. M. Appelqvist, M. Golding, R. Vreeker and N. J. Zuidam, in *Encapsulation and Controlled Release Technologies in Food Systems*, ed. J. M. Lakkis, John Wiley & Sons, Chichester, UK, 2nd edn, 2016, ch. 6, pp. 129–172.
37. P. S. Given Jr., *Curr. Opin. Colloid Interface Sci.*, 2009, **14**, 43.
38. D. J. McClements and J. Rao, *Crit. Rev. Food Sci. Nutr.*, 2011, **51**(4), 285.
39. D. J. McClements, *Curr. Opin. Colloid Interface Sci.*, 2012, **17**, 235.
40. D. J. McClements, *J. Food Sci.*, 2016, **80**(7), N1602.
41. S. M. Melnikov, A. K. Popp, S. Miao, A. R. Patel, L. M. Flendrig and K. P. Velikov, *J. Funct. Foods*, 2017, **28**, 90.
42. H. S. Ribeiro, R. Gupta, K. W. Smith, A. K. Popp and K. P. Velikov, *Soft Matter*, 2016, **12**, 5835.
43. R. V. Tikekar, Y. Pan and N. Nitin, *Food Res. Int.*, 2013, **51**(1), 370.
44. C. C. Berton-Carabin and K. Schroen, *Annu. Rev. Food Sci. Technol.*, 2015, **6**, 263.
45. B. R. Shah, Y. Li, W. Jin, Y. An, L. He, Z. Li, W. Xu and B. Li, *Food Hydrocolloids*, 2016, **52**, 369.
46. A. A. Zardini, M. Mohebbi, R. Farhoosh and S. Bolurian, *J. Food Sci. Technol.*, 2018, **55**(1), 287.
47. E. Nazemiyeh, M. Eskandani, H. Sheikhloie and H. Nazemiyeh, *Adv. Pharm. Bull.*, 2016, **6**(2), 235.
48. A. R. Patel and K. P. Velikov, *Curr. Opin. Colloid Interface Sci.*, 2014, **19**(5), 450.
49. Z.-L. Wan, J. Guoa and X.-Q. Yang, *Food Funct.*, 2015, **6**, 2876.

50. A. R. Patel, Y. C. Hu, J. K. Tiwari and K. P. Velikov, *Soft Matter*, 2010, **6**, 6192.
51. A. R. Patel, E. Drost, T. B. J. Blijdenstein and K. P. Velikov, *ChemPhysChem*, 2012, **13**, 3777.
52. A. Patel, P. Heussen, J. Hazekamp and K. P. Velikov, *Soft Matter*, 2011, **7**, 8549.
53. A. R. Patel, P. C. M. Heussen, J. Hazekamp, E. Drost and K. P. Velikov, *Food Chem.*, 2012, **133**, 423.
54. F. Donsi, P. Voudouris, S. J. Veen and K. P. Velikov, *Food Hydrocolloids*, 2017, **63**, 508.
55. Z. Xiao, W. Liu, G. Zhu, R. Zhou and Y. Niu, *J. Sci. Food Agric.*, 2014, **94**, 1482.
56. A. R. Patel, J. Nijsse and K. P. Velikov, *Soft Matter*, 2011, **7**, 4294.
57. A. R. Patel, E. Drost, R. den Adel, J. Hazekamp and K. P. Velikov, *Soft Matter*, 2012, **8**, 3515.
58. A. R. Patel, C. Remijn, A. I. M. Cabero, P. C. M. Heussen, J. W. M. S. ten Hoorn and K. P. Velikov, *Adv. Funct. Mater.*, 2013, **23**, 4710.
59. A. R. Patel, J. Seijen-ten-Hoorn, P. C. M. Heussen, E. Drost, J. Hazekamp and K. P. Velikov, *J. Colloid Interface Sci.*, 2012, **374**, 150.
60. K. P. Velikov, *WO Pat.* 2009/016018, 2009.
61. E. Dickinson, *Food Biophys.*, 2011, **6**, 1.
62. F. Jiménez-Colmenero, *Food Res. Int.*, 2013, **52**(1), 64.
63. G. Muschiolik and E. Dickinson, *Compr. Rev. Food Sci. Food Saf.*, 2017, **16**, 532.
64. C. M. Beindorff and N. J. Zuidam, *WO Pat.* 2015/032816, 2015.
65. R. J. Baars, Y. M. van Leeuwen, Y. Hendrix, K. P. Velikov, W. K. Kegel and A. P. Philipse, *Colloids Surf., A*, 2015, **483**, 209.
66. L. S. Henson, L. Jublot, J. E. Tapper and N. J. Zuidam, *EP Pat.* 2415357 and *WO Pat.* 2012016764, 2012.
67. C. Y. G. Lemetter, F. M. Meeuse and N. J. Zuidam, *AIChE J.*, 2009, **55**(6), 1487.
68. C. M. Beindorff and N. J. Zuidam, in *Encapsulation Technologies for Active Food Ingredients and Food Processing*, ed. N. J. Zuidam and V. A. Nedović, Springer, 2010, ch. 6, pp. 161–185.
69. M. E. Malone and I. A. M. Appelqvist, *J. Controlled Release*, 2003, **90**, 227.
70. E. Filippidi, A. R. Patel, E. C. M. Bouwens, P. Voudouris and K. P. Velikov, *Adv. Funct. Mater.*, 2014, **24**(38), 5962.

Part C
Food Structure Analysis, Characterisation, and Modelling

CHAPTER 12

Food Structure Analysis Using Light and Confocal Microscopy

CAMILLA ÖHGREN,[a] PATRICIA LOPEZ-SANCHEZ[a] AND NIKLAS LORÉN*[a,b]

[a] Product Design and Perception, RISE Agrifood and Bioscience, Box 5401, SE-402 29, Göteborg, Sweden; [b] Department of Physics, Chalmers University of Technology, SE-412 96, Göteborg, Sweden
*Email: Niklas.loren@ri.se

12.1 Introduction

Microstructural characterization of materials is a key factor in the understanding of food properties and their functionality.[1,2] Food microstructures consist of many different building blocks such as particles, grains, clusters, aggregates, domains of different phases,[3] inclusions, pores, and strands, which all together form a continuous body.[4] Foods having similar structures can be loosely grouped together as foods that have similar textures.

Food microstructures can be heterogeneous or homogeneous, as well as hierarchical, at different length scales.[5] Thus, to obtain a complete image of how a material is built up, the structure must be characterized at different length scales. To be able to fully describe the structure, it needs to be understood how the different parts of the system interact and all together build up a three-dimensional architecture. Structural differences in a material can already be detected by the naked eye. Using a higher magnification but still on an overall level of structure, we can see how small constituents have formed into *e.g.* a network structure, clusters, and different phases, *etc.* At a high resolution, information can be obtained on how macromolecules

Food Chemistry, Function and Analysis No. 18
Handbook of Food Structure Development
Edited by Fotis Spyropoulos, Aris Lazidis and Ian T. Norton

Published by the Royal Society of Chemistry, www.rsc.org

associate into supramolecular assemblies as molecules in a solution or small aggregates in dispersion. Supramolecular assemblies form strands in a network either as aligned molecules or as linked aggregates or particles. The packing of molecules, as well as the geometry and flexibility of strands, has a bearing on the network properties.

Using several different microscopy techniques, it is possible to visualize different features at various structural levels and the dynamic course of events such as the mechanisms taking place in a product when it is produced, used, experienced, or broken down.[6] Examples of such mechanisms are aggregation, phase separation, crystallization, flocculation, and dissolution, which can occur during heating, freezing, melting, mixing, tension, or storage. These mechanisms and the structures they give rise to have an impact on many important physical and functional properties such as water holding, diffusion, and rheological and sensory properties. Knowledge of the relationship between the microstructure and functional properties of materials makes it possible to tailor-make structures for specific purposes within sectors such as foods, paper, paint, pharmaceuticals, and hygiene.[1] Consequently, food microstructures need to be visualized at length scales ranging from a few nanometers, using electron microscopy, up to millimeters, using macroscopy. Light microscopy (LM) and confocal laser scanning microscopy (CLSM)[7] are excellent tools to reveal the food microstructure at the micrometer level. Examples of other powerful tools are X-ray tomography, confocal Raman microscopy, IR microscopy *etc.* However, this chapter focuses on light microscopy and confocal laser scanning microscopy.

This chapter briefly describes the principles and bases for using light microscopy and confocal laser scanning microscopy. It also gives an overview of the use of LM and CLSM to characterize the microstructure at micrometer level in different types of foods. The examples displayed in the chapter involve bread and dough, fibrous vegetable protein structures, plant cell walls, chocolate, and mayonnaise.

12.2 Characterization Methods of Food Structures at the Micrometer Level

12.2.1 Light Microscopy

In the light microscope (LM), the light from a halogen lamp ($\lambda = 400$–800 nm) is focused on the specimen and transmitted through it. A collector lens concentrates the light from the lamp, and a field diaphragm determines the illumination area of the specimen. The condenser consists of the aperture diaphragm and a condenser lens. The aperture diaphragm affects resolution, contrast, and depth of focus, with all these of major importance to image quality. The light intensity is controlled by the condenser lens, and the objective lens determines the magnification of the image. An important parameter of the objectives is the numerical aperture (NA). A high NA gives a high resolution. The depth of focus is the largest distance between two points on the vertical axis where both can be seen in focus. A large NA or a high magnification gives a short depth of focus. The contrast is the difference in light intensity or color between a structure and its surroundings. The contrast is normally enhanced by staining. LM will give information on structure compounds larger than half a micron, the images will be two-dimensional, and staining is required for most samples.

There are different techniques in LM. *Bright field* microscopy can be used for all specimens with enough contrast. The contrast is normally also enhanced by staining, since most food products have bad contrast. The eye will see a magnified and reversed image compared to the specimen. This is the standard technique that requires only Köhler illumination. Increased contrast is introduced using dyes or stains of known specificity for different components in the sample. The violet staining of starch with iodine in Lugol's solution is a long established but very useful method for the food microscopist. Light green is an example of the many stains available for the localization of proteins. Protein also colors yellow by Lugol's solution. Protein-rich products based on meat or fish can advantageously be stained using the so-called Sweden staining, *i.e.* the mixture of aniline blue and orange G. It is then possible to distinguish among the muscle fibers in yellow, the collagen that stains blue, and gelatin in light blue. When samples are fixated and plastic embedded, the fat is block stained with OsO4. Oil red is also a commonly used fat-soluble dye. Anisotropic (crystalline or ordered) structures rotate *polarized light*, which means that they can be detected between two crossed polarizers, while all other structures remain invisible. Thus, the polarizing microscope has one polarizing filter before the specimen, called polarizer, and one filter after the specimen, called the analyzer. *Differential interference contrast* (*DIC*) is a variant of polarized light. A Wollaston prism splits the light into two rays, with a separation smaller than the objective resolution. Both rays pass the specimen, but differences in path length cause interference in the image, which looks like shadowing and gives three-dimensional relief images of structures. *Phase contrast* converts phase shifts in light passing through a transparent specimen to brightness changes in the image. Phase contrast is important when studying living cells and microorganisms, which are not visible without staining and which are killed if staining is used. In *epifluorescence*, autofluorescing substances or those stained with fluorochromes will emit light with a longer wavelength than the illuminating light. It requires a dichroic mirror, which reflects one wavelength but not the other, an excitation filter, and a barrier filter. The lamp is a mercury or xenon illuminator.

LMs require that the material must be very thin to get the light through the sample. If the material is a liquid, a simple smearing of the sample is enough to visualize the structure using LM. However, if the material structure would be destroyed by smearing, cryo or plastic embedding is necessary to obtain thin sections. A quick route to make semithin sections is to freeze the material before sectioning. It is important to freeze the material quickly, preferably in liquid nitrogen, to minimize the creation and sizes of ice crystals, which can destroy and create an artifact structure. In cryo sectioning, the thinnest possible sections to be made is about 5 μm, making it difficult to resolve small objects with this preparation technique. The cryo sectioning technique is not suitable for hard, brittle, porous materials, since it is difficult to make connected sections. By using highly advanced preparation methods properly intended for electron microscopy, such as *e.g.* high-pressure freezing followed by freeze-substitution and embedding, it may be possible to make thinner cryo sections with higher resolution.[8]

The preparation procedure for plastic sections involves five steps: fixation, dehydration, embedding, sectioning, and staining. Fixation is performed in order to denaturate the proteins in such way that the cells and tissues are preserved from osmotic damage and shrinkage and so that they can stand subsequent treatment, such as dehydration, embedding, cutting, and staining, without great changes to their morphology.[9] To avoid extraction of the material, it is important to use a solvent for the fixation compounds that is as similar as possible to the fluid of the sample, *i.e.* similar pH, ionic composition, *etc.* Fixation is usually performed by treatment with a solution of the fixative agent. Sometimes it is necessary not to put samples in fluids before fixation, and then fixation in vapor is a possibility. This is recommended for hard and brittle materials such as snacks and bread. Formaldehyde, glutaraldehyde, and acrolein, all containing aldehyde groups, achieve their effects as protein fixatives *via* the formation of cross-links between various reactive groups present in the polypeptide chains. Glutaraldehyde is also considered to fix the fat membranes. A second fixation step is recommended using Osmium tetroxide, which has an effect on unsaturated lipids and makes cross-links between C=C double bonds and also effects reactive groups in proteins. During dehydration, ethanol is used to replace water, when the embedding medium is not water soluble, and also to disrupt hydrogen bonds in proteins in order to maintain their secondary and tertiary structures. The embedding is performed in different resins, which polymerize relatively slowly so that the plastic solution will reach all spaces inside the material before it gets stiff. Sectioning for LM is performed using glass knives in an ultramicrotome. The sections cut for LM are $\leq$1 μm thick.

The final micrograph from an LM is affected both by the illumination system in the microscope and by the preparation technique used. When LM is used for investigating particulate dispersions or smears spread out on an objective glass, the material will be imaged as projections. Material embedded in plastic or cryo embedding of compounds, which are then cut (into sections), will be visualized as sections. Fractured surfaces or reflection, giving a 3D effect, can be obtained both from smears and sectioned material when the DIC function is used.

12.2.2 Confocal Laser Scanning Microscopy

Fluorescence is the ability of certain atoms and molecules to absorb the light from a particular wavelength and subsequently remit it at longer wavelengths after a short time interval. Each fluorescent molecule or atom can absorb incident light only of certain specific wavelengths, *i.e.* at certain energy levels known as absorption bands.[10] The emission of fluorescence also occurs at certain specific wavelengths, but these wavelengths are longer and have less energy than the incident light due to loss of energy by interactions with the environment of the molecules or atoms before the emission of light. Each fluorescent molecule or atom can repeat the fluorescence process many times before excited state processes bleach the fluorescent species.

The CLSM is used in fluorescence or reflection mode. Most food ingredients are not autofluorescent and therefore require some sort of specific staining to be visible in the CLSM. The available spectrum for fluorescence in the CLSM ranges from a few hundred nanometers up to around 1000 nm. By using fluorochromes that become excited at different wavelengths, several fluorescence processes can be studied simultaneously. Thus, the spatial distribution of food components can be determined simultaneously and independently by labeling them with specific fluorochromes.

In a CLSM, the light from the samples that is coming from outside the focus plane is blocked by the pinhole. In principle, the light in the beam path is focused in the illumination aperture and in the pinhole of the confocal aperture. The point source of light illuminates just a small conical region of the specimen, and a confocal point detector simultaneously detects light from the same illuminated region. A two-dimensional image is generated by scanning the light point source in the illumination aperture and by the confocal point detector in the confocal aperture in unison. This scanning is performed by a system of scanning mirrors.

The basic components in an epifluorescence CLSM, which uses the same objective for the illumination and collection of re-emitted fluorescent light, are: (1) a set of lasers that can emit different light wavelengths; (2) an acusto optical tunable filter (AOTF), controlling the intensity of the laser light pixel by pixel; (3) illumination aperture; (4) a beam splitter, determining which wavelengths should be directed toward the specimen and which wavelengths should go to the detector; (5) scanning mirrors; (6) the objective, focusing the light into the specimen and collecting the emitted light; (7) the pinhole; and (8) the detector, usually a photomultiplier tube (PMT) or hybrid/GASP. Fluorescence light emanating from regions above or below the in-focus plane comes to focus elsewhere and is blocked by the pinhole and almost totally prevented from reaching the detector. It is not possible to block all the light coming from outside the focus plane due to the wave nature of light, but essentially only in-focus information is recorded and used to form the two-dimensional image. The strength of the optical sectioning is a function of the pinhole size and geometry, the objective, and the features of the objects in the specimen. Intuitively, one would like to use a small pinhole to obtain a thin optical section. In many applications, however, the signal-to-noise ratio sets the limits for how small the pinhole can be. This leads to enhanced spatial resolution and the possibility of noninvasive optical sectioning of the specimen. Three-dimensional structural information can be obtained by collecting thin optical sections (images) at different depths in the specimen by moving the specimen in an axial direction. More information about the principle of CLSM can be found in Lorén *et al.*,[7] and Pawley.[11]

Resolution is always a compromise between spatial resolution, temporal resolution, and noise, which involves intensity and spectral resolution, dynamic range, and the signal-to-noise ratio.[12] The objective is the most instrumental component in determining the information content of an image. The role of the objective is to focus the laser beam and to ensure constructive interference at the focal point. The three-dimensional spatial light intensity distribution of the light around the focal point is mathematically described by the point-spread function (PSF). The PSF is closely linked to the spatial resolution of the CLSM, and a more localized PSF will always produce an image with higher resolution.[13] In practice, the most important parameters that a CLSM user must consider are the magnification, the numerical aperture (NA), and the working distance of the objective.

Recently, the CLSM technique has been further improved by the introduction of multispectral detectors, detectors with high quantum efficiency and less noise, high-quality objectives, efficient beam splitters, faster scanning, super-resolution technology, and large arrays of lasers/white lasers.

Several quantitative CLSM techniques can be performed including: (1) image analysis to determine size, shape, phase volume, connectivity, length, *etc.*; (2) fluorescence recovery after photobleaching (FRAP), single particle tracking (SPT), Fluorescence correlation spectroscopy (FCS) and image correlation spectroscopy to quantify diffusion, flow, and binding properties; (3) Förster resonance energy transfer (FRET) to determine whether two fluorophores are within a certain distance of each other, which is useful in biology and chemistry, to mention a few.

FRAP is a commonly used fluorescence microscopy-based technique to determine local molecular transport and interactions at the micrometer scale. A brief high-intensity laser pulse is locally applied to the sample, causing substantial photobleaching of the fluorescent molecules within the illuminated area. This causes a local concentration gradient of fluorescent molecules, leading to diffusional influx of intact fluorophores from the local surroundings into the bleached area. Quantitative information on the molecular transport can be extracted from the time evolution of the fluorescence recovery in the bleached area using a suitable model.[14] FRAP is not so far widely used to investigate foods even if the potential is large. However, there are some examples from different product categories. These involve oil migration in fat crystal networks,[15,16] diffusion of pectin methylesterases in pectin gels and solutions,[17] mechanisms for stabilizing acidified milk drinks,[18] ripening of model cheeses,[19] and diffusion in alginate hydrogels[20] and carrageenan hydrogels,[21] to mention a few.

12.3 Imaging and Dynamics of Foods

For a successful characterization of complex heterogeneous food microstructures, it is important to select different microscopy techniques. A large variety of different preparation techniques for food and other challenging soft materials containing water and fats are also necessary. The selection of techniques is dependent on the length scales of interest in the structure, and if time-dependent phenomena, such as foaming, aggregation, gelation, melting, phase separation, disintegration, migration, *etc.*, shall be visualized. To make this selection properly, it is important to have a good knowledge of the food structures and their properties.

12.3.1 Cereal-based Porous and Heterogeneous Foods

12.3.1.1 Bread—A Porous Heterogeneous System

For the understanding of properties of porous material, such as bread, biscuits, and crisps, it is important to differentiate between the 3D structure, the porosity/heterogeneity of the samples, and the structure of the matrix. This can be done advantageously by combining CLSM and LM. CLSM is very useful in getting an understanding of the 3D structure at a low to medium magnification, in porous cereal material that autofluoresce, which entails no staining; staining can be tricky with some porous materials that can collapse when liquid staining is added. With CLSM, depths of a few millimeters can be analyzed (if the sample is transparent enough with good refractive index matching) by adding several images together into a 3D projected image, while in LM several images from different sections in a depth of some micrometers can be added. High magnification and resolution are, of course, possible to obtain with CLSM, but the advantage with LM when analyzing cereal material is that it is easier to get a clear distinction between the different components of starch, amylose, amylopectin, protein, fat, yeast, fibers, and cell walls, in the same image and at the same time. In Figure 12.1, an example of the structure of white bread is shown in a CLSM image of low

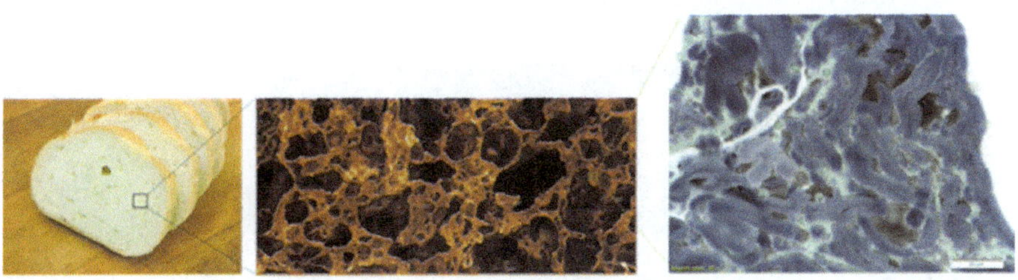

Figure 12.1 From left to right: photo, CLSM, and LM images of white bread. The CLSM image is stained for sodium ions and is shown in bright, and in the LM image, starch is shown in violet, fat in brown, protein in green, yeast cells in green spheres and fiber parts in white. Scale bar is 500 μm in the CLSM image and 20 μm in the LM image.

magnification and in an LM image at high magnification. In the CLSM image, the bread is stained for sodium using CoroNa. The salt was dissolved in the water phase when mixing together the ingredients for the bread prior to baking and is therefore evenly distributed in the bread structure. The image is composed of several images paired together in both the longitudinal and the vertical directions, resulting in an image with size of 9.5×4.4×1.0 mm. Sizes and distribution of pores and lamellas in the bread are here clearly visualized. In the LM image in Figure 12.1, the structure of a separate lamella is shown. This is an image obtained from a plastic embedded sample, which is why good resolution is obtained, and it is possible to resolve the details of starch granules (violet), fat (brown), protein (green), yeast cells (green spheres), and fiber/cell wall (white). In this bread, the granules are intact, and not much amylose has leached out. Somewhat more amylose (blue) is accumulated in the interior of the granules.

Croissant is a bread with a somewhat different structure, as shown with four different microscopy techniques in varying length scales in Figure 12.2. Croissants are prepared using a sheeting technique with a lot of fat between the layers, which gives this special layered structure with large pores in between, shown in the photo and the X-ray image to the left in Figure 12.2. The structure of the separate lamellas is shown with CLSM and LM, and it is demonstrated that all structure components (starch granules, protein, and fat) are distributed in the rolling direction, in contrast to the white bread where no direction is visible. The fat is distributed in large domains, also in the rolling direction. The LM image is prepared using the cryo technique and was cut in 8 μm thickness sections, which is why it does not have the same resolution as the LM image of the white bread in Figure 12.1. By studying ultra-thin (100 nm) plastic sections in transmission electron microscopy (TEM), further information is obtained about the details in the granular structure and components in the surrounding protein phase.

Bakeries are struggling to make gluten-free bread with texture similar to wheat flour–containing bread. In studying the microstructure of different types of gluten-free bread, it is understandable why most of them have not succeeded. To the left in Figure 12.3, an image of a "normal" gluten containing white bread is shown. The starch granules are swelled but only partly

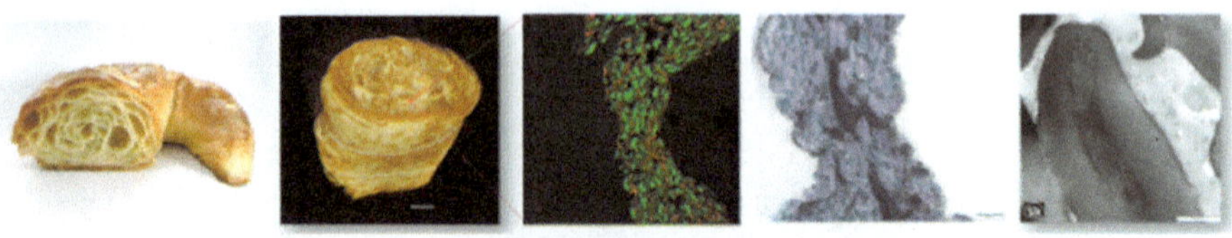

Figure 12.2 From left to right: photo, X-ray tomography, CLSM, LM, and TEM images of croissant. In the CLSM image, starch is shown in green and protein in red, and in the LM image, starch is shown in violet, fat in brown, protein in green, yeast cells in green spheres, and fiber parts in white. Scale bar is 1 cm in the X-ray, 10 μm in the CLSM image, 50 μm in the LM image, and 200 nm in the TEM image.

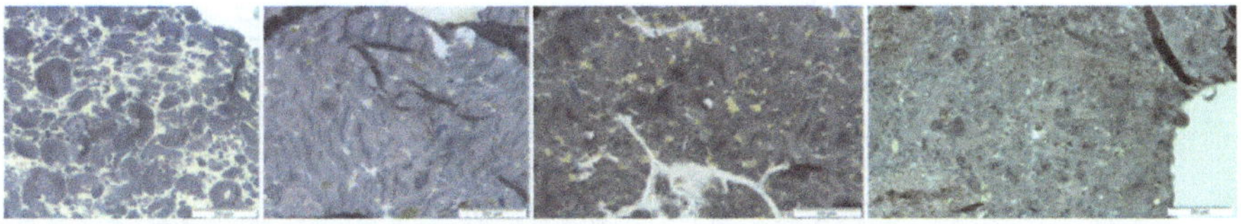

Figure 12.3 From left to right: LM image of white bread, gluten-free white bread, gluten-free bread based on wheat starch and oat, and a commercial gluten-free bread composed of a starch mixture. In the LM image, starch is shown in violet, amylopectin in brown-pinkish, amylose in blue, protein in yellow, and fiber parts in white. Scale bar is 50 μm.

phase separated in amylose and amylopectin, with no leakage to the surrounding protein (gluten) phase (yellow). When baking bread on wheat starch, a compact bread with a texture like sponge cake is obtained. In the micrograph (second image from the left in Figure 12.3), it is also shown that the granules are more densely packed, more swelled, and more separated in amylose (blue) and amylopectin (brown-pinkish), compared to the wheat flour–containing bread. If the wheat starch is mixed with oat flour, a more heterogeneous microstructure with smaller granules, protein, and fibers from the oat is obtained (third image from left in Figure 12.3). The structure is somewhat looser and the texture less spongy, but the protein from the oat does not show the functional properties of gluten forming a continuous phase but is situated in small aggregates (yellow). The image to the right in Figure 12.3 shows a gluten-free bread containing starch from different sources. The structure is very melted with a lot of released amylopectin, which gives the bread a stickier mouthfeel and thereby it is perceived as fresher and juicier. This bread has very likely been baked over a long time at a low temperature.

12.3.1.2 Dough—A Dynamic Porous Heterogeneous System

Except being a porous heterogeneous food system, dough is also a dynamic system, since proofing is ongoing, and thereby the structure is changing continuously during analysis. In a proofed dough, it is of interest, both to characterize the size, form, and equability of the pores on an overall level of structure and to analyze the structure of the lamellas and dynamically study the structural evolution around the growing pores. In Figure 12.4, a proofed dough based on wheat flour is shown, from left to right respectively, in X-ray, LM, and CLSM at different length scales. The X-ray image shows in bright color the area where material is present in the cross section of the whole bun analyzed in a frozen state. Also, the LM image on proofed dough is on a frozen proofed dough where fixation was performed at −20 °C. The dough is formed of a continuous protein phase (green) with a lot of interspersed starch granules (violet). On higher magnification in the CLSM image, it is shown how the protein (red) is built up by a network of varying thicknesses, which also surrounds many of the pores. Yeast is shown as small light green spots, and starch granules are shown in dark green.

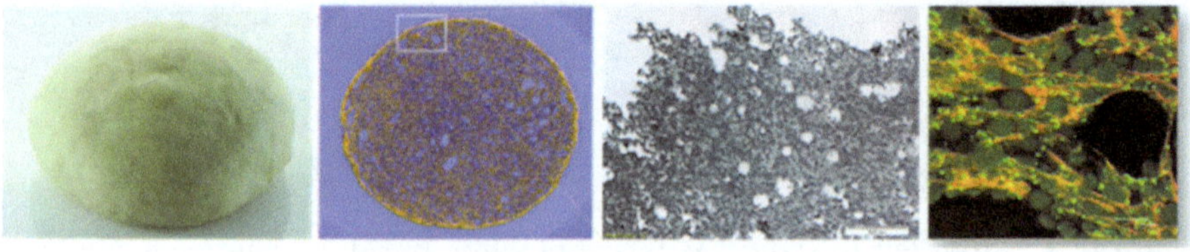

Figure 12.4 From left to right: photo, X-ray, LM, and CLSM images of dough.[25] In the X-ray image, bread material is shown in yellow; in the LM image, starch is shown in violet and protein in green; and in the CLSM image, protein is red, starch granules dark green, and yeast cells light green. Scale bars are 500 μm (LM image) and 30 μm (CLSM image). The diameter of the bun is approximately five centimeters.

The CLSM technique is also advantageous because it is possible to study bulk samples and thereby evaluate structural alteration *in situ*, *e.g.* during expansion. We have seen how pores surrounded by protein threads are restricted in their increase in size and maintain their form, whereas bubbles with a lot of granules around are instead uninhibited in increasing in size, which often results in collapse.

Freezing dough has become a popular and important method for bakeries, allowing improved working hours and making it easier to produce "freshly baked" bread in stores, bars, restaurants, *etc.* However, low-temperature storage is known to negatively affect dough, forming ice crystals that may grow and damage the gluten network during storage.[22–24] The effect of 6 weeks of frozen storage on dough is shown in the CLSM images in Figure 12.5. The fresh reference dough (upper row) has a very well distributed gluten network, but that becomes less homogeneous after frozen storage (lower row), when large areas without gluten appear. On an even higher magnification in the LM images in Figure 12.5, the effect of 6 months of the frozen storage of dough is shown. Following freeze storage, the gaps between the gluten and the starch have considerably increased in size, and the gluten phase has visibly decreased in volume, as if it has lost part of the water.

12.3.1.3 *Fibrous Vegetable Proteins: High Concentration Phase Separated Systems*

The huge trend in altered food intake toward diets with higher vegetarian content has resulted in an increased focus on creating fibrous structures with consistency resembling that of meat. Some vegetable proteins have the functional property to form fibrous structures.[26–28] These proteins originate from different crops, which also contain large amounts of both starch and fibers. A high content of these components makes it difficult to form fibrous structures. Extraction of the protein fraction makes food manufacturing more expensive with the added risk that the functional properties of the proteins might change during the extraction process. Therefore, it is most advantageous to use the crop as it is or just to add a simple fractionation step

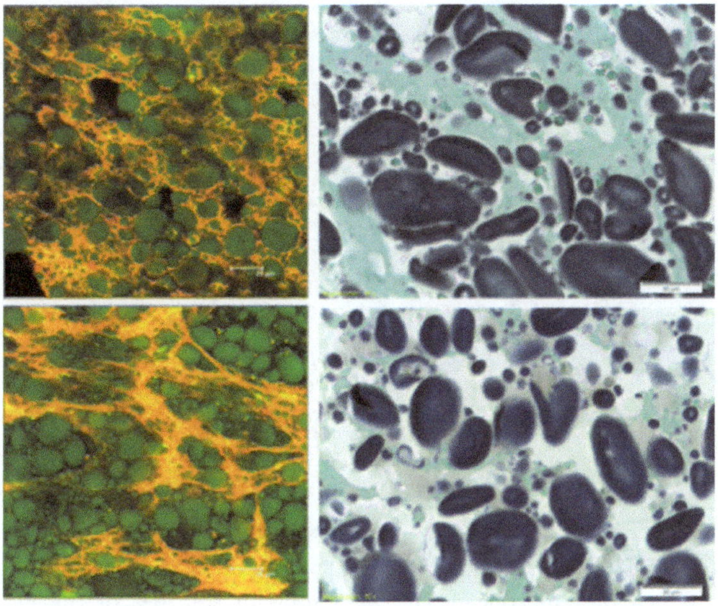

Figure 12.5 Fresh (upper row) and freeze stored (lower row) dough shown in CLSM and LM. In the LM image, starch is shown in violet, protein in green, yeast in dark green spots, and fat in brown, and in the CLSM image protein is red, starch granules dark green, and yeast cells light green. Scale bars are 20 μm (LM image) and 25 μm (CLSM image).

to increase the protein content. In an ongoing study, we are investigating how much of the pea protein can be replaced with starch before losing the ability to form fibrous structures. In Figure 12.6, extruded samples and corresponding LM images of pea protein concentrate (82%) with no starch added (Figure 12.6a) and with 10% (Figure 12.6b) and 15% (Figure 12.6c) of the pea protein replaced with starch are shown. When no extra starch is added to the pea protein concentrate, a rubbery texture is formed, in which a continuous protein phase (yellow) with small parts of starch (violet) and fibers (gray) can be observed. When 10% of the protein is replaced by starch, a desired fibrous texture is formed (see Figure 12.6b). The starch and the protein form a bicontinuous structure. However, when 15% of the protein is replaced by starch (Figure 12.6c), the fibrous structure is lost, the texture is more dough-like, and the LM images show that the continuous phase is no longer the protein but the starch. Phase separation of components at high concentration is an exciting area with a lot of opportunities to create interesting new structures for food textural design.

12.3.2 Plant-based Foods

Confocal scanning laser microscopy has been proven to be a useful technique to visualize plant cells as well as the microstructure of liquid and

Figure 12.6 Photos and LM images of pea protein concentrate extrudates: (a) only pea protein concentrate, 82% protein; (b) 10% of the pea protein concentrate replaced by starch, 72% protein and 10% starch; and (c) 15% of the pea protein concentrate replaced by starch, 67% protein and 15% starch. Scale bar is 50 µm.

semiliquid plant-based foods, such as soups, sauces, and beverages. Stains for cell wall polysaccharides and cell membranes can be used to determine viability and integrity and to localize specific compounds with minimum sample preparation. To get an insight into the overall microstructure of plant suspensions, different dyes are available to identify cell wall polysaccharides. Congo red is a benzidine-based anionic stain that interacts with β-D-glucans, polysaccharides containing continuous β-(1→4)-linked D-glucopyranosyl units, and some hemicelluloses. Calcofluor white binds to cellulose, allowing localization of plant cell walls in the food matrix. Sample preparation for visualization of the microstructure of liquid plant-based foods is minimum; usually a few drops of a diluted dye solution are gently mixed with the sample placed onto a sample holder.

The plant cell wall is made up of primary and secondary cell walls, the latter present in woody plants. Cells are held together by the so-called middle lamella, which contains mainly pectin. The composition of the cell wall depends on plant variety, growing conditions, and ripening stage. In general terms, the primary cell wall surrounds the cell membrane and, for most edible fruits and vegetables, is mainly composed of cellulose, hemicelluloses, and pectin. These polysaccharides are assembled into a complex structure, which is disrupted during food processing and later during digestion in the gastrointestinal tract. Cell walls provide rigidity and mechanical support to the plant and are the source of the structural components in plant-based foods. The texture of plant-based foods is directly related to their microstructure, which in turn is characterized by particle size, shape, and hardness. Therefore, visualization of the microstructure, with minimum sample disruption, is key to understanding, designing, and predicting the textural properties of plant-based foods.

Plant-based foods can be described as suspensions produced from comminuted plant material, containing a liquid phase with mainly salts, soluble sugars, and organic acids, and a dispersed phase formed of all the plant-insoluble solids. In general, conventional processing conditions such as thermal and shear treatments disrupt the plant tissues, breaking cells and releasing plant polysaccharides into the continuous liquid phase. The "break" process is a common manufacturing technique in the production of tomato products. It consists of a mechanical disruption and thermal treatment applied simultaneously. The temperature of the *hot break* varies from 85 °C to 100 °C, and the *cold break* is usually below 70 °C. At temperatures around 60 °C, the cell membranes are denatured, reducing cell turgor and leading to deformed cells with more elongated shapes. Increasing the severity of the thermal treatment leads to further softening of the cell material. Apart from the effect on cell turgor, the applied temperature also influences enzymatic activity, which impacts cell wall hardness and the release of polysaccharides. As revealed by CSLM and light microscopy (Figure 12.7), the

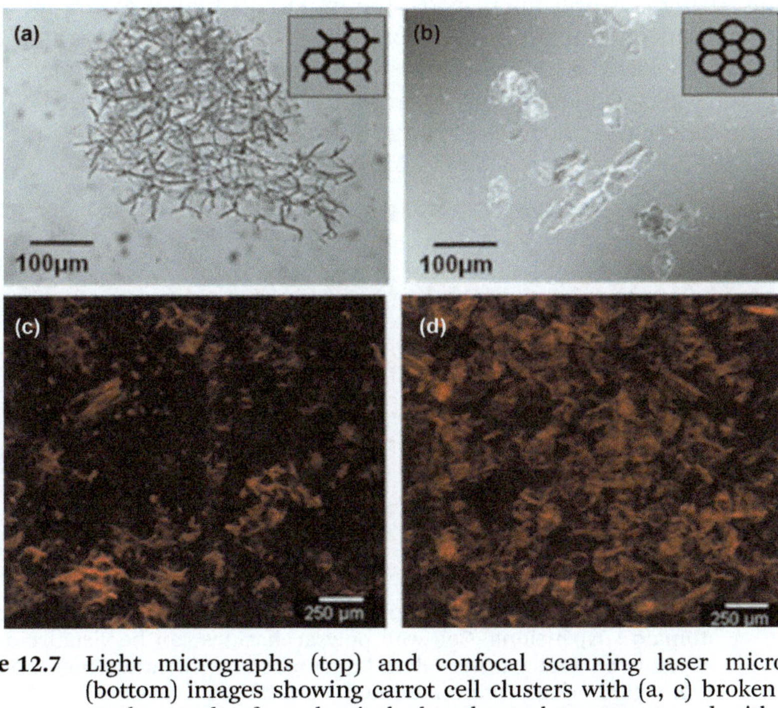

Figure 12.7 Light micrographs (top) and confocal scanning laser microscopy (bottom) images showing carrot cell clusters with (a, c) broken edges as the result of mechanical plus thermal treatment and with (b, d) smooth cell surfaces obtained from thermal plus mechanical treatment. Congo red was the dye used (1 wt% in water).[31]

order of thermal and mechanical treatments during the break process affects particle structure.[29,30] Failure of the tissue occurs across the cell walls when mechanical treatment is applied prior to heating, leading to clusters of cells with broken cell edges on the surface (Figure 12.7a). Cell–cell separation occurs when the thermal step is used prior to the mechanical treatment, due to reduction of intercellular adhesion, as a result of pectin solubilization and leading to particles with smooth surfaces (Figure 12.7b).

Furthermore, confocal micrographs show that the microstructure of carrot suspensions produced of smooth clusters is more homogeneous than those of broken clusters. This observation is likely due to the combination of a reduced particle size, a narrower size distribution, and the smooth surface of the particles, which allow closer particle packing, filling up the voids that were visible in the samples of broken clusters.

High-pressure homogenization (HPH) is a high-shear process widely used in the food industry to manufacture liquid and semiliquid products with homogeneous and smooth textures. In the homogenizer, pressure is used to force a suspension through a small constriction. During the HPH process, the particles are deformed, and they break up due to the different types of flows occurring in the homogenizer, elongational flow toward the constriction, and shear flow within the constriction. It has been shown that HPH processing conditions have different impacts on particle morphologies and microstructure. Using confocal microscopy, the impact of high-pressure (60 MPa) homogenization on tomato suspensions could be visualized. As it can be observed in Figure 12.8, HPH breaks the tomato cells into fragments and disrupts cell wall structure, releasing cell wall polysaccharides (in red) into the continuous phase (Figure 12.8b), rising the viscosity of the tomato products.

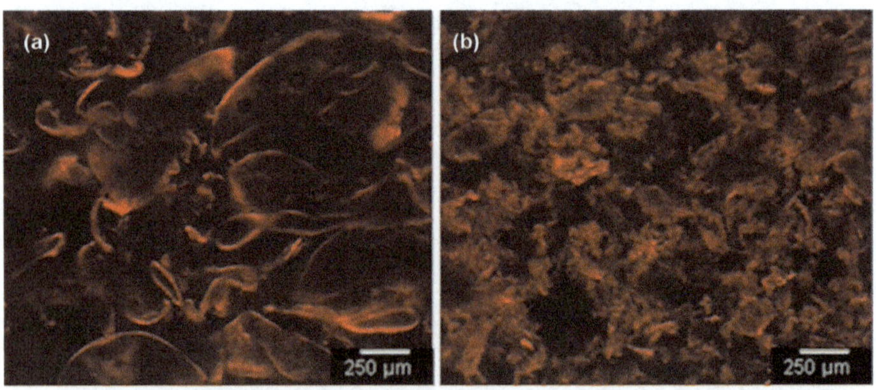

Figure 12.8 Effect of high-pressure homogenization on the microstructure of tomato suspensions. Cell wall polysaccharides can be visualized with Congo red (1 wt% in water) (a) before and (b) after high-pressure homogenization. Figure modified from effect of mechanical and thermal treatments on the microstructure and rheological properties of carrot, broccoli, and tomato dispersions.
Adapted from ref. 29 with permission from Elsevier, Copyright 2011.

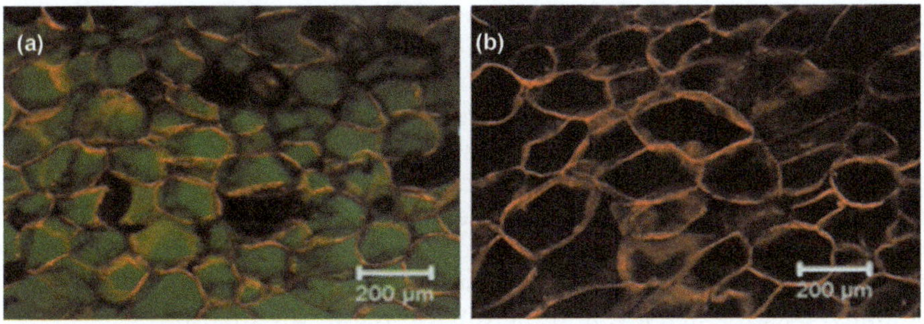

Figure 12.9 Apple parenchyma tissue stained with FDA/P4B (a) after short thermal treatment (70 °C, 5 min) and after prolonged thermal heating (70 °C, 20 min). Cell walls are stained with P4B (in red), and cell membranes are visualized with FDA (in green).
Figure adapted from Liu. Adapted from ref. 33 with permission from the Royal Society of Chemistry.

The plant cell membrane keeps plant cells turgid and significantly impacts on the perceived texture of fruits and vegetables, such as in juiciness and crunchiness. Confocal experiments in which double staining with fluorescein diacetate (FDA)/pontamine fast scarlet (P4B) was used on apple tissues, are depicted in Figure 12.9.[32] FDA action indicates cell viability and cell-membrane integrity. A short thermal treatment (70 °C, 5 min) did not impact on cell membrane (Figure 12.9a); however, after a prolonged thermal treatment (70 °C, 20 min), the membranes are no longer intact, and only the cell wall remains (Figure 12.9b).[33]

12.3.3 Fat-rich Foods

Light microscopy and confocal laser scanning microscopy are powerful tools for characterization (at the micrometer level) of fat-rich food structures. It is, however, important to take care during sample preparation. For nonsmearable samples, light microscopy requires plastic embedding and thin sectioning. During this preparation, there is the risk that part of the fat is washed away. However, CLSM is suitable for fat characterization due to minor sample preparation. For CLSM analysis, the fat-rich food is preferably stained by adding it on a cover glass onto which the staining has been dried. CLSM is in general also very useful for the characterization of food emulsions. The differences in refractive index between the oil and water phases make it difficult to analyze the structure deep into an emulsion system due to the scattering of the light. CLSM, however, normally gives very nice images of the top layers of emulsions.

12.3.3.1 Chocolate

Chocolate has a complex multicomponent hierarchical structure with delicate properties.[34] The structure consists of irregular micrometer-sized cocoa particles and sugar crystals dispersed in a continuous fat phase. The fat phase forms a crystal network with liquid fat between the fat crystals.[35] The hydrophilic sugar crystals need to be covered by an emulsifier, which is normally lecithin, to obtain the desirable rheological properties of chocolate. The chocolate properties are strongly linked to the structure. The most important component of chocolate is cocoa butter. The formed cocoa butter crystals control, for instance, the melting, stability, shelf life, texture, and mouthfeel properties of chocolate. Cocoa butter crystalizes into six different polymorphs. Form V is the most desired polymorph in chocolate due to its stability, packing density, and melting profile in the mouth.

Polarized light microscopy (PLM) utilizes the birefringence of materials. The contrast is between birefringent and nonbirefringent elements. A birefringent material is a material that has two different refractive indexes and so transmits the light in an anisotropic way. A nonbirefringent material has only one refractive index, and light is in this case transmitted isotropically. Cocoa butter structure behavior has previously been studied using polarized light microscopy.[36–38] By taking advantage of the birefringence of fat crystals, this method allows seeing them brightly in contrast to the dark liquid fat. The evolution of the fat crystal network can then be followed as a function of time and temperature for a qualitative or quantitative analysis. Figure 12.10 shows PLM images of untempered cocoa butter stored at two different temperatures for 3 days. The fat crystals are seen as feather-like bright spherulites in the images. Storage temperature influences the crystal growth. Lower temperature leads to more nucleation and formation of many small crystals (see Figure 12.10a). Higher temperature results in fewer and larger crystals (see Figure 12.10b).[39]

Confocal laser scanning microscopy can also be used to determine fat crystal growth in cocoa butter. The fat crystals are visualized indirectly by introducing a few ppm of fluorescent dye (BODIPY FL16) into the melted cocoa butter. Then the fat crystals start to grow and exclude most of the fluorescent dye from the crystal. Therefore, the fat crystals appear as dark objects in the CLSM images. Figure 12.11 shows an example of fat crystal

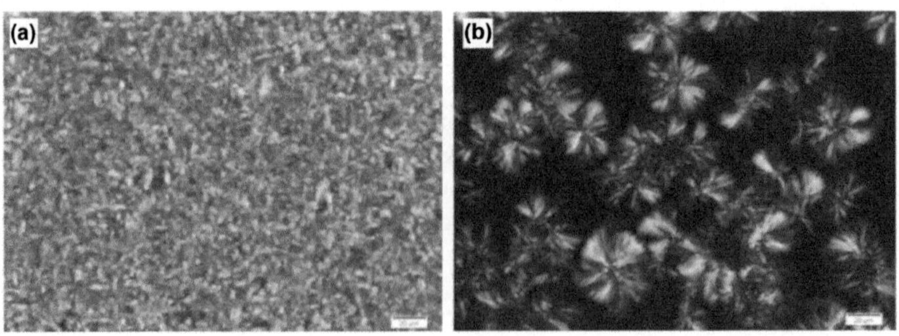

Figure 12.10 Polarized light microscopy images of untempered cocoa butter after 72 hours at (a) 8 °C and (b) 23 °C.

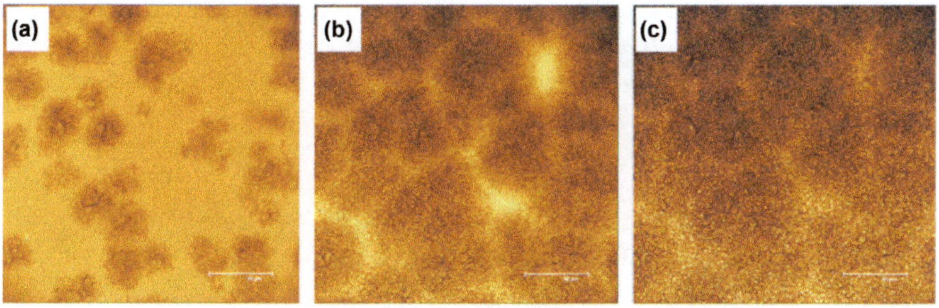

Figure 12.11 CLSM micrographs showing fat crystal growth in seeded cocoa butter: (a) 5 min; (b) 10 min; (c) 20 min, at 14 °C. The scale bar represents 50 μm.
Adapted from ref. 40 with permission from Elsevier, Copyright 2011.

growth in a seeded cocoa butter sample.[40] The samples were seeded with 1 wt% βVI crystals at 34 °C. The image sequence in Figure 12.11 shows how the fat crystals grow and that the structure becomes denser with time. CLSM has been used to show that all chocolate ingredients and precrystalization techniques have a large impact on fat crystalization kinetics and the resulting microstructure.[40]

The migration of fat is a very important phenomenon in chocolate.[41] It is strongly connected to the shelf life properties of chocolate, especially in pralines. The migration of fat to the surface results in fat crystalization at the surface. The fat crystals give rise to fat bloom, which is the grayish haze formed on chocolate. Fat bloom is associated with a bad quality or uneatable chocolate product. Fat bloom is the cause of a lot of chocolate waste. The problem is even more pronounced in pralines with fat-rich filling. In pralines, the fat migrates from the filling into the chocolate shell, dissolves fats in the shell, and then migrates out to the surface. Therefore, it is important to retard fat migration through the chocolate shell. Confocal laser scanning microscopy and fluorescence recovery after photobleaching offer the possibility to determine the local fat migration rate in the chocolate shell.[15,16] Figure 12.12 shows an example of a FRAP measurement in nonseeded badly tempered cocoa butter. The mean intensities in the region-of-interest (ROI) as a function of time are marked as red stars in Figure 12.12. The mean intensities are fitted to an equation (black line) describing the diffusion. The rate of recovery is proportional to the local diffusion rate, which is estimated by the fitting.

12.3.3.2 *Mayonnaise*

Mayonnaise is a common fat-rich food that is suitable for analysis with confocal laser scanning microscopy. Figure 12.13 shows examples of different food emulsions containing different types of stabilizing components. The model food system in Figure 12.13a is a 20% rapeseed oil emulsion

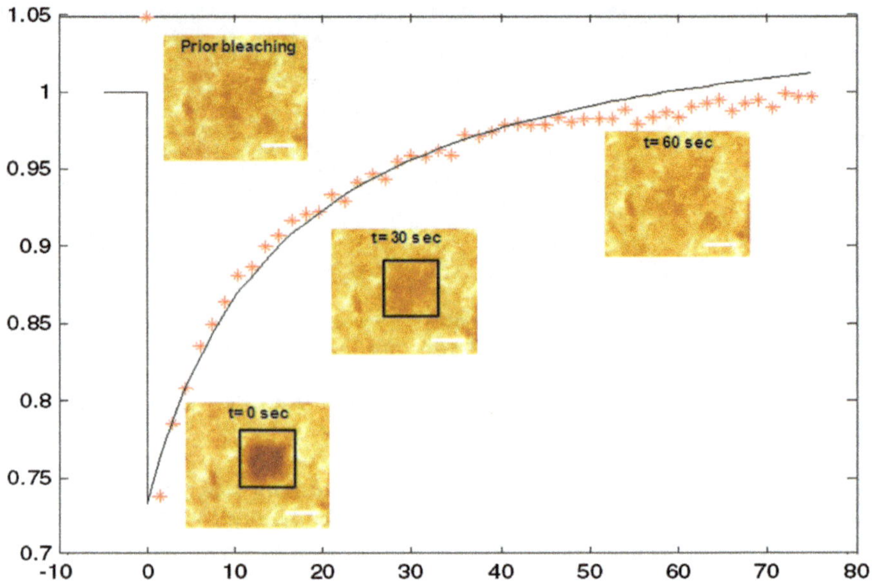

Figure 12.12 FRAP measurement in nonseeded badly tempered cocoa butter. The graph presents a recovery curve, and the CLSM micrographs insets show the intensity of the recovery during the FRAP measurement.[42] The scale bar is 5 μm. The bleached region-of-interest is 7×7 μm.

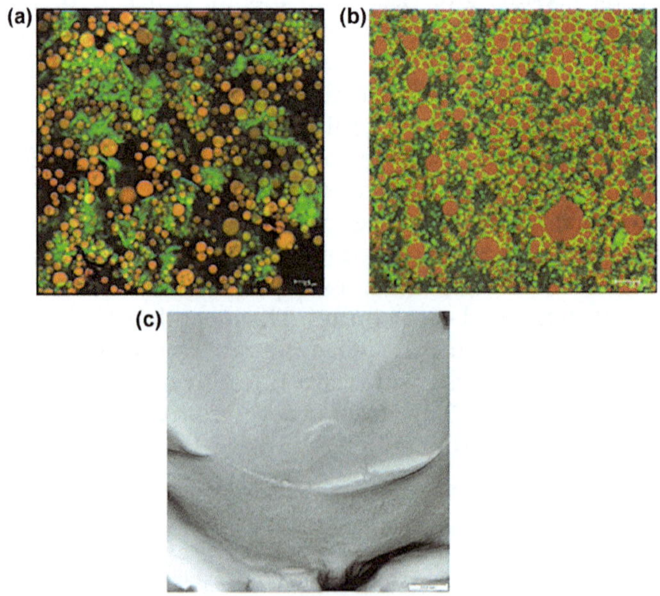

Figure 12.13 CLSM images of (a) food model emulsion with fat in red and microfibrillar cellulose in green and (b) light mayonnaise, 32% fat with fat in red and protein in green. Scale bar is 10 μm. (c) TEM image of a freeze fractured fat droplet in milk. Scale bar is 100 nm.

stabilized with 0.5% microfibrillar cellulose. This emulsion has a consistency resembling that of sour milk. The cellulose gives a steric stabilized emulsion, forming a network-like structure (green) between the fat droplets (red). Instead, in a mayonnaise, Figure 12.13b, the fat drop–stabilizing egg yolk lies around the droplets like an emulsifier. To visualize the fat droplets at a nanometer scale, electron microscopy is required. Instead of plastic embedding and risk of washing away the fat, freeze fracturing is an excellent method. The combination of CLSM, freeze fracturing, and TEM is a powerful tool to determine mayonnaise structure at different length scales.[43]

12.4 Conclusions and Future Perspective

The use of light or confocal scanning laser microscopes, in combination with temperature and deformation stages, allows the *in situ* investigation of dynamic mechanical properties in relation to microstructural changes. This leads to a better understanding of the microscopic mechanisms related to mechanical deformation and fracture behavior to aid product texture design. Particle tracking microrheology is an emerging technique, which utilizes the Brownian motion of fluorescence particles embedded in soft systems to probe local dynamics and measurement of viscoelastic properties at smaller length scales than conventional rheological methods, allowing detection of local inhomogeneities. The use of monoclonal antibodies and molecular probes specific for plant cell wall polysaccharides, in combination with confocal scanning laser microscopy, allows quantitative assessments of cell wall polymers in fruits and vegetables and all forms of biomass. Figure 12.14

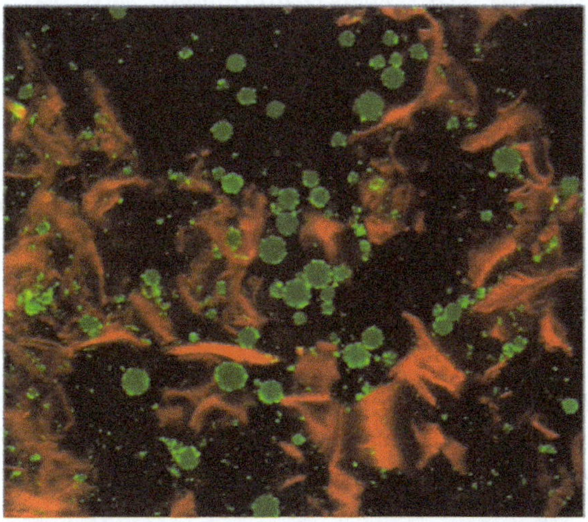

Figure 12.14 CLSM micrograph of tomato purée with olive oil. The cell walls are red, and the pectin is green. The pectin is mainly covering the oil drops. The pectin is stained with LM7. LM7 recognizes a partially methyl-esterified epitope of homogalacturonan that results from a nonblockwise de-esterification process, and it does not bind to unesterified homogalacturonan. The width of the image is 120 µm.

shows an example of tomato purée with added olive oil where pectin is specifically labeled with LM7. The possibilities to analyze solid samples at high magnifications in the light microscope without using chemical fixation preparations is something that can be enabled using cryo techniques mainly customized for electron microscopy. New super-resolution techniques like stimulated emission depletion (STED) and photoactivated localization microscopy (PALM) allow for resolutions in the order of 50 nm, which is very useful for studies of (for instance) plant materials, protein aggregates, living cells, living bacteria, and interfaces in emulsions. Fluorescence lifetime microscopy (FLIM) might open up new ways to discriminate different substances and autofluorescence. Thus, there are a lot of ongoing interesting developments in CLSM and LM, which will open many new exciting possibilities for studies of food microstructure in the future.

Acknowledgements

Annika Krona, RISE, Annika Altskär, RISE, Stephen Hall, LTH and 4D Imaging Lab at Lund University, and Sophia Wassén, RISE, are acknowledged for contributing unpublished images to the book chapter.

References

1. A.-M. Hermansson, M. Langton, and C. Olsson, in *Gums and Stabilisers for the Food Industry*, ed. P. A. Williams and G. O. Phillips, Royal Society of Chemistry, Cambridge, 2004, pp. 3–11.
2. J. M. Aguilera, *J. Food Eng.*, 2005, **67**, 3.
3. I. Norton and W. J. Frith, *Food Hydrocolloids*, 2001, **15**, 543.
4. D. J. McClements, *Understanding and Controlling the Microstructure of Complex Foods*, CRC Press, 2007.
5. N. Lorén, M. Nydén and A.-M. Hermansson, *Adv. Colloid Interface Sci.*, 2009, **150**, 5.
6. P. J. Lillford, *J. Texture Stud.*, 2011, **42**, 130.
7. N. Lorén, M. Langton, and A.-M. Hermansson, in *Understanding and Controlling the Microstructure of Complex Foods*, ed. D. J. McClements, CRC Press, 2007, p. 232.
8. D. Vanhecke, W. Graber and D. Studer, *Methods Cell Biol.*, 2008, **88**, 151.
9. J. Brontë Gatenby, and H. W. Beams, *The Microtomist's Vademecum. A Handbook of the Methods of Animal and Plant Microscopic Technique*, J A Churchill Ltd., London, 1950.
10. B. Herman, *Fluorescence Microscopy*, BIOS Scientific Publishers, Royal Microscopical Society, UK, 1998.
11. J. B. Pawley, *Handbook of Biological Confocal Microscopy*, Springer, 2006.
12. C. J. R. Sheppard and D. M. Shotton, *Confocal Laser Scanning Microscopy*, Bios Scientific Publishers, Royal Microscopical Society, UK, 1997.

13. A. Egner and S. W. Hell, *Trends Cell Biol.*, 2005, **15**, 207.
14. N. Lorén, J. Hagman, J. Jonasson, H. Deschout, D. Bernin, F. Cella, A. Diaspro, J. G. McNally, M. Ameloot, N. Smisdom, M. Nydén, A.-M. Hermansson, M. Rudemo and K. Braeckmans, *Q. Rev. Biophys.*, 2015, **48**, 323.
15. S. Marty, M. Schroeder, K. W. Baker, G. Mazzanti and A. G. Marangoni, *Langmuir*, 2009, **25**, 8780.
16. L. Svanberg, L. Ahrné, N. Lorén and E. Windhab, *Food Res. Int.*, 2011, **44**, 1339.
17. P. Videcoq, K. Steenkeste, E. Bonnina and C. Garnier, *Soft Matter*, 2013, **9**, 5110.
18. S. Jensen, C. Rolin and R. Ipsen, *Food Hydrocolloids*, 2010, **24**, 291.
19. J. Floury, M.-N. Madec, F. Waharte, S. Jeanson and S. Lortal, *Food Chem.*, 2012, **133**, 551.
20. E. Schuster, J. Eckardt, A.-M. Hermansson, A. Larsson, N. Lorén, A. Altskär and A. Ström, *Soft Matter*, 2014, **10**, 357.
21. J. Hagman, N. Lorén and A.-M. Hermansson, *Food Hydrocolloids*, 2012, **29**, 106.
22. M. E. Barcenas and C. M. Rosell, *Food Chem.*, 2006, **95**, 438.
23. V. O. Selomulyo and W. Zhou, *J. Cereal Sci.*, 2007, **45**, 1.
24. J. Yi and W. L. Kerr, *J. Food Eng.*, 2009, **93**, 495.
25. C. Öhgren, N. Fabregat and M. Langton, *LWT–Food Sci. Technol.*, 2016, **68**, 626.
26. S. H. Prudêncio-Ferreira and J. G. Arêas, *J. Food Sci.*, 1993, **58**, 378.
27. K. Liu and F. H. Hsieh, *J. Agric. Food Chem.*, 2008, **56**, 2681.
28. R. Osen, S. Toelstede, F. Wild, P. Eisner and U. Schweiggert-Weisz, *J. Food Eng.*, 2014, **127**, 67.
29. P. Lopez-Sanchez, J. Nijsse, H. C. Blonk, L. Bialek, S. Schumm and M. Langton, *J. Sci. Food Agric.*, 2011, **91**, 207.
30. C. Pickardt, G. Dongowski and H. Kunzek, *Eur. Food Res. Technol.*, 2004, **219**, 229.
31. P. Lopez-Sanchez, PhD thesis, *Microstructure and rheological properties of plant particle suspensions prepared using different physical treatments*, Chalmers University of Technology, 2011.
32. J. Thomas, N. A. Idris and D. A. Collings, *J. Microsc.*, 2017, **268**, 13.
33. D. Liu, P. Lopez-Sanchez, and M. J. Gidley, *Food Funct.*, 2019, **10**, 3008–3017.
34. D. Rousseau, in *Understanding and Controlling the Microstructure of Complex Foods*, ed. D. J. McClements, Woodhead Publishing Limited, Cambridge, 2007, pp. 648–687.
35. F. Peyronel, N. A. Acevedo, D. A. Pink and A. G. Marangoni, in *Crystallization of Lipids – From Molecules to Crystal Networks*, ed. K. Sato, Wiley-Blackwell, UK, 2017.
36. A. Marangoni, *Crystal Growth Design*, 2003, **3**, 95.
37. A. Bowser, *Manufact. Confect.*, 2006, **86**, 115.

38. Y. Kinta and W. Hartel, *J. Am. Oil Chem. Soc.*, 2009, **87**, 19.
39. L. Balestra, Master's thesis, *Characterization of chocolate microstructure: Development and comparison of microscopy techniques to evaluate cocoa butter crystallization*, Chalmers University of Technology, 2013.
40. L. Svanberg, L. Ahrné, N. Lorén and E. Windhab, *J. Food Eng.*, 2011, **104**, 70.
41. G. R. Ziegler, in *Science and Technology of Enrobed and Filled Chocolate, Confectionery and Bakery Products*, ed. G. Talbot, Woodhead Publishing, Cambridge, 2009, pp. 185–210.
42. L. Svanberg, PhD thesis, *Pre-crystallization processing of dark chocolate: Effect on fat crystal structure and product properties*, ETH Zürich, 2012, DOI: 10.3929/ethz-a-007599212.
43. M. Langton, E. Jordansson, A. Altskär, C. Sørensen and A.-M. Hermansson, *Food Hydrocolloids*, 1999, **13**, 113.

CHAPTER 13

Food Structure Characterisation Using Small-angle Scattering Methods

MARTA MARTINEZ-SANZ,[b] AMPARO LOPEZ-RUBIO[b] AND ELLIOT GILBERT*[a]

[a] ANSTO, Australia; [b] Instituto de Agroquímica y Tecnología de Alimentos (IATA), Valencia, Spain
*Email: epg@ansto.gov.au

13.1 Introduction

Whether we consider individual ingredients or whole products, food, at its essence, is a mixture of components of multiple sizes (atomic, molecular, macromolecular) that can exist in various physical states (solution, colloidal, particulate) and that exhibit different energies (translational, rotational, vibrational). Particulate materials may be composed of atoms or molecules that exhibit long-range order (crystalline) or lack such order (amorphous). Food may exhibit properties that are both liquid-like and solid-like, and this behaviour is often time dependent. It is the combination of these attributes of structure, motion, and response to external stressors (*e.g.* temperature, pressure, shear, *etc.*) that informs the suitability of given ingredient, how a food can be developed (*e.g.* processing) and the extent to which that product meets a specific need (*e.g.* physiological). Sanguansri and Augustin stated more than a decade ago: "The future development of food products will require an understanding of the relationship between nano-,

Food Chemistry, Function and Analysis No. 18
Handbook of Food Structure Development
Edited by Fotis Spyropoulos, Aris Lazidis and Ian T. Norton

Published by the Royal Society of Chemistry, www.rsc.org

supramolecular and higher order structures and of functionality on a physical, nutritional and physiological level. . . . [F]ood scientists and technologists will have to be engaged in nanoscience and nanotechnology.''

Structure–property relationships, *i.e.* the connection between food structure and the way a product behaves, is central to materials science and product design,[1] and, as a result, measurements are needed. Food materials science, which includes elements of physics, chemistry, biology, and mathematics, is underpinned by such measurements. Challenges to the manufacturer may arise from consumer-driven changes (*e.g.* lower-fat foods) or inconsistency in supply due to environmental variations. To measure and control food structure, to make predictions of behaviour, and to deliberately engineer consistent or new products, characterisation techniques are essential.[1–3] Scattering methods are indispensable within the arsenal of such methods.

13.2 Scattering Techniques

It is impossible within the scope of a single chapter to describe the full range of scattering techniques, let alone the broad range of applications; indeed, entire books have been written just on a single method. Here, we provide a brief terminology of scattering methods and their application to various length scales. We then focus on how structure can be measured on the nanoscale with a specific approach—small-angle scattering, both of X-rays or neutrons—and present the information that has been derived from such measurements across a diverse range of food systems.

Scattering refers to the change in direction of a source of radiation as a result of its interaction with a material due to some structural variation within; we describe these as inhomogeneities which, when combined with ''contrast'', enables these variations to be discerned. The scattering of X-rays depends on differences between the electronic structure of atoms and, for neutron scattering, on differences between the nuclear structure of atoms. In addition, the directional change can also be accompanied by no energy change (elastic) or an energy change (inelastic). Elastic neutron scattering techniques include small-angle scattering (SANS), ultra-SANS, neutron reflectometry, and neutron diffraction. As a whole, these techniques provide information about structure ranging from the sub-angstrom ($<10^{-10}$ m) to supra-micron size range ($>10^{-5}$ m). This could be packing between chains in fat crystals (neutron diffraction), the structure of a casein micelle (SANS), the conformation of a protein at an interface (neutron reflectometry), or the arrangement of droplets in an emulsion (ultra-SANS). Equivalent X-ray techniques accessing the same size ranges exist. All these techniques provide structural information on a specific length scale. The accessible spatial range using elastic scattering techniques is illustrated in Figure 13.1 for starch, in addition to other, complementary, food-relevant characterisation methods. For a comprehensive review of the application of the range of

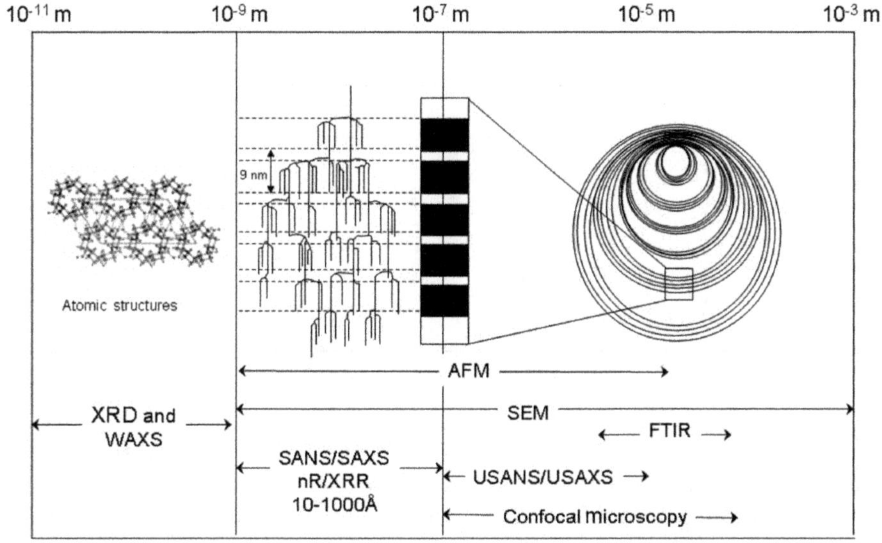

Figure 13.1 The hierarchical structure of starch with relevant elastic scattering techniques and other complementary characterisation methods. Reproduced from ref. 5 with permission from Elsevier, Copyright 2009.

neutron scattering methods to food, the reader should refer to Lopez-Rubio and Gilbert.[4]

The differences in contrast between regions within a material can be appreciated through consideration of atomic composition. While the number of protons in an atomic nucleus defines the elemental type, the number of protons and neutrons defines the elemental isotope. Since neutrons are scattered by the atomic nucleus, this means that the scattering from different isotopes can differ significantly. The most relevant atom for food is hydrogen (one proton in nucleus) and its heavier isotope, deuterium (one proton and one neutron). The extent of neutron scattering is defined by a neutron scattering length whose magnitude effectively defines the size of the nucleus; it is -0.3742×10^{-12} and 0.6671×10^{-12} cm for hydrogen and deuterium, respectively (Table 13.1). This difference is extremely valuable for the study of hydrogen-containing materials and forms the basis of a method known as contrast variation used widely in small-angle scattering. A neutron scattering event may occur when a neutron (dimension $\sim10^{-15}$ m) interacts with an atomic nucleus (approximately 10 times larger) within a material. However, in typical materials, the atoms are separated by $\sim10^{-10}$ m. Thus, if a neutron were the size of a pea, the nucleus would be the size of a tennis ball, and, in a typical material, these tennis balls would be separated by approximately 1 km! As far as a neutron is concerned, materials consist mostly of empty space; consequently, neutrons represent a highly penetrating form of radiation. The latter opens up opportunities for studying materials during realistic processing conditions, since a neutron beam can

Table 13.1 Neutron and X-ray scattering lengths and cross sections[a] for biologically relevant elements. A scattering length, b_i, is associated with a cross section such that $\sigma_i = 4\pi b_i^2$.[b]

Atom	Nucleus	b_{coh} (10^{-12} cm)	σ_{coh} (10^{-24} cm^2)	σ_{incoh} (10^{-24} cm^2)	$b_{X\text{-ray}}$ (10^{-12} cm)
Hydrogen	^{1}H	−0.374	1.76	79.7	0.28
Deuterium	^{2}H	0.667	5.59	2.01	0.28
Carbon	^{12}C	0.665	5.56	0	1.69
Nitrogen	^{14}C	0.940	11.1	0	1.97
Oxygen	^{16}C	0.580	4.23	0	2.25
Phosphorus	^{31}P	0.517	3.31	80	4.22
Sulphur	^{32}S[a]	0.285	1.02	0	4.51

[a]Natural abundance 95%.
[b]coh = coherent; incoh = incoherent.

potentially be transmitted through thick containers. For example, Doutch *et al.* investigated the structural changes occurring during the cooking of starch while simultaneously undergoing a rapid viscoanalysis measurement.[6] For X-rays, the extent of scattering is determined by variations in the electronic (instead of nuclear) structure between atoms in the material. Since the number of protons in an atom is equal to the number of electrons, the X-ray scattering intensity increases linearly with atomic number; consequently, heavier elements in a material will dominate in X-ray scattering. It is also possible to define a scattering cross section, derived from the scattering length, that is a measure of the effective surface area of the target atom presented to the incident radiation; this parameter is proportional to the probability that a scattering event will occur. The difference between the neutron and X-ray scattering lengths and associated cross sections for the most relevant elements found in food is shown in Table 13.1.[7]

13.3 Small-angle Scattering Methods

There is an inverse relationship between scattering angle and the structural variations giving rise to such scattering. Thus, the scattering at small angles gives rise to structural details on length scales greater—from approximately one to several hundred nanometres—than that obtained from conventional diffraction (wide-angle) instruments. This "nano"-size range spans a multitude of disciplines and food systems: proteins (biology and medical sciences), emulsions and microemulsions (polymer and materials science), and phase separation and porosity (materials science and physics). For even larger dimensions, ultra small-angle scattering (USANS/USAXS) yields information up to tens of microns. On these larger size scales, the technique is insensitive to differences between atoms in materials but is, however, sensitive to groups of atoms, enabling molecular structures and their spatial distribution to be determined. As a consequence, small-angle scattering, either with X-rays (SAXS) or neutrons (SANS), has been widely applied to food-based systems. The theory of small-angle X-ray and neutron scattering may be found in a wide range of texts;[8–15] however, for the beginner, a variety of more accessible summaries are available online. The SAS Portal provides an excellent variety of resources on small-angle scattering.[16]

A small-angle scattering experiment conventionally measures the intensity of scattered radiation versus q (equal to $(4\pi/\lambda)\sin\theta$, where 2θ is the scattering angle and λ is the wavelength of the incident radiation). For isotropic scatterers, the data are typically radially averaged to give the scalar quantity q. The measured intensity may be written:

$$I(q) = \frac{d\Sigma}{d\Omega}(q) = (\Delta\rho)^2 V_p^2 N_p P(q) S(q) \tag{13.1}$$

where $d\Sigma/d\Omega(q)$ is the differential scattering cross section, $\Delta\rho$ is the scattering contrast between the species of interest and its surroundings (to be defined), V_p is the volume and N_p the number of species per unit volume, $P(q)$ is the form factor, and $S(q)$ is the structure factor. $d\Sigma/d\Omega(q)$ contains all the information regarding the size, shape, and interactions among the scattering species in the sample. $P(q)$ describes the scattering from an individual scattering species, whereas $S(q)$ describes the scattering between scattering centres. For example, if we consider a protein solution, the protein would be the species of interest with respect to a solvent, $P(q)$ would provide information on the protein size and shape (intraparticle), and $S(q)$ would yield information on the separation between proteins (interparticle).

From eqn (13.1), it is apparent that for small-angle scattering to be observed, there must be a contrast in scattering between the species of interest and its surroundings. The scattering contrast $\Delta\rho$ is equal to $\Delta\rho = \rho_p - \rho_{sm}$, where ρ_p and ρ_{sm} refer to the scattering length densities of the species of interest and solution or surrounding matrix, respectively. The scattering intensity is proportional to $(\Delta\rho)^2$. The scattering length density (SLD) ρ of a molecule with i atoms is defined as $\delta\, N_A\, \Sigma b_i/M$, where δ is the bulk physical density of the molecule, N_A is Avogadro's number, b_i is the scattering length (Table 13.1) of the ith atom in the molecule, and M is the molecular weight. Thus, the contrast may be simply calculated from knowledge of the physical density of the molecule and its atomic composition. Typical scattering length densities are expressed in units of 10^{-6} A^{-2} or 10^{10} cm^{-2}.

Due to differences in the mechanism by which neutrons and X-rays scatter, the intrinsic contrast in a material depends on the radiation used; thus, small-angle neutron and X-ray scattering (SANS or SAXS) provide complementary information. Contrast is a property that can be manipulated *via* a method known as contrast variation (or contrast matching). In the context of using SANS in food materials science, this relies on the different scattering lengths of hydrogen and deuterium. This means that molecules containing hydrogen may have their SLD varied by replacement with deuterium. The SLD for H_2O is negative (-0.56×10^{10} cm^{-2}) and for D_2O is positive (6.38×10^{10} cm^{-2}). Thus, through the preparation of mixtures of H_2O and D_2O, particular components may be strategically contrast-matched so that they effectively become transparent to neutrons. Consider a protein-polysaccharide complex in solution (three components); by selecting the appropriate mixture of solvents, one can "contrast-match" out the contributions from one of the components to "see" only the protein or only the polysaccharide. Figure 13.2 shows the SLDs for water and various biological macromolecules as a function of the deuterium concentration.[11] The range of SLDs that may be achieved through merely mixing normal and heavy water means that a selected biological component in a multicomponent system can be contrast-matched so that it has no contribution to the overall scattering. This approach is not limited to H_2O and D_2O; it is equally possible for non-polar solvents using, for example, normal and deuterated forms of hexadecane. Contrast variation may be achieved based on inherent differences in scattering as a result of chemical composition (as shown in Figure 13.2) or arise from strategic selective deuteration, *e.g.* replacement of hydrogen with deuterium in a fatty acid. It is worthy of note that the solvent scattering may be modified in X-ray scattering by the addition of various contrasting agents such as sucrose, glycerol, or salts. For example, proteins may be contrast-matched in 65% sucrose solutions.[17] In addition, related SAXS methods, known as anomalous SAXS (ASAXS) and resonant soft X-ray

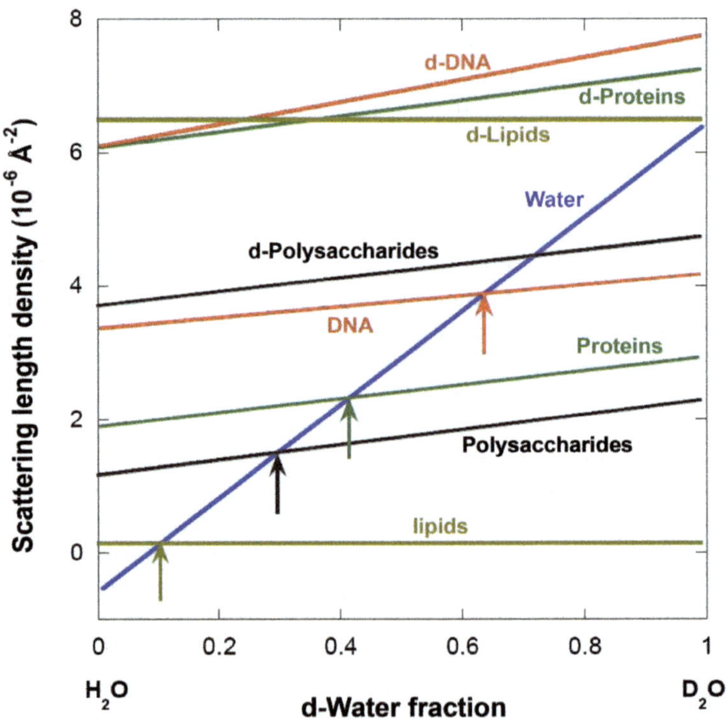

Figure 13.2 Neutron scattering length densities for common food-based materials. Reprinted from ref. 5 with permission from Elsevier, Copyright 2009. (See also ref. 11.)

scattering (RSoXS), use the dependence of the scattering length of a given element, if the energy of the incident X-ray beam is near the absorption edge of this element, to manipulate contrast.[18] The latter is more suitable for lighter elements, but both present technical challenges. Inevitably, the ability to perform both SANS and SAXS on the same sample often reveals even more information due to the complementarity of the techniques. For example, oleogels have been studied with SAXS to reveal that tubular structures of plant-based sterols can entrap oil. When combined with SANS and contrast variation, details of the tubular substructure could be discerned.[19]

To investigate the effect of pH, temperature, pressure, *etc.* on the final characteristics of a food product, it is important to maintain the environment as closely as possible to the relevant conditions. For small-angle scattering, no special preparation techniques are required, *e.g.* dehydration, staining *etc.*; this enables materials to be studied in their "native" state under realistic conditions, including partial hydration, and is appropriate for gel, solutions, aerosols, *etc.* The lack of special preparation also naturally leads to a minimisation of possible sample artefacts. Except in cases where intense synchrotron X-ray beams are used, small-angle scattering is also non-destructive. The beams used for small-angle scattering vary from 10 mm in diameter for SANS, down to the order of microns for synchrotron-based SAXS. Of course, part of the reason for this arises from differences in beam flux (*i.e.* the number of photons or neutrons generated per unit time and unit area). However, these dimensions are somewhat greater than those of molecules; small-angle methods therefore provide bulk information with the scattering representative of the whole sample (as compared, for example, to scanning electron microscopy where only local information is obtained). While this is even more the case for a SANS beam, at modern neutron scattering facilities, measurements using beams down to the size of hundreds of microns are possible.

With all these advantages, why are SAXS and SANS not more widely used? Firstly, one needs to be able to access the appropriate instrumentation (and people who can operate such instruments). For example, at ANSTO, there are two SANS instruments (QUOKKA[20] and BILBY[21]), a USANS instrument (KOOKABURRA[22]), and a lab-based SAXS instrument, all of which have been used to study food materials. However, to access such instruments, experimental proposals must be written that undergo peer review; only then, if beamtime is awarded, can measurements be made. The other major disadvantage is that small-angle scattering techniques yield non-visual information, *i.e.* scattering data yield structural information in "reciprocal space," not in "real space" like microscopy techniques. This means that the scattering data have to be either inverted back to real space or fitted to models describing structures in reciprocal space.

If data are collected over a sufficiently wide q range in the dilute regime, then the data can be Fourier transformed to generate a distance distribution function $P(r)$. This provides information not only on the spatial extent of the particle but also on its shape. This approach is commonly used in the study of protein solutions (*e.g.* Sokolova *et al.*, 2010) using the program GNOM within the ATSAS package;[23] the latter package also includes *ab initio* methods such as shape reconstruction using dummy atom models and tools that utilise known high-resolution protein structures from the Protein Data Bank.

If fitting data to models describing structures in reciprocal space, mathematical models are required, and these must be based on some knowledge of the system (*e.g.* chemical composition, physical density, hydrodynamic radius, X-ray crystal structure); it is therefore essential that information obtained from other techniques is available. Such complementary information must be incorporated into appropriate mathematical models that are physically and chemically robust and meaningful so as to minimise the semi-infinite number of possible alternatives to the nanostructure that can give rise to the scattering data. Conducting SAS measurements with multiple contrasts serves to provide additional constraints in models to reduce such possibilities.

Mathematical form factors $P(q)$ have been developed for many scattering symmetries. Some conventional models are sphere, core–shell, vesicle, cylinder (hollow, core–shell, flexible), ellipsoidal, lamellar, and parallelepiped. These form factors can be combined with a range of structure factors $S(q)$ to describe interparticle interactions, *e.g.* hard sphere, square well, and sticky hard sphere. Modelling can become even more complex if there is a correlation between particle shape and orientation, *e.g.* systems under flow.

Sometimes models are not strictly required, and some structural parameters can be extracted from the data in a straightforward manner. For example, for dilute particles (*i.e.* where $S(q)$ can be considered to be unity), Guinier's law enables the spatial extent of the scattering particle of radius of gyration R_g to be obtained from a plot of the logarithm of intensity against q^2. This parameter is related to the hydrodynamic radius measured in dynamic light scattering. R_g is a single parameter that represents the size of the scattering object. Its interpretation is simple in the case of a spherical scattering particle ($R_g = \sqrt{(3/5)}R$, where R is the radius of the object), but for non-spherical objects, R_g contains contributions from the other axes. If data are collected on an absolute scale (using a direct beam or a known standard sample) and the concentration is known accurately, then it is possible to determine the molecular weight of the particle. If the scattering objects have one or more dimensions that are large compared to the inverse q range, then characteristic slopes may be observable on log–log plots. For example, rod-like objects exhibit q^{-1} power law behaviour, disk-like objects show q^{-2} behaviour, and objects with smooth–sharp interfaces exhibit q^{-4} (Porod) behaviour. From the latter, surface area information may be obtained. Slopes (n) between -3 and -4 characterise rough interfaces of a fractal dimension D, with $n = -(6 - D)$. This is called a surface fractal. In the case of polymer coils, the magnitude of the Porod slope is the inverse of the excluded volume parameter ν. A slope of -2 is a signature of Gaussian chains in a dilute environment, whereas a slope $n = -5/3$ is for fully swollen coils, and a slope of -3 is for collapsed polymer coils. A slope between -2 and -3

is characteristic of "mass fractals" such as branched systems (gels) or networks.

A variety of readily accessible software packages are available, some on commercial platforms and some open-source; they typically have details of the underlying mathematics for each of the models and fitting approaches within their associated documentation (*e.g.* SasView[24]). For a semi-comprehensive listing of software packages, the Software section of the SAS Portal is a good starting point.[16]

13.4 Application of Small-angle Scattering to Characterise Food-relevant Materials

Most foods are complex, heterogeneous systems whose structural organisation is highly relevant for their subsequent nutritional properties. Indeed, many commercial food products are nanostructured composites, in which self-assembly may play a key role.[25] In addition, the individual building blocks of most food products are carbohydrates (principally polysaccharides), proteins, and lipids, with structures on size ranges that can be covered by small-angle scattering techniques. Moreover, the continuous industrial eagerness to develop new products, together with current consumer demands towards healthier food with increased functionality, necessitates an improved knowledge about how these building blocks are structured and interact with one another, along with determining the processing conditions needed for the rational design of novel food products.[4] In the following, we have selected a range of relevant studies in which food components and complex food systems have been characterised with small-angle scattering, providing valuable complementary and, in some cases, unique information about structure and functionality.

13.4.1 Polysaccharides

Polysaccharides are major components in fruits and vegetables, accounting for more than 90% of their dry content.[26] They are also present in food as added ingredients to provide different functionality (*e.g.* thickeners, gelling agents, flocculants, lubricants, bulking and swelling agents, emulsion and suspension stabilisers, adhesives and binders, encapsulants, and additives for low-calorie products). Generally, food polysaccharides can be classified by their native source, as shown in Table 13.2. Besides naturally occurring

Table 13.2 Classification of some of the most important food polysaccharides based on source.

Source	Examples
Higher plants	Cellulose, hemicelluloses, starches, pectins
Seaweed	Agars, alginates, carrageenans
Microorganisms	Xanthans, pullulan, gellan, curdlan

polysaccharides extracted from plants (including seeds, tubers/roots, and fruits) and seaweed, some microorganisms are able to generate polysaccharides as the product of fermentation processes.

The function, and therefore selection, of polysaccharides in food products is associated with their physical properties and structure. Although most food polysaccharides are water soluble or water dispersible, their behaviour is largely conditioned by the nature and extent of intermolecular associations. For instance, polysaccharide chains can be arranged into highly ordered crystalline structures, as is the case for native cellulose microfibrils, which are not soluble in water and have limited moisture accessibility; at the other extreme, polysaccharide chains can exist in solution as fluctuating, disordered coils, interacting with one another only by physical entanglement, as in the case of some amorphous hemicelluloses.

Many polysaccharides are found in nature as partially hydrated systems, with water having a large effect on their structure. To characterise the native structure of food polysaccharides, therefore, and to link it to their specific physical properties, it is essential to utilise techniques that do not require drying. This has been one of the major issues in this research area; for instance, microscopy techniques such as scanning electron microscopy (SEM) and transmission electron microscopy (TEM) have been routinely used to investigate the structure of plant cell wall polysaccharides, but questions arise as to the impact of sample preparation, including drying, on the structure. In this context, small-angle scattering (SAS) techniques offer an excellent approach for the investigation of polysaccharide structure in the native hydrated state, especially when combined with complementary methods such as spectroscopy, microscopy, and rheology.[27–32]

13.4.1.1 Starch

Starch is one of the most abundant plant polysaccharides and is the main component in many raw and processed foods. It is a ubiquitous carbohydrate that provides up to 90% of the human nutritional energy in developing countries. Native starch is found in vegetal resources forming water-insoluble semi-crystalline granules, which present a complex hierarchical architecture characterised by at least four structural levels. The granules (~μm) are composed of concentric semi-crystalline and amorphous growth rings (~120–400 nm). The semi-crystalline rings are at the same time composed of stacks of alternating amorphous and crystalline structures known as lamellae (~9–10 nm). Finally, at the molecular level (~nm), the two polymers comprising starch, *i.e.* linear amylose and branched amylopectin, are organised differently, constituting the amorphous and crystalline lamellar domains. SAS and ultra-SAS methods are able to access this broad range of dimensions. A review of the application of SAS techniques to characterise the structure of starch can be found in the work by Blazek and Gilbert.[27]

The crystalline regions are thought to consist mainly of amylopectin side chains that are organized in double helices,[33] whereas the amorphous regions are mainly composed of linear amylose chains and less ordered amylopectin branch points.[34] Depending on several factors, such as the amylose/amylopectin ratio and the amylopectin chain length, these helices can be arranged differently into A-type (mainly found in cereal starches and related to short double helices) or B-type (long double helices found in tubers and high amylose starches) crystalline unit cells.[35–38]

Water has a great influence on the long-range order of starches, constituting up to 4–7% and 25–27% of the crystalline unit cell in the A and B polymorphs, respectively.[39–41] In fact, increasing the water content within the granule favours chain reorganization, thus increasing molecular order.[42] According to the currently accepted so-called liquid crystalline model, in the presence of a sufficient amount of water, the plasticization of amylose branch points, which act as flexible spacers, allows the double helices to be decoupled from the polymer backbone and aligned into lamellae, resulting in a transition from the nematic (possessing orientational order but lacking positional order) to smectic (aligned in layers) phase. This gives rise to a broad scattering peak in the SAS from hydrated native starches, commonly referred to as the lamellar peak, whose intensity is reduced in dry or low-humidity samples.[27] Additionally, the intensity depends on the SLD difference between the crystalline and amorphous domains in the lamellae and with respect to the surrounding amorphous growth rings.

The paracrystalline model proposed by Cameron and Donald[43] considers the existence of three different regions: (1) crystalline, comprising the amylopectin double helices, (2) amorphous, formed by the amylopectin branch points within the lamellae, and (3) amorphous background, arising from the amorphous growth rings. Although this model might serve as a good approximation, it does not account for the possible involvement of amylose in crystallites[44,45] and the deviations from an ideal lamellar structure. More sophisticated models have also been proposed;[46,47] however, the inclusion of additional parameters in scattering models increases the uncertainty of the fitting results, and therefore additional SANS contrast variation experiments are desirable to perform simultaneous global fitting for the scattering data. Another approach uses a one-dimensional correlation function from Fourier transformation of the scattering data, which provides structural parameters such as the thickness of crystalline and amorphous lamellar regions and the long period distance.[48–51] This method requires the extrapolation of the scattering data to both low and high q, the former being particularly problematic due to the inevitable contribution of longer-range structural features to the scattering. An empirical model based on the combination of a power-law function (to account for the diffuse scattering at low q) with Gaussian/Lorentzian peaks has been successfully applied in a number of studies.[42,52–54] Semi-empirical models, such as fractal[6] and unified[55] approaches, have also been used to fit the scattering data over more extended q ranges.

Native starches from different sources have been characterised by SAXS experiments, revealing a decrease in the intensity of the lamellar peak with increasing amylose contents.[56–61] This was proposed to arise from a decreased SLD difference between crystalline and amorphous regions due to the localization of amylose chains within different regions of the lamellae.[61] Taking advantage of the possibility of creating regions with different SLD, SANS contrast variation experiments have been carried out to study the distribution of water within starch granules[58] and to quantify the molecular densities within different regions in the granules.[62] Furthermore, the combination of SAXS and SANS allowed selected structural regions to be highlighted in dry and hydrated native starches. While the structural features detected in the dry state were consistent with small blocklets or superhelices, in the hydrated state, larger structures consistent with growth rings were detected.[55] SAS has been widely used to characterise the structure of native and genetically modified starches and to evaluate the structural effects of different processing methods and storage conditions.[42,53,63–65]

For most food-related applications, starch is typically consumed after processing. Many food processing methods involve the use of combined heat and humidity, leading to starch gelatinisation. Essentially, the gelatinisation process is initiated by the access of water towards the interior of starch granules, which swell and absorb water. Subsequently, structural changes occur at the different architectural levels, leading to crystallinity losses, disruption of the lamellar structure and amylose molecules leaking out of the starch granules, forming a continuous gel.[66] Due to the existence of SLD contrast between amorphous and crystalline domains, SAXS experiments have been widely used to investigate the structural changes undergone by starch during the gelatinisation process. During the initial stages of gelatinisation, the contrast between the amorphous and crystalline lamellae has been seen to increase with time and temperature as a result of the reorganization of the crystalline regions (from a nematic to a smectic phase), promoted by the presence of moisture that is absorbed into the amorphous lamellae.[43,67–69] After this initial transition, upon increasing the temperature further, the crystalline domains are disrupted, thus decreasing the contrast between the amorphous and crystalline lamellae. After surpassing the crystallites' melting point (the gelatinisation temperature), both the crystalline and lamellar orders are lost, as suggested by the absence of the SAXS lamellar peak.[43] After that, a gel with fractal structure is formed.[50] The gelatinisation of starch can also be induced by high hydrostatic pressure (HHP) treatments, as demonstrated by *in situ* synchrotron SAXS experiments.[51,70] SANS has also been used to perform *in situ* measurements during the Rapid Visco Analysis (RVA) processing of commercial starches,[6] which is a widely used industrial method to evaluate the rheological properties of starch slurries, probing structural transitions in starch related to changes in the viscosity.

The annealing of starch (heating to a temperature between the glass transition and the gelatinisation temperature in the presence of moisture) has been shown to affect gelatinisation behaviour due to decreased water absorption.[71] Additionally, the presence of other compounds, such as lipids, has been demonstrated to affect gelatinisation, inducing the formation of lipid-amylose complexes and limiting the accessibility of moisture towards the granules.[72,73] Following gelatinisation, starch is able to reorganise on cooling *via* a process known as retrogradation. Some studies have investigated the retrogradation of starch by SAS techniques,[74–76] showing the formation of semicrystalline structures that differ significantly from the original granules. While retrogradation of corn starch at low temperatures has been shown to induce the formation of lamellae with a long-range periodicity of up to ca. 20 nm and B-type structures, A-type and V-type structures with no defined periodicity have been observed after retrogradation at high temperatures.[77,78] Conversely, temperature did not show any effect on the retrogradation of wheat starch.[79]

Apart from those methods involving heating, starch is often subjected to other treatments to facilitate its processing or to adapt its properties to target requirements for specific food products. For instance, ultrasonic treatment induces structural changes in the lamellae, promoting a greater swelling of the granules, without affecting the crystal structure.[80] SAS techniques have also been used to assess the structural changes induced by acid and alkaline treatments.[81,82] which are often used to enhance quality characteristics of the colour, flavour, and texture of food products.

From the human nutrition perspective, understanding the factors affecting starch digestibility is essential. After consuming starchy foods, some enzymes in the small intestine are able to digest a fraction of starch, converting it into glucose, while another fraction, known as resistant starch, is not digested, providing several health benefits. The degree of starch gelatinisation in food products is known to affect its digestion rate.[83] This is conditioned by the extent and type of processing to which starch has been subjected in manufacturing the food product and also by the way in which food is prepared prior to its consumption. For instance, the rapid heating rates provided by microwaves lead to disruption of the semi-crystalline lamellae before the onset of gelatinization,[84] and refrigeration of cooked rice for 12 h prior to its consumption promotes the formation of resistant starch.[85] The structural changes taking place during the enzymatic hydrolysis process of starch granules has been investigated by SAS techniques in a relatively limited number of studies.[42,52–54,84,86–88] These studies have shown that the susceptibility of starches to be enzymatically digested is conditioned by the type of crystalline structure (preferential digestion of the amorphous growth rings in A-type starches in native materials), amylose content (high amylose starches present lower digestibility due to decreased accessibility for α-amylase following processing and retrogradation), as well as granule microstructure and nanostructure of the growth rings.

13.4.1.2 Cellulose and Plant Cell Wall Polysaccharides

Plant cell walls (PCWs) are extremely complex structures that may be considered as composite materials in which a semi-crystalline component (cellulose) is embedded in an amorphous multi-component matrix, which is mostly composed of polysaccharides, of which hemicelluloses and pectins are the most abundant. Cellulose is the main structural component of PCWs, providing mechanical strength. It is a linear homopolysaccharide of D-glucose units connected by (1,4)-β-glucosidic linkages that, in natural resources, are frequently organized in a highly ordered manner forming rod-like crystal units.[89,90] These crystalline units are held together in a paracrystalline matrix, intercalated with disordered amorphous domains, forming cellulose microfibrils. While cellulose microfibrils represent the main building blocks in most PCWs, the specific properties of each plant tissue are strongly determined by the matrix cell wall composition and the nature of the interactions established between the different cell wall constituents. Understanding the structural role of the different components in PCWs is not only crucial from a plant biology perspective but also to the food industry; for the latter, it would enable linking the composition and structure of different plant tissues from vegetables and fruits to their particular physical properties which, in turn, determine their processability as well as the application of certain fractions as food additives. Furthermore, the distinct composition and structure of PCWs is also responsible for differences in the digestibility of vegetal products.

SAS techniques offer a tremendous advantage for the study of PCWs since they allow the characterisation of native tissues with minimal sample preparation; in addition, they permit study of the structural role of specific components by means of strategies that limit the complexity of native PCWs, such as sequential component extraction or the use of cell wall analogues based on bacterial cellulose. Although some of the major advances in this field, brought by SAS techniques, are discussed in what follows, an in-depth analysis of their application to characterise the structure of PCWs can be found elsewhere.[30]

As shown in Table 13.3, the SLD contrast between cellulose and other PCW polysaccharides is limited due to their structural similarity. In SAXS, the contrast is expected to be generated between the crystalline cellulose regions and the amorphous domains of cellulose and matrix components. SANS experiments present the advantageous possibility of modifying the solvent–sample neutron contrast by replacing H_2O with different D_2O/H_2O ratios. Moreover, selective deuterium labelling enables the neutron SLD to increase and thus enhance contrast with respect to other PCW components. The ability of SAXS and SANS to highlight distinct structural features has been exploited in several works in which the hierarchical architecture of cellulose and the structural role of diverse PCW polysaccharides have been investigated using model systems based on pure and composite bacterial cellulose hydrogels.[29,31,91–95] Composite hydrogels in which the cellulose component

Table 13.3 Neutron and X-ray scattering length densities of cellulose and matrix plant cell wall polysaccharides. The following physical densities are used: ρ(crystalline cellulose) = 1.60 g cm^{-3} (ref. 98), ρ(paracrystalline cellulose) = 1.51 g cm^{-3} (ref. 99), ρ(amorphous cellulose) = 1.48 g cm^{-3} (ref. 100), ρ(hemicelluloses) = 1.40 g cm^{-3} (ref. 101). The reference neutron SLD value for pectin is from ref. 102.

	Neutron SLD (10^{10} cm^{-2})	X-ray SLD (10^{10} cm^{-2})
Cellulose (crystalline)	1.87	14.46
Cellulose (paracrystalline)	1.77	13.65
Cellulose (amorphous)	1.73	13.38
Cellulose (amorphous, D_2O exchanged)	3.39	—
100% Deuterated cellulose (crystalline)	7.59	13.62
Hemicelluloses[a]	~1.62	~12.65
Pectin	2.47	—

[a]The provided SLD value is based on the estimation previously conducted for tamarind XG and wheat AX[92] but depends on the molecular weight and degree of substitution.

was selectively deuterated have been successfully produced,[94,95] providing an enhanced validation of complex models to fit the experimental data. From the analysis of pure cellulose hydrogels, a model for the multi-scale structure of cellulose, which accounted for the essential role of water at different structural levels, was proposed.[31,93] According to that model, the structure of native cellulose is characterised by (1) the existence of cellulose microfibrils containing an impermeable crystalline core surrounded by a partially hydrated paracrystalline shell, (2) the creation of a strong network of cellulose microfibrils held together by hydrogen bonding to form cellulose ribbons, and (3) the differential behaviour of tightly bound water held within the ribbons compared to bulk solvent. The application of this model to composite hydrogels enabled the identification of a distinct interaction mechanism of cellulose with matrix PCW polysaccharides such as arabinoxylan (AX), xyloglucan (XG), mixed-linkage glucans (MLG), and pectins. Cellulose-AX interactions, which were seen to be limited to the cellulose ribbons' surface, were hypothesized to take place *via* a non-specific adsorption mechanism.[92–94] On the contrary, XG and MLG, although showing distinct interaction mechanisms, were both seen to form two different fractions: (1) interfibrillar domains interacting with the cellulose microfibrils and (2) surface domains, responsible for the cross-linking of ribbons.[94] In the particular case of pectin, SAS techniques were combined with rheological characterisation, showing that, in contrast to conventional models that assume pectin acts as a separated phase in primary cell walls, two different fractions of pectin exist in PCWs: (1) a major fraction that does not interact at the molecular level with cellulose but has a strong impact on the mechanical properties through poroelastic effects and (2) a minor fraction that interacts intimately with cellulose microfibrils at the point of assembly.[91] Furthermore, the presence of calcium leads to the formation of

Ca^{2+}-pectate gels, which were demonstrated to act as load-bearing components in primary PCWs.[29]

Although the conclusions drawn from cited works were based on the characterisation of model cellulose hydrogels, the same approach has been successfully applied to characterise the structure of pristine cotton tissues[96] and cell walls extracted from onion, carrot, and apple,[97] confirming the validity of the proposed hypotheses. These results are relevant to the food industry, from different perspectives: (1) to develop plant-based food additives or products with specific rheological or textural properties based on the composition of the raw material and the distinct structural role of cellulose, hemicelluloses, and pectin; (2) to develop optimized extraction methods to isolate polysaccharides from plant cell walls, used as food ingredients, based on their structural characteristics; and (3) to understand the possible nutritional benefits of plant-based foods based on their composition and bioavailability (*i.e.* some hemicelluloses are strongly linked to cellulose and will most likely not be digested).

13.4.1.3 Gums and Mucilages

Besides providing a matrix in cell walls for vegetal tissues, polysaccharides are essential components in plant exudates, such as gums and mucilages. Gums and mucilages contain mixtures of polysaccharides that are water soluble or water dispersible and at low concentrations form viscous solutions or gels. They are commonly used as food additives, mainly as thickeners and emulsifying agents. Despite the great interest of these polysaccharides in the food industry, very limited work has been carried out on their structural characterisation by SAS techniques.

Gum arabic, a natural composite polysaccharide derived from exudates of *Acacia senegal* and *Acacia seyal* trees, is one of the most commonly used food hydrocolloids that serves as an emulsifier and a long-term stabiliser in food products containing oil–water interfaces. Its structure consists mainly of a highly branched polysaccharide (a β-($1 \rightarrow 3$) galactose backbone with arabinose and rhamnose branches) and a minor fraction of a protein-polysaccharide complex.[103] The structure of aqueous gum arabic solutions has been studied by SAXS and SANS, revealing the presence of a major component comprising polysaccharide spheroidal aggregates, interlaced with a small amount of large coils, likely arising from protein-polysaccharide complexes.[104] A correlation length model was used to fit the SAXS patterns from guar gum aqueous solutions and to determine the effect of additives on the average distance between cross-linking points of the polymeric network,[105] although a detailed analysis of the polysaccharide structure was not carried out.

Seed mucilages are pectin-rich and hemicellulose-rich polysaccharides that help seeds to maintain hydration and survive in arid conditions. A recent study reported on the structural characterisation of two gel-forming arabinoxylan fractions (water extracted and alkali extracted), with nearly identical molecular weight and monosaccharide composition but very different gelation and rheological properties.[32] SANS and SAXS experiments were used to characterise both fractions in non-gelled conditions (using KOH and KOD as solvents). A flexible cylinder model was used to fit the experimental data, and it was concluded that although both fractions presented a random coil conformation and similar degree of chain rigidity in dilute solutions, the conformation of the water-extracted fraction changed from rod-like to sphere-like at high concentrations. Moreover, the same types of scattering patterns were obtained for the fractions in gelled conditions (*i.e.* using water), suggesting that the gelling process was conditioned by the extent of hydrogen bonding within molecular entanglements.

SAS techniques have been used more extensively to study the structure and gelation processes of gums synthesised by microorganisms (*i.e.* extracellular polysaccharides), such as gellan gum, xanthan gum and curdlan. The gelation mechanism of gellan has been comprehensively studied by SAXS. Aqueous solutions (80 °C) and gels (10 °C) at different concentrations were analysed and the data fitted using a broken rod-like chain model.[106] The results suggested that gellan gum consists mainly of single-stranded disordered polymer coils at higher temperatures. On cooling, a considerable proportion of the gellan chains are arranged as short double-stranded helices in the sol state, with gelation originating from the alignment and further aggregation of double helix regions. The same model was used to study the effect of annealing[107] and the addition of different counterions[108–112] on gelation. Furthermore, the SAXS patterns from hybrid gellan-xyloglucan gels were collected and fitted using the semi-empirical Beaucage model.[113] The results were interpreted in terms of the average size of the pores in the gel structure (represented by the calculated radii of gyration), which was seen to increase with the addition of xyloglucan.

Curdlan is a neutral polysaccharide that is not soluble in water but is soluble in alkaline aqueous solutions and polar solvents like dimethyl sulphoxide. Transparent curdlan gels are obtained upon heating of solutions in NaOH above 60 °C. Based on combined SAXS and dynamic light scattering experiments at 25 °C, curdlan molecules associate to form multi-helical structures at NaOH concentrations above 0.05M but form heterogeneously aggregated coils at concentrations below 0.02M; such multi-helical conformations seem to change to a single helical conformation on heating to 60 °C.[114,115] Xanthan gum forms gels by annealing and subsequent cooling of aqueous solutions. Combined SAXS and DSC studies were used to study the nature of the xanthan chains in the sol state and their ability to form aggregates when annealed.[116] A scattering peak was detected in the SAXS as a function of annealing time and interpreted as characteristic of the average size of the junction zones in the gels. The results showed that an ordered structure of single and double helices develops in solution as a result of annealing, with the size of the junction zones in the gels increasing when subsequently formed on cooling. The same peak was also detected in SANS experiments of xanthan solutions in the semi-dilute regime.[117]

13.4.1.3.1 Seaweed Polysaccharides: Carrageenan, Alginate and Agar. Several sulphated polysaccharides abundant in seaweed, such as agar, carrageenan and alginate, are widely used in the food industry as gelling and thickening agents due to their particular gelation characteristics. Carrageenan is a structural polysaccharide in some red algal species, which can be classified into three types, depending on the number of sulphate groups per repeat unit of disaccharide: kappa (κ-) (one sulphate group), iota (ι-) (two sulphate groups), or lambda (λ-) (three sulphate groups). The three types of carrageenan undergo an ion-induced gelation process, which is strongly affected by many factors, mainly the degree of sulphation, the polymer concentration, and the type and concentration of added salts. The gelation of carrageenans has been studied by analysing the SAXS patterns of their K^+-salt forms at different concentrations and temperatures.[118] The results were analyzed by means of molecular models of double helix or associated double helices. The data showed that in the case of κ-carrageenan, gelation takes place through the transition from single chain to double helix and subsequent association; however, the association of double helices does not take place in ι-carrageenan. In contrast, the scattering behaviour of λ-carrageenan was not significantly affected by temperature since it did not form a gel. The effect of various types of monovalent counterions (Na^+, Li^+, K^+, Rb^+, and Cs^+) was also investigated, showing that only Na^+ and Li^+ alter the gelation through aggregation of double helices in κ-carrageenan; on the other hand, ι-carrageenan gelled through the formation of double helices, irrespective of the added ion (Figure 13.3).[119] Aqueous solutions of the Na^+- and Li^+-salts of κ- and ι-carrageenan were also investigated by SAXS, identifying the critical concentrations at which the systems transitioned from a dilute to a semi-dilute regime.[120,121]

Alginate is an anionic polysaccharide composed of 1–4-linked α-L-guluronic acid and β-D-mannuronic acid residues, which is a structural component in marine brown algae (*Phaeophyceae*) and forms gels upon the addition of divalent cations or upon lowering the pH below the pK_a of the uronic acid residues. SAXS data of alginate gels obtained by addition of Ca^{2+} showed that dimerization of chain segments was the principal association mode at low fractional Ca^{2+} saturation of guluronic acid of the alginate,[122,123] while increasing the fractional Ca^{2+} saturation yielded coexisting lateral association modes.[122] Moreover, the gelation kinetics were monitored by time-resolved rheological and SAXS characterization; the results were analysed using a two-component broken rod model showing that, after the initial association of chain segments in junction zones, gelation proceeds *via* the formation of larger bundles.[124] The same approach was used to characterise alginic acid gels,[125] which also showed a two-step gelation process.

Agar is a mixture of two polysaccharides (agarose and agaropectin) and constitutes the main structural component in the cell walls of certain species of red algae. Agarose is the gelling fraction of agar and consists of repeating units (agarobiose) of alternating β-D-galactopyranosil and 3,6-anhydro-β-L-galactopyranosil groups. Agaropectin has a similar structure but contains

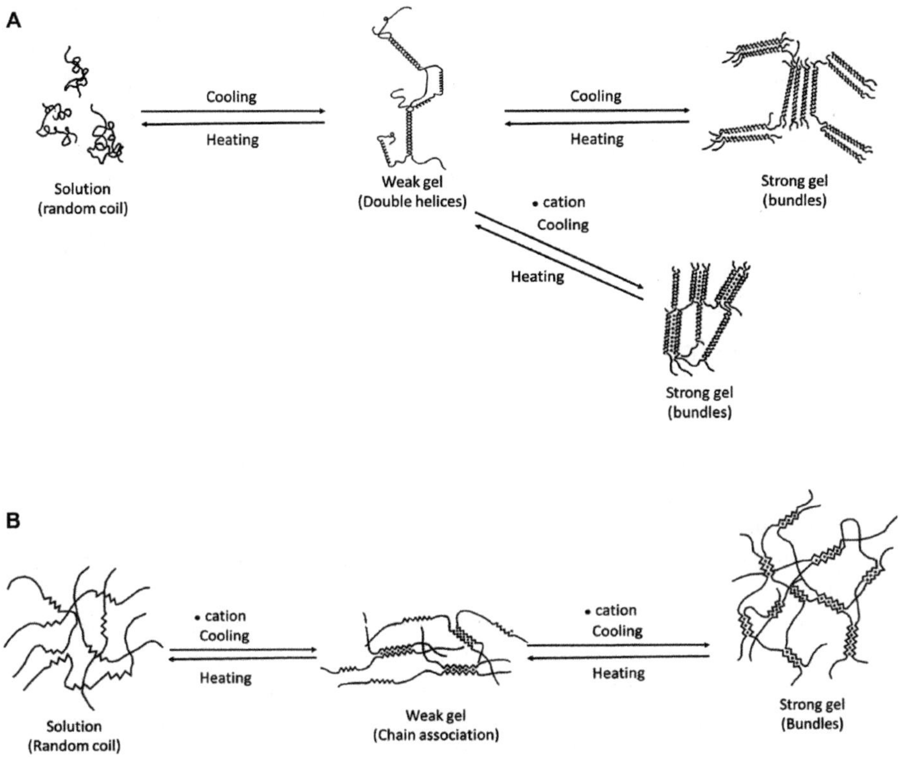

Figure 13.3 Schematic representation of the most common gelation mechanisms of food polysaccharides: (A) gelation through the formation of helical structures; (B) cation-mediated gelation.

5–10% sulphate esters and other residues. Although many studies have been conducted to characterise the rheological properties of agar and gelation mechanisms have been proposed on the basis of these studies, SAS techniques have not been exploited to date to investigate the structural conformation of this polysaccharide; only agarose solutions and gels have been characterised by SAXS[126] and SANS,[127] although more in-depth studies would be needed to fully understand the gelling and melting mechanism.

13.4.1.4 *Interpenetrating Hydrocolloid Networks*

Interpenetrating hydrocolloid networks (IHNs) are blends of two or more hydrocolloids that interact with each other merely by physical entanglements; *i.e.* they are neither chemically nor covalently bonded to each other. These systems have been shown to present improved properties with respect to their individual components and have been mostly studied within the biomedical field,[128,129] despite offering great potential for the development of hydrogels for the food industry. To date, only a reference to the investigation of the water swelling capacity of konjak glucomannan (KGM)-chitosan blends by simultaneous SAXS/WAXS experiments is available in the literature. In that study, SAXS showed that (1) high-molecular-weight chitosan films exhibit low susceptibility to water swelling due to strong intrachain associations and (2) for low-molecular-weight chitosan, the water swelling is restricted at low KGM contents due to strong KGM-chitosan interactions, whereas for higher KGM concentrations, free hydroxyl groups provided by "excess" KGM led to higher swelling.[130]

13.4.2 Proteins

Food proteins constitute a diverse and complex collection of biological macromolecules. Apart from contributing to the nutritional quality of food, proteins also act as integral components to food systems due to their diverse functional properties, largely dictated by changes in their structure or structure-related properties occurring during preparation, processing, and storage. For the optimal use of existing and future food protein sources, a thorough understanding of the nature of the relationships between structure and function is therefore needed. While X-ray crystallography is the most detailed technique, it has severe limitations. For example, it may not be possible to produce protein crystals (this is more common for membrane proteins), and, where they are available, the data are most relevant to the protein in its non-hydrated form, whereas proteins exhibit a variety of dynamic features and functions in aqueous environments. In this sense, SAS techniques are valuable approaches to study protein structure, as they can provide a substantial number of significant geometric parameters consistent with crystallographic results, as well as NMR data, but that can readily show tertiary structural changes occurring under varying conditions and within solutions and gels.[131]

For protein structural characterisation, SAS is normally combined with a number of other characterisation tools (*e.g.* IR spectroscopy, NMR, DSC, TEM, circular dichroism (CD), *etc.*) that can provide complementary information. For instance, size-exclusion chromatography and ion-exchange chromatography, combined with SAXS (SEC-SAXS, IC-SAXS), are reliable methods for the determination of R_g, molecular weight, and the structural properties of proteins with dimensions of 1–10 nm.[132,133] In this section, the ability of SAS to investigate the structure of hydrated proteins is shown, highlighting the usefulness of these techniques to unravel some of the functional properties of these fundamental macromolecules.

13.4.2.1 Milk Proteins

Milk proteins (mainly comprising caseins and whey proteins) have been widely studied both for nutritional and technological reasons. Caseins are phosphorous-containing proteins which, in milk, are present as micelles containing calcium nanoclusters; these proteins coagulate or precipitate at pH 4.6. Whey proteins (mainly formed by α-lactalbumin and β-lactoglobulin) do not contain phosphorous and remain in solution at this pH value.

Although the structure of casein micelles has been broadly studied and there are several good reviews dealing with this topic,[134,135] for some time, there was considerable debate on the actual organisation of the four types of casein proteins (αs1, αs2, β, and κ, at an approximate ratio of 4 : 1 : 4 : 1) with the minerals in the form of colloidal calcium phosphate (CCP) nanoparticles. It is widely accepted that the surface of the casein micelles is covered by a hairy layer of κ-casein, which serves as a stabilizing agent.[136] However, the internal organisation of the micelles, widely explored using SAXS and SANS,[137–139] has been described using either the submicelle model (first proposed by Schmidt[140]) or the nanocluster model (proposed by Holt[141] and De Kruif and Holt[142]). The submicelle model describes casein micelles as being formed by small protein micelles (submicelles) linked together by bridging CCP. In contrast, the nanocluster model depicts the casein micelles as having the CCP nanoparticles randomly distributed within the micelle and having protein molecules surrounding them.[143] This latter model has now been accepted and supported by contrast-matching the calcium phosphate in SANS,[144] by removing the calcium phosphate through acidification,[145] and, most recently, through the use of resonant soft X-ray scattering (RSoXS), by measuring scattering around the Ca L2/L3 absorption edge (~350 eV).[146] RSoXS, together with ASAXS, was used to study the casein structure in iron-fortified milk, indicating that iron was co-located within the CCP nanoclusters.[147] Although casein micelle sizes have been generally described as being naturally polydisperse, ranging from 50 to 230 nm in diameter,[148,149] a recent review has revealed that this polydispersity from a single animal is actually rather small for a colloidal system, and it is related to the "surfactant" properties of κ-casein.[134] Both SAXS and SANS show a characteristic length of 13 to 18 nm within the casein micelles (less clearly observed in SAXS patterns unless in the powder form), which may be interpreted as an interparticle correlation length between CCP nanoparticles.[143] It was found that, even though the casein micelles from various milk origins have different physicochemical properties, as well as hydration and protein composition, the internal structure of the micelle and, more specifically, the correlation length between CCP nanoparticles was the same.[150] The ability of SANS to probe the casein micelle structure is based on the different neutron contrast match points of the components, being around 40–42% D_2O for the proteins and 74–88% for the calcium clusters.[151] SAS techniques, in this sense, have also been used to understand how other components, like sugars, may destabilize casein micelles.[152]

Apart from the structure, the dynamic size, and the stability of casein micelles play important roles in the processing of milk and its conversion to various dairy products, such as cheese, yogurt, and milk protein powder concentrates.[153–156] For instance, during the manufacture of many industrial dairy products, milk may undergo concentration processes using membrane filtration, such as ultrafiltration (UF) and diafiltration (DF).[153] While these processes do not affect the overall micellar state of the caseins,[157] SANS studies have demonstrated that extensive ultrafiltration treatment is able to remove calcium clusters in a similar way as mild diafiltration, as observed by a decrease in the relative height of the characteristic scattering peak associated with the average distance between clusters.[151] This has technological implications, as it is known that CCP plays an important role in determining the cheese-making properties of milk, where losses of insoluble calcium induce the formation of weaker and more flexible gels;[158] it also influences buffering capacity,[159] a key factor during acidification in cheese making. In fact, manipulation of CCP allows for modification of cheese texture properties, such as meltability and stretchability.[160] High-pressure processing also alters casein micelle structure, as demonstrated by neutron scattering.[161] In other cases, the casein micelle structure may be altered by incorporation of other compounds. For instance, ethylenediaminetetraacetic acid (EDTA), which is normally added to milk to improve its foaming properties,[162] is known to disrupt casein micelles. A recent study combining SAXS with NMR, viscometry and DLS demonstrated that treating reconstituted skim milk with transglutaminase prevented dissociation of the casein micelles; upon addition of EDTA, an enhancement in viscosity occurs that is greater than that observed in the absence of transglutaminase treatment.[163] It is also well known that micellar casein powders have poor reconstitution properties. Combining SAXS with differential scanning calorimetry (DSC) and infrared spectroscopy, the structure of various micellar casein powders, containing different salt and protein compositions, was studied; it was observed that the inner micellar structure was hardly affected, thus indicating that the reconstitution ability was rather controlled by higher-order structural changes, such as cross-linking between casein micelles, which could be avoided by adding sodium caseinate to the concentrated milk before spray drying.[164]

Whey proteins are added to a number of food products due to their solubility over a wide range of pH, diverse functionality, nutritional profile, and relatively low cost; they are mainly obtained as a by-product from the cheese industry. Whey proteins have been used, for instance, as fibrils to increase viscosity, as gelling agents, and for stabilizing foams and emulsions.[165] Whey protein fibrils are commonly produced at very acidic pH, as their stability is highly dependent on protein concentration and ionic strength.[166] SAXS studies, in combination with zeta potential measurements, CD, and TEM, demonstrated that the addition of sodium lecithin below its critical micelle concentration did not affect the fibril growth mechanism and protein conformation but decreased the protein susceptibility to pH changes, thus broadening the potential application of nanofibrils as a food ingredient.[167] Protein conformation can be also modified by subjecting food to common preservation processes. For instance, SAXS experiments of β-lactoglobulin exposed to gamma irradiation—normally applied to decontaminate food products—showed that when the proteins were treated in solution, oligomers and aggregates resulting from the cross-linking of dimer subunits were formed, while samples irradiated in the solid state were not modified.[168] Detergent molecules like sodium dodecyl sulphate (SDS), which is a GRAS ingredient used as an emulsifying agent and whipping aid in food products, can also cause protein denaturation and induce a loss of α-helical conformation. However, in globular proteins like β-lactalbumin, SAXS studies, combined with isothermal titration calorimetry (ITC), proved that, depending on the SDS concentration, different aggregation/fibrillation processes take place during protein denaturation and influence the α-helix content of the protein.[169]

Dairy proteins are used as emulsifiers since they can adsorb onto the oil–aqueous droplet interface, forming a strong and cohesive protective film that prevents droplet aggregation. In particular, sodium caseinate has been widely used as an ingredient in the food industry due to its functional properties, which include emulsification, water and fat binding, thickening, and gelation,[170] SAXS has been useful, especially in nanoemulsions stabilized by dairy proteins, providing an average size of oil droplets.[152,171]

When incorporating dairy proteins to food products and subjecting them to common food processing conditions, such as heating, chemical reactions may occur, such as the Maillard reaction, which involves an initial glycation, in which a condensation between an unprotonated amino group from a protein and a carbonyl group of a reducing sugar takes place. The conformational changes of bovine serum albumin (BSA) during a Maillard reaction with dextran under controlled heating was studied by SAXS, fluorescence spectroscopy, DLS, and CD; it was found that conjugates formed, inhibiting BSA aggregation and preserving the secondary structure of the native protein.[172] Moreover, this glycosylation prevented aggregate formation in the presence of polyphenols; due to the interactions established between polyphenols and proteins, this makes these glycosylated proteins promising nanocarriers for polyphenols.[173]

13.4.2.2 Meat Proteins

Meat proteins can be categorised as water soluble (mainly sarcoplasmic enzymes) or soluble in concentrated salt solutions (myofibrillar proteins, the most abundant being myosin and actin). The remaining protein in the muscle flesh comprises connective tissue (collagen and elastin). Synchrotron SAXS has been performed to correlate protein composition with ovine meat quality.[174] The authors combined scattering data with different meat quality parameters such as shear force, collagen content, and solubility and found relationships between the actin/myosin fibril d-spacing and the tenderness of the meat after 1 and 5 days of ageing. The SAXS measurements were also correlated with intramuscular fat content (IMF), showing the potential of SAXS as a complementary tool for meat quality determination.[174] Another recent study on porcine meat using SAXS revealed that rigor bonds, formed during accelerated glycolysis *post mortem*, led to a compressed interfilamental space and loss of sarcomeric water and affected contractile function when the meat was treated in denaturing conditions (38 °C, pH 5.5).[175]

Gelatin, produced from a partial hydrolysis of collagen, is a widely used food protein, mainly employed as a gelling ingredient. The structure of gelatin gels, physically obtained by lowering the temperature and chemically cross-linked to improve their stability, has been studied using SANS and correlated with rheological data. The structural parameters obtained through model fitting the scattering data revealed the presence of small aggregates within the chemically cross-linked network, which hindered the free stretching of the triple helix bundles in the physical gels.[176]

13.4.2.3 Plant Proteins

Soybean proteins are commonly used as ingredients in vegetarian foods as they contain all essential amino acids for human nutrition and are commonly used as gelling, emulsifying and foaming agents.[177] Moreover, given their ability to produce meat-like products with similar texture, flavour, colour and nutritional value, textured soy proteins (and other low-cost vegetable proteins) are promising meat substitutes.[178] It is common practice to use dry protein powders as ingredients in food manufacturing but, usually, subsequent rehydration does not lead to the same protein properties as those seen in the original native states.[179,180] A combination of USAXS, SAXS and SANS was used to study glycinin structure at different levels of relative humidity and over four orders of magnitude in length scale.[180] The enhancement in contrast provided by neutron scattering resulted in Bragg peaks from the protein being well resolved and indicating an expansion of the hexagonal unit cell from the protein as a function of moisture content.[180] Also combining SAXS with FTIR, DSC and NMR, it was possible to study the molecular mobility of glycinin above a certain humidity value (dependent on the technique used), which was related to the glass-to-rubber transition.[181] The long-range order from glycinin is completely lost upon thermal denaturation, which is reflected in the scattering patterns as the disappearance of the Bragg peaks. It was also observed that changes in scattering intensity as a function of moisture were less obvious for the denatured materials, a fact that could indicate a difference in the extent to which water can interact with the protein.[182] Further SAXS studies on glycinin also revealed that extreme acidic (pH 2) and basic (pH 11) conditions lead to the partial denaturation of the protein. However, while acidic pH promotes the dissociation of the hexameric structure, increasing the pH to 11 results in the formation of large unimodal particles, reflected by "featureless" SAXS patterns in both powders and solutions.[183]

SAXS has been used in the structural analysis of gliadin wheat grain proteins. The standard and cross-sectional Guinier approximations were used to obtain the radius of gyration, R_g, and radius of gyration of the cross-section (R_c) respectively of α, γ, and ω-gliadins in aqueous ethanol in addition to employing rod-like and prolate ellipsoid models.[184] When heating these wheat proteins at acidic or basic pH conditions, multi-level structural changes were observed. SAXS indicated that, at acidic conditions (pH 2.5), protein conformations evolved from globular to elongated ellipsoids and, upon solvent evaporation, gliadin tended to aggregate. Alkaline treatment (pH 8.5) did not induce hydrolysis, and the globular conformation was maintained; however, irregular large aggregates assembled during solvent evaporation.[185] SAXS, in combination with AFM, CD and sodium dodecyl sulphate-polyacrylamide gel electrophoresis (SDS – PAGE), has been applied to investigate kafirin, a prolamin from the cereal sorghum, to show that protein concentration and solvent polarity can effectively regulate kafirin assemblies from thick rod-like to slim rod-like structures.[186] The structure of glycoproteins from acacia gum has also been studied by SAXS, in combination with a number of biophysical techniques, to improve the knowledge of these macromolecules and to enhance their emulsifying and flavour-stabilizing properties. SAXS, in combination with TEM, CD and high-performance size exclusion chromatography–multi-angle laser light scattering (HPSEC–MALLS), suggested that the glycoprotein molecular fraction structure from acacia gum is an assembly of ring-like glycoprotein modules ascribed to hydroxyproline (Hyp)–arabinogalactan (AG) subunits.[187]

13.4.2.4 Other Proteins

Interestingly, SAS can also provide useful information about processes occurring during the consumption of specific foods. For instance, the astringency related to the ingestion of food products containing tannins is related to the interactions of these polyphenols with salivary proteins. These proline-rich proteins are divided into acidic and basic types; the only described function for the basic type is to bind tannins and protect the organism against their anti-nutritional effects.[188] The SAXS patterns of solutions containing salivary proteins and polyphenols can be fitted using a combination of the scattering from aggregates (empirical function chosen to match the scattering that takes into account aggregate concentration) and from non-aggregated proteins in solution (power law function). By combining SAXS with DLS and electrospray ionization mass spectrometry (ESI-MS), it was determined that aggregates between the proline-rich proteins and a model tannin had a core–corona structure (with the corona having fewer bound tannins) and that increasing the tannin concentration beyond a threshold value causes a transfer of proteins to the aggregates and a fast rise of the number of proteins per aggregate.[189] A related work used SAXS with DLS, CD, NMR and viscometry to develop a molecular model for astringency produced by polyphenol/protein interactions using β-casein and epigallocatechin gallate (EGCG) as model protein and polyphenol, respectively.[190] Interestingly, glycosylation of these proline-rich proteins avoided the formation of large aggregates; they were subsequently disrupted in the presence of ethanol, as observed by SAXS and DLS, confirming the effect of co-solvents on protein/tannin interactions.[191] As the molecular size of tannins can influence astringency, SAXS has also been used to determine its variation in wine samples depending on their ageing period.[192] Interestingly, SAXS results indicated that the older wines had larger tannin molecules than the younger wines and that wine composition did not have a great impact on tannin molecular size.[191] This increase in tannin size could be related to oxidation reactions taking place during the ageing of wine, as also observed previously using SAXS.[193]

13.4.3 Lipids

Food lipids have key roles in influencing sensory properties, affecting texture, viscosity and mouthfeel. They also impact on the odour and flavour of food products, as these chemicals are hydrophobic. Lipids can affect the appearance of foods by imparting glossy notes, impact satiety and, of course, are an energy and nutrient source, being essential for the solubilisation of hydrophobic vitamins A, D, E and K. Food lipids are largely based on triglycerides of various fatty acids, which can be saturated (typically solid at room temperature) or unsaturated (liquid fats or oils). They are normally classified as simple (fatty acids, hydrocarbons and fatty alcohols) or complex lipids (like glycolipids or phospholipids) depending on their size and structural complexity. SAS has been mainly used to characterise their structuring role, either as constituents of emulsions and emulsion gels, lipid nanocarriers, or in the formation of what are known as edible organogels or oleogels. These non-destructive techniques have also been used to study lipid crystallization behaviour and to ascertain how their structure affects their oxidative stability.

13.4.3.1 Emulsions and Emulsion Gels

Simple food emulsions, which comprise a mixture of two immiscible liquids (generally oil and aqueous phases), are thermodynamically unstable, and surfactants, like phospholipids, are commonly used as stabilizing agents. In these systems, the structural and chemical properties of the boundary layer at the interface between the oil and aqueous phases are of paramount importance in determining stability. Using micro-focus small-angle X-ray scattering (μ-SAXS), the interface created by phospholipids with other amphiphilic molecules was investigated in a water-in-oil emulsion produced by microfluidics and whose thickness was calculated; by adding molecules with different packing parameter, the stiffness of the interfacial layer could be tuned.[194]

Time-resolved synchrotron SAXS has been used to study the structural evolution of emulsion droplets during *in vitro* digestion of mayonnaise. During the course of digestion, highly ordered nanostructures were formed inside the mayonnaise emulsion droplets; these nanostructures act as a controlled release system for food nutrients, especially poorly water-soluble components.[195] Lipid ordering in the course of digestion has been also observed for bile acid/lipid/water systems using combined SAXS/WAXS, in which an increase in the intake of rapeseed oil was seen to cause the formation of multi-lamellar vesicles and lamellar liquid crystals.[196] SAXS has been used to demonstrate that bile salt micelles increase in volume during lipolysis due to the insertion of monoglycerides and fatty acids within their micellar structure.[197]

Although SAXS and SANS are suitable non-invasive tools to investigate turbid emulsions and emulsion gels, the information derived is usually restricted to relatively short length scales but can be expanded with USANS. As an alternative, spin-echo small-angle neutron scattering (SESANS) can provide information at relatively long length scales (~100 nm–10 μm), which can be interpreted in "real space."[198] SESANS has been used to study the effect of temperature cycling on emulsion gels, to characterise the fat droplet cluster structures in these systems, and to provide information on their impact on gel firmness.[199]

13.4.3.2 Fat Crystallization

Lipids consumed in food products are composed of a mixture of naturally occurring triacylglycerols (TAGs) in large proportions (~98%), diacylglycerols (DAGs), monoacylglycerols (MAGs), and free fatty acids (FFAs). Apart from their composition, the molecular features (like degree of unsaturation, position of double bonds, or chain lengths) affect their steric hindrance, miscibility, melting point and crystallization. Due to the importance of fat crystals in food products, affecting texture, consistency, stability, rheology, processing and performance, an enormous amount of literature is already available (*e.g.* ref. 200 and 201).

Synchrotron sources allow the measurement of simultaneous SAXS/WAXS/DSC experiments to follow the thermal monitoring of fat polymorphism even within fat droplets to understand fat crystallization phenomena in food products.[202] The nanostructures of crystallized model triacylglycerol (TAG) systems, specifically pure tristearin (SSS) and pure tripalmitin (PPP), have been studied using USAXS and computer simulations. Scattering profiles observed were interpreted in terms of the surface fractal morphology of grain boundaries surrounding nanoscale voids in the material.[203] Combined SAXS-USAXS has also been used to study isolated milk fat crystals, mainly composed of high-melting triacylglycerols (TAGs). Using Guinier–Porod models, the structural properties of the milk fat crystalline nanoplatelets and their aggregates were characterized.[204]

SAXS combined with WAXS and DSC has also been used to ascertain how incorporation of phytosterols affects the crystallization of milk fat emulsions; this is of relevance in the development of enriched dairy products. The scattering patterns demonstrated the ability of phytosterol to integrate within the milk fat triacylglycerol matrix, leading to an increase in d-spacing packing and system disorder.[205]

13.4.3.3 Oleogels

There is a current trend in the food industry to replace traditional edible saturated fats with "healthier" alternatives, typically unsaturated or polyunsaturated fats. However, saturated fats with their higher melting temperature impact products' texture and pleasurable mouthfeel. The replacement of solid fats therefore leads to formulation challenges; consequently a range of alternative structuring approaches are being pursued.[206,207] For example, mixtures of γ-oryzanol and β-sitosterol have been shown to structure pure edible oil phases. According to SAXS data, the ability of oil structuring of these phytosterols is based on fibril/tubule formation.[208] Structural details of these phytosterol mixtures were obtained through SANS, which revealed that the tubules were composed of an inner shell containing the androsterol group of sitosterol or oryzanol and an outer shell containing the ferulic acid moieties in the oryzanol molecule.[19] As alternative oleogelators, mixtures of oleic acid and sodium oleate are able to structure edible lipid phases. SAS experiments showed that, depending on the molar ratio of both components, either inverse micellar or lamellar phases were formed, with the formation of crystalline space-filling networks when the oleogelators were added above 10 wt%.[209]

13.4.3.4 Edible Oils and Oxidation

The oxidation of edible oil results in primary and secondary oxidation products (such as hydroperoxides, carbonyls and epoxides), which produce undesirable sensory and biological effects; consequently, the suppression of lipid oxidation in food matrices is of great importance. The rate and extent of lipid oxidation in many heterogeneous foods are strongly affected by the physicochemical characteristics of the water–oil interfaces.[210] It has also been observed that association colloids, such as phospholipid inverse micelles and other surface active minor components (such as free fatty acids) present in commercial oils, could increase the rate of oxidation in bulk oils.[211] SAXS is valuable for characterising the size of these phospholipid inverse micelles.[212] In fact, SAXS studies showed that water had a strong impact on the properties of the association colloids formed by phospholipids,[210] with the presence of free fatty acids impacting on the inverse micelle structure of phospholipids in bulk oils and having a pro-oxidative activity.[213] In contrast, addition of a saturated fat, specifically stearin from anhydrous milk, delayed the lipid oxidation of fish oil which, from combined DSC, SAXS and WAXS experiments, was ascribed to restricted oxygen and TAG molecular mobility by the ordered state of milk fat, thereby limiting the propagation of oxidative reactions.[214]

13.4.4 Complex Systems

Although studying the structure of isolated proteins, lipids and polysaccharides undoubtedly provides useful information to understand their behaviour when present in food products, the reality is that natural and processed foods are much more complex systems composed of multi-component mixtures. In the following, some examples of the characterisation of more complex systems will be described where the interactions between different components are essential to generate materials with specific properties that can be exploited in food-related applications.

13.4.4.1 Starch-lipid Complexes

As previously described, starch is one of the polysaccharides most extensively studied by SAS techniques. However, the vast majority of studies available in the literature have been performed on the isolated polysaccharide. Although these studies are undoubtedly meaningful, they do not reproduce the real conditions encountered in food systems, where starch is found associated with other polysaccharides, lipids, proteins and minerals. The interaction of starch with these other components is expected to affect its behaviour during food processing and also its role from a nutritional perspective. For instance, the formation of amylose-lipid complexes is relevant to gelatinisation. Derycke *et al.* showed that the texture of parboiled rice may be related to the level of crystalline amylose-lipid complexes formed during parboiling because these complexes are stable in the cooking process.[215] Rheological characterisation combined with SAXS has been carried out to study the gelation of starch induced by shear forces,[216] showing that stiffer gels are produced in the presence of amylose-lipid complexes. This is due to the release of short amylose fragments from the complexes during heating, which aggregate upon cooling and agglomerate into larger fractal objects when shearing stops. SAXS/WAXS experiments showed that an increased thermal stability of corn starch granules during gelatinisation is induced by the presence of L-alpha-lysophosphatidylcholine (the main phospholipid in corn).[217] A recent study has also investigated the effect of the amylose-lipid complexes in starch blends with microalgae on the gelatinisation process.[72] Although the presence of microalgae was seen to limit water accessibility towards the interior of starch granules, the formation of V-type amylose-lipid complexes counteracted this effect and resulted in the formation of more crystalline and homogeneous lamellar structures. This result is relevant for the food industry since the addition of microalgae, in particular *Spirulina*, may enable the production of starch products with added nutritional value that are more thermodynamically stable and more slowly digested in the gastrointestinal tract.

13.4.4.2 Polyphenol-polysaccharide Interactions

Despite the great controversy on the possible human health benefits of phenolic compounds, several epidemiological studies strongly suggest that long-term consumption of diets rich in plant polyphenols may help prevent cardiovascular diseases, cancer, diabetes, osteoporosis and neurodegenerative diseases.[218,219] The study of polyphenol-polysaccharide interactions is recently gaining in interest, since they have a major impact on the bioaccessibility (release from food matrix) and bioavailability (uptake after digestion) of polyphenols, which ultimately determine their effect on human health. The main source of dietary polyphenols is fruits and vegetables, where these species may be associated with cell wall components, such as cellulose.

Model plant cell wall systems based on bacterial cellulose hydrogels have been used to study the impact of food processing methods, such as hydrothermal, mechanical and drying treatments, on the ability of food polyphenols to interact with cellulose using SAXS and SANS.[220] The polyphenol binding capacity was seen to be strongly affected by the degree of cellulose microfibril aggregation and the availability of hydroxyl groups. A strong decrease in the adsorption of polyphenols occurred after a freeze-drying process, which was attributed to the formation of cellulose interfibrillar aggregates connected by a strong hydrogen bonding network. These structural effects were also highlighted in native apple cell walls subjected to different processing conditions.[221] The adsorption capacity of apple polyphenols was reduced as a result of drying. Furthermore, the presence of pectins and their solubilisation upon oven drying also influenced polyphenol-binding ability. These results show that both the composition and the processing of plant cell walls are crucial for the bioavailability of polyphenols.

13.4.4.3 Protein-polysaccharide Coacervates

Coacervation arises from the separation of colloidal systems into two liquid phases. This phenomenon is relevant to protein/polysaccharide complexes, which mostly originate from interactions between oppositely charged colloids. To reduce the free energy of the system, protein/polysaccharide complexes aggregate until their size and surface properties lead to insolubilization, resulting in a liquid–liquid macroscopic phase separation. At thermodynamic equilibrium, a two-phase system in which a dense fluid (complex coacervate) phase coexists with a dilute (solvent-rich) sub-phase. These systems offer the possibility of widening the functional properties of each individual component and represent interesting alternatives for the development of thickening and gelling agents, foaming and emulsifying agents, and encapsulation systems for the stabilization and release of bioactives or sensitive compounds. In general, it has been observed that the properties of coacervates are strongly affected by the charges of the two different components. For instance, the total amount and distribution of charges carried by pectic polysaccharides have been seen to affect the microstructure of the coacervates formed through interactions with the protein bovine β-lactoglobulin variant A (βLgA).[222] SAXS and SANS revealed a structure of compact particles composed of overlapping protein-binding pectin chains, bridged by βLgA-rich clusters, whose size and distribution were controlled by the overall and local charge densities of pectin. Furthermore, the chain length of the polysaccharide is also relevant to the formation of coacervates, as demonstrated by a recent study on the interactions between βLgA and oligogalacturonic acids (which were used as models for the un-methylesterified segments of pectin), combining SANS with ITC and CD.[223] In another study, SANS from lysozyme/pectin complexes indicated that, on a local scale, proteins were clustered inside the complexes, and, on a larger scale, the complexes formed spherical globules with hydrodynamic radius of 10–50 nm. The compaction of the protein and the size of the globules were driven by electrostatic interactions and could be controlled by changing the charge density of the pectin or pH conditions.[102] Denser or more open structures of whey protein/gum arabic coacervates could be obtained by changing parameters such as pH, protein/polysaccharide ratio and ionic strength.[224] The volume and density of the coacervate phase were the highest at a specific pH value (which differed depending on the protein/polysaccharide ratio) at which, as confirmed by SAXS, the coacervate was dense and structured. On the other hand, more heterogeneous and less structured coacervates were obtained when salt was added to the system.

13.4.5 Whole Foods

Due to the great complexity of whole food systems, in which one typically finds a combination of all three macronutrients, together with micronutrients such as vitamins and minerals, their analysis by SAS is extremely complicated. Only a few studies can be found in the literature where whole foods have been characterised using SAS techniques. Milk, a natural protein-stabilized emulsion containing vitamins, salts, proteins, carbohydrates and fat, has been one of the most studied whole food systems. The transition from normal emulsion through a variety of differently ordered nanostructures during *in vitro* digestion of milk, under simulated *in vivo* conditions of the gastrointestinal tract, has been observed using time-resolved SAXS experiments.[225,226] In another study, the structural effects of ultrasonication on cross-flow ultrafiltration of skim milk were investigated using a custom-designed SAXS cross-flow ultrasound-coupled filtration cell.[227] The study showed that while the internal structure of casein micelles was not affected by the ultrasound treatment, the filtration performance was significantly improved, showing promising results for mitigating the fouling phenomena usually taking place during milk ultrafiltration. The effect of high hydrostatic pressure (HHP) on skim milk was investigated by *in situ* SAXS experiments on samples held in a diamond anvil cell.[228] The results demonstrated that the hierarchical structure of the system was disrupted during processing and that their native casein micelle structure only partially recovered after releasing pressure to ambient.

SAXS has been used to follow the staling process of bread in dry and wet conditions.[229] Significant changes at the nanoscale level were noted during staling; the results suggested that in the dry samples, mainly the gluten phase was affected, while for water suspensions the changes originated mainly in the starch matrix and were suggested to be related to a retrogradation process. Oil migration in multi-component food materials containing cocoa butter has been investigated with μ-SAXS,[230] with important implications for the chocolate fat blooming phenomenon, *i.e.* the formation of white defects on the chocolate surface due to fat crystals. The results suggested a migration mechanism initially driven by capillary forces through pores and cracks in the solid, followed by chemical migration through the fat phase, inducing softening and partial dissolution of crystalline cocoa butter.

13.4.6 Other Food-related Applications

13.4.6.1 Encapsulation

Apart from their role in food structure, proteins and carbohydrates can be utilised to develop encapsulating structures for the protection and/or delivery of bioactive ingredients in food products and, in an associated field, pharmaceuticals.[231–233] An extensive review on the most relevant literature on this area can be found elsewhere.[234] Although microscopy, separation and spectroscopy methods have been typically used to evaluate the stability and release kinetics of encapsulation systems, SAS techniques offer advantages to investigate the native structure of encapsulating systems and their release mechanisms under a wide range of conditions; however, to date, they have been largely unexploited in this area. Some examples include the structural characterisation of soybean lecithin/chitosan nanoparticles loaded with a lipophilic positively charged drug,[235] unilamellar liposomes loaded with antibiotics,[236] and core–shell nanoparticles based on BSA combined with chitosan and poly(sodium-4-styrene)sulphonate for the controlled release of ibuprofen.[237] In a more sophisticated study, the distinct SLD contrast generated when using neutrons and X-rays was exploited to characterise microemulsions embedded in alginate hydrogels, probing the structure of the whole system and of the microemulsion droplets embedded within the gel.[238] In another work, Liu *et al.*[239] investigated the release mechanism of an enzyme encapsulated in freeze-dried chitosan/xanthan gum hydrogels, which was triggered by lowering the pH of the medium. SAXS showed that the addition of a nanoclay into the hydrogels stabilises the system, reducing the pH-induced structural changes in the hydrogels and thus lowering the enzyme release rate. SAS techniques have also been proven useful in characterising the biopolymer interactions in solutions to tune the properties of the subsequently produced encapsulation structures. For instance, SAXS characterisation of chitosan/gelatin solutions for the production of microcapsules through electrospraying showed that clustered networks between both biopolymers were formed due to electrostatic interactions; the formation of nanoscale soluble polyelectrolyte complexes for a particular formulation was observed due to stronger interactions between both biopolymers.[240] Polyelectrolyte complexation was also exploited to develop gelatin/ι-carrageenan encapsulation structures for intestinal-targeted release applications.[28] The interactions of blend solutions were explored by SAXS, highlighting the existence of a particular gelatin/ι-carrageenan ratio at which aggregates (larger than *ca.* 200 nm), with an internal structure of swollen linear chains, were formed through complexation.

13.4.6.2 Packaging Materials

Another potential application of SAS techniques, still largely unexplored, within the food industry is the structural characterisation of polymeric food packaging materials. Most of the available studies report on the structure of petroleum-derived polymeric materials such as polyamides[241] and poly(-ethylene-*co*-vinyl alcohol) (EVOH).[242–244] However, due to the severe environmental issues associated with the use of plastics, current trends in the food packaging sector are geared towards the development of biopolymeric materials, *i.e.* derived from renewable resources. Most of the works on the characterisation of biopolymers for food packaging are related to starch. Some examples include the SAXS/WAXS characterisation of starch-based films to study the structural changes and plasticizer migration during microwave heating in contact with milk[245] and the effect of lignocellulosic fractions extracted from *Posidonia oceanica* biomass[246] and microalgae[247,248] on composite starch films. Some other examples of biopolymeric films that have been characterised by SAXS/WAXS include extruded poly(lactic-acid),[249] blends of aliphatic polyesters with chitosan,[250] extruded wheat gluten films,[251] and chitosan films loaded with cellulose nanocrystals.[252] With the current circular economy policies being promoted around the world, research within the food packaging area is expected to focus on the development of bio-based materials derived from waste biomass. SAS techniques offer great opportunities for the structural characterisation of polysaccharide- and protein-based packaging approaches and investigation of the effect of industrial processing techniques and typical food processing methods on these newly developed packaging materials.

13.5 Conclusions and Future Perspectives

It is apparent that small-angle scattering methods have wide application in obtaining unique and valuable structural information across a diverse range of food-based systems, both naturally occurring and processed. The continued consumer-driven desire for healthy food that is produced sustainably will require the development of new processing technologies or the utilisation of new ingredients. On the nanoscale, small-angle scattering methods have the potential to make significant contributions in these areas particularly in studying processes *in operando*. The challenge has been to bring these methods into the mainstream of food materials science, on par with more conventional methods such as infrared spectroscopy, NMR and microscopy and to make them accessible to physicists, chemists, biologists, and engineers. There is also a need to develop a community of experts in these methods that are willing to share their knowledge but that also have a genuine desire to understand the actual requirements of food scientists working across academia, national research laboratories and industry. As an example, the Neutrons and Food conference series[253] has sought to do exactly this for all neutron-scattering techniques, not just SANS, and to provide a dialogue between the groups. It is also apparent that several synchrotrons have been successful in developing new communities in food science and in demonstrating the benefits of using "large-scale" facilities.[254] However, much work needs to be done to realise these opportunities; it is hoped that this chapter will stimulate the next generation of food materials scientists.

References

1. J. M. Aguilera, P. J. Lillford, Structure–Property Relationships in Foods, in *Food Materials Science: Principles and Practice*, ed. J. M. Aguilera and P. J. Lillford, Springer, New York, 2008, pp. 229–253.
2. P. Belton, *The Chemical Physics of Food*, John Wiley & Sons, 2008.
3. R. Mezzenga, P. Schurtenberger, A. Burbidge and M. Michel, Understanding foods as soft materials, *Nat. Mater.*, 2005, **4**(10), 729.
4. A. Lopez-Rubio and E. P. Gilbert, Neutron scattering: A natural tool for food science and technology research, *Trends Food Sci. Technol.*, 2009, **20**(11–12), 576–586.
5. E. P. Gilbert, A. Lopez-Rubio and M. J. Gidley, Characterisation techniques in food materials science, *Food Mater. Sci. Eng.*, 2012, 52–93.
6. J. Doutch, M. Bason, F. Franceschini, K. James, D. Clowes and E. P. Gilbert, Structural changes during starch pasting using simultaneous Rapid Visco Analysis and small-angle neutron scattering, *Carbohydr. Polym.*, 2012, **88**(3), 1061–1071.
7. V. F. Sears, Neutron scattering lengths and cross sections, *Neutron News*, 1992, **3**(3), 26–37.
8. I. Pilz, O. Glatter, O. Kratky, Small-angle x-ray scattering, in *Methods in Enzymology*, Elsevier, 1979, pp. 148–249.
9. J. Dawkins, R. A. Pethrick, *Modern Techniques for Polymer Characterisation*, John Wiley and Sons, 1999.
10. P. Lindner, T. Zemb, *Neutron, X-ray and Light Scattering*, North-Holland, Amesterdam, 1991.
11. B. Hammouda, The SANS toolbox, NIST Center for Neutron Research, 2008. Available at http://tinyurl.com/SANStoolbox.
12. A. Guiner, G. Fournet, C. Walker, *Small Angle Scattering of X-rays*, J. Wiley & Sons, New York, 1955.
13. G. D. Wignall, Small-angle neutron scattering characterization of polymers, *Phys. Prop. Polym.*, 1993, **424**, e502.

14. (a) H. P. Fink, in *Structure analysis by small-angle X-ray and neutron scattering*, ed. L. A. Feigin and D. I. Svergun, Plenum Press, New York/London, 1987, vol. 8, p. 335; (b) J. Dechant, *Acta Polym.*, 1989, **40**(3), 224.
15. B. Jacrot, The study of biological structures by neutron scattering from solution, *Rep. Prog. Phys.*, 1976, **39**(10), 911.
16. SAS Portal, The home for Small Angle Scattering. Available at http://smallangle.org/.
17. D. I. Svergun and M. H. Koch, Small-angle scattering studies of biological macromolecules in solution, *Rep. Prog. Phys.*, 2003, **66**(10), 1735.
18. M. Ballauff and A. Jusufi, Anomalous small-angle X-ray scattering: analyzing correlations and fluctuations in polyelectrolytes, *Colloid Polym. Sci.*, 2006, **284**(11), 1303–1311.
19. A. Bot, E. P. Gilbert, W. G. Bouwman, H. Sawalha, R. den Adel, V. M. Garamus, P. Venema, E. van der Linden and E. Flöter, Elucidation of density profile of self-assembled sitosterol+oryzanol tubules with small-angle neutron scattering, *Faraday Discuss.*, 2012, **158**(1), 223–238.
20. K. Wood, J. P. Mata, C. J. Garvey, C.-M. Wu, W. A. Hamilton, P. Abbeywick, D. Bartlett, F. Bartsch, P. Baxter, N. Booth, W. Brown, J. Christoforidis, D. Clowes, T. d'Adam, F. Darmann, M. Deura, S. Harrison, N. Hauser, G. Horton, D. Federici, F. Franceschini, P. Hanson, E. Imamovic, P. Imperia, M. Jones, S. Kennedy, S. Kim, T. Lam, W. T. Lee, M. Lesha, D. Mannicke, T. Noakes, S. R. Olsen, J. C. Osborn, D. Penny, M. Perry, S. A. Pullen, R. A. Robinson, J. C. Schulz, N. Xiong and E. P. Gilbert, QUOKKA, the pinhole small-angle neutron scattering instrument at the OPAL Research Reactor, Australia: design, performance, operation and scientific highlights, *J. Applied Crystallogr.*, 2018, **51**(2), 294–314.
21. A. Sokolova, J. Christoforidis, A. Eltobaji, J. Barnes, F. Darmann, A. E. Whitten and L. de Campo, BILBY: time-of-flight small angle scattering instrument, *Neutron News*, 2016, **27**(2), 9–13.
22. C. Rehm, L. d. Campo, A. Brûlé, F. Darmann, F. Bartsch and A. Berry, Design and performance of the variable-wavelength Bonse–Hart ultra-small-angle neutron scattering diffractometer KOOKABURRA at ANSTO, *J. Appl. Crystallogr.*, 2018, **51**(1), 1–8.
23. EMBL, Biological Small Angle Scattering. Available at https://www.embl-hamburg.de/biosaxs/software.html.
24. SasView Model Functions. Available at http://www.sasview.org/sasview/user/models/model_functions.html.
25. A. Donald, Food for thought, *Nat. Mater.*, 2004, **3**, 579.
26. S. Damodaran, K. L. Parkin, *Fennema's Food Chemistry*, CRC Press, 2017.
27. J. Blazek and E. P. Gilbert, Application of small-angle X-ray and neutron scattering techniques to the characterisation of starch structure: A review, *Carbohydr. Polym.*, 2011, **85**(2), 281–293.

28. L. G. Gómez-Mascaraque, B. Llavata-Cabrero, M. Martínez-Sanz, M. J. Fabra and A. López-Rubio, Self-assembled gelatin-ι-carrageenan encapsulation structures for intestinal-targeted release applications, *J. Colloid Interface Sci.*, 2018, **517**, 113–123.
29. P. Lopez-Sanchez, M. Martinez-Sanz, M. R. Bonilla, D. Wang, E. P. Gilbert, J. R. Stokes and M. J. Gidley, Cellulose-pectin composite hydrogels: Intermolecular interactions and material properties depend on order of assembly, *Carbohydr. Polym.*, 2017, **162**, 71–81.
30. M. Martínez-Sanz, M. J. Gidley and E. P. Gilbert, Application of X-ray and neutron small angle scattering techniques to study the hierarchical structure of plant cell walls: A review, *Carbohydr. Polym.*, 2015, **125**, 120–134.
31. M. Martínez-Sanz, D. Mikkelsen, B. Flanagan, M. J. Gidley and E. P. Gilbert, Multi-scale model for the hierarchical architecture of native cellulose hydrogels, *Carbohydr. Polym.*, 2016, **147**, 542–555.
32. L. Yu, G. E. Yakubov, M. Martínez-Sanz, E. P. Gilbert and J. R. Stokes, Rheological and structural properties of complex arabinoxylans from *Plantago ovata* seed mucilage under non-gelled conditions, *Carbohydr. Polym.*, 2018, **193**, 179–188.
33. A. Imberty, A. Buléon, V. Tran and S. Pérez, Recent advances in knowledge of starch structure, *Starch/Stärke*, 1991, **43**(10), 375–384.
34. S. Pérez and E. Bertoft, The molecular structures of starch components and their contribution to the architecture of starch granules: A comprehensive review, *Starch/Stärke*, 2010, **62**(8), 389–420.
35. N. W. H. Cheetham and L. Tao, Variation in crystalline type with amylose content in maize starch granules: an X-ray powder diffraction study, *Carbohydr. Polym.*, 1998, **36**(4), 277–284.
36. R. Hoover, Composition, molecular structure, and physicochemical properties of tuber and root starches: a review, *Carbohydr. Polym.*, 2001, **45**(3), 253–267.
37. P. J. Jenkins, R. E. Cameron and A. M. Donald, A universal feature in the structure of starch granules from different botanical sources, *Starch/Stärke*, 1993, **45**(12), 417–420.
38. H. Salman, J. Blazek, A. Lopez-Rubio, E. P. Gilbert, T. Hanley and L. Copeland, Structure–function relationships in A and B granules from wheat starches of similar amylose content, *Carbohydr. Polym.*, 2009, **75**(3), 420–427.
39. A. Imberty, H. Chanzy, S. Pérez, A. Bulèon and V. Tran, The double-helical nature of the crystalline part of A-starch, *J. Mol. Biol.*, 1988, **201**(2), 365–378.
40. A. Imberty and S. Perez, A revisit to the three-dimensional structure of B-type starch, *Biopolymers*, 1988, **27**(8), 1205–1221.
41. A. Sarko and H. C. Wu, The Crystal Structures of A-, B- and C-Polymorphs of Amylose and Starch, *Starch/Stärke*, 1978, **30**(3), 73–78.

42. A. Lopez-Rubio, A. Htoon and E. P. Gilbert, Influence of Extrusion and Digestion on the Nanostructure of High-Amylose Maize Starch, *Biomacromolecules*, 2007, **8**(5), 1564–1572.
43. R. E. Cameron and A. M. Donald, A small-angle X-ray scattering study of the annealing and gelatinization of starch, *Polymer*, 1992, **33**(12), 2628–2635.
44. L. Lamberts, S. V. Gomand, V. Derycke and J. A. Delcour, Presence of Amylose Crystallites in Parboiled Rice, *J. Agric. Food Chem.*, 2009, **57**(8), 3210–3216.
45. D. Saibene and K. Seetharaman, Amylose involvement in the amylopectin clusters of potato starch granules, *Carbohydr. Polym.*, 2010, **82**(2), 376–383.
46. M. M. Agamalyan, G. A. Evmenenko, A. D. Vilesov, S. Y. Frenkel, V. N. Zgonnik, L. V. Vinogradova and E. Y. Melenevskaya, Small-angle neutron diffraction study of block copolymer superstructure, *Polymer*, 1992, **33**(12), 2542–2547.
47. D. R. Daniels and A. M. Donald, An improved model for analyzing the small angle X-ray scattering of starch granules, *Biopolymers*, 2003, **69**(2), 165–175.
48. M. B. Cardoso and H. Westfahl Jr., On the lamellar width distributions of starch, *Carbohydr. Polym.*, 2010, **81**(1), 21–28.
49. P. Chen, K. Wang, Q. Kuang, S. Zhou, D. Wang and X. Liu, Understanding how the aggregation structure of starch affects its gastrointestinal digestion rate and extent, *Int. J. Biol. Macromol.*, 2016, **87**, 28–33.
50. Q. Kuang, J. Xu, Y. Liang, F. Xie, F. Tian, S. Zhou and X. Liu, Lamellar structure change of waxy corn starch during gelatinization by time-resolved synchrotron SAXS, *Food Hydrocolloids*, 2017, **62**, 43–48.
51. Z. Yang, Q. Gu, E. Lam, F. Tian, S. Chaieb and Y. Hemar, *In situ* study starch gelatinization under ultra-high hydrostatic pressure using synchrotron SAXS, *Food Hydrocolloids*, 2016, **56**, 58–61.
52. J. Blazek and E. P. Gilbert, Effect of Enzymatic Hydrolysis on Native Starch Granule Structure, *Biomacromolecules*, 2010, **11**(12), 3275–3289.
53. H. Chanvrier, S. Uthayakumaran, I. A. M. Appelqvist, M. J. Gidley, E. P. Gilbert and A. López-Rubio, Influence of Storage Conditions on the Structure, Thermal Behavior, and Formation of Enzyme-Resistant Starch in Extruded Starches, *J. Agric. Food Chem.*, 2007, **55**(24), 9883–9890.
54. A. Lopez-Rubio, B. M. Flanagan, A. K. Shrestha, M. J. Gidley and E. P. Gilbert, Molecular Rearrangement of Starch During *In Vitro* Digestion: Toward a Better Understanding of Enzyme Resistant Starch Formation in Processed Starches, *Biomacromolecules*, 2008, **9**(7), 1951–1958.
55. J. Doutch and E. P. Gilbert, Characterisation of large scale structures in starch granules via small-angle neutron and X-ray scattering, *Carbohydr. Polym.*, 2013, **91**(1), 444–451.

56. I. Bocharnikova, L. A. Wasserman, A. V. Krivandin, J. Fornal, W. Baszczak, V. Y. Chernykh, A. Schiraldi and V. P. Yuryev, Structure and thermodynamic melting parameters of wheat starches with different amylose content, *J. Therm. Anal. Calorim.*, 2003, **74**(3), 681–695.
57. P. J. Jenkins and A. M. Donald, The influence of amylose on starch granule structure, *Int. J. Biol. Macromol.*, 1995, **17**(6), 315–321.
58. S. S. Kozlov, A. Blennow, A. V. Krivandin and V. P. Yuryev, Structural and thermodynamic properties of starches extracted from GBSS and GWD suppressed potato lines, *Int. J. Biol. Macromol.*, 2007, **40**(5), 449–460.
59. S. S. Kozlov, A. V. Krivandin, O. V. Shatalova, T. Noda, E. Bertoft, J. Fornal and V. P. Yuryev, Structure of starches extracted from near-isogenic wheat lines, *J. Therm. Anal. Calorim.*, 2007, **87**(2), 575–584.
60. J. S. Sanderson, R. D. Daniels, A. M. Donald, A. Blennow and S. B. Engelsen, Exploratory SAXS and HPAEC-PAD studies of starches from diverse plant genotypes, *Carbohydr. Polym.*, 2006, **64**(3), 433–443.
61. V. P. Yuryev, A. V. Krivandin, V. I. Kiseleva, L. A. Wasserman, N. K. Genkina, J. Fornal, W. Blaszczak and A. Schiraldi, Structural parameters of amylopectin clusters and semi-crystalline growth rings in wheat starches with different amylose content, *Carbohydr. Res.*, 2004, **339**(16), 2683–2691.
62. A. M. Donald, K. L. Kato, P. A. Perry and T. A. Waigh, Scattering Studies of the Internal Structure of Starch Granules, *Starch/Stärke*, 2001, **53**(10), 504–512.
63. N. Aravind, M. Sissons, C. M. Fellows, J. Blazek and E. P. Gilbert, Optimisation of resistant starch II and III levels in durum wheat pasta to reduce *in vitro* digestibility while maintaining processing and sensory characteristics, *Food Chem.*, 2013, **136**(2), 1100–1109.
64. H. Chanvrier, I. A. Appelqvist, A. R. Bird, E. Gilbert, A. Htoon, Z. Li, P. J. Lillford, A. Lopez-Rubio, M. K. Morell and D. L. Topping, Processing of novel elevated amylose wheats: functional properties and starch digestibility of extruded products, *J. Agric. Food Chem.*, 2007, **55**(25), 10248–10257.
65. A. Htoon, A. K. Shrestha, B. M. Flanagan, A. Lopez-Rubio, A. R. Bird, E. P. Gilbert and M. J. Gidley, Effects of processing high amylose maize starches under controlled conditions on structural organisation and amylase digestibility, *Carbohydr. Polym.*, 2009, **75**(2), 236–245.
66. J. W. Donovan, Phase transitions of the starch–water system, *Biopolymers*, 1979, **18**(2), 263–275.
67. P. J. Jenkins, R. E. Comerson, A. Donald, W. Bras, G. E. Derbyshire, G. R. Mant and A. Ryan, *In situ* simultaneous small and wide angle X-ray scattering: A new technique to study starch gelatinization, *J. Polym. Sci., Part B: Polym. Phys.*, 1994, **32**(8), 1579–1583.
68. R. Vermeylen, V. Derycke, J. A. Delcour, B. Goderis, H. Reynaers and M. H. J. Koch, Gelatinization of Starch in Excess Water: Beyond the

Melting of Lamellar Crystallites. A Combined Wide- and Small-Angle X-ray Scattering Study, *Biomacromolecules*, 2006, **7**(9), 2624–2630.
69. R. Vermeylen, V. Derycke, J. A. Delcour, B. Goderis, H. Reynaers and M. H. J. Koch, Structural Transformations during Gelatinization of Starches in Limited Water: Combined Wide- and Small-Angle X-ray Scattering Study, *Biomacromolecules*, 2006, **7**(4), 1231–1238.
70. R. Gebhardt, M. Hanfland, M. Mezouar and C. Riekel, High-Pressure Potato Starch Granule Gelatinization: Synchrotron Radiation Micro-SAXS/WAXS Using a Diamond Anvil Cell, *Biomacromolecules*, 2007, **8**(7), 2092–2097.
71. H. Jacobs, N. Mischenko, M. H. J. Koch, R. C. Eerlingen, J. A. Delcour and H. Reynaers, Evaluation of the impact of annealing on gelatinisation at intermediate water content of wheat and potato starches: A differential scanning calorimetry and small angle X-ray scattering study, *Carbohydr. Res.*, 1998, **306**(1), 1–10.
72. M. Martínez-Sanz, M. J. Fabra, L. G. Gómez-Mascaraque and A. López-Rubio, Structural effects of microalgae additives on the starch gelatinisation process, *Food Hydrocolloids*, 2018, **77**, 257–269.
73. S. Zabar, U. Lesmes, I. Katz, E. Shimoni and H. Bianco-Peled, Structural characterization of amylose-long chain fatty acid complexes produced via the acidification method, *Food Hydrocolloids*, 2010, **24**(4), 347–357.
74. K. J. I'Anson, M. J. Miles, V. J. Morris, S. G. Ring and C. Nave, A study of amylose gelation using a synchrotron X-ray source, *Carbohydr. Polym.*, 1988, **8**(1), 45–53.
75. J. J. Mueller, C. Gernat, W. Schulz, E. C. Mueller, W. Vorwerg and G. Damaschun, Computer simulations of x-ray scattering curves: Gelation and crystallization process in amylose solutions, *Biopolymers*, 1995, **35**(3), 271–288.
76. T. Suzuki, A. Chiba and T. Yarno, Interpretation of small angle X-ray scattering from starch on the basis of fractals, *Carbohydr. Polym.*, 1997, **34**(4), 357–363.
77. K. Shamai, H. Bianco-Peled and E. Shimoni, Polymorphism of resistant starch type III, *Carbohydr. Polym.*, 2003, **54**(3), 363–369.
78. K. Shamai, E. Shimoni and H. Bianco-Peled, Small Angle X-ray Scattering of Resistant Starch Type III, *Biomacromolecules*, 2004, **5**(1), 219–223.
79. Z. Shiran, S. Eyal and B. P. Havazelet, Development of Nanostructure in Resistant Starch Type III During Thermal Treatments and Cycling, *Macromol. Biosci.*, 2008, **8**(2), 163–170.
80. J. Zhu, L. Li, L. Chen and X. Li, Study on supramolecular structural changes of ultrasonic treated potato starch granules, *Food Hydrocolloids*, 2012, **29**(1), 116–122.
81. J. Cai, Y. Yang, J. Man, J. Huang, Z. Wang, C. Zhang, M. Gu, Q. Liu and C. Wei, Structural and functional properties of alkali-treated high-amylose rice starch, *Food Chem.*, 2014, **145**, 245–253.

82. S. Wang and L. Copeland, Effect of alkali treatment on structure and function of pea starch granules, *Food Chem.*, 2012, **135**(3), 1635–1642.
83. J. Holm, I. Lundquist, I. Björck, A.-C. Eliasson and N.-G. Asp, Degree of starch gelatinization, digestion rate of starch *in vitro*, and metabolic response in rats, *Am. J. Clin. Nutr.*, 1988, **47**(6), 1010–1016.
84. D. Fan, L. Wang, W. Chen, S. Ma, W. Ma, X. Liu, J. Zhao and H. Zhang, Effect of microwave on lamellar parameters of rice starch through small-angle X-ray scattering, *Food Hydrocolloids*, 2014, **35**, 620–626.
85. M. Roberts, Eat rice cold for fewer calories, 2015. Available at https://www.bbc.com/news/health-32019176.
86. L. Cai, Y.-C. Shi, L. Rong and B. S. Hsiao, Debranching and crystallization of waxy maize starch in relation to enzyme digestibility, *Carbohydr. Polym.*, 2010, **81**(2), 385–393.
87. A. K. Shrestha, J. Blazek, B. M. Flanagan, S. Dhital, O. Larroque, M. K. Morell, E. P. Gilbert and M. J. Gidley, Molecular, mesoscopic and microscopic structure evolution during amylase digestion of maize starch granules, *Carbohydr. Polym.*, 2012, **90**(1), 23–33.
88. F. Zeng, Q.-Y. Gao, Z. Han, X.-A. Zeng and S.-j. Yu, Structural properties and digestibility of pulsed electric field treated waxy rice starch, *Food Chem.*, 2016, **194**, 1313–1319.
89. D. P. Delmer and Y. Amor, Cellulose biosynthesis, *Plant Cell*, 1995, **7**(7), 987–1000.
90. M. Mutwil, S. Debolt and S. Persson, Cellulose synthesis: a complex complex, *Curr. Opin. Plant Biol.*, 2008, **11**(3), 252–257.
91. P. Lopez-Sanchez, M. Martinez-Sanz, M. R. Bonilla, D. Wang, C. T. Walsh, E. P. Gilbert, J. R. Stokes and M. J. Gidley, Pectin impacts cellulose fibre architecture and hydrogel mechanics in the absence of calcium, *Carbohydr. Polym.*, 2016, **153**, 236–245.
92. M. Martínez-Sanz, P. Lopez-Sanchez, M. J. Gidley and E. P. Gilbert, Evidence for differential interaction mechanism of plant cell wall matrix polysaccharides in hierarchically-structured bacterial cellulose, *Cellulose*, 2015, **22**(3), 1541–1563.
93. M. Martínez-Sanz, M. J. Gidley and E. P. Gilbert, Hierarchical architecture of bacterial cellulose and composite plant cell wall polysaccharide hydrogels using small angle neutron scattering, *Soft Matter*, 2016, **12**(5), 1534–1549.
94. M. Martínez-Sanz, D. Mikkelsen, B. M. Flanagan, M. J. Gidley and E. P. Gilbert, Multi-scale characterisation of deuterated cellulose composite hydrogels reveals evidence for different interaction mechanisms with arabinoxylan, mixed-linkage glucan and xyloglucan, *Polymer*, 2017, **124**, 1–11.
95. M. Martinez-Sanz, D. Mikkelsen, B. M. Flanagan, C. Rehm, L. de Campo, M. J. Gidley and E. P. Gilbert, Investigation of the micro- and nano-scale architecture of cellulose hydrogels with plant cell wall polysaccharides: A combined USANS/SANS study, *Polymer*, 2016, **105**, 449–460.

96. M. Martínez-Sanz, F. Pettolino, B. Flanagan, M. J. Gidley and E. P. Gilbert, Structure of cellulose microfibrils in mature cotton fibres, *Polymer*, 2017, **175**, 450–463.
97. P. Lopez-Sanchez, M. Martinez-Sanz, M. R. Bonilla, F. Sonni, E. P. Gilbert and M. J. Gidley, Nanostructure and poroviscoelasticity in cell wall materials from onion, carrot and apple: Roles of pectin, *Food Hydrocolloids*, 2020, **98**, 105253.
98. A. P. Heiner and O. Teleman, Interface between Monoclinic Crystalline Cellulose and Water: Breakdown of the Odd/Even Duplicity, *Langmuir*, 1997, **13**(3), 511–518.
99. M. Ioelovich, A. Leykin and O. Figovsky, Study of cellulose para-crystallinity, *BioResources*, 2010, **5**(3), 1393–1407.
100. W. Chen, G. C. Lickfield and C. Q. Yang, Molecular modeling of cellulose in amorphous state. Part I: model building and plastic deformation study, *Polymer*, 2004, **45**(3), 1063–1071.
101. C. V. Cerclier, A. Guyomard-Lack, F. Cousin, B. Jean, E. Bonnin, B. Cathala and C. Moreau, Xyloglucan–Cellulose Nanocrystal Multi-layered Films: Effect of Film Architecture on Enzymatic Hydrolysis, *Biomacromolecules*, 2013, **14**(10), 3599–3609.
102. I. Schmidt, F. Cousin, C. Huchon, F. Boué and M. A. V. Axelos, Spatial Structure and Composition of Polysaccharide-Protein Complexes from Small Angle Neutron Scattering, *Biomacromolecules*, 2009, **10**(6), 1346–1357.
103. R. C. Randall, G. O. Phillips and P. A. Williams, The role of the proteinaceous component on the emulsifying properties of gum arabic, *Food Hydrocolloids*, 1988, **2**(2), 131–140.
104. D. Yael, C. Yachin and Y. R. Rachel, Structure of gum arabic in aqueous solution, *J. Polym. Sci., Part B: Polym. Phys.*, 2006, **44**(22), 3265–3271.
105. C. Berlangieri, G. Poggi, S. Murgia, M. Monduzzi, L. Dei and E. Carretti, Structural, rheological and dynamics insights of hydroxypropyl guar gel-like systems, *Colloids Surf., B*, 2018, **168**, 178–186.
106. Y. Yuguchi, M. Mimura, S. Kitamura, H. Urakawa and K. Kajiwara, Structural characteristics of gellan in aqueous solution, *Food Hydrocolloids*, 1993, **7**(5), 373–385.
107. H. Yoshida and M. Takahashi, Structural change of gellan hydrogel induced by annealing, *Food Hydrocolloids*, 1993, **7**(5), 387–395.
108. H. Fukada, K. Takahashi, S. Kitamura, Y. Yuguchi, H. Urakawa and K. Kajiwara, Thermodynamics and structural aspect of the gelling process in the gellan gum/metal salt aqueous solutions, *J. Therm. Anal. Calorim.*, 2002, **70**(3), 797–806.
109. Y. Izumi, N. Kikuta, K. Sakai and H. Takezawa, Phase diagrams and molecular structures of sodium-salt-type gellan gum, *Carbohydr. Polym.*, 1996, **30**(2), 121–127.
110. Y. Yoshiaki, U. Hiroshi and K. Kanji, Structural characteristics of crosslinking domain in gellan gum gel, *Macromol. Symp.*, 1997, **120**(1), 77–89.

111. Y. Yuguchi, M. Mimura, H. Urakawa, S. Kitamura, S. Ohno and K. Kajiwara, Small angle X-ray characterization of gellan gum containing a high content of sodium in aqueous solution, *Carbohydr. Polym.*, 1996, **30**(2), 83–93.
112. Y. Yuguchi, H. Urakawa and K. Kajiwara, The effect of potassium salt on the structural characteristics of gellan gum gel, *Food Hydrocolloids*, 2002, **16**(3), 191–195.
113. C. F. de Souza, I. C. Riegel-Vidotti, M. B. Cardoso, L. Ono, N. Lucyszyn, A. F. Lubambo, C. V. Sens, A. Grein-Iankovski and M. R. Sierakowski, Nanometric organisation in blends of gellan/xyloglucan hydrogels, *Carbohydr. Polym.*, 2014, **114**, 48–56.
114. T. Tada, T. Matsumoto and T. Masuda, Structure of molecular association of curdlan at dilute regime in alkaline aqueous systems, *Chem. Phys.*, 1998, **228**(1), 157–166.
115. T. Tada, T. Matsumoto and T. Masuda, Dynamic viscoelasticity and small-angle X-ray scattering studies on the gelation mechanism and network structure of curdlan gels, *Carbohydr. Polym.*, 1999, **39**(1), 53–59.
116. F. X. Quinn, T. Hatakeyama, M. Takahashi and H. Hatakeyama, The effect of annealing on the conformational properties of xanthan hydrogels, *Polymer*, 1994, **35**(6), 1248–1252.
117. M. Milas, M. Rinaudo, R. Duplessix, R. Borsali and P. Lindner, Small angle neutron scattering from polyelectrolyte solutions: from disordered to ordered xanthan chain conformation, *Macromolecules*, 1995, **28**(9), 3119–3124.
118. Y. Yuguchi, T. T. Thu Thuy, H. Urakawa and K. Kajiwara, Structural characteristics of carrageenan gels: temperature and concentration dependence, *Food Hydrocolloids*, 2002, **16**(6), 515–522.
119. Y. Yuguchi, H. Urakawa and K. Kajiwara, Structural characteristics of carrageenan gels: various types of counter ions, *Food Hydrocolloids*, 2003, **17**(4), 481–485.
120. B. Denef, N. Mischenko, M. H. J. Koch and H. Reynaers, Small-angle X-ray scattering of κ- and ι-carrageenan in aqueous and in salt solutions, *Int. J. Biol. Macromol.*, 1996, **18**(3), 151–159.
121. N. Mischenko, B. Denef, M. H. J. Koch and H. Reynaers, Influence of ionic effects on the ordering and association phenomena in dilute and semidilute carrageenan solutions, *Int. J. Biol. Macromol.*, 1996, **19**(3), 185–194.
122. B. T. Stokke, K. I. Draget, O. Smidsrød, Y. Yuguchi, H. Urakawa and K. Kajiwara, Small-Angle X-ray Scattering and Rheological Characterization of Alginate Gels. 1. Ca–Alginate Gels, *Macromolecules*, 2000, **33**(5), 1853–1863.
123. W. Zheng-Yu, J. W. White, K. Mikio, S. Shozaburo and N. Tsunenori, A small-angle X-ray scattering study of alginate solution and its Sol–Gel transition by addition of divalent cations, *Biopolymers*, 1995, **35**(2), 227–238.

124. Y. Yuguchi, H. Urakawa, K. Kajiwara, K. I. Draget and B. T. Stokke, Small-angle X-ray scattering and rheological characterization of alginate gels. 2. Time-resolved studies on ionotropic gels, *J. Mol. Struct.*, 2000, **554**(1), 21–34.
125. K. I. Draget, B. T. Stokke, Y. Yuguchi, H. Urakawa and K. Kajiwara, Small-Angle X-ray Scattering and Rheological Characterization of Alginate Gels. 3. Alginic Acid Gels, *Biomacromolecules*, 2003, **4**(6), 1661–1668.
126. M. Djabourov, A. Clark, D. Rowlands and S. Ross-Murphy, Small-angle x-ray scattering characterization of agarose sols and gels, *Macromolecules*, 1989, **22**(1), 180–188.
127. N. Fatin-Rouge, K. J. Wilkinson and J. Buffle, Combining Small Angle Neutron Scattering (SANS) and Fluorescence Correlation Spectroscopy (FCS) Measurements To Relate Diffusion in Agarose Gels to Structure, *J. Phys. Chem. B*, 2006, **110**(41), 20133–20142.
128. G. D'Arrigo, C. Di Meo, E. Geissler, T. Coviello, F. Alhaique and P. Matricardi, Hyaluronic acid methacrylate derivatives and calcium alginate interpenetrated hydrogel networks for biomedical applications: physico-chemical characterization and protein release, *Colloid Polym. Sci.*, 2012, **290**(15), 1575–1582.
129. D. A. Rennerfeldt, A. N. Renth, Z. Talata, S. H. Gehrke and M. S. Detamore, Tuning mechanical performance of poly(ethylene glycol) and agarose interpenetrating network hydrogels for cartilage tissue engineering, *Biomaterials*, 2013, **34**(33), 8241–8257.
130. A. Lopez-Rubio, P. Tarancón, L. G. Gómez-Mascaraque, M. Martínez-Sanz, M. J. Fabra, J. C. Martínez and S. Fiszman, Development of glucomannan-chitosan interpenetrating hydrocolloid networks (IHNs) as a potential tool for creating satiating ingredients, *Food Hydrocolloids*, 2016, **60**, 533–542.
131. H. Pessen, T. F. Kumosinski and H. M. Farrell, Investigation of differences in the tertiary structures of food proteins by small-angle X-ray scattering, *J. Ind. Microbiol.*, 1988, **3**(2), 89–103.
132. E. Mathew, A. Mirza and N. Menhart, Liquid-chromatography-coupled SAXS for accurate sizing of aggregating proteins, *J. Synchrotron Radiat.*, 2004, **11**, 314–318.
133. Y. Watanabe, Assessment study of ion-exchange chromatography combined with solution X-ray scattering measurement for protein characterization, *J. Chromatogr. A*, 2018, **1539**, 103–107.
134. C. G. De Kruif, The structure of casein micelles: a review of small-angle scattering data, *J. Appl. Crystallogr.*, 2014, **47**(5), 1479–1489.
135. C. Holt, J. A. Carver, H. Ecroyd and D. C. Thorn, Invited review: Caseins and the casein micelle: their biological functions, structures, and behavior in foods, *J. Dairy Sci.*, 2013, **96**(10), 6127–6146.
136. L. Day, J. K. Raynes, A. Leis, L. H. Liu and R. P. W. Williams, Probing the internal and external micelle structures of differently sized casein micelles from individual cows milk by dynamic light and small-angle X-ray scattering, *Food Hydrocolloids*, 2017, **69**, 150–163.

137. C. David, F. Pignon, T. Narayanan, M. Sztucki, G. Gésan-Guiziou and A. Magnin, Spatial and temporal *in situ* evolution of the concentration profile during casein micelle ultrafiltration probed by small-angle X-ray scattering, *Langmuir*, 2008, **24**(9), 4523–4529.
138. C. G. de Kruif, T. Huppertz, V. S. Urban and A. V. Petukhov, Casein micelles and their internal structure, *Adv. Colloid Interface Sci.*, 2012, **171–172**, 36–52.
139. S. Hansen, R. Bauer, S. B. Lomholt, K. B. Quist, J. S. Pedersen and K. Mortensen, Structure of casein micelles studied by small-angle neutron scattering, *Eur. Biophys. J.*, 1996, **24**(3), 143–147.
140. D. G. Schmidt, in *Developments in Dairy Chemistry. I. Proteins*, ed. P. F. Fox, Applied Science Publishers, Barking, 1982, pp. 61–86.
141. C. Holt, *The Biological Function of Casein: Research Reviews*, Hannah Research Institute, 1994, pp. 60–68.
142. C. G. H. De Kruif, in *Advanced Dairy Chemistry*, ed. P. F. M. Fox and P. L. H. McSweeney, Kluwer Academic, New York, 2003, p. 233.
143. J. P. Mata, P. Udabage and E. P. Gilbert, Structure of casein micelles in milk protein concentrate powders via small angle X-ray scattering, *Soft Matter*, 2011, **7**(8), 3837–3843.
144. C. Holt, C. De Kruif, R. Tuinier and P. Timmins, Substructure of bovine casein micelles by small-angle X-ray and neutron scattering, *Colloids Surf., A*, 2003, **213**(2–3), 275–284.
145. S. Marchin, J.-L. Putaux, F. Pignon and J. Léonil, Effects of the environmental factors on the casein micelle structure studied by cryo transmission electron microscopy and small-angle X-ray scattering/ultrasmall-angle x-ray scattering, *J. Chem. Phys.*, 2007, **126**(4), 01B617.
146. B. Ingham, G. Erlangga, A. Smialowska, N. Kirby, C. Wang, L. Matia-Merino, R. Haverkamp and A. Carr, Solving the mystery of the internal structure of casein micelles, *Soft Matter*, 2015, **11**(14), 2723–2725.
147. B. Ingham, N. Kirby, C. Wang, M. Brady and A. Carr, Elemental fingerprinting of mineral species in iron-fortified milk: anomalous small-angle X-ray scattering and resonant soft X-ray scattering studies, *J. Synchrotron Radiat.*, 2018, **25**(Pt 4), 1106–1112.
148. E. Bijl, R. de Vries, H. van Valenberg, T. Huppertz and T. van Hooijdonk, Factors influencing casein micelle size in milk of individual cows: Genetic variants and glycosylation of κ-casein, *Int. Dairy J.*, 2014, **34**(1), 135–141.
149. C. G. de Kruif and T. Huppertz, Casein micelles: size distribution in milks from individual cows, *J. Agric. Food Chem.*, 2012, **60**(18), 4649–4655.
150. B. Ingham, A. Smialowska, N. M. Kirby, C. Wang and A. J. Carr, A structural comparison of casein micelles in cow, goat and sheep milk using X-ray scattering, *Soft Matter*, 2018, **14**(17), 3336–3343.
151. M. Alexander, M. P. Nieh, M. A. Ferrer and M. Corredig, Changes in the calcium cluster distribution of ultrafiltered and diafiltered fresh skim

milk as observed by Small Angle Neutron Scattering, *J. Dairy Res.*, 2011, **78**(3), 349–356.
152. C. Huck-Iriart, J. Montes-de-Oca-Avalos, M. L. Herrera, R. J. Candal, C. L. Pinto-de-Oliveira and I. Linares-Torriani, New insights about flocculation process in sodium caseinate-stabilized emulsions, *Food Res. Int.*, 2016, **89**(Pt 1), 338–346.
153. G. Brans, C. G. P. H. Schroën, R. G. M. van der Sman and R. M. Boom, Membrane fractionation of milk: state of the art and challenges, *J. Membr. Sci.*, 2004, **243**(1–2), 263–272.
154. A. Logan, A. Leis, L. Day, S. K. Øiseth, A. Puvanenthiran and M. A. Augustin, Rennet gelation properties of milk: Influence of natural variation in milk fat globule size and casein micelle size, *Int. Dairy J.*, 2015, **46**, 71–77.
155. R. Niki, G. Y. Kim, T. Kimura, K. Takahashi, K. Kohyama and K. Nishinari, Physical-properties and microstructure of rennet gels from casein micelles of different sizes, *Milchwissenschaft – Milk Sci. Int.*, 1994, **49**(6), 325–329.
156. R. P. W. Williams, L. D'Ath and B. Zisu, Role of protein aggregation in heat-induced heat stability during milk powder manufacture, *Dairy Sci. Technol.*, 2008, **88**(1), 121–147.
157. M. D. E. Mp, Functional milk proteins: production and utilisation, in *Advanced Dairy Chemistry*, ed. P. F. M. Fox and P. L. H. McSweeney, Kluwer Academic, New York, 2003, pp. 1175–1228.
158. J. Choi, D. S. Horne and J. A. Lucey, Effect of insoluble calcium concentration on rennet coagulation properties of milk, *J. Dairy Sci.*, 2007, **90**(6), 2612–2623.
159. J. Lucey and P. Fox, Importance of calcium and phosphate in cheese manufacture: A review, *J. Dairy Sci.*, 1993, **76**(6), 1714–1724.
160. J. Choi, D. S. Horne, M. E. Johnson and J. A. Lucey, Effects of the concentration of insoluble calcium phosphate associated with casein micelles on the functionality of directly acidified cheese, *J. Dairy Sci.*, 2008, **91**(2), 513–522.
161. R. Tromp, T. Huppertz and J. Kohlbrecher, Casein micelles at non-ambient pressure studied by neutron scattering, *Food Biophys.*, 2015, **10**(1), 51–56.
162. B. R. Ward, S. J. Goddard, M.-A. Augustin and I. R. McKinnon, EDTA-induced dissociation of casein micelles and its effect on foaming properties of milk, *J. Dairy Res.*, 1997, **64**(4), 495–504.
163. E. Lam, C. Holt, P. Edwards, I. McKinnon, D. Otter, N. Li and Y. Hemar, The effect of transglutaminase treatment on the physico-chemical properties of skim milk with added ethylenediaminetetraacetic acid, *Food Hydrocolloids*, 2017, **69**, 329–340.
164. E. P. Schokker, J. S. Church, J. P. Mata, E. P. Gilbert, A. Puvanenthiran and P. Udabage, Reconstitution properties of micellar casein powder: Effects of composition and storage, *Int. Dairy J.*, 2011, **21**(11), 877–886.

165. F. J. Rossier-Miranda, K. Schroën and R. Boom, Mechanical characterization and pH response of fibril-reinforced microcapsules prepared by layer-by-layer adsorption, *Langmuir*, 2010, **26**(24), 19106–19113.
166. L. M. Sagis, C. Veerman, R. Ganzevles, M. Ramaekers, S. G. Bolder and E. van der Linden, Mesoscopic structure and viscoelastic properties of β-lactoglobulin gels at low pH and low ionic strength, *Food Hydrocolloids*, 2002, **16**(3), 207–213.
167. R. A. Mantovani, J. Fattori, M. Michelon and R. L. Cunha, Formation and pH-stability of whey protein fibrils in the presence of lecithin, *Food Hydrocolloids*, 2016, **60**, 288–298.
168. C. L. Oliveira, L. de la Hoz, J. C. Silva, I. L. Torriani and F. M. Netto, Effects of gamma radiation on beta-lactoglobulin: oligomerization and aggregation, *Biopolymers*, 2007, **85**(3), 284–294.
169. Y. Sun, P. L. Oseliero Filho and C. L. P. Oliveira, α-Lactalbumin and sodium dodecyl sulfate aggregates: Denaturation, complex formation and time stability, *Food Hydrocolloids*, 2017, **62**, 10–20.
170. P. G. Maher, M. A. Fenelon, Y. Zhou, M. Kamrul Haque and Y. H. Roos, Optimization of β-casein stabilized nanoemulsions using experimental mixture design, *J. Food Sci.*, 2011, **76**(8), C1108–C1117.
171. J. M. Montes de Oca-Avalos, R. J. Candal and M. L. Herrera, Colloidal properties of sodium caseinate-stabilized nanoemulsions prepared by a combination of a high-energy homogenization and evaporative ripening methods, *Food Res. Int.*, 2017, **100**, 143–150.
172. S. Xia, Y. Li, Q. Zhao, J. Li, Q. Xia, X. Zhang and Q. Huang, Probing Conformational Change of Bovine Serum Albumin-Dextran Conjugates Under Controlled Dry Heating, *J. Agric. Food Chem.*, 2015, **63**(16), 4080–4086.
173. S. Xia, Y. Li, Q. Xia, X. Zhang and Q. Huang, Glycosylation of bovine serum albumin via Maillard reaction prevents epigallocatechin-3-gallate-induced protein aggregation, *Food Hydrocolloids*, 2015, **43**, 228–235.
174. J. M. Hoban, D. L. Hopkins, N. Kirby, D. Collins, F. R. Dunshea, M. G. Kerr, K. Bailes, J. J. Cottrell, B. W. Holman, W. Brown and E. N. Ponnampalam, Application of small angle X-ray scattering synchrotron technology for measuring ovine meat quality, *Meat Sci.*, 2016, **117**, 122–129.
175. J. Liu, M. Schwartzkopf and A. Arner, Rigor bonds cause reduced sarcomeric volume in skinned porcine skeletal muscle under PSE-like conditions, *Meat Sci.*, 2018, **139**, 91–96.
176. Z. Yang, Y. Hemar, L. Hilliou, E. P. Gilbert, D. J. McGillivray, M. A. Williams and S. Chaieb, Nonlinear Behavior of Gelatin Networks Reveals a Hierarchical Structure, *Biomacromolecules*, 2016, **17**(2), 590–600.
177. S. Utsumi, Y. Matsumura and T. Mori, Structure–Function Relationships of Soy Proteins, in *Food Proteins and Their Applications*, ed., S. D. A. Paraf, Marcel Dekker, New York, 1997, pp. 257–291.

178. P. Kumar, M. K. Chatli, N. Mehta, P. Singh, O. P. Malav and A. K. Verma, Meat analogues: Health promising sustainable meat substitutes, *Crit. Rev. Food Sci. Nutr.*, 2017, **57**(5), 923–932.
179. J. M. S. Aguilera and D. W. Stanley, *Microstructural Principles of Food Processing and Engineering*, Aspen Publishers, Gaithersburg, 1999.
180. C. S. Kealley, M. M. Elcombe, R. Wuhrer and E. P. Gilbert, Application of small-angle scattering to study the effects of moisture content on a native soy protein, *J. Appl. Crystallogr.*, 2008, **41**(3), 628–633.
181. C. S. Kealley, M. K. Rout, M. R. Dezfouli, E. Strounina, A. K. Whittaker, I. A. Appelqvist, P. J. Lillford, E. P. Gilbert and M. J. Gidley, Structure and molecular mobility of soy glycinin in the solid state, *Biomacromolecules*, 2008, **9**(10), 2937–2946.
182. M. G. Huson, E. V. Strounina, C. S. Kealley, M. K. Rout, J. S. Church, I. A. Appelqvist, M. J. Gidley and E. P. Gilbert, Effects of thermal denaturation on the solid-state structure and molecular mobility of glycinin, *Biomacromolecules*, 2011, **12**(6), 2092–2102.
183. A. Sokolova, C. S. Kealley, T. Hanley, A. Rekas and E. P. Gilbert, Small-angle X-ray scattering study of the effect of pH and salts on 11S soy glycinin in the freeze-dried powder and solution states, *J. Agric. Food Chem.*, 2010, **58**(2), 967–974.
184. R. Urade, N. Sato and M. Sugiyama, Gliadins from wheat grain: an overview, from primary structure to nanostructures of aggregates, *Biophys. Rev.*, 2018, **10**(2), 435–443.
185. J. Xiao, C. Shi, L. Zhang, Y. Li, J. Qi, Y. Wang and Q. Huang, Multilevel structural responses of beta-conglycinin and glycinin under acidic or alkaline heat treatment, *Food Res. Int.*, 2016, **89**, 540–548.
186. J. Xiao, Y. Li, J. Li, A. P. Gonzalez, Q. Xia and Q. Huang, Structure, morphology, and assembly behavior of kafirin, *J. Agric. Food Chem.*, 2015, **63**(1), 216–224.
187. D. Renard, E. Lepvrier, C. Garnier, P. Roblin, M. Nigen and C. Sanchez, Structure of glycoproteins from Acacia gum: an assembly of ring-like glycoproteins modules, *Carbohydr. Polym.*, 2014, **99**, 736–747.
188. A. Bennick, Interaction of Plant Polyphenols with Salivary Proteins, *Crit. Rev. Oral Biol. Med.*, 2002, **13**, 184–196.
189. F. Canon, F. Pate, V. Cheynier, P. Sarni-Manchado, A. Giuliani, J. Perez, D. Durand, J. Li and B. Cabane, Aggregation of the salivary proline-rich protein IB5 in the presence of the tannin EgCG, *Langmuir*, 2013, **29**(6), 1926–1937.
190. E. Jöbstl, J. O'Connell, J. P. A. Fairclough and M. P. Williamson, Molecular model for astringency produced by polyphenol/protein interactions, *Biomacromolecules*, 2004, **5**(3), 942–949.
191. C. Pascal, C. Poncet-Legrand, B. Cabane and A. Vernhet, Aggregation of a proline-rich protein induced by epigallocatechin gallate and condensed tannins: effect of protein glycosylation, *J. Agric. Food Chem.*, 2008, **56**(15), 6724–6732.

192. J. M. McRae, N. Kirby, H. D. Mertens, S. Kassara and P. A. Smith, Measuring the molecular dimensions of wine tannins: comparison of small-angle X-ray scattering, gel-permeation chromatography and mean degree of polymerization, *J. Agric. Food Chem.*, 2014, **62**(29), 7216–7224.
193. C. Poncet-Legrand, B. Cabane, A.-B. N. Bautista-Ortín, S. Carrillo, H. L. N. Fulcrand, J. Pérez and A. Vernhet, Tannin oxidation: intra-versus intermolecular reactions, *Biomacromolecules*, 2010, **11**(9), 2376–2386.
194. E. Di Cola, K. Torbensen, I. Clemente, F. Rossi, S. Ristori and A. Abou-Hassan, Lipid-Stabilized Water-Oil Interfaces Studied by Microfocusing Small-Angle X-ray Scattering, *Langmuir*, 2017, **33**(36), 9100–9105.
195. S. Salentinig, H. Amenitsch and A. Yaghmur, *In Situ* Monitoring of Nanostructure Formation during the Digestion of Mayonnaise, *ACS Omega*, 2017, **2**(4), 1441–1446.
196. Y. Nonomura, K. Nakayama, Y. Aoki and A. Fujimori, Phase behavior of bile acid/lipid/water systems containing model dietary lipids, *J. Colloid Interface Sci.*, 2009, **339**(1), 222–229.
197. O. Rezhdo, S. Di Maio, P. Le, K. C. Littrell, R. L. Carrier and S. H. Chen, Characterization of colloidal structures during intestinal lipolysis using small-angle neutron scattering, *J. Colloid Interface Sci.*, 2017, **499**, 189–201.
198. W. G. Bouwman, T. V. Krouglov, J. Plomp, S. V. Grigoriev, W. H. Kraan and M. T. Rekveldt, SESANS studies of colloid phase transitions, dairy products and polymer fibres, *Phys. B*, 2004, **350**, 140–146.
199. A. Bot, F. Duval and W. Bouwman, Effect of processing on droplet cluster structure in emulsion gels, *Food Hydrocolloids*, 2007, **21**(5–6), 844–854.
200. D. M. Tang and A. G. Marangoni, Quantitative study on the microstructure of colloidal fat crystal networks and fractal dimensions, *Adv. Colloid Interface Sci.*, 2006, **128**, 257–265.
201. *Crystallization and Solidification Properties of Lipids.* ed. N. Widlak, R. W. Hartel and S. Narine, AOCS Press, 2001.
202. D. Kalnin, G. Garnaud, H. Amenitsch and M. Ollivon, Monitoring fat crystallization in aerated food emulsions by combined DSC and time-resolved synchrotron X-ray diffraction, *Food Res. Int.*, 2002, **35**(10), 927–934.
203. F. Peyronel, B. Quinn, A. G. Marangoni and D. A. Pink, Ultra Small Angle X-Ray Scattering for Pure Tristearin and Tripalmitin: Model Predictions and Experimental Results, *Food Biophys.*, 2014, **9**(4), 304–313.
204. P. R. R. Ramel Jr., F. Peyronel and A. G. Marangoni, Characterization of the nanoscale structure of milk fat, *Food Chem.*, 2016, **203**, 224–230.
205. L. M. Zychowski, A. Logan, M. A. Augustin, A. L. Kelly, A. Zabara, J. A. O'Mahony, C. E. Conn and M. A. Auty, Effect of Phytosterols on the

Crystallization Behavior of Oil-in-Water Milk Fat Emulsions, *J. Agric. Food Chem.*, 2016, **64**(34), 6546–6554.
206. A. J. Gravelle, C. Blach, J. Weiss, S. Barbut and A. G. Marangoni, Structure and properties of an ethylcellulose and stearyl alcohol/stearic acid (EC/SO:SA) hybrid oleogelator system, *Eur. J. Lipid Sci. Technol.*, 2017, **119**(11), 1700069.
207. A. G. Marangoni, N. Acevedo, F. Maleky, E. Co, F. Peyronel, G. Mazzanti, B. Quinn and D. Pink, Structure and functionality of edible fats, *Soft Matter*, 2012, **8**(5), 1275–1300.
208. A. Bot, Y. S. J. Veldhuizen, R. den Adel and E. C. Roijers, Non-TAG structuring of edible oils and emulsions, *Food Hydrocolloids*, 2009, **23**(4), 1184–1189.
209. C. V. Nikiforidis, E. P. Gilbert and E. Scholten, Organogel formation via supramolecular assembly of oleic acid and sodium oleate, *RSC Adv.*, 2015, **5**(59), 47466–47475.
210. B. C. Chen, A. Han, D. J. McClements and E. A. Decker, Physical Structures in Soybean Oil and Their Impact on Lipid Oxidation, *J. Agric. Food Chem.*, 2010, **58**(22), 11993–11999.
211. K. Kittipongpittaya, A. Panya, D. J. McClements and E. A. Decker, Impact of Free Fatty Acids and Phospholipids on Reverse Micelles Formation and Lipid Oxidation in Bulk Oil, *J. Am. Oil Chem. Soc.*, 2013, **91**(3), 453–462.
212. R. Subramanian, S. Ichikawa, M. Nakajima, T. Kimura and T. Maekawa, Characterization of phospholipid reverse micelles in relation to membrane processing of vegetable oils, *Eur. J. Lipid Sci. Technol.*, 2001, **103**(2), 93–97.
213. K. Kittipongpittaya, A. Panya, L. Cui, D. J. McClements and E. A. Decker, Association Colloids Formed by Multiple Surface Active Minor Components and Their Effect on Lipid Oxidation in Bulk Oil, *J. Am. Oil Chem. Soc.*, 2014, **91**(11), 1955–1965.
214. T. Truong, S. Janin, B.-Z. Li and B. Bhandari, Effect of Crystalline Structure on Oxidation of Fish Oil in Stearin: Fish Oil Mixtures, *Food Bioprocess Technol.*, 2016, **9**(5), 792–800.
215. V. Derycke, G. E. Vandeputte, R. Vermeylen, W. De Man, B. Goderis, M. H. J. Koch and J. A. Delcour, Starch gelatinization and amylose–lipid interactions during rice parboiling investigated by temperature resolved wide angle X-ray scattering and differential scanning calorimetry, *J. Cereal Sci.*, 2005, **42**(3), 334–343.
216. J. A. Putseys, C. J. Gommes, P. Van Puyvelde, J. A. Delcour and B. Goderis, *In situ* SAXS under shear unveils the gelation of aqueous starch suspensions and the impact of added amylose–lipid complexes, *Carbohydr. Polym.*, 2011, **84**(3), 1141–1150.
217. E. Hernández-Hernández, C. A. Ávila-Orta, B. S. Hsiao, J. Castro-Rosas, J. A. Gallegos-Infante, J. Morales-Castro, L. Araceli Ochoa-Martínez and C. A. Gómez-Aldapa, Synchrotron X-ray scattering analysis of the

interaction between corn starch and an exogenous lipid during hydrothermal treatment, *J. Cereal Sci.*, 2011, **54**(1), 69–75.
218. I. C. Arts and P. C. Hollman, Polyphenols and disease risk in epidemiologic studies, *Am. J. Clin. Nutr.*, 2005, **81**(1), 317S–325S.
219. C. Hui, X. Qi, Z. Qianyong, P. Xiaoli, Z. Jundong and M. Mantian, Flavonoids, Flavonoid Subclasses and Breast Cancer Risk: A Meta-Analysis of Epidemiologic Studies, *PLoS One*, 2013, **8**(1), e54318.
220. D. Liu, M. Martinez-Sanz, P. Lopez-Sanchez, E. P. Gilbert and M. J. Gidley, Adsorption behaviour of polyphenols on cellulose is affected by processing history, *Food Hydrocolloids*, 2017, **63**, 496–507.
221. D. Liu, P. Lopez-Sanchez, M. Martinez-Sanz, E. P. Gilbert, M. J. Gidley, Adsorption isotherm studies on the interaction between polyphenols and apple cell walls: effects of variety, heating and drying, *Food Chemistry*, 2019, **282**, 58–66.
222. A. Y. Xu, L. D. Melton, T. M. Ryan, J. P. Mata, A. Rekas, M. A. K. Williams and D. J. McGillivray, Effects of polysaccharide charge pattern on the microstructures of β-lactoglobulin-pectin complex coacervates, studied by SAXS and SANS, *Food Hydrocolloids*, 2018, **77**, 952–963.
223. A. Y. Xu, L. D. Melton, T. M. Ryan, J. P. Mata, G. B. Jameson, A. Rekas, M. A. Williams and D. J. McGillivray, Sugar-coated proteins: the importance of degree of polymerisation of oligo-galacturonic acid on protein binding and aggregation, *Soft Matter*, 2017, **13**(14), 2698–2707.
224. F. Weinbreck, R. H. Tromp and C. G. de Kruif, Composition and Structure of Whey Protein/Gum Arabic Coacervates, *Biomacromolecules*, 2004, **5**(4), 1437–1445.
225. S. Salentinig, S. Phan, J. Khan, A. Hawley and B. J. Boyd, Formation of Highly Organized Nanostructures during the Digestion of Milk, *ACS Nano*, 2013, **7**(12), 10904–10911.
226. S. Salentinig, S. Phan, A. Hawley and B. J. Boyd, Self-Assembly Structure Formation during the Digestion of Human Breast Milk, *Angew. Chem.*, 2015, **127**(5), 1620–1623.
227. Y. Jin, N. Hengl, S. Baup, F. Pignon, N. Gondrexon, M. Sztucki, G. Gésan-Guiziou, A. Magnin, M. Abyan, M. Karrouch and D. Blésès, Effects of ultrasound on cross-flow ultrafiltration of skim milk: Characterization from macro-scale to nano-scale, *J. Membr. Sci.*, 2014, **470**, 205–218.
228. Z. Yang, Q. Gu, W. Banjar, N. Li and Y. Hemar, *In situ* study of skim milk structure changes under high hydrostatic pressure using synchrotron SAXS, *Food Hydrocolloids*, 2018, **77**, 772–776.
229. P. Stanisław, O. Elżbieta and B. Magdalena, Investigation of the staling process of bread using the small-angle X-ray scattering (SAXS) method, *Flavour Fragrance J.*, 2006, **21**(1), 37–41.
230. S. K. Reinke, S. V. Roth, G. Santoro, J. Vieira, S. Heinrich and S. Palzer, Tracking Structural Changes in Lipid-based Multicomponent Food

Materials due to Oil Migration by Microfocus Small-Angle X-ray Scattering, *ACS Appl. Mater. Interfaces*, 2015, **7**(18), 9929–9936.
231. J. Boetker, T. Rades, J. Rantanen, A. Hawley and B. J. Boyd, Structural Elucidation of Rapid Solution-Mediated Phase Transitions in Pharmaceutical Solids Using *in Situ* Synchrotron SAXS/WAXS, *Mol. Pharmaceutics*, 2012, **9**(9), 2787–2791.
232. T.-H. Nguyen, T. Hanley, C. J. H. Porter and B. J. Boyd, Nanostructured liquid crystalline particles provide long duration sustained-release effect for a poorly water soluble drug after oral administration, *J. Controlled Release*, 2011, **153**(2), 180–186.
233. S. Phan, W.-K. Fong, N. Kirby, T. Hanley and B. J. Boyd, Evaluating the link between self-assembled mesophase structure and drug release, *Int. J. Pharm.*, 2011, **421**(1), 176–182.
234. L. G. Gómez-Mascaraque, M. J. Fabra, J. L. Castro-Mayorga, G. Sánchez, M. Martínez-Sanz, A. López-Rubio, Nanostructuring Biopolymers for Improved Food Quality and Safety, in *Biopolymers for Food Design*, ed. A. M. Grumezescu and A. M. Holban, Academic Press, 2018, ch. 2, pp. 33–64.
235. Y. Gerelli, M. T. D. Bari, A. Deriu, L. Cantù, P. Colombo, C. Como, S. Motta, F. Sonvico and R. May, Structure and organization of phospholipid/polysaccharide nanoparticles, *J. Phys.: Condens. Matter*, 2008, **20**(10), 104211.
236. I. Colzi, A. N. Troyan, B. Perito, E. Casalone, R. Romoli, G. Pieraccini, N. Škalko-Basnet, A. Adessi, F. Rossi, C. Gonnelli and S. Ristori, Antibiotic delivery by liposomes from prokaryotic microorganisms: *Similia cum similis* works better, *Eur. J. Pharm. Biopharm.*, 2015, **94**, 411–418.
237. N. Varga, M. Benkő, D. Sebők and I. Dékány, BSA/polyelectrolyte core–shell nanoparticles for controlled release of encapsulated ibuprofen, *Colloids Surf., B*, 2014, **123**, 616–622.
238. E. Josef, K. Barat, I. Barsht, M. Zilberman and H. Bianco-Peled, Composite hydrogels as a vehicle for releasing drugs with a wide range of hydrophobicities, *Acta Biomater.*, 2013, **9**(11), 8815–8822.
239. H. Liu, K. Nakagawa, D. I. Kato, D. Chaudhary and M. O. Tadé, Enzyme encapsulation in freeze-dried bionanocomposites prepared from chitosan and xanthan gum blend, *Mater. Chem. Phys.*, 2011, **129**, 488–494.
240. E. Atay, M. J. Fabra, M. Martínez-Sanz, L. G. Gomez-Mascaraque, A. Altan and A. Lopez-Rubio, Development and characterization of chitosan/gelatin electrosprayed microparticles as food grade delivery vehicles for anthocyanin extracts, *Food Hydrocolloids*, 2018, **77**, 699–710.
241. A. López-Rubio, R. Gavara and J. M. Lagarón, Unexpected partial crystallization of an amorphous polyamide as induced by combined temperature and humidity, *J. Appl. Polym. Sci.*, 2006, **102**(2), 1516–1523.
242. N. Alipour, U. W. Gedde, M. S. Hedenqvist, S. Yu, S. Roth, K. Brüning, A. Vieyres and K. Schneider, Structure and properties of polyethylene-based and EVOH-based multilayered films with layer thicknesses of 150 nm and greater, *Eur. Polym. J.*, 2015, **64**, 36–51.

243. A. Lopez-Rubio, P. Hernandez-Munoz, R. Catala, R. Gavara and J. M. Lagaron, Improving packaged food quality and safety. Part 1: Synchrotron X-ray analysis, *Food Addit. Contam.*, 2005, **22**(10), 988–993.
244. A. López-Rubio, J. M. Lagarón, P. Hernández-Muñoz, E. Almenar, R. Catalá, R. Gavara and M. A. Pascall, Effect of high pressure treatments on the properties of EVOH-based food packaging materials, *Innovative Food Sci. Emerging Technol.*, 2005, **6**(1), 51–58.
245. C. Huang, J. Zhu, L. Chen, L. Li and X. Li, Structural changes and plasticizer migration of starch-based food packaging material contacting with milk during microwave heating, *Food Control*, 2014, **36**(1), 55–62.
246. I. Benito-González, A. López-Rubio and M. Martínez-Sanz, Potential of lignocellulosic fractions from Posidonia oceanica to improve barrier and mechanical properties of bio-based packaging materials, *Int. J. Biol. Macromol.*, 2018, **118**, 542–551.
247. M. J. Fabra, M. Martínez-Sanz, L. G. Gómez-Mascaraque, J. M. Coll-Marqués, J. C. Martínez and A. López-Rubio, Development and characterization of hybrid corn starch-microalgae films: Effect of ultrasound pre-treatment on structural, barrier and mechanical performance, *Algal Res.*, 2017, **28**, 80–87.
248. M. J. Fabra, M. Martínez-Sanz, L. G. Gómez-Mascaraque, R. Gavara and A. López-Rubio, Structural and physicochemical characterization of thermoplastic corn starch films containing microalgae, *Carbohydr. Polym.*, 2018, **186**, 184–191.
249. G. Stoclet, J. M. Lefebvre, R. Séguéla and C. Vanmansart, *In-situ* SAXS study of the plastic deformation behavior of polylactide upon cold-drawing, *Polymer*, 2014, **55**(7), 1817–1828.
250. V. M. Correlo, L. F. Boesel, M. Bhattacharya, J. F. Mano, N. M. Neves and R. L. Reis, Properties of melt processed chitosan and aliphatic polyester blends, *Mater. Sci. Eng., A*, 2005, **403**(1), 57–68.
251. R. Kuktaite, T. S. Plivelic, Y. Cerenius, M. S. Hedenqvist, M. Gällstedt, S. Marttila, R. Ignell, Y. Popineau, O. Tranquet, P. R. Shewry and E. Johansson, Structure and Morphology of Wheat Gluten Films: From Polymeric Protein Aggregates Toward Superstructure Arrangements, *Biomacromolecules*, 2011, **12**(5), 1438–1448.
252. F. A. Corsello, P. A. Bolla, P. S. Anbinder, M. A. Serradell, J. I. Amalvy and P. J. Peruzzo, Morphology and properties of neutralized chitosan-cellulose nanocrystals biocomposite films, *Carbohydr. Polym.*, 2017, **156**, 452–459.
253. Neutrons and Food 5. Available at http://www.ansto.gov.au/Events/Neutronsandfoodconference2018/index.htm.
254. There's SCIENCE in your food! Available at http://www.synchrotron.org.au/component/content/article?id=885:there-s-science-in-your-food.

CHAPTER 14

Applications of Scanning Electron Microscopy and Atomic Force Microscopy to Food Structure Characterisation

BRYONY JAMES* AND JENNY MALMSTRÖM

The University of Auckland, New Zealand
*Email: b.james@auckland.ac.nz

14.1 Introduction

As summarised in previous chapters, the structural building blocks of food materials span length scales of many orders of magnitude. Increasingly, the influence of microstructure on the functional properties of food such as texture, flavour, digestibility, and nutrition is being recognised. Tailored food structures will take on an even more important role as the drive for personalised nutrition influences consumers' food choices.

Macrostructural features, such as the air cells in bread and cakes[1,2] or the parenchyma cells in pome fruits,[3] can be seen with little or no supplementary magnification. Features at the microscopic level are amenable to structural analysis by the light-based microscopy techniques described earlier in this book. To look more closely again, the food scientist might turn to electron microscopy. Scanning electron microscopy (SEM), in its many iterations, reveals details at the micrometre level, such as the cellular structures of fruit- and vegetable-based foods,[4,5] the fat structures and

Food Chemistry, Function and Analysis No. 18
Handbook of Food Structure Development
Edited by Fotis Spyropoulos, Aris Lazidis and Ian T. Norton

Published by the Royal Society of Chemistry, www.rsc.org

protein distribution in dairy-based foods,[6,7] the droplet size in emulsions,[8,9] and the distribution of phases in multi-component, multi-scale food materials such as chocolate.[10,11] Where higher magnification is required, such as the casein micellar structure in dairy products[12] or the structure of plant cell walls,[13] it might be necessary to use the higher magnifications offered by transmission electron microscopy (TEM). Recent advances in atomic force microscopy (AFM) have also seen this technique maturing from a relatively simple imaging technique for surface microstructure to a powerful technique for mapping physicochemical interactions at surfaces.

As with all microscopy techniques, the researcher must first answer the question, "What information do I need to determine," as this guides the bulk of the effort of any microscopy, which is the sample preparation. This chapter therefore limits itself to the underlying operating principles for the various modes of operation for SEM (and, more briefly, AFM) and to the sample handling philosophy that needs to be adopted for each. There is no recipe book for sample preparation; rather, an understanding of how an image will be formed is needed in order to ensure that sample preparation reveals rather than conceals or, even worse, introduces artefacts that are compelling but misleading.

14.2 Components of the Scanning Electron Microscope

SEM has been applied to the study of food microstructure since the early 1960s. Applications of conventional SEM to a variety of food materials have been comprehensively reviewed elsewhere.[14–16] More recent advances in SEM techniques include those enabled by environmental SEM (ESEM, also referred to as variable pressure SEM and occasionally low vacuum SEM).

The basic components of an SEM are shown in Figure 14.1, and, whilst an understanding of the electron optics of the column is not entirely necessary for the interpretation of microstructural images, an understanding of the interactions of the beam and the specimen is crucial.

14.3 The Compromises of High-resolution Imaging

Figure 14.2 shows the parameters of the electron beam as it creates the *probe* that forms the image. These four factors – accelerating voltage, beam current, convergence angle, and spot size – are under the control of the operator. Understanding the interaction of these parameters allows an SEM user to manage the trade-offs that are part and parcel of acquiring high-quality and relevant microstructural images.

In microscopy, the concept of resolution is more important than magnification. High-magnification images require high resolution to be of any value. Resolution is defined as the smallest distance between two features that can be observed where the two features are distinguishable. High-resolution images require that the diameter of the probe at the sample surface (d_p, in common usage referred to as *spot size*) is small. Looking at

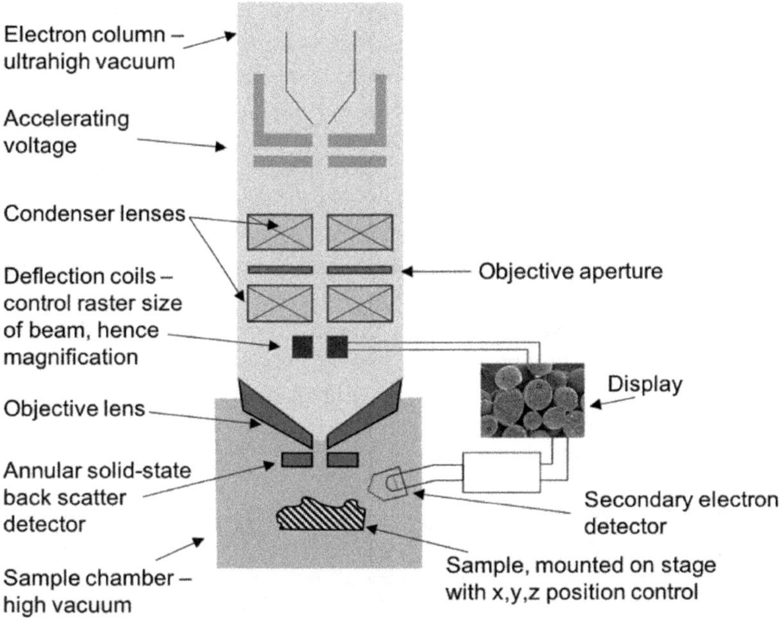

Figure 14.1 Basic components of a conventional scanning electron microscope.

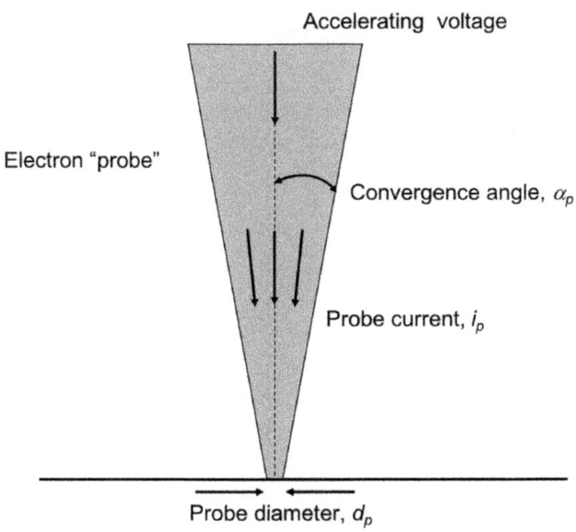

Figure 14.2 The electron beam as it hits the sample.

Figure 14.2 would suggest that increasing the convergence angle α_p would decrease spot size and hence increase resolution; the user controls α_p by altering aperture size. However, increasing α_p also increases lens aberrations, leading to the beam creating a *disk of confusion* rather than a discrete point. Increasing accelerating voltage decreases spot size and increases probe current i_p. A minimum i_p is necessary for adequate sample brightness, so a higher accelerating voltage seems like an easy solution; however, higher accelerating voltage also means greater beam penetration into the sample, reducing the surface sensitivity of images. A higher accelerating voltage also increases the electron *dose* to the sample, which can result in sample charging and in beam damage (Figure 14.3). This last factor is particularly relevant for food microstructural analysis where the sample is frequently highly susceptible to damage from the heating created by the electron beam.

Different electron sources produce significantly different probe diameters for a given probe current. Field emission electron sources produce greater image brightness at lower spot sizes and are considered high-resolution instruments. Thermionic electron sources are lower-resolution instruments but less costly to purchase and run. So high-resolution imaging is a compromise between spatial resolution, surface-sensitivity, and beam damage; these factors are summarised in Figure 14.4.

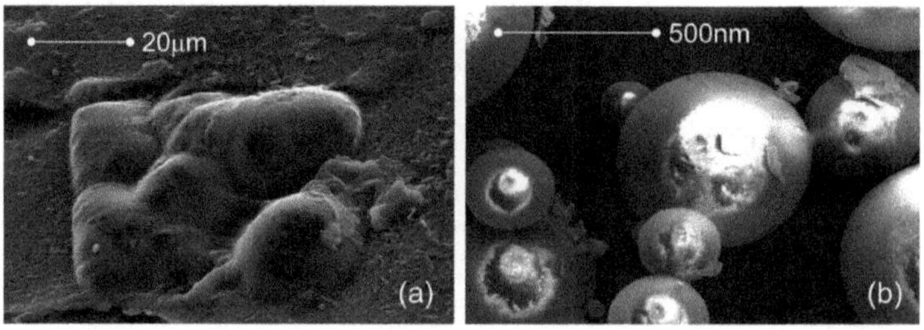

Figure 14.3 When an electron dose to a sample is inappropriately high, (a) the sample might suffer thermally induced beam damage, or (b) a static charge may build up on the sample, obscuring detail and introducing image artefacts.

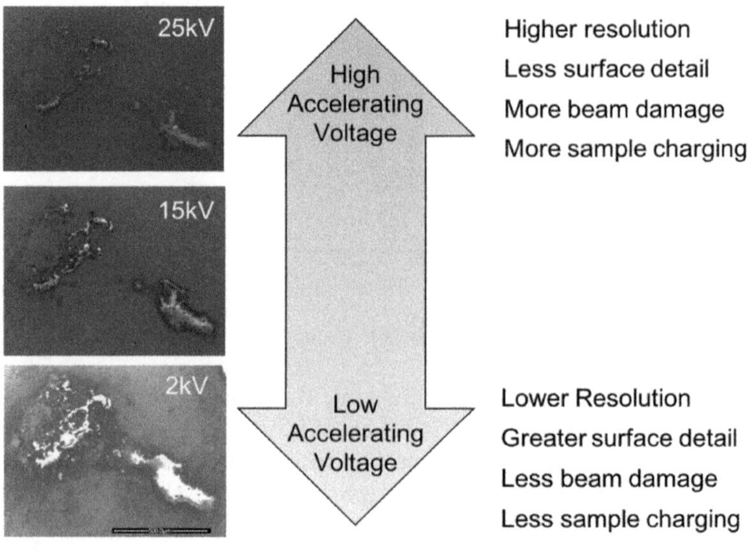

Figure 14.4 Summary of the compromises in imaging parameters.

14.4 Electron-specimen Interactions and Image Formation

Once the probe is formed and strikes the sample, interactions between the electrons of the beam and the atoms of the sample create the signal electrons that form the image. These signal electrons can be either secondary electrons (SE) or back-scattered electrons (BSE), depending on whether they were formed by elastic or inelastic collisions (Figure 14.5). The beam penetrates the sample to a distance dictated by the accelerating voltage and the material of the specimen. There are numerous models to predict the beam penetration, but a good starting point is the Kanaya–Okayama (KO) range:[17]

$$\text{Penetration depth } (\mu\text{m}) = (0.276\, A\, V^{1.67})/(z^{0.89}\, \rho) \tag{14.1}$$

where A = atomic weight (g mol^{-1}), V = accelerating voltage (kV), Z = atomic number, and ρ = density (g cm^{-3}).

The KO range makes several assumptions but gives an SEM user a rough guide to beam penetration, though this is not the same as the volume from which a signal electron might be generated. Since secondary electrons are generated from inelastic collisions, they are low energy and escape only from the near surface of the sample. Back-scattered electrons are generated from elastic collisions and conserve nearly all the energy of the primary beam; as such, they can escape from deeper within the sample. These two points have a direct impact on the resolution and surface sensitivity of an SEM image, as shown in Figure 14.5.

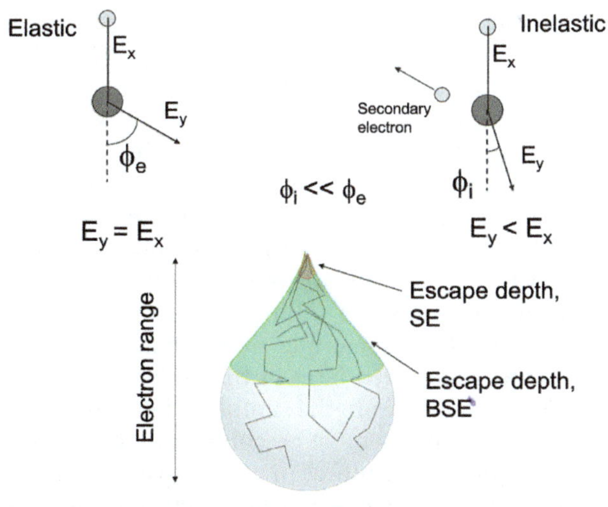

Figure 14.5 Image-forming electrons can be generated from elastic collisions (resulting in back-scattered electrons [BSE]) or from inelastic collisions (resulting in secondary electrons[SE]). The interaction depth of the primary beam with the sample depends on both accelerating voltage and the material of which the sample is made. SE escape from a smaller volume than BSE. This is apparent during imaging, as SE images are higher resolution (lower effective spot size) and more surface sensitive.

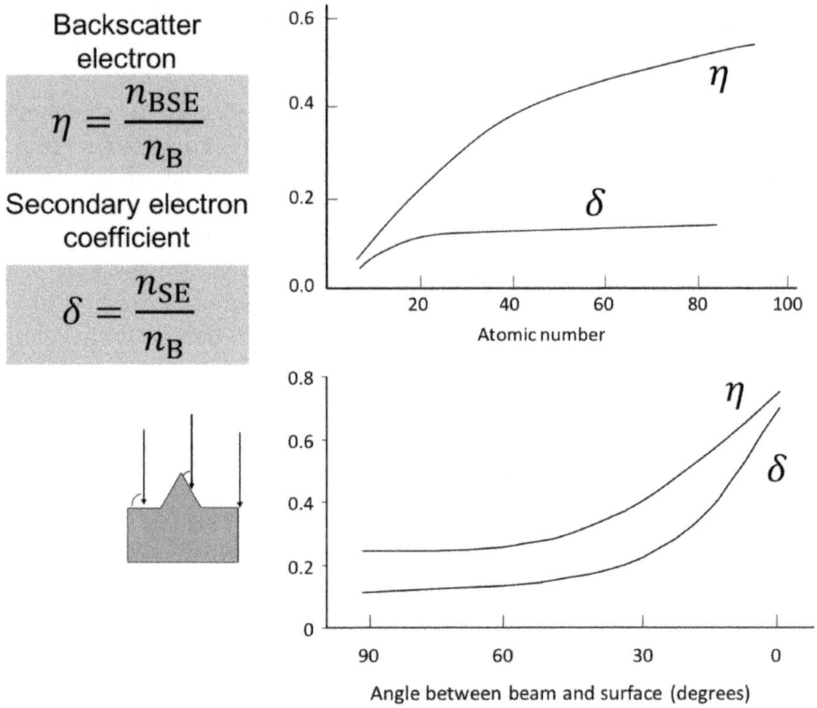

Figure 14.6 Variation in back scatter and secondary electron coefficient with both sample topography and atomic number.

Once the signal electrons are generated, they are collected by a suitable detector (usually an Everhart-Thornley detector for SE and a solid-state detector for BSE). The number of electrons generated depends on the topography of the sample and its atomic number, as shown in Figure 14.6. Atomic number contrast effects are not frequently useful for food microstructure analysis where light elements are the bulk of the material (one exception being when heavy metal labels are used to highlight specific components). Whilst both BSE and SE vary in similar fashion with angle, SE are generally used for topographical imaging due to their greater surface sensitivity. There are reasons for opting to image topography with BSE, including reducing the effect of sample charging and working with variable pressure instruments, as covered later in the chapter.

Electrons are always ejected from the sample at in a lower number than they are injected;[18] as such, samples must be conducting to avoid the build-up of static charge on the surface, resulting in charging artefacts. Frequently insulating samples are sputter-coated with gold, platinum, or gold-palladium to render them sufficiently conducting for SEM imaging. These coaters operate at vacuum conditions requiring samples to be vacuum compatible; this issue is discussed in the next section. Imaging using BSE can reduce charging artefacts as BSE are emitted from deeper within the sample.

14.5 Lens Aberrations

Electromagnetic lenses are not free of imperfections. These result in lens aberrations that limit image resolution by preventing the beam from being focussed in a clearly defined focal point in the image plane. Spherical aberration means that portions of the beam furthest from the centre of the lens are focussed more strongly than those at the axis of the lens. This aberration can be corrected to some extent by using apertures, but these, in turn, can introduce diffraction effects. Chromatic aberrations occur since the electron beam has an energy spread, and portions of the beam of lower energy get more strongly focussed than those of higher energy. In general, there is little that SEM manufacturers can do about chromatic aberration, but more recent, ultra-high-resolution instruments have introduced monochromators.[19]

The aberration of most immediate relevance to the jobbing SEM operator is astigmatism, which is a lack of symmetry in the lens, resulting in an elongated spot. This can be addressed by applying correcting fields from stigmators, which, although it takes some practice, is a relatively easy operation. Figure 14.7 illustrates this with an SEM image of rice.

14.6 Overcoming the Limitations Imposed by Vacuum

Electron beam instruments require high vacuum to ensure adequate electron range; the mean free path of an electron at atmospheric pressure is less than 100 nm. Achieving a workable range of tens of centimetres requires pressures between 10^{-3} and 10^{-7} Torr. Conventional SEM instruments maintain the sample chamber at low pressure, and these vacuum requirements impose constraints on the sample, meaning it must have low vapour

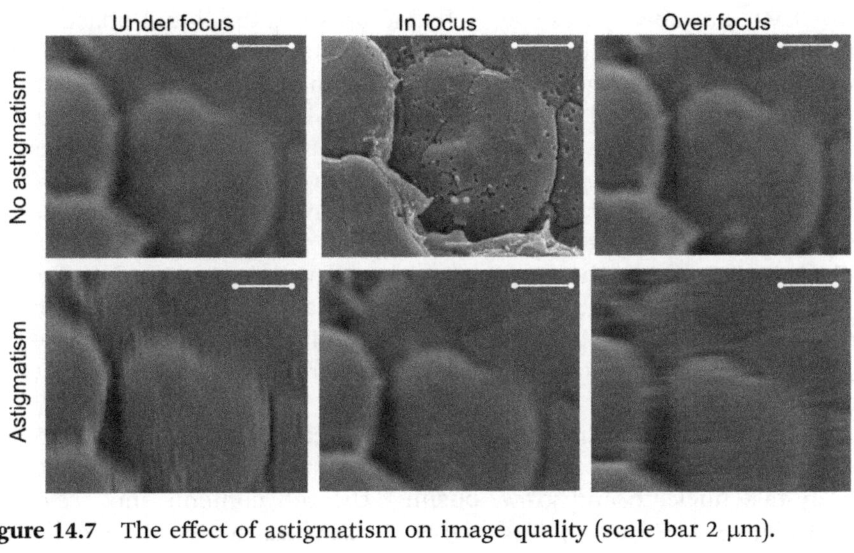

Figure 14.7 The effect of astigmatism on image quality (scale bar 2 μm).

pressure. For many foods, this means steps must be taken to reduce water activity prior to imaging (or prior to sputter-coating).

14.6.1 Drying

With any dehydration technique, care must be taken to limit shrinkage and preserve the structure as close to its native state as possible. Frequently, fixation using glutaraldehyde followed by osmium tetroxide is used to prevent conformation changes in the components of a food material during drying. Even fixed samples will shrink, though different drying techniques result in different shrinkage.[20] The most commonly used technique for drying biological tissues, including plant and meat-based food materials, is critical point drying (CPD).[21] During CPD, water is displaced from the specimen by immersing in increasing concentrations of ethanol. The ethanol is then removed from the sample at the critical point by controlling pressure and temperature, so removing sample damage from surface tension by never allowing a liquid–gas interface to form. Freeze drying is an alternative technique whereby the sample is plunged into a cryogen (such as liquid nitrogen slush or liquid nitrogen–cooled Freon or propane). The aim is to freeze the sample as rapidly as possible in order to avoid or limit ice crystal growth. The frozen samples are then dried under vacuum conditions allowing the ice to sublime and again avoiding a gas–liquid interface. An excellent comparison of CPD *versus* freeze drying is provided in Pathan *et al.*,[20] where the impact of the drying technique on cabbage, broccoli, and peas illustrates the extent to which surface and internal structures can be preserved.

14.6.2 Cryo-SEM

Cryo-SEM[22] has advantages over dehydration since the liquid phase can be left intact or only partially removed, thus maintaining the sample closer to its native state. Challenges of cryo-SEM are similar to those of freeze drying, inasmuch as ice crystal formation is a critical consideration. Figure 14.8 illustrates a typical cryo-SEM arrangement. The most common cryogen for cryo-SEM is liquid nitrogen slush, necessitating a separate *slushing station* and an arrangement for transferring the sample under vacuum to the preparation and sample chamber of the SEM.

Once transferred to the preparation chamber, the sample might be fractured, and surface water can be sublimed away in a process often called *etching*, due to its similarity to metallographic etching. The sample will usually then be sputter-coated prior to transfer to the sample chamber for imaging.

Cooling rate is the chief consideration in cryo-SEM imaging, since the formation of ice crystals results in distinctive artefacts. These are formed as ice crystals nucleate and grow, pushing the cell contents into *ice-crystal ghosts* (Figure 14.8 [inset]). The higher the cooling rate is, the smaller the

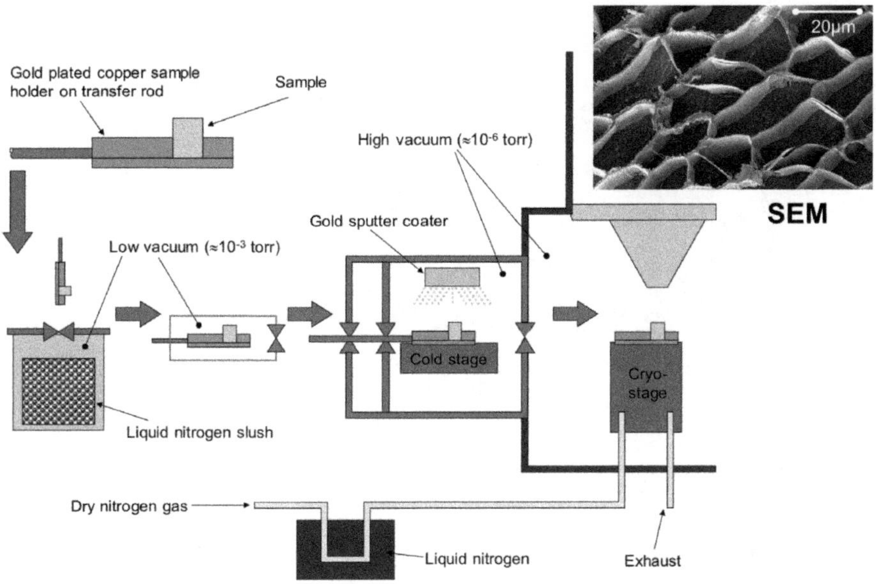

Figure 14.8 Schematic of cryo-SEM; inset: image of ice crystal ghosts in tomato soup.

ice-crystals will be, and use may be made of high-pressure freezing to create glassy ice for no ice crystal formation at all;[22] however, access to high-pressure freezers for many food microstructure studies is not often practical, as the equipment is expensive and specialised. Cryo-protectants also slow ice crystal formation, and therefore foods high in sugar will result in less ice crystal growth.

14.6.3 Environmental SEM

One of the more recent developments in SEM is that of environmental SEM, sometimes referred to as variable pressure SEM or, somewhat confusingly, as low vacuum SEM.[23] Initially developed in the 1970s and 1980s[24,25] and now commonly found in microanalysis labs, the ESEM removes the need for the sample to be at high vacuum. This allows wet samples to be imaged wet, preserving their native state and allowing dynamic processes to be followed. It is important to note that the sample chamber is not at atmospheric pressure; this requires samples to be cooled for full access to the relevant part of the water phase diagram, keeping wet samples wet (Figure 14.9). As such, samples are usually mounted on a Peltier stage to control temperature.

The operation of the ESEM relies on having a controlled pressure of an imaging gas (usually water vapour) in the chamber. This gas scatters the primary beam but, at typical ESEM pressures of 5 to 20 Torr, not completely. A spot size similar to that of conventional SEMs can be achieved, meaning that ESEM does not sacrifice resolution.

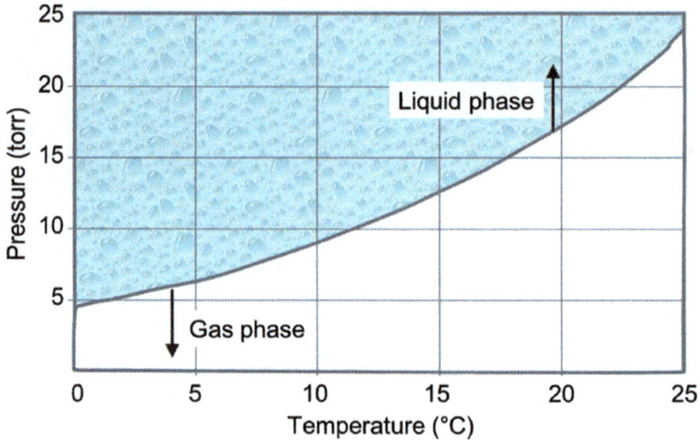

Figure 14.9 Typical operating pressures and temperatures for environmental SEM mean that liquid water can be present in the sample.

As secondary electrons are emitted from the sample, they interact with the imaging gas, ionising it, and creating a cascade of environmental SE, thus amplifying the signal (Figure 14.10). An additional advantage of this ionisation cascade is that the positive ions accelerate toward the sample, supressing any sample charging and so enabling the imaging of insulating specimens. Back scatter electrons, being the result of elastic collisions, have greater energy than SE and, as such, can be directly detected without relying on scattering events. The new generation of benchtop SEMs, which operate with less stringent vacuum requirements than conventional SEMs, frequently have BSE detectors and no SE detectors. Different imaging gases promote different ratios of SE and BSE and can thus have a direct impact on image contrast.[26]

A consideration of Figure 14.10 reveals some features of ESEM that a microscopist needs to be aware of to maximise image quality. Firstly, the sample thickness must be minimised in order to most closely control the temperature; since only the underside of the sample is in contact with the Peltier stage. Secondly, the gas in the chamber becomes part of the imaging system. The usable portion of the beam is controlled by features of the gas, including its ionisation potential and the distance of the sample from the pole piece of the microscope. This beam gas path length (BGPL) can have a profound effect on image quality at higher pressures. The accelerating voltage is also critical, both for the reasons of resolution mentioned previously and for its impact on scattering the beam. The various parameters influencing the usable portion of the beam are summarised:[23]

$$I_s \propto I_p e^{-(kpd/V)} \tag{14.2}$$

where I_s = proportion of beam that forms image, I_p = primary beam current, k = gas-related constant, p = pressure, d = beam gas path length (BGPL), and V = accelerating voltage.

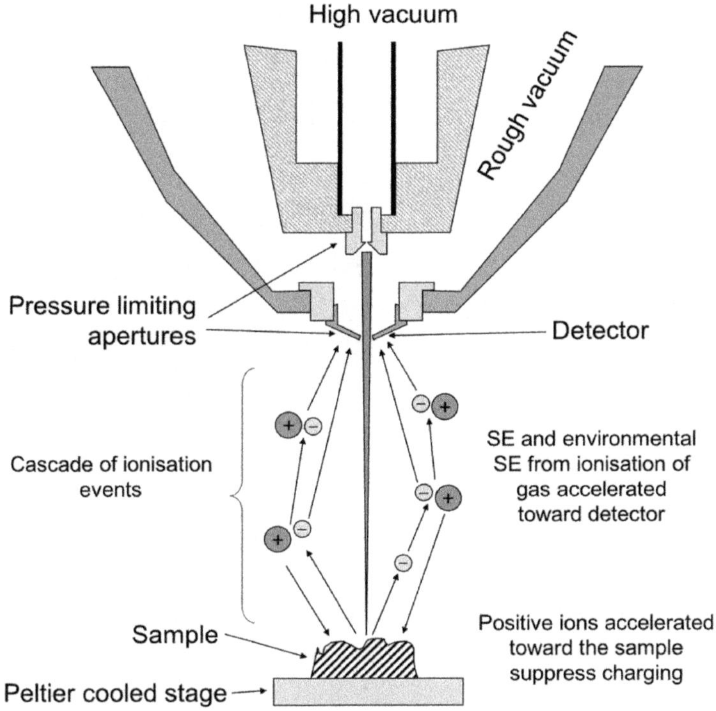

Figure 14.10 The cascade of scattering events that occurs between the emitted SE and the gas present in the chamber serves to amplify the signal and supress sample charging.

From a practical perspective, this interaction with the imaging gas can appear to slow the image formation and thus demand a slower scan speed than usual. This can be problematic when following dynamic processes where there will be a trade-off between image quality and missing a change in the sample.

The ability to image samples close to their native state means that mechanical properties can be examined directly using an *in situ* straining stage (Figure 14.11). Also dehydration and, to some extent, rehydration can be followed, but the challenge is that the presence of water conceals features that might be of interest, since water is not transparent to the electron beam in the same way it is to light. This means that, when examining freshly fractured surfaces of hydrated materials, for example, a combination of ESEM and cryo-SEM might be needed.

14.6.4 Examples

An understanding of how images are formed during SEM guides the approach to sample preparation. A few examples are given here in order to illustrate an approach that might be taken for a particular food material.

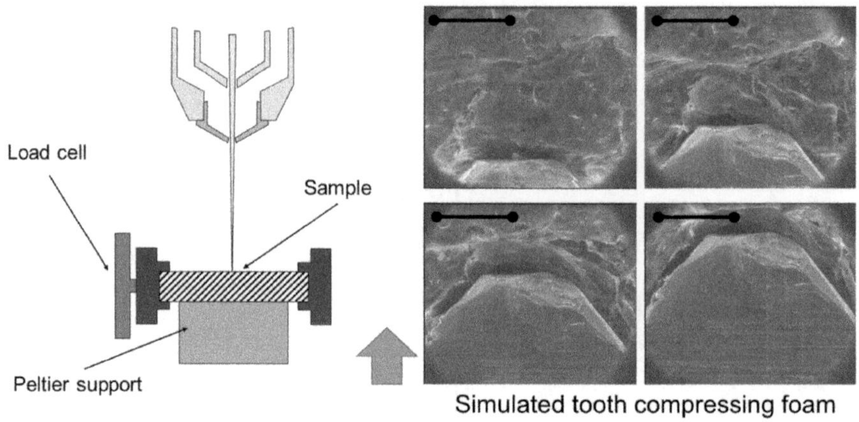

Figure 14.11 Since samples for ESEM imaging do not have to be dehydrated or coated, it is possible to conduct dynamic studies during imaging. A straining stage needs to have suitable geometry to allow a hydrated sample to be kept cool, whilst minimising the beam gas path length (scale bar 200 μm).

This is not intended to be an exhaustive instruction manual, simply a starting point for microstructural investigations.

14.6.4.1 Chocolate

As mentioned elsewhere in this book, chocolate is a complex multi-phase material. Both polarised light optical microscopy and confocal laser scanning microscopy offer valuable insights into the structures relevant to the performance of chocolate.[27,28] Scanning electron microscopy, with its exceptional depth of field, also offers an opportunity to examine surface structures related to the appearance of chocolate.

In a study of the fat bloom of chocolate,[10] a modified sample stub allowed a shallow, cast sample of chocolate to be examined both in ESEM and cold-stage X-ray photoelectron spectroscopy (XPS). If a surface-sensitive technique is to be used on the same sample, subsequent to SEM imaging, it is essential that the surface is not coated by, for example, gold. As such cryo-SEM would not have been suitable for this correlative study. ESEM imaging conditions were optimised to limit beam damage of the sample; nitrous oxide was used as the imaging gas to enable the Peltier stage to be held at −5 °C. In ESEM, beam damage can occur not only from thermal heating of the sample but through the presence of the positive ions released during the imaging process.[29,30] Conducting the SEM imaging in the ESEM also allowed the effect of heating and cooling to be followed in real time as the Peltier stage could easily be cycled between −5 °C and +40 °C (Figure 14.12). It is worth noting that even with optimum imaging conditions, beam damage to such a delicate sample is all but inevitable, and the food microscopist needs to be able to identify truth from artefact.

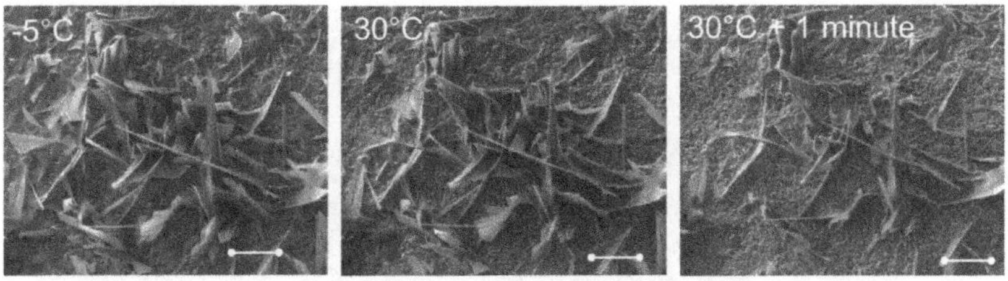

Figure 14.12 Impact of sample heating on fat bloom crystals on chocolate (scale bar 20 μm).

Correlation of light and fluorescence microscopy, with both ESEM and cryo-SEM, offers a suite of techniques that allows the surface structures of chocolate to be investigated across relevant scales of length and time.[31]

14.6.4.2 Emulsions

Emulsions are common structures for many liquid and semi-solid foods. Elsewhere in this book, the power of freeze-fracturing to examine the emulsion structure of mayonnaise without the need for fixing the sample was illustrated. Cryo-SEM is an obvious complementary technique to this approach as it offers the ability to examine the fracture face with increased depth of field. In addition, ESEM offers a novel approach to examining emulsion microstructure. Not only can a sample be kept in its native state (liquid or semi-solid) by controlling temperature and pressure, but the control of charge compensation, through the positive ions of the imaging gas cascade, offers a novel contrast mechanism. Charge contrast imaging (CCI) describes the effect when charge compensation is deliberately reduced by altering the gas pressure in the chamber to the bare minimum to avoid sample charging.[32] This results in contrast are due to the marginally differing electrical properties in the two phases of the sample (Figure 14.13). Coupled with the ability to alter temperature (within an admittedly narrow range), this would enable dynamic studies of emulsion structural changes with heating.

14.6.4.3 Plant-based foods

Comparing SEM techniques to those described elsewhere in this book for plant-based materials illustrates again the strong complementary nature of these methods. Light-based microscopies, coupled with appropriate stains, can reveal details of cell walls, including indications of their composition. Coupled with cryo-SEM, structural details at high resolution can be revealed and differences in mechanical properties inferred. For example, Figure 14.14 shows the different fracture paths for cooked and raw potatoes, with

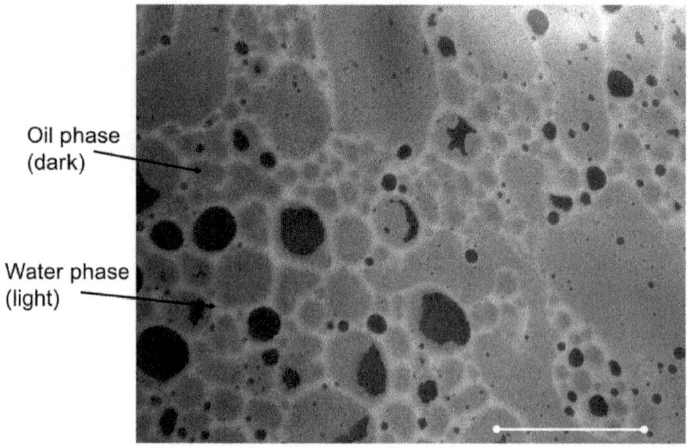

Figure 14.13 Low-fat mayonnaise imaged using charge contrast imaging in ESEM (scale bar 20 μm).

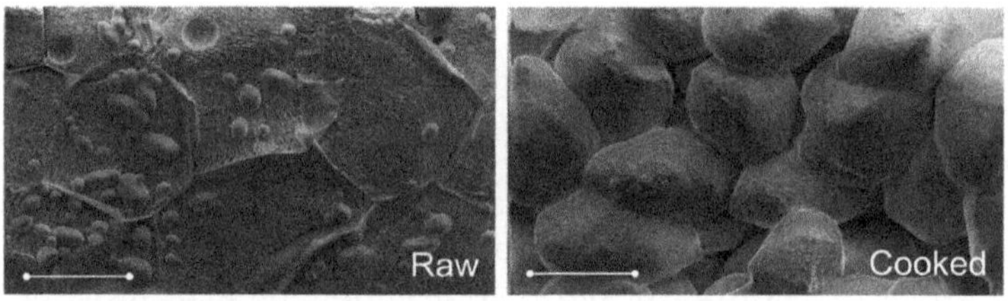

Figure 14.14 Cryo-fractured surfaces of raw and cooked potato imaged using cryo-SEM (scale bar 200 μm).

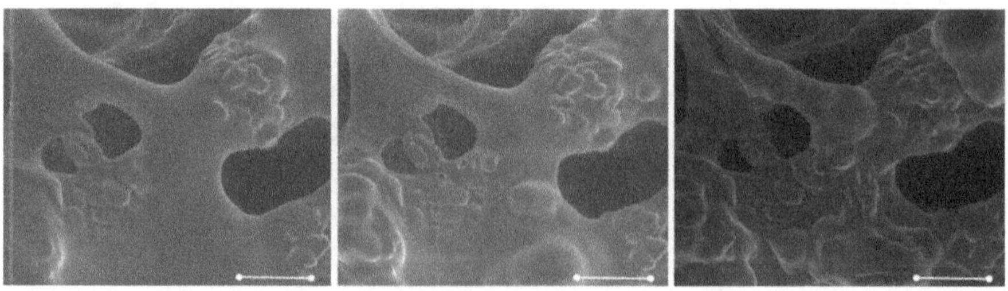

Figure 14.15 Sequence of images of dehydrating banana flesh imaged using ESEM (scale bar 100 μm).

cooked potatoes failing between the cells and raw failing across the cells, as the intercellular adhesion is greater than the cell wall strength. ESEM offers the ability to conduct *in situ* tests on plant material at near native conditions, such as the images of dehydrating banana flesh, shown in Figure 14.15.

14.7 Atomic Force Microscopy

Whereas SEM is a microscopy technique that has been applied to food materials since the 1960s, atomic force microscopy (AFM) is a far more recent addition to the food microscopist's toolbox. The resolution of AFM can be better than that of SEM, and it also offers a "3D" view of the surface of a sample. Possibly where AFM really excels is when it is used as a probe for information beyond simple topography. As with all microscopy techniques, there are challenges and advantages unique to AFM. Whilst this section aims to give a brief insight into the potential applications of AFM to food materials, it is worth, as with SEM, coming to grips with the fundamental mechanism of image formation in order to apply the technique most appropriately.

Atomic force microscopy (AFM) is a member of the scanning probe microscopy (SPM) family. This means that it uses a sharp tip to scan the surface of a sample in a raster pattern to build up an image. As the imaging does not rely on light or radiation, the resolution limit does not have the same inherent limit. Scanning tunnelling microscopy (STM), a predecessor of AFM, is based on the principle of applying a potential difference between a sharp tip and the surface and measuring the tunnelling current to keep a constant tip-surface distance. The tip is moved vertically, whilst scanning across a sample surface, to maintain a constant tunnelling current; as a result, an atomic resolution image of the surface is built up. STM can be applied only to conductive or semiconductive materials and is traditionally an ultra-high-vacuum technique, so it is generally not easily applicable to food applications or soft matter. In response to these limitations, AFM was developed by Binning *et al.* in 1986.[33] In AFM, the sharp tip is placed at the end of a flexible cantilever, and the detection is based on the forces between the tip and the sample. These forces vary with the distance from the surface, with repulsive forces originating from orbital overlap dominating close to the surface (Figure 14.16).

AFM can be used in ambient conditions or in physiologically relevant liquids (water and buffers) and can create high-resolution images of a wide range of surfaces, including single molecules immobilised on flat surfaces such as mica.[34–36] The resolution is mainly limited by the size of the tip and noise from the surroundings.

The two primary operating modes in AFM are the contact and tapping modes. For contact mode, the cantilever is operated in the repulsive end of the force spectrum. The direct interaction between tip and sample often makes this mode unsuitable for soft matter. Tapping, or intermittent contact, mode, greatly reduces the stresses on the samples, as the cantilever is driven to oscillate at its resonance frequency, tapping the surface as it scans. It is also possible to infer extra information from the phase shift in tapping mode AFM. The phase shift holds information about material properties, such as surface chemistry and compliance.[35] It is important to keep in mind that the vertical resolution of AFM is much better than the lateral one

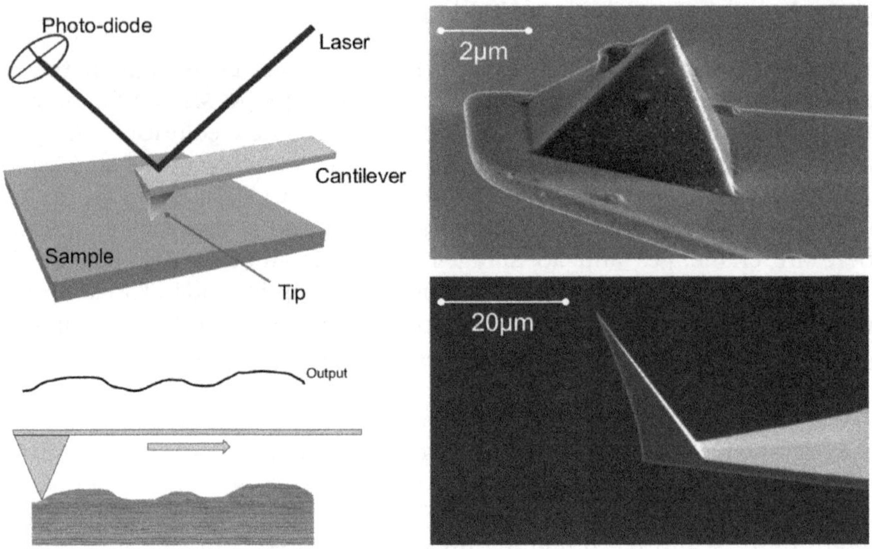

Figure 14.16 Schematic illustration of an AFM setup. The deflection of the cantilever, as it scans the surface, is detected by a position-sensitive detector monitoring the position of a reflected laser beam. The feedback system compensates to maintain a constant distance from the surface, or a constant oscillation amplitude in tapping mode, by moving the stage up and down with a piezo element. Inset images show SEM images of two of the many available designs of AFM tip.

(vertical resolution <0.5 Å, while the lateral resolution is limited by the tip size). When imaging soft surfaces, care must be taken not to compress surface features by the force of the tip.

AFM is becoming increasingly used in food-related fields, in particular to study the smaller structural building blocks of food, such as biopolymers,[37,38] emulsion systems,[39–42] biological systems such as microorganisms,[43–45] and fungi.[46,47] A particularly important contribution from AFM to food science is the information it can provide about the individual molecules that make up the food material.[48] The following sections present a few examples of how AFM has been applied to the smaller structural entities in food materials, though applications of the technique are multiplying rapidly.

14.7.1 Polysaccharides

A lot of the detailed information about how polysaccharides assemble has been derived from AFM imaging and probing of these molecules.[49] Early examples of the application of AFM in this field was imaging of acetan,[50] xanthan,[51–53] starch,[54–58] and pectin.[48,59–61] One of the main difficulties in imaging polysaccharides is the amount of water they couple, which makes imaging in ambient conditions very challenging due to capillary forces.

Therefore, most polysaccharides have been imaged under alcohols or other non-aqueous solvents; imaging in water generally leads to removal of polysaccharide from the surface.[49] The supramolecular structure of polysaccharides, elucidated by AFM, has been correlated with rheological properties to derive structure–function relationships. Such molecular understanding is vital for the creation of new texture-based products.[62]

While imaging of dilute polysaccharides on surfaces may be informative to understand structure–function relationships, polysaccharides are generally present in foods as gels or compacted into particles. The deformation properties and rheology of such gels influence how we sense certain foods, and here too AFM has contributed to our understanding. In the case of gels and particles, imaging may not be the most important contribution of AFM but instead its ability to measure local mechanical properties and produce maps of them. Traditionally, this has been done through performing force curves and fitting the approach part of the curve with a model, such as the Hertz model.[63] For this kind of modelling to be accurate, it is beneficial to use a controlled tip geometry, such as a colloidal tip.[64] It is also paramount to have a well calibrated cantilever and optical detection system.[64–66] Recently, nanomechanical testing based on tapping mode has been developed in AFM, making such characterisation easier. By using both amplitude and frequency modulation, it is possible to derive mechanical properties of a material while imaging.[67] The main advantages are that resolution is better and mechanical properties can be overlaid on the topographical image. Drawbacks are the need for calibration on a known material for each cantilever and the need to image in repulsive mode (not suitable for soft gels).

14.7.2 Lipids

Lipids and lipid bilayers are important structures in biology and for food materials. Phospholipid membranes are lipid self-assemblies, generally with cholesterol and proteins incorporated. Monoglycerides also self-assemble into globules or drops, often in response to lipolytic digestion. Digestion is directly affected by the stability of these various self-assemblies. An understanding of the driving forces for assembly and for the structures formed is important for developing functionalities in food, such as the controlled release of nutrients or flavours or the modulation of oxidation reactions.[68] When it comes to studying lipid formations, TEM and AFM are often complementary techniques. One of the main benefits with AFM is the ability to image under physiological conditions in order to study dynamic effects such as fusion or the rupture of vesicles.[69] A particularly relevant area of research for food microstructure studies is how flavonoids interact with lipid metaphases.[70] Such metaphases, in turn, have been observed by AFM, and the development of high-speed AFM has even made it possible to look at the temporal changes of cholesterol-rich domains in lipid bilayers.[71]

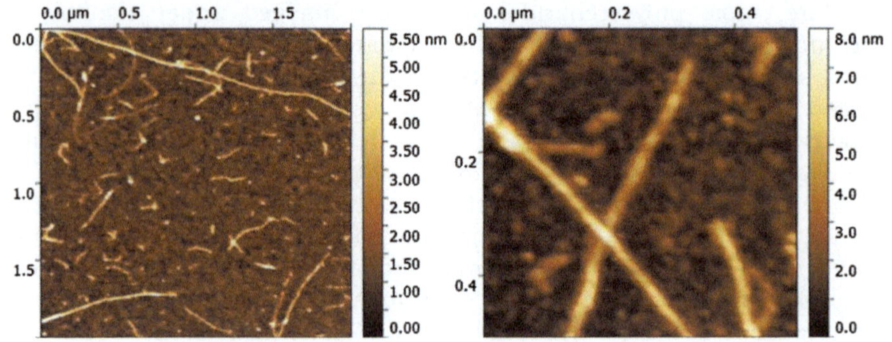

Figure 14.17 Amyloid fibres imaged using AFM (prepared in accordance with Jayawardena *et al.*[74])

14.7.3 Proteins

In food, the stabilisation of interfaces by protein aggregates is of importance for texture, flavour, and stability. One type of protein aggregate, amyloid fibrils, has been particularly well studied using AFM (Figure 14.17). One study by Jordens *et al.*[72] is a good example of correlating nanoscale characteristics with macroscopic properties, such as interfacial shear rheology. High-resolution AFM imaging has been applied to amyloid fibres or sequences[72,73] of various origin, such as haemoglobin,[74] bovine serum protein,[75] and Tau protein.[76] While amyloid fibres are generally studied for non-food applications, such as material science and drug delivery, the fact that many of the source proteins are from food products makes the insights into nucleation and growth of such fibres highly relevant to food science and food microstructural engineering.[77] Such knowledge may be particularly important to the future of creating meat-like textures from vegetarian protein sources.

The ability to "activate" the AFM tip by attaching macromolecules or other active species has opened a new avenue of food structure-related research, where the interaction of, for example, proteins with a solid surface can be directly measured.[78,79]

14.7.4 Plant-based Foods: Cell Wall Structure

Almost from its inception, food scientists embraced AFM as a tool for studying the ultrastructure of plant cell walls. Whilst sample preparation to isolate cell walls can be intrusive, subsequent imaging can be under a variety of fluid conditions, preserving structures. The high spatial resolution of AFM can reveal details of the cellulosic structures in cell walls,[80,81] and the ability to directly probe mechanical properties has been applied to a variety of plant materials.[82]

14.8 Conclusions and Future Perspective

Structural length scales from the low nanometre to the macroscopic scale are relevant to food properties. So when studying microstructure, it is essential to consider a range of characterisation techniques. Identifying and probing all of the relevant size scales require that several complementary techniques are used.

With the advent of benchtop SEMs, the techniques of microstructural analysis will become more commonly available in food research labs. The ease of use of this new generation of instruments is an advantage. However, it is also a risk. When it is nearly as easy to snap an SEM image as it is to snap a selfie, it is too easy for inexperience to lead to misinterpreting images or, worse, attributing information to mere artefact.

This chapter introduced only two tools available to the researcher wishing to study food microstructure, SEM and AFM. Whichever technique is applied, however, the researcher requires an understanding of the fundamentals of image formation to determine the correct experimental approach. This philosophy, guiding sample preparation and imaging technique, taken in concert with a broad knowledge of possible techniques, will result in the best possible outcome for researchers interested in food microstructure.

References

1. C. Ramírez, J. C. Germain and J. Aguilera, *J. Food Sci.*, 2009, **74**, 65–72.
2. M. E. Bárcenas and C. M. Rosell, *Food Hydrocolloids*, 2005, **19**, 1037–1043.
3. M. González-Fésler, D. Salvatori, P. Gómez and S. M. Alzamora, *J. Food Eng.*, 2008, **87**, 323–332.
4. Y. Deng and Y. Zhao, *J. Food Eng.*, 2008, **85**, 84–93.
5. X. Chen, Y. Chiu, S. Lin and B. James, *Drying Technol.*, 2006, **24**, 965–972.
6. S. W. Yang, P. Watkinson, G. Gillies and B. J. James, *Int. Dairy J.*, 2016, **62**, 19–27.
7. L. Ong, R. R. Dagastine, S. E. Kentish and S. L. Gras, *LWT–Food Sci. Technol.*, 2011, **44**, 1291–1302.
8. V. Klang, N. B. Matsko, C. Valenta and F. Hofer, *Micron*, 2012, **43**, 85–103.
9. S. Frasch-Melnik, I. T. Norton and F. Spyropoulos, *J. Food Eng.*, 2010, **98**, 437–442.
10. B. J. James and B. G. Smith, *LWT – Food Sci. Technol.*, 2009, **42**, 929–937.
11. Y. Kinta and T. Hatta, *J. Am. Oil Chem. Soc.*, 2007, **84**, 107–115.
12. S. Marchin, J.-L. Putaux, F. Pignon and J. Léonil, *J. Chem. Phys.*, 2007, **126**, 045101.
13. D. N. Sila, T. Duvetter, A. De Roeck, I. Verlent, C. Smout, G. K. Moates, B. P. Hills, K. K. Waldron and M. Van Loey, *Trends Food Sci. Technol.*, 2008, **19**, 309–319.
14. K. Groves, in *Food Microstructures: Microscopy, Measurement and Modelling*, ed. V. J. Morris and K. Groves, Woodhead Publishing Ltd, Cambridge, UK, 2013, pp. 386–428.
15. J. M. Aguilera, D. W. Stanley, *Microstructural Principles of Food Engineering*, Aspen Publications Inc., Gaithersberg, USA, 1999.

16. D. N. Holcomb, M. Kalab, *Studies of Food Microstructure*, Scanning Electron Microscopy Inc., O'Hare, USA, 1981.
17. K. Kanaya and S. Okayama, *J. Phys., D*, 1972, **5**, 43–58.
18. J. I. Goldstein, D. E. Newbury, P. Echlin, D. C. Joy, C. E. Lyman, E. Lifshin, L. Sawyer, J. R. Michael, in *Scanning Electron Microscopy and X-ray Microanalysis*, Springer, Boston, USA, 3rd edn, 2003, pp. 99–193.
19. R. Young, T. Templerton, L. Roussel, I. Gestmann, G. van Veen, T. Dingle and S. Henstra, *Microsc. Today*, 2008, **7**, 24–28.
20. A. K. Pathan, J. Bond and R. E. Gaskin, *Mater. Today*, 2010, **12**, 32–43.
21. Brief Introduction to Critical Point Drying, Leica Microsystems. Available at https://www.leica-microsystems.com/science-lab/brief-introduction-to-critical-point-drying/ (accessed February 2019).
22. P. Echlin, *Low Temperature Microscopy and Analysis*, Springer, Boston, USA, 1st edn, 1992.
23. D. Stokes, *Principles and Practice of Variable Pressure/Environmental Scanning Electron Microscopy (VP-ESEM)*, Wiley, Chichester, UK, 2008.
24. G. D. Danilatos and V. N. E. Robinson, *Scanning*, 1979, **2**, 72–82.
25. G. D. Danilatos, *Microsc. Res. Tech.*, 1993, **25**, 354–361.
26. A. L. Fletcher, B. L. Thiel and A. M. Donald, *J. Microsc.*, 1999, **196**, 26–34.
27. M. A. E. Auty, M. Twomey, T. P. Guinee and D. M. Mulvihill, *J. Dairy Res.*, 2001, **68**, 417–427.
28. L. Svanberg, L. Ahrné, N. Lorén and E. Windhab, *Food Res. Int.*, 2011, **44**, 1339–1350.
29. C. P. Royall, B. L. Thiel and A. M. Donald, *J. Microsc.*, 2001, **204**, 185–195.
30. A. M. Donald, *Nat. Mater.*, 2003, **2**, 511–516.
31. H. Zhao and B. J. James, *J. Food Eng.*, 2019, **249**, 9–14.
32. D. J. Stokes, B. L. Thiel and A. M. Donald, *Langmuir*, 1998, **14**, 4402–4408.
33. G. Binnig, C. F. Quate and C. Gerber, *Phys. Rev. Lett.*, 1986, **56**, 930–933.
34. S. Scheuring, D. Fotiadis, C. Möller, S. A. Müller, A. Engel and D. J. Müller, *Single Mol.*, 2001, **2**, 59–67.
35. R. García and R. Pérez, *Surf. Sci. Rep.*, 2002, **47**, 197–301.
36. H. G. Hansma, K. J. Kim, D. E. Laney, R. A. Garcia, M. Argaman, M. J. Allen and S. M. Parsons, *J. Struct. Biol.*, 1997, **119**, 99–108.
37. V. J. Morris, A. R. Kirby and A. P. Gunning, *Scanning*, 1999, **21**, 287–292.
38. V. J. Morris, A. P. Gunning, A. R. Kirby, A. R. Mackie, P. J. Wilde, in *Hydrocolloids*, ed. K. Nishinari, Elsevier Science, Amsterdam, 2000, pp. 99–109.
39. A. P. Gunning, A. R. Mackie, P. J. Wilde and V. J. Morris, *Langmuir,*, 2004, **20**, 116–122.
40. R. R. Dagastine, G. W. Stevens, D. Y. C. Chan and F. Grieser, *J. Colloid Interface Sci.*, 2004, **273**, 339–342.
41. S. Mettu, C. Wu and R. R. Dagastine, *J. Colloid Interface Sci.*, 2018, **517**, 166–175.

42. R. F. Tabor, F. Grieser, R. R. Dagastine and D. Y. C. Chan, *J. Colloid Interface Sci.*, 2012, **371**, 1–14.
43. L. S. Dorobantu, G. G. Goss and R. E. Burrell, *Micron*, 2012, **43**, 1312–1322.
44. H. H. P. Fang, K. Y. Chan and L. C. Xu, *J. Microbiol. Methods*, 2000, **40**, 89–97.
45. Q. Liu and H. Yang, *Trends Food Sci. Technol.*, 2018, **87**, 73–83.
46. Y. F. Dufrêne, *Yeast*, 2010, **27**, 465–471.
47. H. Ma, L. A. Snook, C. Tian, S. G. W. Kaminskyj and T. E. S. Dahms, *Mycol. Res.*, 2006, **110**, 879–886.
48. A. R. Kirby, A. P. Gunning and V. J. Morris, *Trends Food Sci. Technol.*, 1995, **6**, 359–365.
49. N. I. Abu-Lail and T. A. Camesano, *J. Microsc.*, 2003, **212**, 217–238.
50. A. R. Kirby, A. P. Gunning, V. J. Morris and M. J. Ridout, *Biophys. J.*, 1995, **68**, 360–363.
51. T. A. Camesano and K. J. Wilkinson, *Biomacromolecules*, 2001, **2**, 1184–1191.
52. A. R. Kirby, A. Patrick Gunning and V. J. Morris, *Carbohydr. Res.*, 1995, **267**, 161–166.
53. H. Li, M. Rief, F. Oesterhelt and H. E. Gaub, *Adv. Mater.*, 1998, **10**, 316–319.
54. P. Liu, L. Chen, A. Corrigan Penny, L. Yu and Z. Liu, *Int. J. Food Eng.*, 2008, **4**(7), 8.
55. P. M. Baldwin, J. Adler, M. C. Davies and C. D. Melia, *J. Cereal Sci.*, 1998, **27**, 255–265.
56. P. M. Baldwin, M. C. Davies and C. D. Melia, *Int. J. Biol. Macromol.*, 1997, **21**, 103–107.
57. Z. Liu, S. Chen, Z. Ouyang, Y. Guo, J. Hu and M. Li, *J. Vacuum Sci. Technol. B: Microelectron. Nanometer Struct.*, 2001, **19**, 111–114.
58. P. E. Marszalek, H. Li and J. M. Fernandez, *Nat. Biotechnol.*, 2001, **19**, 258–262.
59. M. L. Fishman and P. H. Cooke, *Carbohydr. Res.*, 2009, **344**, 1792–1797.
60. A. Gromer, A. R. Kirby, A. P. Gunning and V. J. Morris, *Langmuir*, 2009, **25**, 8012–8018.
61. A. N. Round, N. M. Rigby, A. J. MacDougall, S. G. Ring and V. J. Morris, *Carbohydr. Res.*, 2001, **331**, 337–342.
62. T. Funami, *Food Sci. Technol. Res.*, 2010, **16**, 13–22.
63. K. D. Costa, *Dis. Markers*, 2004, **19**, 139–154.
64. M. R. Bonilla, J. R. Stokes, M. J. Gidley and G. E. Yakubov, *Soft Matter*, 2015, **11**, 1281–1292.
65. H. Schillers, C. Rianna, J. Schäpe, T. Luque, H. Doschke, M. Wälte, J. J. Uriarte, N. Campillo, G. P. A. Michanetzis, J. Bobrowska, A. Dumitru, E. T. Herruzo, S. Bovio, P. Parot, M. Galluzzi, A. Podestà, L. Puricelli, S. Scheuring, Y. Missirlis, R. Garcia, M. Odorico, J. M. Teulon, F. Lafont, M. Lekka, F. Rico, A. Rigato, J. L. Pellequer, H. Oberleithner, D. Navajas and M. Radmacher, *Sci. Rep.*, 2017, **7**, 5117.

66. W. Megone, N. Roohpour and J. E. Gautrot, *Sci. Rep.*, 2018, **8**, 6780.
67. O. G. Jones, *Curr. Opin. Food Sci.*, 2016, **9**, 77–83.
68. L. Sagalowicz, M. Michel, I. Blank, O. Schafer and M. E. Leser, *Curr. Opin. Colloid Interface Sci.*, 2017, **28**, 87–95.
69. E. Reimhult, M. Zäch, F. Höök and B. Kasemo, *Langmuir*, 2006, **22**, 3313–3319.
70. A. Sadeghpour, D. Sanver, M. Rappolt, in *Advances in Biomembranes and Lipid Self-Assembly*, ed. A. Iglič, A. Garcia-Sáez and M. Rappolt, Academic Press, London, UK, 2017, vol. 25, pp. 95–123.
71. G. R. Heath, S. D. Evans and S. D. Connell, *Biophys. J.*, 2016, **110**, 571a.
72. S. Jordens, P. A. Rühs, C. Sieber, L. Isa, P. Fischer and R. Mezzenga, *Langmuir*, 2014, **30**, 10090–10097.
73. G. Wei, Z. Su, N. P. Reynolds, P. Arosio, I. W. Hamley, E. Gazit and R. Mezzenga, *Chem. Soc. Rev.*, 2017, **46**, 4661–4708.
74. N. Jayawardena, M. Kaur, S. Nair, J. Malmstrom, D. Goldstone, L. Negron, J. A. Gerrard and L. J. Domigan, *Biomolecules*, 2017, **7**, 37.
75. I. Usov, J. Adamcik and R. Mezzenga, *ACS Nano*, 2013, **7**, 10465–10474.
76. J. Adamcik, A. Sánchez-Ferrer, N. Ait-Bouziad, N. P. Reynolds, H. A. Lashuel and R. Mezzenga, *Angew. Chem., Int. Ed.*, 2016, **55**, 618–622.
77. A. K. Buell, in *International Review of Cell and Molecular Biology*, ed. M. Sandal, Academic Press, London, UK, 2017, vol. 329, pp. 187–226.
78. A. T. T. Tran and B. J. James, *Colloids Surf., A*, 2012, **414**, 104–114.
79. A. T. T. Tran, D. A. Patterson and B. J. James, *J. Food Eng.*, 2012, **112**, 38–49.
80. S.-Y. Ding and M. E. Himmel, *J. Agric. Food Chem.*, 2006, **54**, 597–606.
81. C. Paniagua, S. Posé, V. J. Morris, A. R. Kirby, M. A. Quesada and J. A. Mercado, *Ann. Bot.*, 2014, **114**, 1375–1383.
82. H. Vogler, D. Felekis, B. J. Nelson and U. Grossniklaus, *Plants*, 2015, **4**, 167–182.

CHAPTER 15

Modelling and Computer Simulation Approaches to Understand and Predict Food Structure Development: Structuring by Gelation and Self-association of Biomolecules

GEORGIOS DALKAS* AND STEPHEN R. EUSTON*

Heriot-Watt University, School of Engineering and Physical Sciences, Institute of Mechanical, Process and Energy Engineering, John Muir Building, EH14 4AS, Edinburgh, UK
*Emails: G.Dalkas@hw.ac.uk; S.R.Euston@hw.ac.uk

15.1 Introduction

Computer-based molecular methods have become a powerful tool to facilitate the interpretation of experimental data at an atomic or molecular level. Numerous high-quality molecular graphics packages are available that allow researchers to visualise, predict, and manipulate macromolecular structures in a 3D atomic-scale model, providing insight into the interactions and structural features at the atomic scale. Computational techniques play an

Food Chemistry, Function and Analysis No. 18
Handbook of Food Structure Development
Edited by Fotis Spyropoulos, Aris Lazidis and Ian T. Norton

Published by the Royal Society of Chemistry, www.rsc.org

important role in several areas such as drug design, homology modelling, functional site location, protein–ligand and protein–protein docking simulations, and molecular dynamics (MD) simulations.[1] For instance, MD simulations have been used extensively to study the fluctuations and conformational changes of proteins and nucleic acids since the late 1970s.[1] These methods are now used routinely to investigate the structure, dynamics, and thermodynamics of biomolecules and their complexes. However, the application of computer simulation techniques in food science is limited by the complexity of many natural and manufactured foods and therefore has lagged behind other science fields.

The intent of this chapter is to outline recent applications of computational methods to model a phenomenon of great importance in food systems, namely structure formation through the self-association or gelation of biomolecules. Starting with a description of the *in silico* methods that can be used for simulations, the chapter will move on to look at simulation in a number of important food applications such as the structuring of edible fats and oils and the structure formation in self-associated or gelled proteins and polysaccharides.

15.2 Basics of Computer-based Molecular Methods

Since computer-based molecular methods are relatively new in food structure development, we will start this chapter by outlining the theoretical concepts behind these approaches.

15.2.1 Molecular Modelling

Molecular modelling studies usually involve three stages (Figure 15.1).[2] Firstly, a model is selected to describe the intra- and intermolecular interactions in a system. Several computational methods have been developed for accurately characterizing molecular conformations and interactions, such as

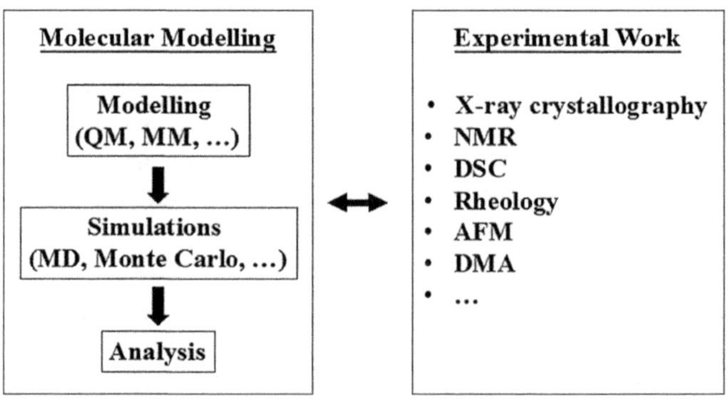

Figure 15.1 A schematic representation of three-stage molecular modelling and its relationship with experimental studies.

quantum mechanics (QM) and molecular mechanics (MM). The use of QM is an approach for accurately characterizing molecular conformations and interactions; however, its application is limited to small systems because of the high computational cost. QM deals with the electrons in a system, and even though some of the electrons can be ignored, the approach is usually limited to relatively small systems. Therefore, the use of this approach for complex food structures is usually not feasible. On the other hand, MM is often used for complex molecular systems at the atomic level by ignoring electronic motions and calculates the energy of a system as a function of the nuclear positions only. In contrast to QM, MM modelling is able to perform calculations on systems containing significantly high numbers of atoms, such as those required to model food systems, in the timescale of nanoseconds to microseconds.

The second stage is the calculation itself, using energy minimization, MD, Monte Carlo, or virtual docking simulations, with the aim of obtaining sufficient information to predict accurately the properties of the system. This stage is the more time-consuming step in the molecular modelling study. Finally, in the last stage of molecular modelling, the simulations must be analysed, as well as checked that they have been carried out properly, and the computational results must be evaluated with reliable experimental data (Figure 15.1).

15.2.2 The MM Force Fields

To be able to carry out a successful simulation of a biomolecule, the interactions among all atoms (bonded and non-bonded) must be defined. The potential energy function that describes the interactions between the atoms is typically called the *force field*. The basic form of an MM force field (energy function) consists of two components that describe intramolecular and intermolecular interactions: the bonded and non-bonded components:

$$E_{\text{ff}} = E_{\text{bonded}} + E_{\text{non-bonded}} \tag{15.1}$$

The bonded component comprises the energy of bond stretching, valence angle bending, and torsional contributions, while the non-bonded component describes pairwise interactions among atoms that are separated by more than three bonds or that are not part of a single molecule; such as the electrostatic and van der Waals (vdW) interactions among all pairs of atoms in the system (Figure 15.2). Some force fields may employ additional terms, such as cross terms that reflect the coupling between individual bond stretches, bends, torsions, *etc.*, and extra terms for hydrogen bonding, which resemble Lennard-Jones potentials but are of shorter range, such as the 10–12 potential. The force fields are empirical, and parameters are usually obtained from QM calculations and experimental data. Therefore, the accuracy of an MM simulation is closely connected to the quality of the chosen force field.

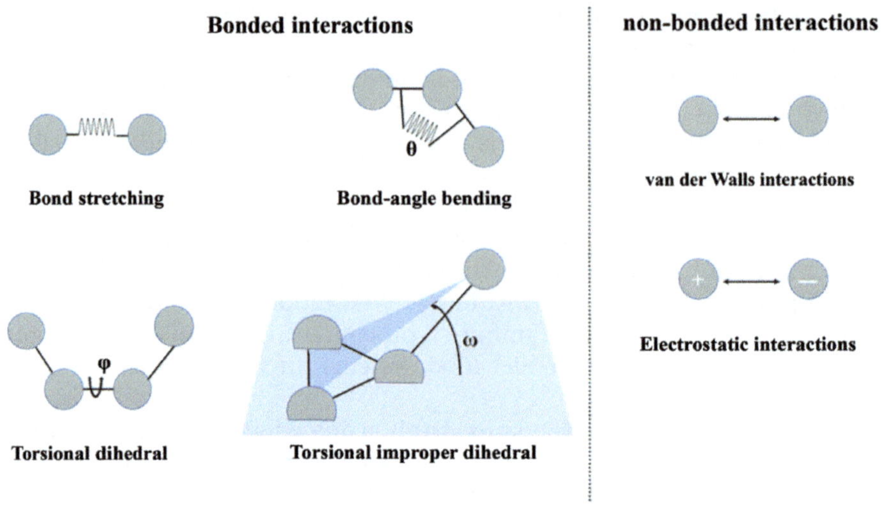

Figure 15.2 A schematic representation of force field energy terms.

15.2.3 Most Used Existing Force Fields

Many groups have developed various force fields for a wide range of biological systems. Some typical examples are the AMBER,[3] CHARMM,[4] GROMOS,[5] MMFF,[6] OPLS,[7] MM2-4,[8,9] and CVFF.[10] These force fields currently differ in the ways they are parameterised. This means that parameters are not necessarily interchangeable and that not all force fields allow for the representation of all atom types, but simulations performed using modern force fields are normally equivalent.[11]

15.2.4 Docking Simulations

When considering two molecules with 3D conformations in atomic detail, it is important to know whether the molecules bind to one another, the conformation adopted by the molecules when they do bind, and how strong the binding affinity of the complex is. Docking simulation explores the behaviour of small molecules in the binding site of a target receptor and is one of the most important applications of molecular modelling approaches in structural biology. Whilst each ligand containing one or several single bonds exists at each moment in many different conformers, only low-energy conformers are found in a significant proportion.[12] Importantly, the biological activity of a ligand will depend on its adopting one conformer state from an ensemble of all the low-energy conformers. Docking packages probe the potential binding sites for a receptor-ligand pair by searching the conformation space with genetic algorithms. By employing an affinity scoring function, it is possible to rank the candidate poses as the sum of various molecular interactions between ligand and receptor groups (Figure 15.2), such as bonded (stretching, bending, torsion, improper), electrostatic interactions, van der Waals interactions, π-stacking, hydrogen bonds and hydrophobic contacts. It is then possible to rank these conformations based on binding energy and similarity of binding site, allowing identification of the most likely binding sites and lowest energy conformation.

15.2.5 Molecular Dynamics Simulations

It is known that molecules are not rigid and that at room temperature their kinetic energy is large enough to allow the atoms to move.[12] This means that the absolute positions of atoms in a molecule are not fixed and that the relative location of substituents on a single bond vary over time. Knowledge of the molecule's dynamics can provide important insights into the structure–function relationship. Several computational approaches have been developed to describe the dynamics of molecules by following the precise position of each atom in time along with the corresponding energies. MD simulation is one of the principal tools in the theoretical study of biological systems that can compute the time-dependent behaviour of a molecular system. MD is a conformation space search procedure where the atoms of a molecule are given an initial velocity and are then allowed to evolve in time according to the laws of Newtonian mechanics. By assuming Newtonian dynamics, the trajectories of atoms can be calculated using modified versions of Newton's second law of motion. Depending on the simulated temperature of the system, the macromolecule can then overcome barriers found in the potential energy surface and explore different conformational states. Before the initiation of the MD simulation, it is necessary to carry out energy minimization of the investigated system. This allows the energy to equilibrate between the atoms in order to eliminate any unfavourable contacts that may have occurred as a result of the model building process and to gently adjust the initial structure to the molecular mechanics force field prior to the production of MD simulations.

The speed and the efficiency of MD simulations are determined in large part by the intrinsic properties of the simulated systems, such as the number of atoms in the system, the complexity of the selected potential energy function, and the use of different boundary conditions. The efficient performance of MD simulations of complex molecular systems has been made possible by advances in software that include the fine-tuning of energy calculations. In addition, the present generation of computers benefit greatly from advances in hardware parallelisation or the use of graphical processing units (GPUs) to speed up the process and reduce computation time. Some of the most popular MD codes, such as AMBER,[3] CHARMM,[13] GROMACS,[14] and NAMD,[15] are compatible with the Message Passing Interface (MPI) software, which has facilitated the development of distributed memory calculations. Combined with software advances such as domain decomposition algorithms, MPI has allowed very large-scale simulations to be carried out on large numbers of computer cores simultaneously. This has driven the adoption of high-performance computational clusters with many tens or hundreds of thousands of processors as regional, national, or supranational facilities. For example, the UK high-end cluster, ARCHER, has over 118 000 processors, whilst PRACE (Partnership for Advanced Computing in Europe) provides access to the high-end HPC facilities of five EU member countries.

15.3 Structure Formation in Triglyceride Systems

Oils and fats are chemically defined as triacylglycerols (TAGs) that are liquid and solid, respectively, at room temperature. Liquid oils are predominantly composed of mono and/or polyunsaturated fatty acids, while solid fats show mainly higher proportions of saturated and in some cases *trans*-fatty acids.[16] TAGS are important structurants in many products, including obvious examples of fat-continuous foods such as margarines, spreads, butter, and chocolate, to name a few, but also baked goods. Much of their structuring capability relies on their capacity to solidify not only into regular crystalline structures but also into amorphous structures when cooled from the melt. Although TAGS are relatively simple molecules, comprising the trihydric alcohol glycerol esterified with three fatty acid side chains, they have proven to be rather difficult to simulate in food-relevant systems, and there are but few simulations of their structure and properties. The solid, crystalline state of triglycerides is of most relevance to food systems, but the crystallization of triglycerides is complicated by extensive structural polymorphism, where the structure can become trapped in a kinetically metastable state that only very slowly decays to the thermodynamically most stable crystal structure.

Early simulations of cooling from the melt using all atom molecular dynamics were unsuccessful, as prohibitively long simulation times prevented the system reaching the thermodynamically stable crystalline state, and only the partially ordered metastable structure could be explored.[17] Much of the literature in this area has been reviewed previously.[18]

Since these early attempts, significant progress has been made using coarse-grained simulation techniques that have already been widely used to simulate cell membrane structure and dynamics.[19] Early attempts to model the C10 triglyceride tridecanoin[20,21] were able to model the liquid–solid phase transition temperature and to reproduce physical properties of the melt and solid phases. The same research group have extended their model to longer acyl chain lengths (C16 tripalmitin and C18 tristearin) and mixtures of the two and have successfully been able to simulate the crystallization of these longer acyl chain TAGS.[22] Using this model, the highly ordered crystalline structure of the α-phase of pure tristearin and tripalmitin crystals was confirmed and shown to contain a high proportion of TAGS in the tuning fork conformation. This behaviour is also observed for mixed tristearin-tripalmitin systems cooled from the melt. Similarly, simulations of the melting of a fully formed α-phase of the mixed system reproduces a melting temperature that is in qualitative agreement with the experimental behaviour.

15.3.1 Molecular Modelling and Computer Simulation Approaches to Characterise the Structure and Complexation of Oleogels

In recent years, the possibility of formulating food products with an improved nutritional profile without solid fats, containing predominantly mono and/or polyunsaturated fatty acids, has received great attention from the scientific and industrial communities. Traditionally, the necessary functionality, desirable texture, as well as enhanced solid structure of most lipid-based food products such as margarines and spreads, are obtained by adding a proportion of crystallizing TAG in the polyunsaturated lipid phase of the food product. However, the crystalline TAGs contain saturated fatty acids (SaFa) that raise blood cholesterol level, which in turn has a negative effect in terms of cardiovascular diseases.[23] This adverse health effect can be reversed by replacing these high-SaFa TAG building blocks with healthier alternative structurants that provide the same level of solid structure. The quest to find suitable alternative structurants with good organoleptic properties (melt-in-mouth) and to provide the same physical characteristics, such as hardness, texture, crispiness, and spreadability is a major technological challenge for food manufacturers.

Among different methods that can be explored for the reduction of saturated fatty acid content, oleogelation is a promising approach that can help to address this challenge. Oleogelation is the formation of liquid oils into a gel-like structure which has features (rheological properties, viscoelasticity, firmness, *etc.*) of a solid fat without containing large amount of saturated fats.[24] Oleogels are the structured oils prepared by oleogelation of liquid oil using oleogelators, such as monoacylglycerides, wax esters, lecithin, and sorbitan tristearate, mixtures of low-molecular-weight organogelators (LMOGs), long-chain fatty acids, or fatty alcohols and their mixtures.[16,24,25] The specific functionality of oleogelators depends on solvent, processing, and the presence of other materials. This efficient structuring in oleogels is typically achieved by the self-assembly of oleogelators that organize themselves into a 3D structure, where the oil liquid phase is entrapped in a thermo-reversible gel network.[26] The possibility of gelling ≥ 90 wt% of liquid oil at a relatively lower mass fraction of oleogelators makes oleogelation a very efficient way to improve the nutritional profile of food products.[25] However, one of the main bottlenecks that holds back the full-scale commercial exploitation of oleogelation methods is the identification of ideal oleogelators with the desired properties. Most of those that are currently investigated have been found through serendipity and have to be optimised empirically. A deep understanding of the structuring mechanisms and the underlying behaviour of such building blocks in lipid phases is required to allow structuring by design, since many of the most interesting systems are built through mixed materials.

Although much work has been done using neutron scattering measurements and atomic force microscopy (AFM),[27,28] the 3D structure of oleogelating systems is not yet clear, and therefore modelling and computer simulations are ideal techniques to gain insight into these structures and to help identify novel oleogelators.[29–31] One of the most appealing edible oleogel systems that has been studied using computational methods is that of the phytosterol β-sitosterol and the phytosterol ester γ-oryzanol.[29,30] In a recent published work, Dalkas *et al.*[29] have employed docking and atomistic MD simulations, supported by complementary experiments using AFM, rheology, and spectroscopy, to investigate the self-assembled structure of β-sitosterol and γ-oryzanol molecules into cylindrical tubules in an organic solvent. Earlier Fourier transform infrared (FTIR) spectroscopy suggested that tubules are built up from stacked dimers of β-sitosterol and γ-oryzanol stabilised by a hydrogen bond.[32] Simulation results agreed with the FTIR findings that a hydrogen bond is formed between the hydroxyl group of the β-sitosterol and the carbonyl group of the ester bond in γ-oryzanol.[29,32] In addition, docking simulations reveal that the sterane group of sitosterol is stacked with the sterane group of oryzanol and that the presence of a methyl group, orthogonal to the cyclopentane ring of the sterane group in oryzanol, prevents parallel stacking of the dimer (Figure 15.3). The authors speculate that the formation of this wedge-shaped dimer may cause β-sitosterol and γ-oryzanol to form a helical tubule when successive dimers self-associate (Figure 15.4), as confirmed by experimental studies.[33]

With small-angle neutron scattering and AFM images[27,28] confirming the presence of 10 nm fibrils as the building blocks of the oleogel, further research was carried out using MD simulations to investigate the structure of these fibrils. Tubules were built from the dimer structure (Figure 15.3) by sequentially stacking dimers into a tubule structure (Figure 15.4). MD results showed that the initial self-assembled tubular configuration of sitosterol and oryzanol maintained its tubular structure during the simulation time, stabilised by a cooperative network of interactions. The formation of hydrogen bonds between the hydroxyl group of the sitosterol and the carbonyl group of the oryzanol was observed, as previously suggested by molecular energy minimisation calculations and FTIR spectroscopy,[28,34] with these interactions forming a cooperative network that stabilises the tubule.

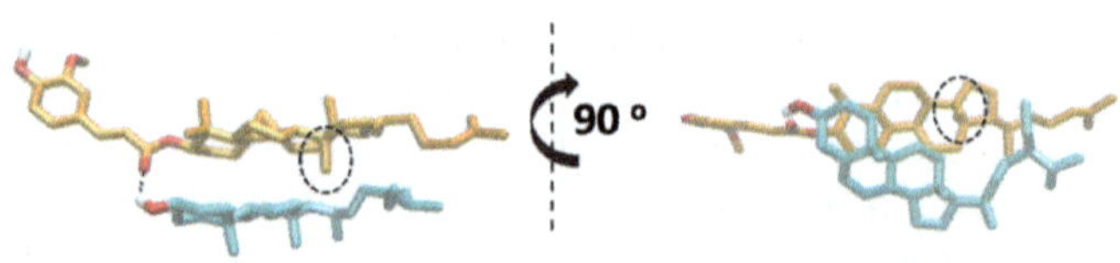

Figure 15.3 Views differing by 90° of the lowest energy conformation for the molecular models of sitosterol (cyan carbons) bound to oryzanol (orange carbons). Oxygens of oryzanol are coloured with red and hydrogens with white. Potential hydrogen bond is indicated with dashed line, and the oryzanol's methyl group, which prevents parallel stacking of the sterane groups, is indicated with dashed circle. Reproduced from ref. 29 with permission from American Chemical Society, Copyright 2018.

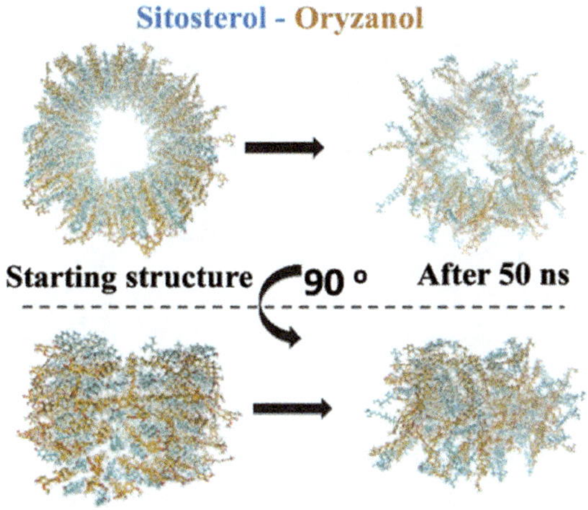

Figure 15.4 Snapshots of the sitosterol-oryzanol organogel at the beginning and at the end of the course of the MD simulation. Sitosterol is shown with cyan carbons and oryzanol with orange carbons.
Reproduced from ref. 29 with permission from American Chemical Society, Copyright 2018.

The ability of sitosterol and oryzanol to self-associate into tubules is not necessarily enough on its own for a gel to form. High-resolution AFM of the tubules also indicated that these associate with one another through surface–surface interactions to form bundles.[27] MD simulations of two interacting tubules identified an interfibril network of vdW interactions, π–π contacts that form between the ferulic acid groups of oryzanol through the aromatic, hydroxyl, methoxy groups that cause the tubules to stick together and consequently to form bundles (Figure 15.5). This mechanism is consistent with models of tubule structure based on small-angle neutron scattering proposing that the sterol dimers are oriented in the tubule such that the ferulate group of oryzanol sticks out from the tubule surface.[28]

Sitosterol–oryzanol gels have not yet been employed widely to structure foods such as margarines and spreads because of the disruption of the gelling process in the presence of water. This is a particular problem for spreads and margarine, which are water-in-oil emulsions. It is believed that this is due to the tendency of water to bond to the hydroxyl group of the sitosterol, promoting the formation of hydrate crystals and breaking up the cooperative hydrogen-bonded network between sitosterol and oryzanol.[30] Inspired by the work of Sawalha *et al.*[35] showing that by reducing the water activity of the aqueous phase with salt or sugar, the effect of water may be mitigated somewhat, Matheson *et al.*[30] replaced the water with an alternative polar phase, glycerol that does not have the same H-bonding propensity as water, and by using experimental and computational work, they investigated

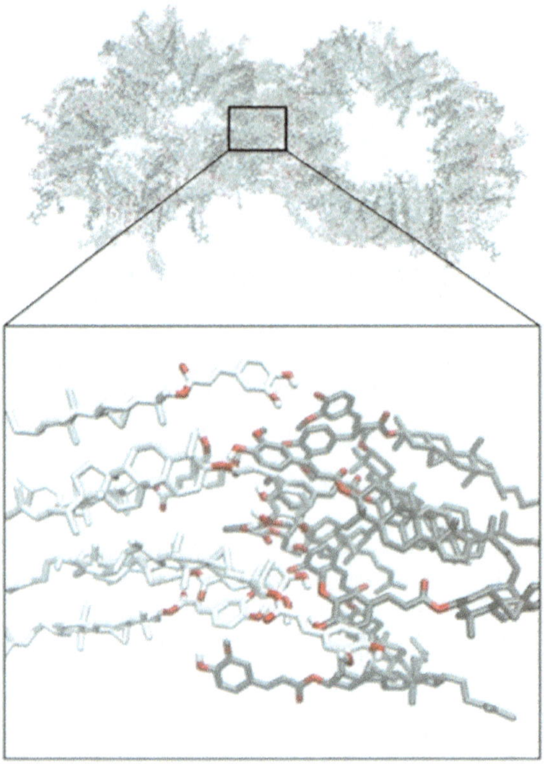

Figure 15.5 Hydrogen bonds, vdW interactions, and π–π contacts between the ferulic acid groups of oryzanol in the interface of two tubules. Reproduced from ref. 29 with permission from American Chemical Society, Copyright 2018.

the stability of the oleogel system. Using MD simulations they have observed that sitosterol–oryzanol tubules were more stable in the presence of glycerol than water, with the cooperative H-bonding network absent in the presence of water, which was in agreement with experimental outcomes.[30]

Another type of oleogel that was studied using computational methods was the oryzanol-lecithin system. Phytosterols fall into the larger family of steroids, along with bile acids and their salts. Bile acids and their salts have been shown to form oleogels when mixed at the correct molar ratio with lecithin, which commonly refers to a range of phospholipids prevalent in the membranes of both plant and animal cells, and is often used as an emulsifier in foods.[36,37] Taking into account the similarity in structure between bile acids and bile salts with oryzanol, MD simulations and experimental studies were carried out by Matheson *et al.*[31] to explore whether oryzanol would form oleogels with lecithin. A combination of light scattering, spectroscopy, and MD simulations showed that lecithin and oryzanol were found to interact strongly and that the mechanism for this is the formation of

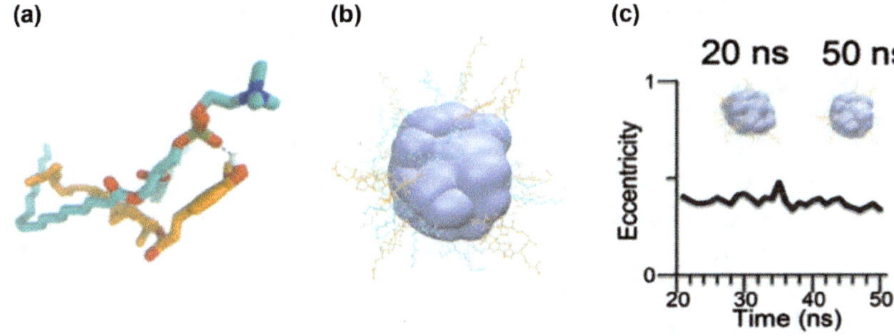

Figure 15.6 (a) Molecular docking simulation of lecithin (cyan carbons, red oxygen, blue nitrogen, and khaki phosphorus) and oryzanol (orange carbons, red oxygen; crucial hydrogen is white), showing the presence of an intermolecular hydrogen bond in dashed line. (b) Snapshot corresponding to a configuration of the mixed system at the end of the simulation time (lecithin's lipid head group and oryzanol's ferulic acid group are rendered as a mauve surface, lecithin's tails as cyan lines and oryzanol's ring system and alkyl chain as orange lines). (c) Eccentricity of the reverse micelle over the last 30 ns of the simulation.
Reproduced from ref. 31 with permission from the Royal Society of Chemistry.

mixed reverse micelles, whereby the oryzanol appears to sit between lecithin molecules in such a manner that the packing ratio of lecithin molecules in the micelle is not significantly altered, and thus the micelles remain spherical (Figure 15.6). The micelles are then able to interact and form a continuous network, trapping the oil in a thermoreversible gel. Importantly, the oryzanol-lecithin oleogel is able to solubilize water, probably by taking the water molecule within the hydrophilic cores of the reverse micelles.

15.4 Molecular Modelling and Computer Simulation Approaches to Study Carbohydrate Gelation

Carbohydrates are major components of foods, and many polysaccharides are utilised as gelling agents.[38] Several computational studies of carbohydrate food systems have investigated the interactions between carbohydrates and water or other solvents, providing atomic-level information on the behaviour of these molecules, as well as the conformational, energetic, and dynamic properties of the complex food systems.[18] Most often these works have used MD simulations with explicit water molecules as solvent, but in some cases oligosaccharides in other solvents or in mixed solvents have been studied.[39,40] Recent studies of computational approaches as applied to carbohydrate gelling agents are discussed here.

Huang *et al.*[41] used MD simulations to probe the solvation and gelation behaviour of methylcellulose (MC), a food-safe additive. MC is derived from cellulose, when one or more hydroxyl groups on the glucose monomer are substituted with methyl groups. The MC monomer has three functional groups available for substitution. Natural cellulose is insoluble in water because of an extensive intermolecular hydrogen-bonding network formed by the hydroxyl groups and facilitated by the molecular orientation of the glucose monomers due to the β1–4 glycosidic bond. Similarly, MC with substituted hydrophobic functional groups is also insoluble in water because of the strong hydrophobic interactions. However, when one or two hydroxyl groups are hydrophobically substituted, the modified MC is water soluble.[41] MC forms a thermoreversible gel upon heating, and while this transition has been extensively studied, the gel-forming mechanism is not yet clearly understood. Simulations of multiple methylcellulose homo-oligomers in water at low concentrations ($\approx$4.5 wt%) at 25 °C and 90 °C revealed that the three substitution positions have different hydrogen-bond-forming preferences.[41] MD data analysis revealed that the O6 position forms the most hydrogen bonds, followed by the O2 and O3 positions. Computational results revealed that dimethylated and trimethylated MCs form loosely aggregated structures at 25 °C and that most MCs form precipitate-like structures at 90 °C. By using atomistic MD simulations, they have shown that the models form closer-packed clusters around the experimentally determined gelation temperature than at room temperature and that, by using contact maps, they were able to probe the gel-formation mechanism. They reported that hydrophobic interactions were mainly responsible for inducing MC gelation rather than hydrogen bonding, which is in agreement with the proposed mechanisms in literature. In another study, Tusch *et al.*[42] have employed computational simulations to study the behaviour of helically structured amylose in water/DMSO mixtures. Amylose is a linear polysaccharide of α-(1→4)-linked glucose units and the principal component of starch. It can also exist in one of two helical structures or as a double-helical form. Using an amylose helix model of 55 glucose residues, Tusch *et al.*[42] found that increasing the percentage of DMSO results in increasing helical stability, which agreed with their experimental findings. They mentioned that water interacts with all the OH groups at the helix surface and then penetrates the helix coil, while DMSO initially forms single hydrogen bonds to part of the hydroxyl groups of the amylose and therefore allows a longer conservation of intramolecular hydrogen bonds than those necessary to maintain the helix. Lopez *et al.*[43] have used a non-polar solvent for the MD simulations, and the results showed that the amylose chain undergoes a conformational change to a V-amylose (helical form bound to an inclusion compound), which was very stable after the formation.

The hemicellulose polysaccharides play an important role in the swelling behaviour of the primary plant cell wall, and MD simulations provided important information about the interactions of hemicellulose polysaccharides with water.[44–49] The gelation of xyloglucan in ethanol and water mixtures has been simulated by Umemura and Yuguchi.[50] Xyloglucan in aqueous solution can form a gel with the addition of alcohol, such as methanol or ethanol. In this work, they studied the adhesive property of alcohol to xyloglucan and proposed the mechanism of the gelation by MD simulations of a xyloglucan in water/methanol and water/ethanol solutions. They showed the interactions between xyloglucan and the water/alcohol solvent, where the water interacts with xyloglucan in a hydrophilic way and with alcohol molecules in hydrophobic way.

Within the family of plant cell wall polysaccharides, the pectins (extracted from fruits and traditionally used as a gelling agent) are a group of acidic polysaccharides containing α-(1→4)-linked galacturonic acid units, with distinct structural and functional domains.[51,52] Rhamno-galactorunans I (RG-I) are the second most abundant polysaccharide of the pectin family, after homogalacturonan (HG). RG-I is one of the major components of ginseng pectin, which has been credited with tonic and medical properties in Asia for thousands of years.[53] Makshakova *et al.*[51] studied the 3D structures and dynamic features of the constituents of RG-I with MD simulations. They found that the extended threefold helical structure of the RG-I linear backbone is the most energetically favourable structure and that branching helps to stabilise a conformer of the backbone twisted along the 1→2 glycosidic linkage, triggering the orientation of the side chains without altering the extended backbone chain conformation.[51]

Atomistic MD simulations were used by Hecht *et al.*[54] to investigate the association and aggregate structure of several alginate chains and for different chain compositions under various Ca^{2+} concentrations. Alginate is a polysaccharide that is abundant in nature, and due to its ability to retain water and its gelling and stabilizing properties, it is used by food manufacturers as a thickening, stabilising, emulsifying, and gelling agent in several food products.[55] Alginates are linear copolymers composed of (1→4)-linked α-L-guluronic (G) and β-D-mannuronic (M) residues of varying sequences, and they aggregate to produce gels in the presence of multivalent cations in aqueous solvent. Its stiffness and gel structure are determined by the ion type and the alginate sequence composition. Simulations by Hecht *et al.*[54] showed that Ca^{2+} ions facilitate chain aggregation and gelation, mainly in a semi-flexible worm-like chain model, and that the aggregation was affected by the alginate monomer composition. In addition, MD data showed that greater fractions of M monomers increase the chain flexibility of heteropolymer chains, and differences in chain G acid composition lead to different interchain association mechanisms (lateral association, zipper mechanism, and entanglement), while the M residues promote chain association.

The combination of experimental liquid-state NMR techniques together with MD simulations has played a critical role in conformational and dynamics analysis of monosaccharides, disaccharides, or polysaccharides.[56] Olsson *et al.*[57] have investigated the conformational flexibility and dynamics of two (1→6)-linked disaccharides that are related to the action of the glycosyl transferase GnT-V, by using NMR, NOE, and T-ROE spectroscopy, conformation-dependent coupling constants, and MD simulations. The dynamic model that was produced in this study from the computer simulations was in very good agreement with the experimental observations.[57]

15.5 Molecular Modelling and Computer Simulation Approaches to Study Protein Gelation

Proteins gels have been used as foods for many years, with examples including cheese, yogurt, tofu, to name a few. The gelation mechanism can occur through acidification or heating. Early computational studies of protein gelation have been reviewed elsewhere,[18,58,59] and here we concentrate on more recent studies over the past few years.

For a protein molecule to form a heat-set gel, the concentration of the protein should be above a certain critical gelation concentration, and the protein should be heated above its denaturation temperature. The first stage of gelation is the thermal unfolding of the protein, followed by aggregation of the individual molecules into aggregates that grow and form a space-filling network that traps water within its structure to form a soft-solid gel. The thermal unfolding of proteins has received significant attention from MD modellers, since protein stability under heating is important in not only the food industry but also other biotechnology sectors. Euston and co-workers,[60,61] using conventional and accelerated molecular dynamics techniques (essential dynamics and replica exchange MD [REMD]), were able to probe the unfolding event that occurs during heating of the bovine protein whey β-lactoglobulin (β-lac). The simulation results were broadly in agreement with experimental studies in that α-helical structures were lost during the early stages of heating, the β-barrel (calyx) structure that defines the lipid-binding pocket unfolded more slowly, and the β-sheet making up calyx remained relatively intact. These structural changes expose the hydrophobic core, as well as the disulphide bonds and free sulphydryl in β-lac, which allows the molecules to aggregate through hydrophobic interactions, which are strengthened by subsequent intramolecular disulphide bonding. Zeiler and Bolhuis[62] have used REMD to probe the mechanisms of exposure of the free sulphydryl and disulphide groups that take part in intramolecular thiol-disulphide interchange reactions. They have shown that, whilst unfolding of the α-helix exposes the free sulphydryl group, the opposite is seen for the buried disulphide bond, thus suggesting an order in which thiol groups contribute to the disulphide bonding between molecules in the aggregates.

There have been very few simulation studies of heat-induced protein gelation over the past five years that are of interest to the food community. For previous studies, earlier review articles should be consulted.[18,58,59] One aspect of protein gelation relevant to foods does deserve mention, in part because it bears a remarkable similarity to the oleogels systems discussed earlier. These are the fibrillar gels that are formed from hydrolysed proteins including β-lactoglobulin,[63] ovalbumin,[64] gluten,[65] conglycininin, and glycinin,[66] to name a few.

It has been known for some time that under certain conditions proteins are able to form long fibrillar aggregates[67] that share similarities in structure with amyloid fibrils in disease states. The formation of these fibrils usually requires the protein to be denatured (*e.g.* by heating) and partially

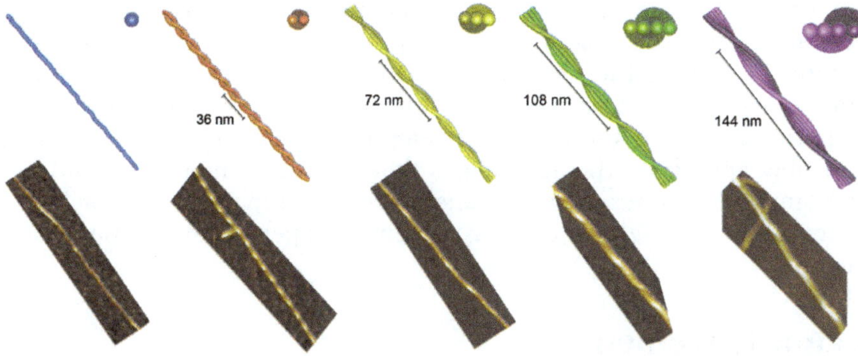

Figure 15.7 Linear relation between the number of filaments and periodicity of β-lactoglobulin fibrils. The top images are coarse-grain molecular dynamics reconstructions of the AFM images below.
Reproduced from ref. 72 with permission from Springer Nature, Copyright 2010.

hydrolysed, for example under acidic conditions.[68] At high enough concentrations, the resulting fibrils can form a transparent hydrogel or used, as they have been, to stabilize foams[69] and emulsions.[70] The structure of these fibrils has been observed using AFM, which has revealed an underlying helical ultrastructure that is similar to the fibrils formed from oleogelating plant sterols, although the interactions and formation mechanisms differ. For protein fibrils, peptides are believed to form through cross β-sheet formation between individual β-strands on separate peptides. These propagate to form a protofibril. Once the protofibrils reach a critical length, they align and twist together into a fibril of periodic pitch.[67] Lai *et al.*[71] have shown, using all atom MD simulations, that aligned planar fibrils made from peptide amphiphiles will spontaneously twist into a chiral helix. AFM images of fibrillar protein aggregates consistently show differing helical pitches (repeat lengths). Adamcik *et al.*[72] have used coarse-grained MD simulation to show how these are able to form. The simulations reveal that when increasing numbers of protofibrils twist into a helix, this gives rise to an increasing pitch length (Figure 15.7). The driving force for the adoption of helical structures has not been determined definitively, with entropic effects, hydrophobic interactions, and chirality all proposed to play a role.[67] The predominant form of the helix formed is left-handed, and this is exclusively so for fibrils from β-lactoglobulin, strongly suggesting a role for chirality.

15.6 Conclusions and Future Perspective

Molecular modelling and molecular simulations have experienced rapid progress over the past few years for many applications dealing with complex multi-component food systems. Throughout this chapter, it was demonstrated how computational approaches have been used to predict structure and physicochemical properties of various food systems at the molecular level or how several computational methods can help to explore and understand properties that are important in processing functionality. Some limitations, such as the timescale and the length scale used in the simulations, are expected to be overcome with the development of more powerful computing hardware and advanced computing algorithms. Once scientists more fully appreciate the advantages that molecular modelling and computer simulations offer to food science, it is hoped that they will make a far greater use of these techniques to complement experimental studies on food structure.

Acknowledgements

This work was supported by the BBSRC DRINC project Edible Oleogels for Reduction of Saturated Fat [BBSRC Grant No.: BB/M027449/1] and made use of time on ARCHER granted *via* the UK High-end Computing Consortium for Biomolecular Simulation, HECBioSim (http://hecbiosim.ac.uk), supported by EPSRC (grant no. EP/L000253/1).

References

1. M. Karplus and J. A. McCammon, *Nat. Struct. Biol.*, 2002, **9**, 646.
2. A. R. Leech, *Molecular Modelling: Principles and Applications*, Pearson Education, 2nd edn, 2001.
3. D. S. C. D. A. Case, T. E. Cheatham, III, T. A. Darden, R. E. Duke, T. J. Giese, H. Gohlke, A. W. Goetz, D. Greene, N. Homeyer, S. Izadi, A. Kovalenko, T. S. Lee, S. LeGrand, P. Li, C. Lin, J. Liu, T. Luchko, R. Luo, D. Mermelstein, K. M. Merz, G. Monard, H. Nguyen, I. Omelyan, A. Onufriev, F. Pan, R. Qi, D. R. Roe, A. Roitberg, C. Sagui, C. L. Simmerling, W. M. Botello-Smith, J. Swails, R. C. Walker, J. Wang, R. M. Wolf, X. Wu, L. Xiao, D. M. York, P. A. Kollman, *AMBER 18*, University of California, San Francisco, 2018.
4. A. D. MacKerell, D. Bashford, M. Bellott, R. L. Dunbrack, J. D. Evanseck, M. J. Field, S. Fischer, J. Gao, H. Guo, S. Ha, D. Joseph-McCarthy, L. Kuchnir, K. Kuczera, F. T. Lau, C. Mattos, S. Michnick, T. Ngo, D. T. Nguyen, B. Prodhom, W. E. Reiher, B. Roux, M. Schlenkrich, J. C. Smith, R. Stote, J. Straub, M. Watanabe, J. Wiorkiewicz-Kuczera, D. Yin and M. Karplus, *J. Phys. Chem. B*, 1998, **102**, 3586–3616.
5. M. M. Reif, P. H. Hunenberger and C. Oostenbrink, *J. Chem. Theory Comput.*, 2012, **8**, 3705–3723.
6. T. A. Halgren, *J. Comput. Chem.*, 1999, **20**, 730–748.
7. W. L. Jorgensen, D. S. Maxwell and J. Tirado-Rives, *J. Am. Chem. Soc.*, 1996, **118**, 11225–11236.
8. N. L. Allinger, K. H. Chen, J. H. Lii and K. A. Durkin, *J. Comput. Chem.*, 2003, **24**, 1447–1472.
9. N. L. Allinger, Y. H. Yuh and J. H. Lii, *J. Am. Chem. Soc.*, 1989, **111**, 8551–8566.

10. P. Dauber-Osguthorpe, V. A. Roberts, D. J. Osguthorpe, J. Wolff, M. Genest and A. T. Hagler, *Proteins*, 1988, **4**, 31–47.
11. A. Hospital, J. R. Goni, M. Orozco and J. L. Gelpi, *Adv. Appl. Bioinform. Chem.*, 2015, **8**, 37–47.
12. H. Höltje and G. Folkers, in *Molecular Modeling: Basic Principles and Applications*, ed. H. Höltje and G. Folkers, Wiley, Weinheim, 2008.
13. B. R. Brooks, R. E. Bruccoleri, B. D. Olafson, D. J. States, S. Swaminathan and M. Karplus, *J. Comput. Chem.*, 1983, **4**, 187–217.
14. H. J. C. Berendsen, D. van der Spoel and R. van Drunen, *Comput. Phys. Commun.*, 1995, **91**, 43–56.
15. J. C. Phillips, R. Braun, W. Wang, J. Gumbart, E. Tajkhorshid, E. Villa, C. Chipot, R. D. Skeel, L. Kale and K. Schulten, *J. Comput. Chem.*, 2005, **26**, 1781–1802.
16. A. R. Patel, *Alternative Routes to Oil Structuring*, Springer, 2015.
17. S. B. Engelsen, J. W. Brady and J. W. Sherbon, *J. Agric. Food Chem.*, 1994, **42**, 2099–2107.
18. S. R. Euston, in *Food Microstructures: Microscopy, Measurement and Modelling*, ed. V. Morris and K. K. Groves, Woodhead Publishing, London, 2013, pp. 336–385.
19. S. J. Marrink and D. P. Tieleman, *Chem. Soc. Rev.*, 2013, **42**, 6801–6822.
20. A. Brasiello, L. Russo, C. Siettos, G. Milano and S. Crescitelli, in *Computer Aided Chemical Engineering*, ed. S. Pierucci and G. B. Ferraris, Elsevier, 2010, vol. 28, pp. 625–630.
21. A. Brasiello, S. Crescitelli and G. Milano, *Phys. Chem. Chem. Phys.*, 2011, **13**, 16618–16628.
22. A. Pizzirusso, A. Brasiello, A. De Nicola, A. G. Marangoni and G. Milano, *J. Phys. D: Appl. Phys.*, 2015, **48**, 494004.
23. J. A. Nettleton, I. A. Brouwer, J. M. Geleijnse and G. Hornstra, *Ann. Nutr. Metab.*, 2017, **70**, 26–33.
24. A. G. Marangoni, *J. Am. Oil Chem. Soc.*, 2012, **89**, 749–780.
25. A. R. Patel and K. Dewettinck, *RSC Food Funct.*, 2015, **7**, 20–29.
26. A. G. Marangoni and N. Garti, in *Edible Oleogels*, Elsevier, 2011, pp. 1–17.
27. A. B. Matheson, V. Koutsos, G. Dalkas, S. Euston and P. Clegg, *Langmuir*, 2017, **33**, 4537–4542.
28. A. Bot, E. P. Gilbert, W. G. Bouwman, H. Sawalha, R. den Adel, V. M. Garamus, P. Venema, E. van der Linden and E. Flöter, *Faraday Discuss.*, 2012, **158**, 223.
29. G. Dalkas, A. B. Matheson, H. Vass, A. Gromov, G. O. Lloyd, V. Koutsos, P. S. Clegg and S. R. Euston, *Langmuir*, 2018, **34**, 8629–8638.
30. A. Matheson, G. Dalkas, R. Mears, S. R. Euston and P. S. Clegg, *Soft Matter*, 2018, **14**, 2044–2051.
31. A. B. Matheson, G. Dalkas, A. Gromov, S. R. Euston and P. S. Clegg, *Food Funct.*, 2017, **8**, 4547–4554.
32. R. den Adel, P. C. M. Heussen and A. Bot, *J. Phys.: Conf. Ser.*, 2010, **247**, 012025.

33. A. Bot, R. den Adel and E. C. Roijers, *J. Am. Oil Chem. Soc.*, 2008, **85**, 1127–1134.
34. M. Pernetti, K. F. van Malssen, E. Flöter and A. Bot, *Curr. Opin. Colloid Interface Sci.*, 2007, **12**, 221–231.
35. H. Sawalha, R. den Adel, P. Venema, A. Bot, E. Floter and E. van der Linden, *J. Agric. Food Chem.*, 2012, **60**, 3462–3470.
36. S.-H. Tung, Y.-E. Huang and S. R. Raghavan, *J. Am. Chem. Soc.*, 2006, **128**, 5751–5756.
37. C.-W. Njauw, C.-Y. Cheng, V. A. Ivanov, A. R. Khokhlov and S.-H. Tung, *Langmuir*, 2013, **29**, 3879–3888.
38. S. Banerjee and S. Bhattacharya, *Crit. Rev. Food Sci. Nutr.*, 2012, **52**, 334–346.
39. A. Vishnyakov, A. P. Lyubartsev and A. Laaksonen, *J. Phys. Chem. A*, 2001, **105**, 1702–1710.
40. A. Vishnyakov, G. Widmalm and A. Laaksonen, *Angew. Chem., Int. Ed. Engl.*, 2000, **39**, 140–142.
41. W. Huang, I. S. Dalal and R. G. Larson, *J. Phys. Chem. B*, 2014, **118**, 13992–14008.
42. M. Tusch, J. Kruger and G. Fels, *J. Chem. Theory Comput.*, 2011, **7**, 2919–2928.
43. C. A. Lopez, A. H. de Vries and S. J. Marrink, *Carbohydr. Res.*, 2012, **364**, 1–7.
44. J. Hanus and K. Mazeau, *Biopolymers*, 2006, **82**, 59–73.
45. K. Mazeau and L. Charlier, *Cellulose*, 2012, **19**, 337–349.
46. M. Umemura and Y. Yuguchi, *Carbohydr. Res.*, 2005, **340**, 2520–2532.
47. Q. Zhang, H. Brumer, H. Agren and Y. Tu, *Carbohydr. Res.*, 2011, **346**, 2595–2602.
48. J. Sauter and A. Grafmuller, *J. Chem. Theory Comput.*, 2015, **11**, 1765–1774.
49. J. Sauter and A. Grafmuller, *J. Chem. Theory Comput.*, 2017, **13**, 223–236.
50. M. Umemura and Y. Yuguchi, *Cellulose*, 2009, **16**, 361–371.
51. O. N. Makshakova, T. A. Gorshkova, P. V. Mikshina, Y. F. Zuev and S. Perez, *Carbohydr. Polym.*, 2017, **158**, 93–101.
52. Y. Li, Q. Zhao and Q. Huang, *Carbohydr. Polym.*, 2014, **101**, 544–553.
53. H. Shi, L. Yu, Y. Shi, J. Lu, H. Teng, Y. Zhou and L. Sun, *Molecules*, 2017, **22**, 1016.
54. H. Hecht and S. Srebnik, *Biomacromolecules*, 2016, **17**, 2160–2167.
55. I. A. Brownlee, C. J. Seal, M. Wilcox, P. W. Dettmar and J. P. Pearson, *Applications of Alginates in Food*, 2009, **13**, 211–228.
56. G. Widmalm, *Carbohydr. Res.*, 2013, **378**, 123–132.
57. U. Olsson, E. Sawen, R. Stenutz and G. Widmalm, *Chemistry*, 2009, **15**, 8886–8894.
58. S. R. Euston, *Curr. Opin. Colloid Interface Sci.*, 2004, **9**, 321–327.
59. E. Dickinson and S. R. Euston, *Adv. Colloid Interface Sci.*, 1992, **42**, 89–148.

60. S. R. Euston, S. Ur-Rehman and G. Costello, *Food Hydrocolloids*, 2007, **21**, 1081–1091.
61. S. R. Euston, *Food Hydrocolloids*, 2013, **30**, 519–530.
62. R. N. W. Zeiler and P. G. Bolhuis, *Mol. Simul.*, 2014, **41**, 1006–1014.
63. C. Akkermans, P. Venema, A. J. van der Goot, H. Gruppen, E. J. Bakx, R. M. Boom and E. van der Linden, *Biomacromolecules*, 2008, **9**, 1474–1479.
64. C. Lara, J. Adamcik, S. Jordens and R. Mezzenga, *Biomacromolecules*, 2011, **12**, 1868–1875.
65. D. M. Ridgley, K. C. Ebanks and J. R. Barone, *Biomacromolecules*, 2011, **12**, 3770–3779.
66. C. H. Tang and C. S. Wang, *J. Agric. Food Chem.*, 2010, **58**, 11058–11066.
67. J. Adamcik and R. Mezzenga, *Macromolecules*, 2011, **45**, 1137–1150.
68. W. S. Gosal, A. H. Clark, P. D. A. Pudney and S. B. Ross-Murphy, *Langmuir*, 2002, **18**, 7174–7181.
69. Z. Wan, X. Yang and L. M. Sagis, *Langmuir*, 2016, **32**, 8092–8101.
70. Z. M. Gao, J. J. Zhao, Y. Huang, X. L. Yao, K. Zhang, Y. P. Fang, K. Nishinari, G. O. Phillips, F. T. Jiang and H. Yang, *LWT - Food Sci. Technol.*, 2017, **76**, 1–8.
71. C. T. Lai, N. L. Rosi and G. C. Schatz, *J. Phys. Chem. Lett.*, 2017, **8**, 2170–2174.
72. J. Adamcik, J. M. Jung, J. Flakowski, P. De Los Rios, G. Dietler and R. Mezzenga, *Nat. Nanotechnol.*, 2010, **5**, 423–428.

Part D
Future Opportunities and Challenges in Food Structure Development

CHAPTER 16

Development of Food Structure via Sustainable Processing Systems

JAMES ATUONWU,[a] SAVVAS TASSOU[a] AND TIM FOSTER*[b]

[a] Brunel University London, UK; [b] University of Nottingham, UK
*Email: Tim.Foster@nottingham.ac.uk

16.1 Introduction

Food processing can be defined as the transformation of raw ingredients and intermediates into products intended for human consumption, with the purpose of improving the digestibility, bioavailability of nutrients and energy, taste, appearance, safety, storability, and distribution.[1] Two key elements stand out from this definition: food product property transformation (which in most cases relates to structure) and food preservation (intended to keep the desired properties for as long as possible for the maximum benefit to the final consumer). Structural changes also occur during preservation processes. Food structure development is therefore a crucial aspect of food processing that must be factored into design and operation.

The food and beverage market is extremely competitive. Therefore, designing for quality improvement alone is not sufficient to assure the success of any technology. The improvements must be achievable at a commercially viable process cost per production unit with as low as possible carbon footprint. There is also the issue of waste minimisation and, more importantly, valorisation. To accomplish these, sustainable processing unit operations and systems that consume less energy and water per unit output than

Food Chemistry, Function and Analysis No. 18
Handbook of Food Structure Development
Edited by Fotis Spyropoulos, Aris Lazidis and Ian T. Norton

Published by the Royal Society of Chemistry, www.rsc.org

conventional processes, at the same time minimising quality depreciation and optimising waste, are necessary. The purpose of this work is to examine the interrelation between the mechanisms of various sustainable food processing operations and the structures of their various products.

16.2 Sustainable Food Processing Systems: An Overview

The global population has consistently been on the rise and is predicted to reach 9.7 billion by 2050.[2] This growth, coupled with changing lifestyles and income levels, places increasing demands on global food security in general and on food processing facilities in particular. As the food and drink industry is a significant energy and water consumer and hence an emissions producer, the need for sustainable food processing systems becomes increasingly important. The food and drink manufacturing industry is quite diverse with many subsectors such as meat processing, fish and seafood, poultry, dairy, brewery, bakery, sugar, confectionery, rendering, distilling, malting, soft drinks, animal feed, amongst others. Each subsector has very specific processing technologies,[3] and so the energy demand patterns differ. In general, however, the main processing techniques applied throughout the food and drink sector can be classified into the following categories: materials reception and preparation, size reduction, mixing and forming, separation techniques, product processing technology, heat processing, concentration by heat, processing by removal of heat, post-processing operations, and utility processes.[4] Each of these, are further divided into different unit operations.[4] In principle, any of these aspects can be made more sustainable by deploying appropriate technologies and systems. The heating-related operations (*e.g.* heat processing, concentration by heat, and some aspects of product processing and separation techniques) form the most energy-demanding aspects of food processing and so possess the greatest sustainability improvement potentials. Steam and hot water generation (for indirect heating) and direct heating together constitute about 81% of the total energy demand of the entire UK food process industry, with motors, refrigeration, and compressed air systems contributing 12%, 5%, and 2%, respectively.[3] The majority of this energy demand (~67%) is currently served by natural gas, followed by electricity and a minor amount of oil and coal.

In view of the foregoing, any technology that accomplishes its intended transformation and/or preservation effect under milder heating conditions than the conventional means could be classified as a sustainable processing technology. The milder temperatures, apart from reducing heat energy input, also imply that less energy and cooling water would be necessary for cooling the product. Furthermore, these technologies enable the production of ready-to-eat meals with the quality of fresh chilled meals but without the need for chilled or frozen storage. Extension of the shelf life of fresh products also contributes to the avoidance of food waste in the supply chain.[5] Some of these technologies are also post-packaging technologies that prevent recontamination of the product and significantly reduce the volume of packaging material as repacking under aseptic conditions is not necessary. Post-package processing systems also have the advantage of preventing the fouling of heating surfaces and thus saving water and other inputs that would have been used for cleaning operations. The following have been identified[1,5] as sustainable food processing technologies:

A. Novel pasteurisation and sterilisation technologies
   i. High-pressure processing
   ii. Microwave volumetric heating
   iii. Radiofrequency heating
   iv. Ohmic heating
   v. Pulsed electric field processing
B. Nano- and microtechnologies
   i. Membrane processes

A common characteristic of the novel pasteurisation and sterilisation technologies is the direct volumetric interaction of the applied energy (*e.g.* electricity) with the food, without any heat exchanger surface heating and conductive heat transfer limitations. This is a critical factor in the energy performance improvements derivable from them. In some cases (*e.g.* HPP, RF, and MW heating), post-package processing is possible. Membrane processes do not require heat in performing many food processing operations that would ordinarily require heat. They are thus energy efficient. In the sections that follow, we introduce the principles of these technologies, their current applications in the food industry, a review of the impacts of the technologies on various food structures, and possible future research directions in this regard.

16.3 Sustainable Food Processing Technologies: Principles and Applications

16.3.1 High-pressure Processing HPP

A typical HPP system consists essentially of a high-pressure vessel, sealing closures (plugs, wedges, and frames or yokes), a pressure-generating system, a pressure-transmitting medium reservoir, and a material handling system. The food sample to be processed is, in most cases, packaged in a flexible, airtight hermetically-sealed container, *e.g.* plastic pouches loaded onto a carrier basket and then into the high-pressure vessel.[6] The vessel is filled with the pressure-transmitting medium, usually deaerated water, by a low-pressure pump. The plugs are closed and secured with a wedge, and the system is thereafter pressurised. After the desired pressure is attained and held for a specified time, depending on the intended purpose, it is

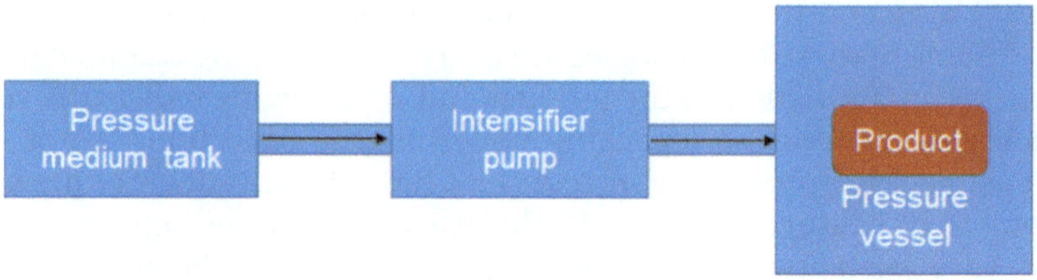

Figure 16.1 Process flow diagram of a high-pressure processing system.

depressurised, the pressurising medium is drained or recirculated, and the basket and its contents are unloaded. Figure 16.1 shows a process flow diagram of a HPP system indirectly pressurised by pumping extra pressure medium to the sealed vessel *via* an intensifier, as is common in the food industry. HPP has been widely deployed across a variety of food sectors such as in the cold pasteurisation of fruits and vegetables, meat products, juices and beverages, among others. Apart from pasteurisation, other current industrial applications of HPP include freezing, thawing, extraction, and pretreatment to enhance subsequent drying and frying processes.[7] HPP does not destroy bacterial spores, unless combined with high temperatures.[8] Hence, HPP-only sterilisation processes are not yet commercially available.[9] New frontiers have been opened by HPP, leading to the development of niche foods. HPP has been identified as one of the most commercially promising emerging food processing technologies.[10]

16.3.2 Radiofrequency and Microwave Processing

Radiofrequency (RF) and microwave (MW) heating exploits the direct interaction of electrical energy in different forms with dielectric materials in foods (mainly water but also fats and sugars). In RF heating, which occurs in the frequency range of 10–50 MHz, a capacitor is formed by passing or placing the food material between two metallic electrodes connected to an oscillator that generates alternating electric fields within the food. The rapid movements of the water molecules in the food (dipole rotation) to align with the rapidly changing polarity of the fields create massive molecular collisions/friction and subsequent heat generation. Sugars and fats are less polar than water and so produce less dipole rotation effects. If there are free ions (charged atoms) or ionic species in the food product, heat is generated *via* the oscillatory migration of ions by a process known as ionic conduction. Frequencies commonly employed for RF heating are 13.56 ± 0.00678, 27.12 ± 0.16272, and 40.68 ± 0.02034 MHz. RF heating requires a very uniform bed of material for efficient operation.[11] RF heating has been successfully applied in thawing, tempering, pasteurisation, cooking, roasting, and in post-baking drying operations.[12]

Microwave heating works by the same principles (dipole rotation and ionic conduction) but is different in a number of ways. Constructionally, there is

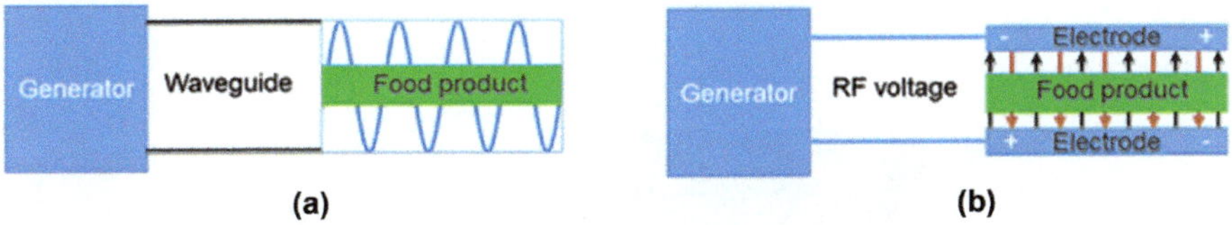

Figure 16.2 Schematic of (a) microwave and (b) RF heating of foods.

no capacitive arrangement, but a closed cavity within which electromagnetic waves, generated in a power source (usually magnetrons), are delivered (see Figures 16.2a and b for comparison). These waves bouncing off the walls are delivered into the food material to drive the dipole rotation/ionic conduction processes. The waves occur due to the much higher frequencies employed (in the range 300 MHz–300 GHz). Common frequencies employed are 896/915 $\pm$ 13 MHz and 2450 $\pm$ 50 MHz. Due to the significantly higher frequencies of microwaves, the heating rate is higher, but the penetration depth is lower than that of RF heating. It, however, requires less floor space than RF heating as large capacitors are not needed. Microwave heating in stand-alone mode or in combination with other energy sources is mainly applied industrially in pasteurization, sterilization, drying, and tempering of high-value foods. Other applications include baking, thawing, blanching, extraction, reheating, amongst others.[13] Due to the direct interaction of electrical signals with the dielectric food components, RF and microwave heating are selective, volumetric, and not limited by the normally slow conductive heat transfer dynamics of conventional systems. The fast heating that results reduces processing time significantly and hence energy intensity in many applications while yielding products of improved quality. Both processes can also be performed on pre-packaged foods.

16.3.3 Ohmic Heating

Ohmic heating systems are similar to RF heaters in construction as they consist of metallic electrodes between which the food is placed (batch operation) or passed (continuous operation within pipes) (Figure 16.3). Unlike RF heaters, the electrodes are connected to a low-frequency power source (usually the mains frequency of 50/60 Hz), with the food in contact with and forming a continuous electrical path between the electrodes.

Due to the conductivity of the food (preferably in the range 0.1–10 S m^{-1}),[14] a current flows through it, generating a volumetric Joule heating effect within. Because of the continuous electrical path requirement, ohmic heating is most suitable for fluid foods such as milk, juices, and other beverages. Fluids containing particulate matter can also be processed. When pumpable, the fluid foods can be processed in continuous flow systems. Solid foods immersed in liquid media can also be processed, but non-uniform heating becomes significant. In principle, uniform heating can be attained only in homogeneous foods. Non-homogeneous foods would require more complex

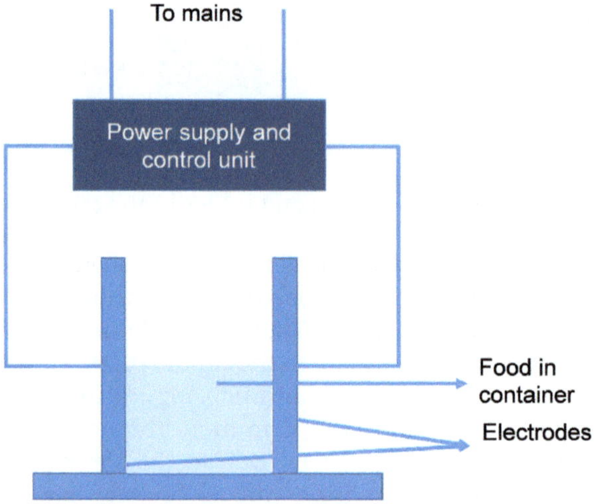

Figure 16.3 Schematic of batch ohmic heater.

operations, such as stirring, intermittent heating, or a combination with other energy sources, to improve heating uniformity. The voltage gradients (volt per distance) usually employed in ohmic heating are in the range 10–90 V cm^{-1}. Ohmic heating is mainly applied in pasteurisation and sterilisation processes and has been shown to be more energy efficient than HPP and microwave heating.[15] Other applications include blanching and thawing.[14] More information on OH from a flexible process point of view can be found in Chapter 18 (Section 18.2.2) of this book.

16.3.4 Pulsed Electric Field Processing (PEF)

In pulsed electric field (PEF) processing, the processed food material is placed between or passed through electrodes, just as in ohmic heating. However, very high-voltage gradients (typically 18 kV cm^{-1} or higher for pasteurisation processes[16] but sometimes less, *e.g.* for pre-drying or pre-peeling operations) are applied in short pulses (μs range) at repetition rates and durations that allow for sufficient treatment of all volume elements of the food. Such high-voltage gradients (for pasteurisation processes) ensure that the voltages across microbial cells (typically in the size range of 2–100 μm) exceed the critical membrane potential (about 200 mV for many cells),[17] puncturing the cell membrane. Small pores of size range 50–100 nm are formed, leading to the loss of barrier function and ultimate destruction of the microbial cells. This process, also known as electroporation, is fast, mild, energy-efficient, and targeted specifically on cell membranes, while retaining vital nutritional and sensory components of foods at extended shelf life. A typical PEF system consists of a high-voltage DC power source, a large capacitor bank, a high-voltage electronic switch, a low-voltage signal generator, and a treatment

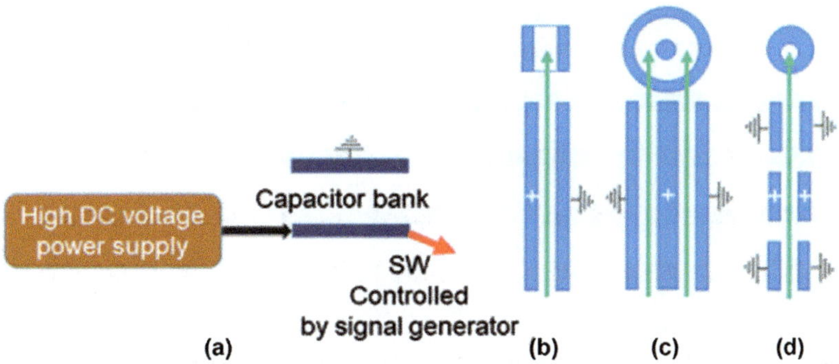

Figure 16.4 (a) Continuous PEF system with power supply system and different treatment chamber configurations, such as (b) parallel plate, (c) coaxial, and (d) collinear. In operation, SW from (a) connects in a pulsed manner to the +electrode in (b), (c), or (d).

chamber housing the electrodes (Figure 16.4) with suitable fluid handling systems.

Mains power is increased by a step-up transformer and converted to DC by suitable electronic circuits, which is used to charge the capacitor bank. The electronic switch connected to the capacitor switches on only when triggered by a pulse from the signal generator. For the very short period when this pulse is active, the capacitor discharges its stored energy *via* the switch to the treatment chamber electrodes, leading to the flow of short pulses of current through the food. Some ohmic heating may occur due to the current flow. The dielectric strength of the food must be high enough to prevent "corona" breakdown due to the high electric fields. Critical operating parameters include pulse width, pulse shape, pulse repetition rate, total number of pulses, specific energy, operating temperature, and voltage gradient. PEF processing is mainly applied in the pasteurisation of foods to extend refrigerated shelf life, although sterilisation is possible for high-acid products.[18] It has also been shown to be an effective pre-treatment technique that enhances drying and extraction processes.[19] PEF was first commercially applied in 2006 in the United States in the preservation of fruit juices. Since then, more industrial-scale processing equipment for liquid and solid products have been developed in Europe and other parts of the world.[20] The technology is still emerging and expected to be widely applied industrially in the very near future.[1] Some further applications of PEF are discussed at Section 18.2.1.2 of this book.

16.3.5 Membrane Food Processes

In membrane processing, liquid food is pressurised through specially designed semipermeable membranes to accomplish particle separation on the basis of their molecular size and shape. This advanced separation technique easily lends itself to a wide variety of food processes such as reduction of

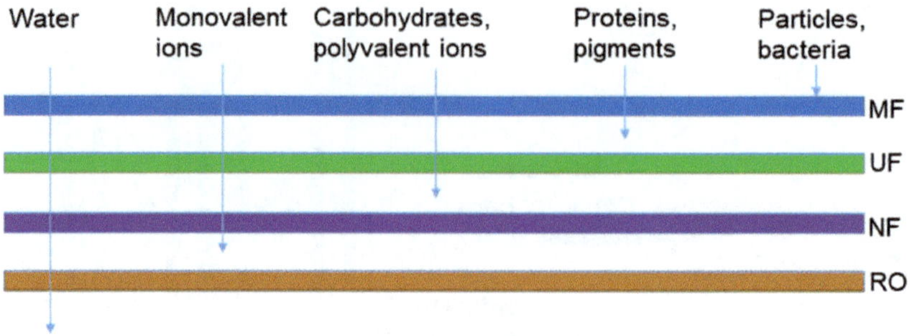

Figure 16.5 Schematic showing different membrane processes (microfiltration MF, ultrafiltration UF, nanofiltration NF, and reverse osmosis RO) based on pore-size differences.

microbial load, demineralization, desalination, stabilization, concentration, de-acidification, purification, clarification, fractionation, amongst others. A detailed review of numerous food processing applications of membrane processes is presented in Kotsanopoulos and Arvanitoyannis.[21] Based on membrane pore size, membrane processes have been classified into four groups:[22,23] microfiltration, ultrafiltration, nanofiltration, and reverse osmosis. Another membrane process, electrodialysis, uses ion exchange membranes through which ions are transferred to different solutions under the influence of electric fields.[21] In general, the applied pressure energy decreases with pore size, and because heat is not required, the process is energy efficient. Fouling occurs during operation, making cleaning operations a crucial aspect of membrane food processes. Figure 16.5 shows a schematic of different membrane processes.

16.4 Food Structure Development in Sustainable Processes: Current Status, Future Research Directions

The various sustainable food processing technologies affect different structural aspects of various food types at the microscopic and macroscopic levels, depending on how the technologies are deployed. A review of the literature on the effects of different applications of HPP, microwave, RF, ohmic, PEF, and membrane processes on the structural properties of different kinds of food products is presented in this section.

16.4.1 HPP and Food Structure

The high pressures exerted during HPP have been observed to have distinct structural effects on different food substances. These effects are essentially based on the Le Chatelier principle, which states, "If a dynamic equilibrium is disturbed by changing the conditions, the position of equilibrium moves to counteract the change." Hence the pressurisation stage of HPP will favour processes that lead to reductions in sample effective volume. If the pressure-induced changes in structure are irreversible, the product comes out structurally modified even after the depressurisation stage. For carbohydrates, little pressure-induced structural changes are observed on simple sugars, formed with high-energy covalent bonds, whereas, complex carbohydrates (such as starch), consisting of chains of sugars linked by low-energy bonds can be modified by HPP.[24] This could result in the swelling of starch and ultimately gelatinization, approaching thermally obtainable levels.[25] In a case study involving potato, wheat, and tapioca starch suspensions,[26] it was observed that starch gelatinisation rates were highly dependent on pressure. Studies on the structural parameters of amylose and amylopectin starches[27] showed HPP-induced changes in branch frequency, intrinsic viscosity, and radius of gyration for all the treated starches. Data obtained from size exclusion chromatography and Fourier transform infrared spectroscopy showed that a-1,6-glycosidic bonds are more frequently split in pressurized amaranth, Hylon VII, and waxy maize starch, while in sorghum and maize starches, the a-1,4 bonds are most commonly split. Liu *et al.*[28] observed no change in the A-type crystalline pattern of sorghum starch after pressure treatments at 120–480 MPa. However, treatment at 600 MPa produced a pattern similar to B-type crystalline structures. HPP treatment also resulted in sorghum starch granules with rough surfaces. Measured amylose content, water absorption capacity, alkaline water retention, pasting temperature, and thermostability increased with increasing pressure levels, while the oil absorption capacity, swelling power, relative crystallinity, and viscosity decreased. Compared with native starch, HPP-modified sorghum starch samples had lower *in vitro* hydrolysis, reduced amount of rapidly digestible starch, as well as increased levels of slowly digestible starch and resistant starch, indicating that HPP treatment is an effective modification method for altering the *in vitro* digestibility and physicochemical properties of sorghum starch.[28] HPP also impacts on proteins. For proteins, which generally exist as macromolecules in aqueous environments in equilibria among different conformations: oligomers, monomers, partially denatured monomers, and denatured monomers, HPP causes non-covalent (hydrogen, ionic, and hydrophobic) bond interactions at molecular level. This leads to secondary and tertiary structure modifications and hence denaturation, aggregation, gelation, fragmentation, or gelatinization.[29] HPP can also trigger the formation of disulfide bonds as demonstrated during HPP treatment of skimmed milk at 200 MPa.[30] These HPP effects could induce texture and colour modifications, especially in meat and meat products. Moreover, reactions that depend on protein structure can be modified by pressure, changing its function, for instance, by modifying active sites of enzymes and masking or better access of epitopes.[24] The pressure-induced secondary and tertiary protein structural changes also apply to the structure of food allergens and antigens, which are almost always proteins. In a review of various works in this regard, Pottier *et al.*[24] shows that HPP indeed significantly alters IgE binding, which reflects allergenicity, and IgG binding, which reflects antigenicity in foods; in many cases, the allergenicity is reduced, improving the health attributes of foods. In an *in vitro* protein digestion study on meats, Kaur *et al.*[31] showed fewer proteins or higher molecular weight peptides in HPP-treated meats than in untreated meats, especially at higher pressures where greater changes in the myofibrillar structure of the meat was observed. This suggests that HPP improves the digestibility of proteins. Lipid systems are the most pressure-sensitive biological components in foods,[32] as they are governed by hydrophobic interactions that are highly susceptible to high pressure. The adiabatic heat of compression of lipid systems is much higher (about 8–9 °C per 100 MPa) than that of water (about 3 °C per 100 MPa) and other food components. Hence, for high pressures, the holding temperature may be high enough to cause thermally induced structural modifications, provided the holding time is sufficiently high. Their melting temperatures also increase by more than 10 °C per 100 MPa such that lipids existing in liquid state at room temperature will crystallize under pressure.[33] Depending on the target pressure, operating temperature, holding time, compression rate, compression cycle, and chemical composition of the lipid, this crystallisation may be irreversible or reversible by decompression.[34] A detailed review on the mechanisms of crystallisation in various lipids under different pressurisation conditions is available.[34] HPP-processed lipid-containing products such as meat and fish have been found to be more prone to certain chemical structural modifications, such as oxidation, above a threshold of 300–500 MPa.[35] The threshold pressure depends on the operating temperature, time, as well as the composition of lipid and non-lipid fractions of the treated food. This is very important as the secondary products of lipid oxidation lead to product rancidity and off flavours.[36] For foods of marine origin, the high presence of polyunsaturated fatty acids promotes the initiation of radical mechanisms that accelerate oxidation in subsequent storage periods. The liberation of radicals or precursors of radicals due to the breakdown of cell membranes induced by pressure has been given as one of the most important causes of increased lipid degradation.[35] A comprehensive review of HPP-induced structural modifications in various foods such as fruits and vegetables, meat, dairy products and ingredients, as well as individual biopolymers such as carbohydrates and proteins, is available.[37]

16.4.2 RF/Microwave Processing and Food Structure

RF and microwave processes operate primarily by thermal effects. Hence similarities exist between the effects of these and conventional heating technologies on food structure. Examples of such effects include secondary, tertiary, or quaternary structural modifications of proteins (denaturation), structural changes leading to gelation and gelatinisation of starches, among others. RF and microwaves especially, however, also show significantly different heating patterns from conventional heating technologies and hence could yield different structures that could be either positive or negative depending on the specific application. Microwaves are characterised by inverse temperature gradients: higher internal temperatures (sometimes up to boiling and superheating levels) and lower surface temperatures. In the microwave baking of flour-, wheat-, and starch-based products, the low surface temperature prevents the much desired browning and crust formation. Evaporated vapour from the interior tends to condense on the cooler surface, preventing browning and crisping reactions and leading to a soggy, gummy product.[38] Also, the internal boiling and steam generation causes structural changes that manifest in product toughening. Besides moisture loss, the internally generated condensable steam replaces the non-condensable gases (air) in the product voids. When the steam condenses by cooling, a vacuum may be created in the voids, causing their collapse and the formation of a more compact and hence tougher structure.[39] Amylose extraction from starch granules during the boiling phase and its rapid crystallization during cooling are thought to be the main cause of toughening of starch-based microwave-baked products.[40] In the well known microwave cooking and pasteurisation process, MicVac,[41] a special valve is created on the packaging that releases oxygen and the majority of the steam during the cooking and pasteurisation stage. The remaining steam condenses and causes an under-pressure in the package, yielding a vacuum-packed product. For short-time microwave pasteurisation and sterilisation processes under carefully controlled temperatures, no objectionable changes in product structure are expected.[42] RF and, to a limited degree, microwave processing are, however, very useful in post-baking drying operations of foods such as biscuits as they minimise the so-called checking phenomenon, which sometimes occurs in products processed in conventional baking ovens. Checking is an undesirable structural transformation that arises due to the build-up of stresses caused by differential moisture content between the outer surface and the core of the product, ultimately leading to breakages.[43]

While the absence of crust formation is a problem for baking operations, it comes in handy for microwave drying processes as crust formation is responsible for the increased heat and mass transfer resistance that occurs in the so-called falling rate period in conventional hot air drying.[44] To prevent wet product surfaces, microwave drying is usually combined with airflow (microwave-convective) or reduced pressure (microwave-vacuum), which also permits drying at lower temperatures. The main advantage of microwave drying lies in the fact that microwaves penetrate the product, preferentially heating the water within to evaporating temperature and setting up temperature, moisture, and pressure gradients, simultaneously in one direction (interior to exterior). A more porous structure is created,[45] providing an easier path for vapour escape, without the limiting effects of crust formation and shrinkage that occur in hot air drying. During the drying of starchy product pellets, the material undergoes state transition from a glassy to a rubbery state, provided the temperature exceeds the glass transition temperature. If microwave heating continues, superheated steam is eventually generated in the glassy matrix and if the vapour pressure is sufficient to overcome the resistance of the rubber-like matrix, microwave expansion or puffing occurs.[46] If after expansion, the microwave energy is turned off, the pellet cools down and reverts to a glassy state but remains puffed, creating a porous structure in the final product. Compared with other heating technologies, microwave expansion of snacks has been reported as faster and more efficient, while eliminating the need for additional fat.[47,48] A review of studies relating to the effects of microwave processing on the structural properties of various food components is available in Jiang *et al.*[49]

16.4.3 Ohmic Heating and Food Structure

Apart from thermal effects, electroporation, a non-thermal electric field effect by which pores are formed in the food structure at the cell/tissue membrane level, has been proposed as one of the mechanisms of ohmic processing.[50] Although the nature of ohmic electroporation in plant tissue is rather complex and not well understood,[51] its effect on certain food process behaviours are probably linked to structural modifications due to this non-thermal effect. For instance, Halden *et al.*[52] observed an increase in beet dye diffusion during ohmic heating. Kulshrestha and Sastry[53] showed that the permeability of plant cell membranes during ohmic heating is higher than that due to a thermal system delivering the same product temperature. Praporscic *et al.*[54] observed high levels of membrane destruction and mechanical softening of tissues even at a moderate temperatures, <50 °C. When combined with osmotic dehydration, ohmic heating was found to cause structural changes such as cell shape, cell size, and middle lamella thickness in apples.[55] Ohmic pre-treatment has also been found in many studies[56–58] to increase the drying rates of fruits, vegetables, and other products due to tissue structural modifications. As these drying rates are in many cases significantly higher than those of conventional thermal treatment at the same temperature,[59] the increase is likely due to the non-thermal electroporation effect. Lower-frequency operation has been shown to favour these structural changes.[60,61] More research is needed, using imaging and other techniques to unravel the nature of these structural changes and their dependence on process parameters.

16.4.4 Pulsed Electric Field Processing and Food Structure

Pulsed electric field (PEF) treatment works primarily by non-thermal effects – electroporation – and so, by default, alters the structure, at the cellular/tissue level, of food products subjected to it. Parasitic ohmic heating may also occur, depending on the magnitude of the operating current, pulse width, and total number of pulses. As in ohmic heating, the nature of the non-thermal structural modifications and the controllability thereof in diverse food products are not well understood. However, several studies and some industrial applications show the peculiar beneficial effects of PEF pretreatment, which could be linked to structural modifications owing to its cell disintegration capabilities. For instance, PEF processing enhances various kinds of mass transport processes such as extraction, drying, marinating, cutting, and peeling. In industrial potato French fry production applications, PEF softens whole potatoes, thereby improving cut quality, the controlled release of sugar, and colour uniformity, while reducing product breakages, frying times, and oil uptake.[62] Significant improvements in drying rates have been observed.[63–67] Improved extraction has also been demonstrated.[68–70] In addition, PEF treatment has been shown to greatly enhance chemical structural modifications such as the acetylation of starch products,[71,72] hence reducing the chemical treatment required for the modification. This is very important as it promotes the clean label concept. Comprehensive reviews of various studies relating to PEF-induced structural modifications of starches and proteins in various food products as well as their impacts on product functionalities are available in Zhu[73] and Han *et al.*,[74] respectively. The general consensus is that PEF induces structural and hence functional changes in food products particularly beyond certain product-dependent threshold voltage gradients (electric field strengths). Most of these modifications are attributed to non-thermal effects, while some require significant ohmic heating.

16.4.5 Membrane Processing and Food Structure

Little is available in the literature on the specific effects of the different membrane process technologies on the structure of processed foods. However, being highly precise separation/concentration processes, membrane processes alter the functional properties of raw and partly processed food materials to produce enhanced functionality products.[75] These functionality changes are linked to structural changes, but more research is needed to unravel the details.

16.5 Concluding Remarks and Future Perspectives

Electricity-driven mild processing techniques have no doubt shown great potential for sustainable food processing in term of energy efficiency and decarbonisation, while improving product quality with less dependence on chemical additives compared to their conventional counterparts. The various studies reviewed in this work show that these technologies have the capacity to engineer structures that create some desired functional properties in starches, proteins, and lipids in various food products. Such structural modifications in some cases are due to normal thermal effects like their conventional counterparts but, more importantly, to non-thermal effects such as pressure and electric field parameters. Enhanced thermal effects such as superheating and the unique temperature distribution regimes encountered, for instance, in microwave processing also aid structural modifications. The mechanisms and effects of these non-thermal, enhanced, and unique thermal variables of the sustainable processing techniques on various food structure modification possibilities are not readily available in the literature. Neither are the interaction between the structural transformations and the sustainability parameters of energy and water efficiency, product shelf life, and overall carbon footprints. Such studies and the data derived from them would aid process and product designers to deliberately design processes and operating regimes to engineer the desired structural and hence functional transformations in food products, while preserving the sustainability advantages of these mild processing techniques.

References

1. Sustainable, Safe and Nutritious Food, Food processing technologies. European Union, Business Innovation Observatory, Contract No. 190/PP/ENT/CIP/12/C/N03C01. Available at https://ec.europa.eu/docsroom/documents/13425/attachments/1/translations/en/renditions/native (accessed 19/5/19).
2. World population projected to reach 9.7 billion by 2050. Available at http://www.un.org/en/development/desa/news/population/2015-report.html (accessed 17/7/18).
3. Industrial decarbonisation and energy efficiency roadmaps to 2050, Food and Drink. Available at https://assets.publishing.service.gov.uk/government/uploads/system/uploads/attachment_data/file/416672/Food_and_Drink_Report.pdf (accessed 19/5/19).
4. Best available techniques (BAT) reference document in the food, drink and milk industries. Available at http://eippcb.jrc.ec.europa.eu/reference/BREF/FDM/FDM_02-10-2018BW.pdf (accessed 19/5/19).
5. Technological options for feeding 10 billion people. Options for sustainable food processing. State-of-the-art report K STOA 2013/122. Available at https://dspace.uevora.pt/rdpc/bitstream/10174/18263/118/062%20-%20513.533HLI_EN.pdf (accessed 19/5/19).
6. J. C. Atuonwu and S. A. Tassou, *Innovative Food Sci. Emerging Technol.*, 2018, **47**, 214.
7. Y. Tao, D. Sun, E. Hogan, A. L. Kelly, in *Emerging Technologies for Food Processing*, ed. D. Sun, Academic Press, Amsterdam, 2nd edn, 2014, ch. 1, pp. 3–24.
8. H. Zhang and G. S. Mittal, *Food Rev. Int.*, 2008, **24**, 330.
9. R. Sevenich, F. Bark, E. Kleinstueck, P. C. Crews, J. Hradeckyc, K. Reineke, *et al.*, *Food Control*, 2015, **50**, 539.
10. C. Jermann, T. Koutchma, E. Margas, C. Leadley and V. Ros-Polski, *Innovative Food Sci. Emerging Technol.*, 2015, **31**, 14.
11. Microwave *vs* radiofrequency heating. Available at https://ferriteinc.com/wp-content/uploads/2018/02/Microwave-and-Radio-Frequency-Heating.pdf (accessed 19/5/19).

12. P. Piyasena, C. Dussault, T. Koutchma, H. S. Ramaswamy and G. B. Awuah, *Crit. Rev. Food Sci. Nutr.*, 2003, **43**, 587.
13. J. C. Atuonwu and S. A. Tassou, *J. Food Eng.*, 2018, **234**, 1.
14. H. Jaeger, A. Roth, S. Toepfl, T. Holzhauser, K. Engel, D. Knorr, *et al.*, *Trends Food Sci. Technol.*, 2016, **55**, 84.
15. J. C. Atuonwu, C. Leadley, A. Bosman, S. A. Tassou, E. Lopez-Quiroga and P. J. Fryer, *Innovative Food Sci. Emerging Technol.*, 2018, **50**, 174.
16. J. Dunn, in *Pulsed Electric Fields in Food Processing: Fundamental Aspects and Applications*, ed. G. V. Barbosa-Canovas and H. Zhang, Technomic Publishing, Lancaster, 2001, ch. 1, p. 2.
17. J. Teissie and M. Rols, *Biophys. J.*, 1993, **65**, 409.
18. N. Yang, K. Huang, C. Lyu and J. Wang, *Food Control*, 2016, **61**, 28.
19. Pulsed electric field processing of foods, New Food Magazine. Available at https://www.newfoodmagazine.com/article/1594/pulsed-electric-field-processing-of-foods/ (assessed 19/5/19).
20. R. Buckow, S. Ng and S. Toepfl, *Compr. Rev. Food Sci. Food Saf.*, 2013, **12**, 455.
21. K. V. Kotsanopoulos and I. S. Arvanitoyannis, *Crit. Rev. Food Sci. Nutr.*, 2015, **55**, 1147.
22. O. Rodriguez-Gonzalez, R. Buckow, T. Koutchama and V. M. Balasubramamian, *Compr. Rev. Food Sci. Food Saf.*, 2015, **14**, 536.
23. S. Mikhaylin and L. Bazinet, *Adv. Colloid Interface Sci.*, 2016, **229**, 34.
24. L. Pottier, G. Villamonte and M. de Lamballerie, *Curr. Opin. Food Sci.*, 2017, **16**, 21–27.
25. H. Simonin, C. Guyon, M. Orlowska, M. de Lamballerie and A. Le Bail, *LWT–Food Sci. Technol.*, 2011, **44**, 779.
26. B. A. Bauer and D. Knorr, *J. Food Eng.*, 2005, **68**, 329.
27. A. Szwengiel, G. Lewandowicz, A. R. Gorecki and W. Blaszczak, *Food Chem.*, 2018, **240**, 51.
28. H. Liu, H. Fan, R. Cao, C. Blanchard and M. Wang, *Int. J. Biol. Macromol.*, 2016, **92**, 753.
29. T. Rahaman, T. Vasiljevic and L. Ramachandran, *Trends Food Sci. Technol.*, 2016, **49**, 24.
30. H. A. Patel, H. Singh, S. H. Anema and L. K. Creamer, *J. Agric. Food Chem.*, 2006, **54**, 3409.
31. L. Kaur, T. Astruc, A. Vénien, O. Loison, J. Cui, M. Irastorza and M. Boland, *Food Funct.*, 2016, **7**, 2389.
32. N. Rivalain, J. Roquain and G. Demazeau, *Biotechnol. Adv.*, 2010, **28**, 659.
33. J. C. Cheftel and J. Culioli, *Meat Sci.*, 1997, **46**, 211.
34. M. Zulkurnain, F. Maleky and V. M. Balasubramaniam, *Food Eng. Rev.*, 2016, **8**, 393.
35. I. G. Medina-Meza, C. Barnaba and G. V. Barbosa-Canovas, *Innovative Food Sci. Emerging Technol.*, 2014, **22**, 1.
36. H. Simonin, F. Duranton and M. de Lamballerie, *Compr. Rev. Food Sci. Food Saf.*, 2012, **11**, 285.

37. T. Bolumar, D. Middendorf, S. Toepfl, V. Heinz, in *High Pressure Processing of Food, Principles, Technology and Applications*, ed. V. M. Balasubramaniam, G. V. Barbosa-Cánovas and H. L. M. Lelieveld, Springer, New York, 2016, ch. 23, pp. 509–537.
38. R. F. Schiffmann, in *Thermally-Generated Flavours: Maillard, Microwave and Extrusion Processes*, ed. T. H. Parliament, M. J. Morello and R. J. McGorrin, American Chemical Society, Washington, DC, 1994, pp. 386–394.
39. S. Mizrahi, *J. Food Sci.*, 2012, **71**, R57.
40. M. Uzzan, E. Kesselman, O. Ramon, I. J. Kopelman and S. Mizrahi, *J. Food Eng.*, 2006, **74**, 279–284.
41. In-pack cooking and pasteurisation. Available at https://www.micvac.com/retail-solutions/method/ (accessed 10/10/18).
42. J. Ahmed, H. Ramaswamy, in *Handbook of Food Preservation*, ed. M. S. Rahman, 2nd edn, Taylor & Francis, Boca Raton, FL, 2007, ch. 28, pp. 691–712.
43. Radio frequency heating and post-baking. Available at http://www.strayfield.co.uk/images/radio_frequency.pdf (accessed 10/10/2018).
44. T. T. H. Tran, J. G. Avila-Acevedo and E. Tsotsas, *Drying Technol.*, 2016, **34**, 1185.
45. H. M. Torringa, E. J. V. Dijk, P. V. Bartels, Microwave puffing of vegetables: Modeling and measurements, in *Proceedings of 31st microwave power symposium*, Manassas, VA, 1996, pp. 16–19.
46. J. D. Gutierrez-Cano, I. E. Hamilton, J. M. Catala-Civera, J. Bows and F. L. Penaranda-Foix, *J. Food Eng.*, 2018, **232**, 21.
47. E. Y. Lee, K. I. Lim, J. Lim and S. Lim, *Cereal Chem.*, 2000, **77**, 769.
48. C. I. Moraru and J. L. Kokini, *Compr. Rev. Food Sci. Food Saf.*, 2003, **2**, 147.
49. H. Jiang, Z. Liu and S. Wang, *Crit. Rev. Food Sci. Nutr.*, 2018, **58**, 2476.
50. W. C. Wang and S. K. Sastry, *Innovative Food Sci. Emerging Technol.*, 2002, **3**, 371.
51. R. Kaur, K. Gul and A. K. Singh, *Cogent Food Agricult.*, 2016, **2**, 1159000.
52. K. Halden, A. A. De Alwis and P. J. Fryer, *Int. J. Food Sci. Technol.*, 1990, **25**, 9.
53. S. A. Kulshrestha and S. K. Sastry, *Innovative Food Sci. Emerging Technol.*, 2010, **11**, 78.
54. I. Praporscic, N. I. Lebovka, S. Ghnimi and E. Vorobiev, *Biosyst. Eng.*, 2006, **93**, 199.
55. J. Moreno, R. Simpson, D. Estrada, S. Lorenzen, D. Moraga and S. Almonacid, *Innovative Food Sci. Emerging Technol.*, 2011, **12**, 562.
56. T. Zhong and M. Lima, *Bioresour. Technol.*, 2003, **87**, 215.
57. N. I. Lebovka, M. V. Shynkaryk and E. Vorobiev, *Drying Technol.*, 2006, **24**, 601.
58. F. Icier, *J. Food Process Eng.*, 2010, **33**, 661.
59. W. C. Wang and S. K. Sastry, *J. Food Process Eng.*, 2000, **23**, 299.
60. M. Lima and S. J. Sastry, *J. Food Eng.*, 1999, **41**, 115.

61. S. Salengke and S. K. Sastry, *Drying Technol.*, 2005, **23**, 551.
62. A. Jansen, Pioneering with pulsed electric field. Available at http://foodfortext.nl/wp-content/uploads/2014/03/HFI01_2015_06_09_PEF.pdf (Accessed 20/5/19).
63. P. Arevalo, M. O. Ngadi, M. I. Bazhal and G. S. V. Raghavan, *Drying Technol.*, 2004, **22**, 1233.
64. T. K. Gachovska, A. A. Adedeji, M. Ngadi and G. S. V. Raghavan, *Drying Technol.*, 2008, **26**, 1244.
65. A. Wiktor, M. Iwaniuk, M. Sledz, M. Nowacka, T. Chudoba and D. Witrowa-Rajchert, *Drying Technol.*, 2013, **31**, 112.
66. Y. Won, S. C. Min and D. Lee, *Drying Technol.*, 2015, **33**, 926.
67. A. Wiktor, M. Nowacka, M. Dadan, K. Rybak, W. Lojkowski, T. Chudoba, *et al.*, *Drying Technol.*, 2016, **31**, 112.
68. E. Puertolas, O. Cregenzan, E. Luengo, I. Alvarez and J. Raso, *Food Chem.*, 2013, **136**, 1330.
69. A. Zderic and E. Zondervan, *Chem. Eng. Res. Des.*, 2016, **109**, 586.
70. S. El-Kantar, N. Boussetta, N. Lebovka, F. Foucart, H. N. Rajha, R. G. Maroun, N. Louka and E. Vorobiev, *Innovative Food Sci. Emerging Technol.*, 2018, **46**, 153.
71. J. Hong, R. Chen, X. A. Zeng and Z. Han, *Food Chem.*, 2016, **192**, 15.
72. J. Hong, X. A. Zeng, R. Buckow, Z. Han and M. S. Wang, *Food Hydrocolloids*, 2016, **54**, 139.
73. F. Zhu, *Trends Food Sci. Technol.*, 2018, **75**, 158.
74. Z. Han, M. Cai, J. Cheng and D. Sun, *Trends Food Sci. Technol.*, 2018, **75**, 1.
75. V. Dhineshkumar and D. Ramaswamy, *J. Applied Biotechnol. Bioeng.*, 2017, **3**, 399.

CHAPTER 17

Food Structure Development/ Production Through Flexible Processes: The Use of Electric Fields to Enable Food Manufacturing

RICARDO N. PEREIRA, ANTÓNIO A. VICENTE AND JOSÉ A. TEIXEIRA*

CEB – Centre of Biological Engineering, University of Minho – Campus de Gualtar, 4710-057, Braga, Portugal
*Email: jateixeira@deb.uminho.pt

17.1 Electric Field Processing: An Overview

The use of electric field in food processing is gaining increasing attention in the last few years. Thus a question arises as to why electrotechnologies are now on the front line of innovative food processing. First, it is important to understand the historical background and to point out that the principle of using electricity to process food is not novel as it dates back to when electricity was used to pasteurize milk through a process that becomes known as Electropure.[1] The Electropure process showed satisfactory reduction of microorganisms in milk but did not show enough technical and economical robustness for long-term production, impairing its commercial application. Since that time and into the early beginnings of the 21st century, this

Food Chemistry, Function and Analysis No. 18
Handbook of Food Structure Development
Edited by Fotis Spyropoulos, Aris Lazidis and Ian T. Norton

Published by the Royal Society of Chemistry, www.rsc.org

technology was somehow unnoticed from the applied point of view, but research and scientific evidence on its use continued to occur. The development of power supply technology and inert materials, as well as the reduced price of electricity brought the opportunity to revive the electricity-based processing technologies traduced by the increase of available equipment by manufacturers and relevant scientific studies. Presently, sister technologies such as pulsed electric fields (PEF) and ohmic heating (OH) are thriving in the food industry. Table 17.1 shows a list some PEF and OH equipment suppliers and examples of some of their applications.

These technologies present specific singularities but share a common aspect: to be effective in their application, a given food structure must behave as a semiconductor, allowing the passage of electrical current. When this condition is fulfilled, electric field processing brings competitive advantages such as high energy efficiency, nonthermal microbiological and enzymatic inactivation, disintegration of cellular structures (*i.e.* electroporation effects), and the possibility of instantaneous and direct volumetric heating. For example, one of the major advantages of OH is that almost all the electrical energy delivered to the food is dissipated into internal heat; OH efficiencies range from 90% to 95%, while in microwave heating is about 50%.[2] These features bring an environmentally friendly nature to electric processing since the use of water as well of fossil energy can be drastically reduced. For all these reasons, processing technologies based on electro-thermodynamic principles are now aligned with a holistic vision on food, giving rise to minimally processed, nutritionally adequate, safe, and healthy food products without overlooking environmental concerns.

The applications of electric fields in food processing are vast and include thermal and nonthermal unit operations such as pasteurization, sterilization, blanching, thawing, cell disintegration, or permeabilization, extraction, drying, and protein functionalization. Important information about these emerging technologies, regarding technical and procedural aspects, equipment requirements and layout, impact on food properties (regarding safety and quality), potential applications, and research needs is remarkably reported in recent scientific literature.[3–11] Electric field technologies were recently considered part of the restricted group of food processing technologies with great potential for future use, this interest mainly being driven by SMEs (small and medium-sized enterprises) in close collaboration with research centers that seek development of innovative niche products.[12] Despite this thorough scientific evaluation, more information is needed regarding these technologies, such as scale-up and economic viability (case-by-case analysis). Moreover, it is also essential to build on the fundamental knowledge concerning the interaction of electric fields with food macronutrients at a structural level. For example, regarding OH, an important research need has been pointed out, *i.e.* the differentiation between thermal and electrical effects and their influence on the physicochemical properties of foods. PEF technology is also struggling with constraints related to the standardization of processing protocols that would allow a better comprehension of the different factors that

Table 17.1 Manufacturers of PEF and OH equipment (not an exhaustive list).

Technology	Company	Country	Website	Applications (examples)
PEF	DIL/ELEA	Germany	www.elea-technology.com	Potato processing, dairy products, juices, and smoothies
	Energy Pulse Systems Portugal	Portugal	www.energypulsesystems.pt	Processing of grapes Apple juice extraction
	Pulsemaster	Netherlands	www.pulsemaster.us	Wide range of solid and liquid products (*e.g.* potato, fruit, meat, dairy and egg products)
	Arc Aroma Pure AB	Sweden	https://www.arcaroma.com	Olive oil (Olivecept®), wine processing (Winecept®)
OH	Emmepiemme srl	Italy	https://www.emmepiemme-srl.com/	Fruit puree (peach, pear, strawberry); fruit pieces (peach, pear, strawberry, pineapple, apricot); vegetable (cream) or whole or pieces of vegetables; egg products; dairy cream and cheese paste; algae; creamy and vegetable sauces; viscous product such as jams and syrups
	Yanagiya Machinery Co., Ltd.	Japan	http://ube-yanagiya.com/	Soymilk products
	Alfa-Laval	Netherlands	http://www.alfalaval.nl/	Fruits, vegetables, prepared foods, cheese, liquid eggs, ready-to-eat dishes, sauces, and juices
	C-Tech	United Kingdom	www.ctechinnovation.com	Pasteurization and sterilization (pressurized system) of dairy products, fruit juices and pieces, food products with high viscosity or with particulate inclusions such as soups, stews, and jams.

influence the outcomes of PEF treatments as well as their complex interactions with raw material properties.[13]

A better understanding of how electric fields interact at a structural level with food macronutrients such as lipids, polysaccharides, and proteins can bring novel processing electroheating protocols aimed at the development of innovative food products that can change food perception. The development of a novel food sensory experience is actually a main driving force of new product design pathways, together with tailor-made foods for niche markets and functional foods (through nano- and microencapsulation of nutraceuticals or bioactive molecules).[12] Electro-based technologies are now emerging in these domains. This chapter intends to give an overview of the most recent developments of electroheating technologies giving particular emphasis on the flexibility that the effects of electric fields can bring on food processing.

17.2 Flexible Processes: Thermal and Nonthermal Processing

The application of an external electric field to a given food material, although simple, can be done in a very flexible way by combining thermal and non-thermal processing for different purposes. Different unit operations involving heat or mass transfer, such as diffusion of solutes, pressing operations, as well as drying and freezing processes, can be enhanced through electro-technologies. Currently, several food processing technologies based on the application of electric fields are being exploited on the academic and commercial levels. In this sense, several common types of electrotechnologies are found in the literature such as ohmic heating, pulsed ohmic heating, moderate electric fields, pulsed electric fields, and high-voltage electrical discharge. For a better understanding of these technologies, it is important to adopt a clear definition of each and to distinguish them regarding their main properties and applications. Table 17.2 summarizes the main characteristics of electric-based technologies and their typical operational parameters.

17.2.1 High-pulsed Electric Technologies

High-voltage electrical discharge (HVED) and PEF are generally recognized as "nonthermal" technologies since they consist of an application of electric field pulses of high intensity (in the kilovolt region) for a very short time (in the range of micro- or nanoseconds), thus avoiding significant heat dissipation within the material.

17.2.1.1 High-voltage Electrical Discharge

In the case of HVED, an electrical discharge flows from an electrode with a high-voltage potential into a neutral fluid generating localized plasma channels, which in turn give rise to electrical phenomena such as corona (electrical discharges in gases, known as electrical wind) or electrohydraulic

Table 17.2 Electric field processing and its main features.[a] [13,17,18,24,56,57]

Designations	Electric field processing characteristics	Main effects
Pulsed electric fields	• $E_F = 20\ \text{kV cm}^{-1}$ to $50\ \text{kV cm}^{-1}$ (microbial inactivation) • $E_F = 0.5–1\ \text{kV cm}^{-1}$ (tissue softening) • Square or exponential wave (unipolar or bipolar) • $\tau = 4–20\ \mu s$ • $t = 1–2400\ \mu s$ • $f = 1–2000$ Hz • $T_{max} = 10–60\ °C$ (possible preheating or ohmic heating-induced effect) • E_{input} from 2 to 20 kJ kg^{-1} (plant and vegetable tissue softening)	• Electroporation • Nonthermal inactivation of microbial cells, • Extraction of thermolabile compounds • Disintegration and softening vegetable/fruit tissues
High-voltage electric discharges	• $E_F = 20–60\ \text{kV cm}^{-1}$ • $\tau = 2–300\ \mu s$ • $t = 1000–3000\ \mu s$ • $f = 0.5–1000$ Hz • E_{input} from 35 to 800 kJ kg^{-1} (tissue softening) • $T_{max} = 20\ °C$ (possible preheating)	• Disintegration and mechanical damage of vegetables and fruit tissues • Nonthermal extraction of biocompounds
Ohmic heating	• $E_F < 1000\ \text{V cm}^{-1}$ • Square or exponential wave • t (no theoretical upper limit) • $f = 0.06–25$ kHz • T_{max} (no theoretical upper limit)	• Internal and volumetric fast heating • Thermal and electrical inactivation of microorganisms • Thermal extraction of biocompounds
Pulsed ohmic heating	• $E_F < 1000\ \text{V cm}^{-1}$ • Bipolar pulses (delay > 15 μs) • t (no theoretical upper limit) • $f = 0.06–10$ kHz • T_{max} (no theoretical upper limit)	• Internal and volumetric fast heating • Thermal and electrical inactivation of microorganisms • Thermal extraction of biocompounds • Avoidance of electrochemical reactions at low frequencies
Moderate electric fields	• $E_F < 1000\ \text{V cm}^{-1}$ • Square or exponential wave • t (no theoretical upper limit) • $f = 0.06–25$ kHz • Performed at sublethal temperatures (<60 °C) or when thermal effects are simulated with a reference	• Nonthermal inactivation of some microorganisms (vegetative cells) • Extraction of thermolabile biocompounds • Controlling activity of some enzymes and bacteria (*e.g.* reducing lag phase of fermentation) • Effects of functional properties of globular proteins

[a]*Legend:* E_F, electric fields intensity; τ, pulse width; t, treatment time; f, frequency; T_{max}, maximum temperature; E_{input}, energy input.

discharge and arcing (when immersed in liquid).[14,15] Application of HVED results in shock waves, cavitation, UV-light emission, generation of radicals, which contribute to cellular inactivation and disintegration and thus the extraction of intracellular compounds. Industrial applicability of this technology is still scarce, largely in part because of the lack of standardization about the positive impacts in food properties and scale-up studies.[15] Other major drawbacks are related to operational problems – such as the need of batch mode operation and high pulse generators – and issues related to the disintegration of food particles by shock waves and nonspecific extraction, as well as the eventual contamination of the treated food by chemical products of electrolysis.[15–17]

17.2.1.2 Pulsed Electric Fields

In the case of PEF, the application of a pulsed electric field with a given geometry (*e.g.* monopolar or bipolar square, exponential decay, sinusoidal waves) occurs between two immersed parallel electrodes, and depending on the treatment conditions (*e.g.* pulse number and duration and electric field intensity and temperature), it determines an accumulation of electrical charges in the cell membrane, causing its reversible or irreversible breakdown due to the formation of pores – a phenomenon commonly known as electroporation.[8] Through the application of this electroporation effect, PEF has been finding success by assisting several unit operations in food processing, such as drying, osmotic dehydration, freeze-drying, freezing, thawing, and extraction.[13]

PEF assisted-processing brings several advantages when compared to conventional treatments allowing the following aspects:[18] (1) increased mass transfer; (2) improved extraction yield of bioactive compounds or molecules; (3) reduced severity of conventional extraction parameters regarding processing time, treatment temperature and excessive use of organic solvents, thus preserving the degradation of heat-sensitive compounds (*e.g.* flavors, vitamins, and proteins); (4) nonthermal microbial inactivation; and (5) reduction of environmental impact (*i.e.* possibility of reducing water consumption and fostering the use of electrical energy instead of fossil energy). Several commercial applications of this technology can be found in the processing of vegetables, and an outstanding example of this is the production of French fries. PEF can be used to replace the preheating process (60 °C, 30 min) before frying with capacity ranges up to 50 ton h^{-1}, improving aspects such as higher cut quality and product length.[19]

However, cold pasteurization by PEF is still a challenging topic not only because of the reduced effectiveness regarding inactivation of the more resistant forms of microorganisms (*i.e.* spores and enzymes) but also due to the lack of fundamental knowledge about its effect on the structural aspects of macronutrients, such as proteins, that can have a protective effect in complex food media.[20,21] Barba *et al.*[18] compiled several studies regarding the use of PEF on the inactivation of pathogenic and food spoilage microorganisms, indicating varying inactivation results and a wide range of processing conditions (*i.e.* electric field intensity, number of pulses, pulse width, treatment time and temperature). The standardization of PEF, particularly at industry scale conditions, regarding nonthermal inactivation, seems difficult to obtain due to the diversity of published literature, which refers to the use of different electrical protocols, the complexity of the composition media and the temperature range of processing. However, the synergistic interactions between PEF and thermal treatments (even at sublethal temperature) can increase significantly the levels of microbial inactivation.[21,22] This combination is even more important in respect to the inactivation of enzymes. Studies performed on the inactivation of food enzymes such as lipoxygenase, polyphenoloxidase, pectin-methylesterase, and peroxidase by PEF alone have shown poor efficacy of the technology, reinforcing the need for a combination with heat treatment and further refrigeration for a reduced enzyme activity during storage.[23] However, it is important to note that due to the lack of standardization in relation to the experimental apparatus and the processing parameters applied in PEF, the results of the different researchers are often heterogeneous. Other challenges to the industrial application of PEF are related to the investment cost and the negative impact on the textural properties of solid foods at elevated electric field intensity and processing time. For example, an industrial PEF system that combines a high field strength (up to 40 $kV\,cm^{-1}$) and an energy input of about ~100 $kJ\,kg^{-1}$ (similar to the ones observed in heat processing) can represent an investment of €2 million.[8] For these reasons, PEF as a technology is not yet at the state that can replace heat processing in food manufacturing since the latter can still guarantee a certain level of confidence necessary regarding food safety.

17.2.2 Ohmic Heating

OH appears as one of the electroprocesses with the most promising future regarding its industrial application since its main processing effect is thermal. In fact, OH represents the genesis of electrical processing since its first application on the electric pasteurization of milk in the 1800s. Since then, it had a long time of maturation through which sister technologies (such as PEF) were developed and in a wide number of applications in food technology, as well as in other scientific fields such as biotechnology and modern organic chemistry.[24] OH is based on the direct passage of electric current through a food material with a given electrical conductivity. Contrary to PEF, in the case of OH, the electric field is of low to moderate intensity (<1 $kV\,cm^{-1}$) and is applied in a continuous way (in a sinusoidal or square waveform) for a sufficient time scale (not restricted to seconds or limited in time) that will allow the increase of kinetic and vibration energy of charged molecules, thus resulting in the generation of volumetric internal heat between the equipment's electrodes in accordance with Joule's law. Depending on the electric power supplied and the composition of the media, this volumetric heat can be generated in a very fast and efficient way, once is not limited to heat transfer phenomena (*i.e.* conduction and convection). This allows the achievement of either high temperatures and short heating times (HTST principle) or ultra-high-temperature (UHT) regimes. Several pieces of scientific literature can be found that describe the characteristics of this technology.[5,25,26]

The most verified advantages of OH lie on the absence of hot surfaces and the ability of fast and uniform heating, which contributes to a better retention of food product quality when compared with traditional conventional heat exchangers, without compromising safety. Even when compared to other emergent electrothermal technologies such as microwaves, radiofrequency, and induction heating, OH can provide enhanced heating uniformity. Another advantage of OH, as a purely thermal process, in essence has no limitations in the inactivation of bacterial spores or enzymes contrary to nonthermal methods such as high-pressure or PEF.[25] OH finds several industrial applications regarding HTST processing of solid-liquid mixtures such as fruit and vegetable products (*e.g.* syrups, purees, and pulps), as well as protein-rich foods as in the case of dairy and egg products.[5] Some examples of fruit and vegetables processing companies claiming the use OH include Sicoly (France, https://www.sicoly.com) and Campbell (Canada, https://www.canadianpackaging.com/packpress/campbell-uses-carton-pack-for-soups-in-france/), respectively. Motivated by the reduction costs of power generators and electricity, the number of manufacturers of ohmic heaters has increased drastically in recent years offering several possibilities of applications and designs.

The effectiveness of OH will be always dependent on a comprehensive knowledge of process factors related to food material and equipment, such as (1) the electrical conductivity of multiphase food products, (2) the geometry and size of solids on solid-liquid mixtures, (3) the flow conditions (if a continuous process), (3) the distribution of the electric field, and consequent distribution of temperature (identification of cold and hot spots). This encompasses important optimization and validation procedures at laboratory scale, as well as important design considerations prior to scale-up. Currently the use of OH in combination with other techniques is being discussed following a hurdle strategy in microbial inactivation and food preservation. Some of these approaches include (1) the use of "in-pack" sterilization using conductive films or packages;[27] (2) PEF cooking of food materials through a combination of intensive PEF treatment and induced ohmic heating;[28,29] and (3) a combination of OH pretreatment with high-pressure thermal processing for the development of high-quality products and efficient design process.[30]

Undoubtedly, heat effects are still crucial to guarantee food safety during processing, which justifies the observed novel trends of combining OH with so-called nonthermal technologies such as PEF or high-pressure processing. To our understanding, OH is gaining visibility and is now being applied in innovative food applications (*e.g.* development of protein functional systems at the nano- and microscales), taking competitive advantage of the synergy between electrical and thermal effects in properties of food biocomponents. However, as for PEF, there still is a lack of systematic and comparable studies, particularly regarding nonthermal effects – once microbial inactivation can be enhanced due to nonthermal effects provided by the application of moderate electric fields (this is outlined in the following sections). It is then important to make a clear distinction between thermal and nonthermal inactivation effects, which sometimes is difficult to demonstrate.[5,25] In the recent past, several studies have proved the nonthermal effects of moderate electric fields on the inactivation of some microorganims and also enzymes. This potential benefit, which should be carefully evaluated through a case-by-case analysis, can bring an opportunity to underprocess food without compromising its safety. Nevertheless, it is important to highlight that, at the moment, the OH industrial applications are straightforward, since microbiological safety is ensured by utilizing the well-established thermal requirements through pasteurization/sterilization operational protocols.

17.2.2.1 Moderate Electric Fields

The research devoted to OH research allowed the scientific recognition that low- to moderate-intensity electric fields can *per se* interact with food and biological matrices in different ways, opening applications in other areas. In this sense, moderate electric fields (MEF) has emerged, putting the emphasis on the study of electrical field effects combined or not with thermal phenomena that occur during OH application,[24,30] MEF is defined as an electrical treatment with electric fields intensity less than or equal to 1000 $V\,cm^{-1}$.[30,31] This way, MEF is clearly distinct from PEF, since the electric field strength levels are considerably smaller and are normally applied through an arbitrary waveform (with variable frequency) not being restricted to pulses and reduced treatment times.

Some of the examples that illustrate this versatility are related to the following: the role of the MEF and OH in extraction procedures aiming at the valorization of bioresources such as agrofood waste and biomasses;[24] the use of electric field in downstream and upstream processing of microalgae, which are now emerging as an alternative food protein source;[32] the innovative combination with hydrodistillation processes – ohmic-assisted hydrodistillation (OAHD);[33] as well as applications related to organic synthesis.[34] More recently, the effects of electric field with varying electrical frequency have been reported as crucial parameters on the activation or deactivation of important food enzymes depending on the treatment temperature.[35–37] Castro *et al.* (2007)[38] suggested for the first time an electrical disturbance on specific enzymes, reporting enhanced inactivation of enzymes such as polyphenoloxidase and lipoxygenase, eventually due to interaction of their prosthetic groups with applied electric field. Samaranayake and Sastry (2016)[36,37] have shown that the activity of pectin methylesterase and α-amylase can be tuned when electric field and electrical frequency are properly combined; they have shown that frequencies below 60 Hz and electric fields of low intensity (<10 V cm^{-1}) can contribute to enhance enzyme electrophoretic motion, thus having an activating effect. MEF performed at 60 °C under an electric field intensity of 1 V cm^{-1} and electrical frequencies ranging from 1 to 60 Hz can increase α-amylase activity up to 41%, when compared with a control treatment at the same temperature but without electric field.[39] The use of MEF at sublethal thermal conditions can also represent an exciting future in food bioprocesses, once they can also be used as "activators" in living bacteria cells by accelerating fermentation processes.[40–42]

The study of the interaction between electric field processing and biomolecules has also been extended to important food macronutrients such as milk proteins. Research has previously shown the positive impact of OH on the preservation of the native structure of whey proteins that benefits from a less aggressive and fast thermal treatment due to direct and volumetric heating.[43] In particular, β-lactoglobulin and α-lactalbumin can better preserve their native structure (*i.e.* by 10% and 8%, respectively) when compared to conventional thermal treatments at temperatures of 85 °C.[44] Fundamental research on this topic has been conducted suggesting that globular proteins such as β-lg when subjected to electroheating suffer distinctive conformational disturbances that change their dynamic behavior during denaturation.[45] It has been shown that OH at denaturation temperature of 90 °C and neutral pH changes the distribution of protein secondary structural elements of β-lg, displaying higher content of β-sheet helix and random coil than a conventional thermal treatment performed under the same thermal conditions. These results somehow corroborate previous works showing the influence of OH on gelation properties of globular whey proteins and derived ingredients, such as whey protein isolate.[46,47] OH assembles potential to control the size and morphology of protein functional networks, at the nano- and microscale ranges, by combining electrical thermal processing with environmental factors – *i.e.* electrical conductivity, pH, ionic strength, among others.[48] OH is proven to reduce the size of protein aggregates during denaturation process, resulting in development of whey protein gels with a more brittle but uniform structure.[44] These events result from the applied electric field but are also potentiated by fast heating rates that can result in the development of protein structures with distinctive morphologies, resembling fibrillar aggregates.[49] Even in caseins, recognized by their highly stable behavior toward mild thermal processing, OH has introduced slight structural alterations when performed at 95 °C – *i.e.* reorientation of hydrophobic residues and redistributions of charges – which were sufficient to give rise to acidified gels with less water-holding capacity.[50] Electric field processing appears as a promising tool to design and assemble food proteins as novel biomaterials to be used in production of functional foods – *i.e.* development of protein hydrogels for drug delivery and acidified gels as textural enhancers.

Another emergent field of application is related to the development of edible coatings and films based on proteins (*e.g.* whey) and polysaccharides (*e.g.* starch and chitosan), where electric field effects (as mentioned for proteins) are being used to change the barrier, physical, and structural properties of the produced films for food packaging purposes.[51–54] However, the way in which electric fields interact at the molecular level with polysaccharides is still undisclosed and needs to be investigated.

17.2.2.2 Pulsed Ohmic Heating

Another common designation for OH is pulsed ohmic heating (POH), which is somehow ambiguous in the scientific literature. Some authors mention POH as a PEF-induced ohmic heating treatment, where PEF pulse conditions are designed to allow heat dissipation by increasing energy input.[29,55] But POH was first introduced as a novel technique by Samaranayake,[56] being referred as an ohmic heating technique that uses insulated gate bipolar transistor (IGBT) pulses. These IGBT pulses allowed the replacement of the conventional OH – *i.e.* sinusoidal waveform at low-frequency conditions (50–60 Hz) and relatively low electric fields – being those pulses very effective in minimizing undesirable electrochemical reactions during processing. Through this application, it is possible to explore the effects of electric field at a low-frequency range avoiding the chemical contamination of food.

17.3 Novel Perspectives Toward Food Structure

From an application point of view, the increase of electric-based niche applications at small scale can be expected to be taking advantage of electroporation and electroheating (ohmic heating) effects during preprocessing (*e.g.* disintegration and softening of vegetable and fruit materials), as well as the effects on unit operations involving heat exchange and mass transfer. Together with the impacts of these emergent processing techniques in living cells and biological systems, they also influence chemical, physical, and consequently structural properties of food systems. The role of these different biological and physical pathways and their corresponding impact on food structure development are illustrated in Figure 17.1. The interaction of electric process variables (*i.e.* electric field intensity, current density, frequency, waveform, Joule effect, among others) with food at the molecular, cellular, and tissue levels needs a comprehensive knowledge and needs to be thoroughly controlled. Electric field in a pulsed or continuous mode imposes changes at the molecular level that have important consequences for the nano-, micro-, and macroproperties of food. At the molecular level, the impact that electric fields may have on the allergenic or sensitizing potential of some food proteins is now under debate.[3,5,58] It is important to highlight that the allergenic potential of food products that have been processed through electric-based techniques should also be properly evaluated and not automatically presumed. The role of electrically induced changes

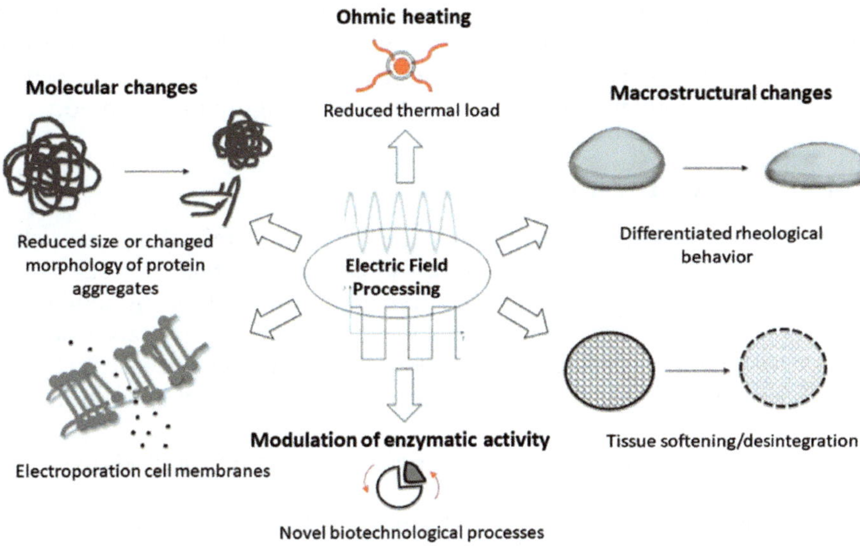

Figure 17.1 Pathways of food structure development through electric field processing.

(*i.e.* electrochemical reactions, electroporation phenomena, as well as Maillard reactions during electroheating) in the balance between reducing the existent allergenicity potential and triggering the formation of neoallergenic structures is unknown.[5,59] The effects of MEF can also be used to tailor the production of nanostructures, including changing their morphology, in a way to encapsulate bioactive compounds and their intended functional delivery. Whey proteins, in particular β-lg or β-lg-rich ingredients such as whey protein isolate, make for a good model to demonstrate that molecular disorders during electrothermal denaturation can give rise to gels with different rheological properties – from nano to macro effects.

Understanding which electrical parameters allow enhanced microbial inactivation or which rule enzyme activity will also bring benefits in terms of process efficacy and design while improving organoleptic properties of food. Regarding OH, the knowledge of electrical conductivity of food material and the identification of cold spots to a proper modelling and validation of a process is crucial. The influence of electrothermal processing in the death kinetics of microorganisms and enzymes should consider the complexity, composition, and structure of food materials. It would be interesting to adjust the thermal processes taking into account the effects of inactivation brought about by the electric field in order to guarantee the necessary microbiological safety and at the same time reduce to the minimum the thermal impact of a given process.

Taking into consideration these recent findings, the use of electric field processing at its different levels shows the potential to change the structural and functional aspects of important food molecules and foods themselves bringing an opportunity to redefine traditional processing and biotechnological processes such as fermentation and enzymatic hydrolysis. All these aspects can decisively contribute to change the color, flavour, and texture of processed food products, and this should be thoroughly investigated.

17.4 Research Needs

Electric field processing is definitely widening and establishing new challenges to the food industry. However, several transversal aspects for all electric-based technologies need to be investigated. Several recommendations are being outlined,[5,13,18,30] and to our knowledge many others can be added. Their achievement can be crucial for a better exploitation of these technologies at commercial level. These research needs can be listed as follows:

- standardization and appropriate report of research methodologies for better comparability of results
- improvement of the current state of knowledge about the interaction of electric fields with living cells – *e.g.* to identify critical factors that may contribute to reversible or irreversible (cell lysis) electroporation but that can also serve as a stimulus to enhance the activity of microorganisms
- evaluation of physically and chemically induced changes in food micronutrients (*e.g.* vitamins) and in macronutrients such as proteins and polysaccharides, along with the corresponding impact on the physicochemical, structural (*e.g.* texture), and nutritional properties of food products
- research on modelling, equipment components, and process design aiming at scaling up the electric field systems with competitive processing costs – *e.g.* low energy requirements, continuous operation, and easy implementation and maintenance
- evaluation of the impact of electric-based technologies on health and well-being by focusing aspects related to *e.g.* allergenic potential, digestibility (about bioavailability and bioaccessibility of food bioactive compounds) and cytotoxicity of food products.
- research on the possibility of complementing preservation – the hurdle concept – combining electric-based technologies with other conventional or novel technologies (*e.g.* high-pressure processing) seeking to extend food shelf-life.

17.5 Conclusions

In view of these considerations, it is clear that for a successful scale-up of these electric-based technologies in any given food process, research needs first to focus on fundamental science. Multidisciplinary teams, combining expertise from different science fields such as food and biological engineering, molecular biology, and biophysics, are necessary to understand the impact that novel food processing brings on dynamic properties of living cells and structural food molecules, such as proteins. But in order to offer an attractive alternative in an industrial setting, these technologies must also prove that they can be economically feasible providing competitive advantages. The clarification of many of these fundamental questions is pointless without the industry's active involvement. The establishment of partnerships between academia and industry is vital, once they can support a right balance between fundamental science and industry interests through the validation of case studies performed at high technology-readiness levels (relevant industrial environment at the pilot and semi-industrial scales). This will provide the necessary critical mass and technological conditions to support the development and resilient growth of electric fields processing in the food industry.

Acknowledgements

This study was supported by the Portuguese Foundation for Science and Technology (FCT) under the scope of the strategic funding of UID/BIO/04469/2019 unit and BioTecNorte operation (NORTE-01-0145-FEDER-000004) funded by the European Regional Development Fund under the scope of Norte2020–Programa Operacional Regional do Norte.

References

1. A. K. Anderson and R. Finkelstein, *J. Dairy Sci.*, 1919, **2**, 374–406.
2. S. K. Sastry, B. F. Heskitt, S. S. Sarang, R. Somavat and K. Ayotte, in *Ohmic Heating in Food Processing*, ed. H. S. Ramaswamy, M. Marcotte, S. Sastry and K. Abdelrahim, CRC Press, 2004, pp. 7–14.
3. L. P. Cappato, M. V. S. Ferreira, J. T. Guimaraes, J. B. Portela, A. L. R. Costa, M. Q. Freitas, R. L. Cunha, C. A. F. Oliveira, G. D. Mercali, L. D. F. Marzack and A. G. Cruz, *Trends Food Sci. Technol.*, 2017, **62**, 104–112.
4. A. Goullieux and J.-P. Pain, in *Emerging Technologies for Food Processing*, Academic Press, San Diego, 2nd edn, 2014, pp. 399–426.
5. H. Jaeger, A. Roth, S. Toepfl, T. Holzhauser, K. H. Engel, D. Knorr, R. F. Vogel, N. Bandick, S. Kulling, V. Heinz and P. Steinberg, *Trends Food Sci. Technol.*, 2016, **55**, 84–97.
6. M. C. Knirsch, C. Alves dos Santos, A. A. Martins de Oliveira Soares Vicente and T. C. V. Penna, *Trends Food Sci. Technol.*, 2010, **21**, 436–441.
7. J. G. Lyng, in *Ohmic Heating in Food Processing*, ed. H. S. Ramaswamy, M. Marcotte, S. Sastry and K. Abdelrahim, CRC Press, 2014, pp. 285–308.
8. S. Toepfl, V. Heinz and D. Knorr, in *Emerging Technologies for Food Processing*, ed. D.-W. Sun, Academic Press, London, 2005, pp. 69–97.
9. A. A. Vicente, R. N. Pereira, T. C. V. Penna and M. Knirsch, in *Ohmic Heating in Food Processing*, CRC Press, 2014, pp. 93–104.
10. E. Vorobiev and N. Lebovka, *Extraction from Foods and Biomaterials Enhanced by Pulsed Electric Energy*, Elsevier, 2016.

11. G. V. Barbosa-Canovas, U. R. Pothakamury, M. M. Gongora-Nieto and B. G. Swanson, *Preservation of Foods with Pulsed Electric Fields*, Elsevier Science, 1999.
12. H. De Vries, M. Mikolajczak, J.-M. Salmon, J. Abecassis, L. Chaunier, S. Guessasma, D. Lourdin, S. Belhabib, E. Leroy and G. Trystram, *Innovative Food Sci. Emerging Technol.*, 2018, **46**, 122–130.
13. J. Raso, W. Frey, G. Ferrari, G. Pataro, D. Knorr, J. Teissie and D. Miklavčič, *Innovative Food Sci. Emerging Technol.*, 2016, **37**, 312–321.
14. C. Delsart, N. Grimi, N. Boussetta, C. Miot Sertier, R. Ghidossi, M. Mietton Peuchot and E. Vorobiev, *Food Res. Int.*, 2015, **77**, 718–724.
15. M. Dalvi-Isfahan, N. Hamdami, A. Le-Bail and E. Xanthakis, *Food Res. Int.*, 2016, **89**, 48–62.
16. A. Jan, M. Sood, S. A. Sofi and T. Norzom, *Int. J. Food Sci. Nutr.*, 2017, **2**, 171–180.
17. N. Boussetta and E. Vorobiev, *C. R. Chim*, 2014, **17**, 197–203.
18. F. J. Barba, O. Parniakov, S. A. Pereira, A. Wiktor, N. Grimi, N. Boussetta, J. A. Saraiva, J. Raso, O. Martin-Belloso, D. Witrowa-Rajchert, N. Lebovka and E. Vorobiev, *Food Res. Int.*, 2015, **77**, 773–798.
19. S. Toepfl, *Stewart Postharvest Rev.*, 2012, **2**, 1–7.
20. O. Martín, B. L. Qin, F. J. Chang, G. V. Barbosa-Cánovas and B. G. Swanson, *J. Food Process Eng.*, 1997, **20**, 317–336.
21. P. Sharma, I. Oey, P. Bremer and D. W. Everett, *Int. Dairy J.*, 2014, **39**, 146–156.
22. M. Amiali, M. O. Ngadi, J. P. Smith and G. S. V. Raghavan, *J. Food Eng.*, 2007, **79**, 689–694.
23. A. Van Loey, B. Verachtert and M. Hendrickx, *Trends Food Sci. Technol.*, 2001, **12**, 94–102.
24. C. M. R. Rocha, Z. Genisheva, P. Ferreira-Santos, R. Rodrigues, A. A. Vicente, J. A. Teixeira and R. N. Pereira, *Bioresour. Technol.*, 2018, **254**, 325–339.
25. S. K. Sastry, in *Ohmic Heating in Food Processing*, ed. H. S. Ramaswamy, M. Marcotte, S. Sastry and K. Abdelrahim, CRC Press, 2014.
26. R. N. Pereira and A. A. Vicente, *Food Res. Int.*, 2010, **43**, 1936–1943.
27. S. Jun and S. Sastry, *J. Food Process Eng.*, 2005, **28**, 417–436.
28. J. Blahovec, P. Kouřím and M. Kindl, *Food Bioprocess Technol.*, 2015, **8**, 2027–2035.
29. J. Blahovec, E. Vorobiev and N. Lebovka, *Food Eng. Rev.*, 2017, **9**, 71–81.
30. S. Sastry, *Food Sci. Technol. Int.*, 2008, **14**, 419–422.
31. L. F. Machado, R. N. Pereira, R. C. Martins, J. A. Teixeira and A. A. Vicente, *J. Food Eng.*, 2010, **96**, 520–527.
32. P. Geada, R. Rodrigues, L. Loureiro, R. Pereira, B. Fernandes, J. A. Teixeira, V. Vasconcelos and A. A. Vicente, *Renewable Sustainable Energy Rev.*, 2018, **94**, 656–668.
33. M. Gavahian and A. Farahnaky, *Trends Food Sci. Technol.*, 2018, **72**, 153–161.

34. V. L. M. Silva, L. M. N. B. F. Santos and A. M. S. Silva, *Chem. Eur. J.*, 2017, **23**, 7853–7865.
35. B. Brochier, G. D. Mercali and L. D. F. Marczak, *LWT–Food Sci. Technol.*, 2016, **74**, 396–403.
36. C. P. Samaranayake and S. K. Sastry, *Food Chem.*, 2016, **199**, 265–272.
37. C. P. Samaranayake and S. K. Sastry, *J. Food Eng.*, 2016, **186**, 17–26.
38. I. Castro, B. Macedo, J. A. Teixeira and A. A. Vicente, *J. Food Sci.*, 2004, **69**, C696–C701.
39. C. P. Samaranayake and S. K. Sastry, *LWT*, 2018, **90**, 448–454.
40. H.-Y. Cho, A. E. Yousef and S. K. Sastry, *Biotechnol. Bioeng*, 1996, **49**, 334–340.
41. L. Loghavi, S. K. Sastry and A. E. Yousef, *Biotechnol. Prog.*, 2008, **24**, 148–153.
42. L. Loghavi, S. K. Sastry and A. E. Yousef, *Biotechnol. Prog.*, 2009, **25**, 85–94.
43. R. N. Pereira, J. A. Teixeira and A. A. Vicente, *J. Agric. Food Chem.*, 2011, **59**, 11589–11597.
44. R. M. Rodrigues, A. J. Martins, O. L. Ramos, F. X. Malcata, J. A. Teixeira, A. A. Vicente and R. N. Pereira, *Food Hydrocolloids*, 2015, **43**, 329–339.
45. R. M. Rodrigues, A. A. Vicente, S. B. Petersen and R. N. Pereira, *Innovative Food Sci. Emerging Technol.*, 2019, **52**, 1–7.
46. G. de Figueiredo Furtado, R. N. C. Pereira, A. A. Vicente and R. L. Cunha, *Food Res. Int.*, 2018, **103**, 371–379.
47. R. N. Pereira, R. M. Rodrigues, E. Altinok, O. L. Ramos, F. Xavier Malcata, P. Maresca, G. Ferrari, J. A. Teixeira and A. A. Vicente, *Food Res. Int.*, 2017, **99**, 435–443.
48. O. L. Ramos, R. N. Pereira, A. Martins, R. Rodrigues, C. Fucinos, J. A. Teixeira, L. Pastrana, F. X. Malcata and A. A. Vicente, *Crit. Rev. Food Sci. Nutr.*, 2017, **57**, 1377–1393.
49. R. N. Pereira, R. M. Rodrigues, Ó. L. Ramos, F. Xavier Malcata, J. A. Teixeira and A. A. Vicente, *Food Bioprocess Technol.*, 2016, **9**, 576–587.
50. T. C. P. Moreira, R. N. Pereira, A. A. Vicente and R. L. da Cunha, *Food Res. Int.*, 2019, **116**, 628–636.
51. C. C. S. Coelho, M. A. Cerqueira, R. N. Pereira, L. M. Pastrana, O. Freitas-Silva, A. A. Vicente, L. M. C. Cabral and J. A. Teixeira, *Carbohydr. Polym.*, 2017, **174**, 1181–1191.
52. R. N. Pereira, B. W. S. Souza, M. A. Cerqueira, J. A. Teixeira and A. A. Vicente, *Biomacromolecules*, 2010, **11**, 2912–2918.
53. B. W. S. Souza, M. A. Cerqueira, A. Casariego, A. M. P. Lima, J. A. Teixeira and A. A. Vicente, *Food Hydrocolloids*, 2009, **23**, 2110–2115.
54. B. W. S. Souza, M. A. Cerqueira, J. T. Martins, A. Casariego, J. A. Teixeira and A. A. Vicente, *Food Hydrocolloids*, 2010, **24**, 330–335.
55. N. El Darra, N. Grimi, E. Vorobiev, N. Louka and R. Maroun, *Food Bioprocess Technol.*, 2013, **6**, 1281–1289.

56. C. P. Samaranayake, S. K. Sastry and H. Zhang, *J. Food Sci.*, 2005, **70**, e460–e465.
57. E. Vorobiev and N. Lebovka, in *Innovative Food Processing Technologies*, ed. K. Knoerzer, P. Juliano and G. Smithers, Woodhead Publishing, 2016, pp. 31–56.
58. R. N. Pereira, J. A. Teixeira, A. A. Vicente, L. P. Cappato, M. V. da Silva Ferreira, R. da Silva Rocha and A. G. da Cruz, *Curr. Opin. Food Sci.*, 2018, **22**, 95–101.
59. R. N. Pereira, R. M. Rodrigues, Ó. L. Ramos, A. C. Pinheiro, J. T. Martins, J. A. Teixeira and A. A. Vicente, *J. Agric. Food Chem.*, 2018, **66**, 11227–11233.

CHAPTER 18

Development of Healthy Food Structures: Reduction of Sugar, Salt, and Fat

BETTINA WOLF

University of Birmingham, UK
Email: B.Wolf@bham.ac.uk

18.1 Introduction

The development of the physical structure of a food is a key part of the manufacturing process as it plays a vital role in the final product quality. It is closely related to the perception of a food and how it breaks down during the digestion process, delivering the nutritional load to the human body. The product may be designed for a certain flavour and texture experience or for the targeted release of a bioactive during the passage of the gastrointestinal tract. In recent years, the reduction of the health-sensitive nutrients sugar, salt, and fat has become an additional focus of new product design. Consuming too much sugar can lead to weight gain and related health problems, as well as tooth decay.[1] Salt is one of the most common sources of sodium in the diet, and excessive intake of sodium has been linked to the development of hypertension. This subsequently increases the risk of developing cardiovascular and renal diseases.[2] A diet high in fat is associated with the development of noncommunicable cardiometabolic disease although the fat, or rather the fatty acid distribution of the triacyly glycerides are to be considered. Unsaturated fatty acids are linked with health benefits.[3] One of the biggest challenges for existing product development is to deliver to the consumer an

Food Chemistry, Function and Analysis No. 18
Handbook of Food Structure Development
Edited by Fotis Spyropoulos, Aris Lazidis and Ian T. Norton

Published by the Royal Society of Chemistry, www.rsc.org

uncompromised experience of eating the product. Maintaining the physical structure of a food formulated with healthier ingredients or a physical structure that follows the same breakdown pathway from intake to swallow is the holy grail of microstructure-based product reformulation. Any associated changes with regard to flavour experience then often can be taken care of by the addition of small-molecular-weight ingredients that will not impact on the food structure.

The principle strategies for sugar, salt, and fat reduction in foods are summarised in Figure 18.1. The practical application of some of these strategies can be very similar for either of these three health-sensitive nutrients, although the impact on food texture and other physical properties regulated by the presence (or absence) of the specific nutrient may be very different. Gradual reduction, also known as health by stealth, relies on accustoming the consumer to a less sweet, salty, or fatty/-rich product and the consequences the reduction has on texture and flavour perception.[4] As with all reduction strategies, there is a limit to the level of the possible reduction, which depends on the nutrient and the overall impact it has on the food structure. The replacement approach can also be applied to sugar, salt, and fat. However, to successfully replace either of these nutrients, their roles in the food product need to be understood and replicated as closely as possible. Potential and practically applied replacers are mentioned in the subsequent sections covering each of these nutrients individually. The exploitation of sensory interactions for the formulation of healthier foods without compromising flavour and mouthfeel experience is by far the least generic approach and requires deep insights into the product-specific perceptions.

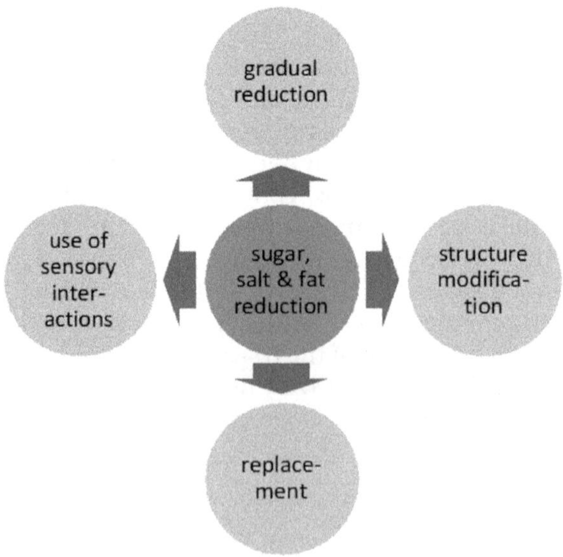

Figure 18.1 Principle strategies for sugar, salt, and fat reduction in foods.

Structure modification includes a variety of approaches that have been demonstrated for model systems, such as cubes of gels or commercial products, for example chocolate confectionery or crisps. This chapter focuses on microstructure approaches, covering the colloidal to small particle length scale. An example for a structure approach at larger length scale is the inhomogeneous distribution of salt in bread. This can be achieved by preparing a high and a low salt content dough and placing these as alternating layers into the bread pan.[5] This idea has also been applied to fat[6] and sugar.[7] The change in the spatial distribution of a sensory stimulus is noticed by consumers and triggers enhanced perception of this stimulus. This means that the overall amount of the compound in a product that triggers this stimulus, such as sugar, salt, and fat, can be reduced without compromising consumer perception. The effect of this approach has been also demonstrated for liquids by pumping solutions of varying salt concentration directly into the mouth of subjects and asking them to continuously rate the taste intensity,[8] However, the way that this type of experiment is designed could have affected the findings of the study itself, as others have not been able to conclude that temporal profiles of salt delivery had an impact on the overall saltiness perception.[9] The taste enhancement approach has recently also been applied to emulsion systems as discussed in later parts under the w/o/w emulsions heading.

This chapter is organised as follows. Initially, reduction approaches specific to sugar, salt, and fat are introduced, omitting microstructure approaches. The latter are then introduced in the penultimate main section before concluding and discussing future perspectives. While the subject is beyond the scope of this chapter, one cannot overlook that salt and sugar in particular often fulfil the added role of a preservative by creating a low-water-activity environment. Microbial stability is a factor also to consider in fat reduction should it be achieved through aqueous-based reformulation.

18.2 Sugar

Sugar as a food ingredient includes different forms while consumers often think of sugar as the sucrose they use at home as a sweetener, a terminology used here unless otherwise stated. As a food ingredient, sugar assumes many roles, including the provision of bulk, density, viscosity, colour, and sweet taste. It enhances flavour, acts as humectant, thus controlling texture but also water activity, contributing to preservation, and performs as an anticoagulant. This multifaceted role of sugar is an indication of the challenges in (re)formulating foods with lowered levels of sugar or even foods without sugar.

In recent years, the functionality of sugar[10] and sugar reduction strategies in foods have been covered in a number of reviews and book chapters, for example, as applied to dairy food[11] or bakery, confectionery, and gel-based products,[12] and from how industry can help to achieve government reduction targets[13] to technological challenges and health consequences.[14] Applying replacement strategy to sugar reduction, the sweetness functionality is typically replaced by a high-intensity sweetener. These tend to be of

Table 18.1 Sugar replacement in different product categories, modified from literature.[12,16]

Product category	Sweeteners	Bulking agents
Biscuits and cakes	Erythritol, isomalt, maltitol, mannitol, rebaudioside A, sorbitol, stevioside, sucralose, xylitol	Cocoa fibre, inulin, maltodextrin, polydextrose
Chocolate	Neotame, rebaudioside, sap-based sugar, stevia, sucralose, thaumatin	Inulin, polydextrose
Ice cream	Erythritol, fructose, honey, trehalose, xylitol	Inulin

low molecular weight, and therefore higher-molecular-weight bulking agents are combined with these to try to mimic the food structure and thus the texture experience of the consumer. The use of high-molecular-weight bulk sweeteners is also a possibility, although their relative degree of sweetness is much lower than that of sugar and high-intensity sweeteners. A comprehensive overview of bulk sweeteners and sugar from production to physicochemical and sensory properties has been carried out by Kruger.[15] Table 18.1 provides an overview of the sugar replacers that have been reported in connection with the three product categories discussed in more depth in this chapter: biscuits (cookies) and cakes, chocolate, and ice cream.

18.2.1 Biscuits and Cakes

While many different types of biscuit and cake formulations exist, the role sugar plays in the final product structure remains similar as the same main structuring components are present, albeit at different ratios as shown in Figure 18.2. The structure roles of sugar may be direct, such as retaining a soft and moist texture in the final cake, muffin, or brownie product, or indirect by affecting the functionality of other major structuring ingredients during the manufacture process. The latter is mainly due to the sugar taking up water that is then not available to the other ingredients to develop undesired functionality. It has been recognised that the replacement of sugar in biscuits and cakes presents by enlarge a structure challenge. Taste and colour on the other hand can be adjusted by the addition of an intensive sweetener and the amount of reducing sugar in the recipe.[17]

18.2.1.1 Biscuits

The structure of a biscuit develops in stages beginning with the dough mixing process until it is set during the baking process, as illustrated in Figure 18.3.[16] Structure development goes hand in hand with a number of phase transitions, and it is argued[16] that if these phase transitions are followed by sugar replacers, it should be possible to formulate a structurally equivalent sugar-reduced biscuit. Taste and colour then can be dealt with by the addition of an intensive sweetener and the amount of reducing sugar,

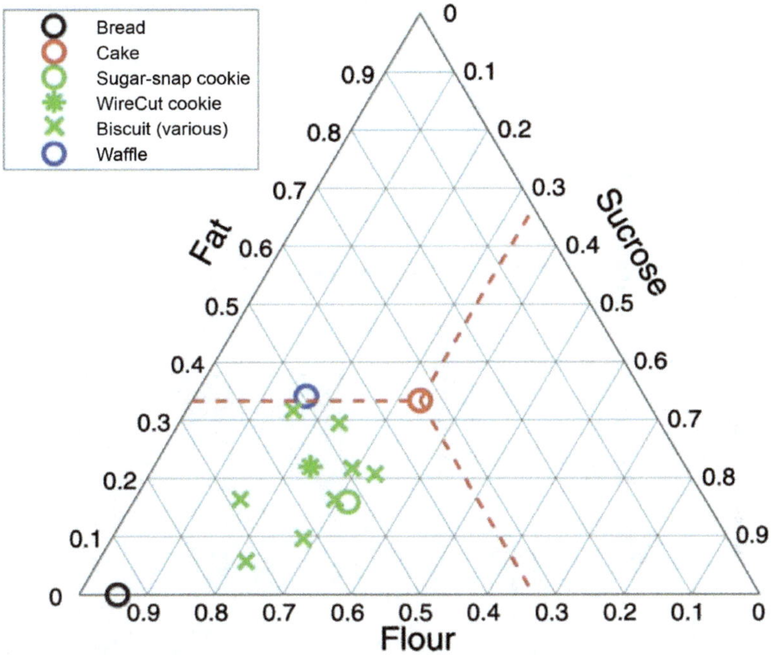

Figure 18.2 Flour : fat : sucrose ratios of common bakery products. Reproduced from ref. 16 with permission from Taylor & Francis Group, LLC, Copyright 2018 The Authors.

respectively.[17] The functionality of sugar, except for the early stages of the mixing process, comes from its dissolved state acting as a plasticizer/co-solvent, moderating the viscosity and viscoelastic behaviour of the dough and phase transitions of the biopolymers and water.[16]

Replacing sugar in a biscuit formulation therefore requires a material that can assume the role of co-solvent/plasticiser and of adding sweetness. This is in practice achieved with various levels of success by formulating with a combination of ingredients. To replicate the structure of a biscuit, bulking agents providing little or no sweetness are applied, such as polyols, oligosaccharides, and dextrins.[16] Sweetness levels then are regulated by the addition of intensive sweeteners, such as saccharin, cyclamate, aspartame, to name only a few.[17] Levels of sugar reduction that can be achieved are limited by the product structure. It is clear that the focus of achieving further sugar reduction based on current biscuit manufacturing technology should be on developing an in-depth understanding of the interactions between sucrose and the other ingredients of a biscuit formulation, in particular the flour.

18.2.1.2 Cakes

Differently from biscuits, cake recipes also contain egg, imparting the characteristic moistness and springiness in the final product. The functionality of sucrose in cake is largely similar to that in biscuits except for the

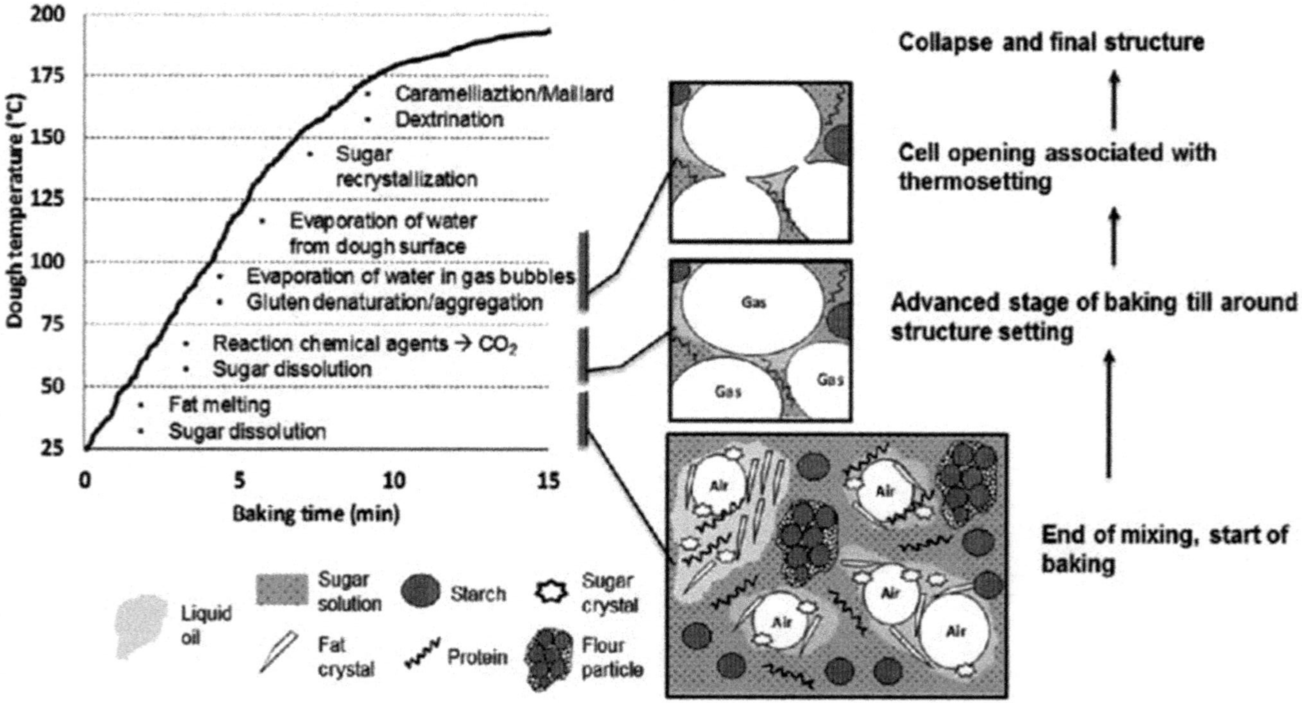

Figure 18.3 Structure of a biscuit developing during baking.
Reproduced from ref. 16 with permission from Taylor & Francis Group, LLC, Copyright 2018 The Authors.

additional interplay with the egg protein, delaying its denaturation, although to a lesser extent than starch gelatinisation, controlling oven rise, structure fixation, and collapse.[18] Adjusting the moisture and sugar levels in the formulation enables the simultaneous occurrence of starch gelatinisation and egg protein for optimal cake volume.[19] However, an increase in the gelatinisation temperature of starch from 57 °C to 92 °C has been reported for high sugar-to-water ratios, up to 1 : 1, resulting in high volume and uniform fine crumb structure, leading to collapse following baking.[20]

18.2.2 Chocolate

The sucrose level in chocolate is around 40–50%, depending on type. This high sugar fraction influences the material properties during processing, as well as the sensorial texture and flavour properties of the final chocolate product. In the final product, it is present as crystalline sugar, milled to a particle size of less than around 30 μm to avoid a gritty mouthfeel. The hydrophilic nature of sugar demands the addition of surfactants to chocolate to prevent aggregation of the sugar particles in the continuous lipophilic cocoa butter phase and to control the unfavourable rheological properties due to the immobilisation of cocoa butter, which is then not available to contribute to the chocolate flow. The sugar particles affect the material properties of chocolate not only directly but also indirectly by providing nucleation sites for cocoa butter crystallisation.[21] A very dense sugar phase, on the other hand, may retard fat crystallisation due to reduced triacyglycerol mobility, confinement, and increased tortuosity.[21] Post initial bite and accelerated by the mechanical breakdown of the ingested chocolate piece by chewing, a large fraction of the sugar will dissolve during the oral phase of consumption, thereby releasing its sweet taste and contributing to the overall flavour experience. It is also possible that the increased viscosity of the dissolved sugar fraction, mixed with saliva, contributes to the oral texture perceived by the consumer.

Sugar replacement in chocolate has recently been reviewed,[12] and the high intensity and bulk sweeteners that have been applied in research or commercially available chocolate have been included in Table 18.1. A recently reported structure approach to sugar reduction in chocolate is discussed in Section 18.5.

18.2.3 Ice Cream

A typical ice cream formulation contains 12–16% sugar, not including lactose. While traditionally sucrose would have been the major type of sugar used in ice cream, nowadays a variety of sugars are applied in a view to create the desired flavour and mouthfeel experience. Sugar not only imparts sweetness but also interacts with the volatile aroma components, thus contributing to the overall flavour of the ice cream.[10] Ice cream softness is affected by the sugar component of the formulation as the level of sugar controls the amount of ice. The less sugar that is present in the formulation, the higher is the ice content and thus the harder the product. The ice content may also be affected by the type of sugar in the formulation. A lower-molecular-weight sugar, such as dextrose, will decrease the freezing point further than sucrose at the same concentration, and thus the ice content in the final product. The molecular weight of the sugar also affects the viscosity of the unfrozen ice cream matrix and thereby the mouthfeel perception that is often described as creaminess. A higher-molecular-weight sugar imparts a higher matrix viscosity that has been linked to creamier texture and warmer eating products, but due to the impact on ice content, they are more difficult to scoop.[22]

Based on a recent review paper on sugar reduction in foods with a section on ice cream,[12] it is obvious that sweeteners can be used successfully as partial sugar replacers. Complete substitution would lead to significant modifications of the physical properties of ice cream, beyond consumer acceptance.

18.3 Salt

Salt as food ingredient is synonymous with sodium chloride. Its salty taste, as well as the detrimental effect to human health, is attributed to the sodium ion. In addition to imparting salty taste, the functionality of salt includes enhancing flavour and aiding processing and preservation. Process functionality encompasses making processing easier as well as promoting interactions with other ingredients for structure development; for example, salt assists gluten development in bread making and enhances water holding capacity in meat products.[23] Over the last decade, salt reduction strategies for foods have been the subject of many reviews and book chapters.[24]

The replacement strategy for salt reduction includes substitution by non-sodium mineral salts or salty compounds. Most successfully applied is potassium chloride, although product-specific critical concentration thresholds exist above which a bitter taste is perceived, although the addition of bitter blockers may overcome this limitation.[24d] This particular approach of replacement with potassium chloride has recently been reviewed including a useful focus on the patent literature.[25] Glutamates, compounds with a strong umami taste, are also frequently applied with the aim to reduce salt. Another option includes flavour enhancers to mask the less salty taste of the product.[24d] A comprehensive overview of commercially available products to improve the palatability of salt-reduced products has recently been published.[26] As noted, salt also fulfils a number of structure functionality roles, and the largest product categories concerned include cheese, processed meat, and baked products (bread). In the case of cheese and processed meat, salt is also critical for preservation through water activity control.

18.3.1 Cheese

Salt plays a number of roles during cheese manufacture.[23,27] It has been suggested that reducing salt levels would result in a very different product, and the best approach would be to select cheeses that are already only lightly salted and modify their manufacturing process.[23] Salt regulates a number of steps during the conversion of milk to cheese, including the activity of the starter culture microorganism, enzyme activity, and water content. Salt levels and points of addition during the manufacturing process depend on the type of cheese. In the final product, salt also assumes a preservative role. Processed cheese manufacture additionally requires the presence of emulsifying salts, and these are traditionally sodium salts. Surveying recent reviews on the functionality and reduction of salt in cheese[27] and processed cheese[28] clearly paints a picture of partial substitution, for example with monosodium glutamate, hydrolysed vegetable protein, or potassium chloride,[27] being the most promising option, allowing for partial salt reduction without too significant changes in cheese quality. While a structural approach by means of commercially available hollow salt particles is mentioned by Bae and co-authors,[27] (see Section 18.5) for detail, no specific reference to the application in cheese has been made until now.

18.3.2 Processed Meat

Salt plays a vital role in the manufacture and product quality of processed meat. In addition to promoting the solubilisation of myofibrillar proteins, which is essential to obtain a cohesive processed meat product and thus its desired texture, it acts as preservative, tastant, and flavour enhancer. In a review on salt reduction in processed meat,[26] it is stated that the salt replacement strategy with lower-sodium ingredient options can be successfully applied provided that the salt-soluble meat proteins are replaced by soy or milk proteins, gums, or starches delivering product cohesiveness. One example of the challenge of formulating a salt-reduced meat product with acceptable texture, taste, and flavour has been provided by Gaudette and Pietrasik.[29] They applied a small number of commercially available salt replacers to smoked turkey sausages and restructured ham only to conclude that texture, taste, and flavour are impacted differently in either product by the same salt replacer and that the application of salt replacers may be more successful in more complex formulations as negative taste and flavour associations may be suppressed. This, however, will not resolve the challenge of maintaining an acceptable texture.

18.3.3 Baked Products (Bread)

Salt is added into formulations for baked products to ensure the correct texture, taste, and flavour of the finished product. It impacts on the crust colour, controls the rate of yeast fermentation and enzyme activity, in addition to developing a cohesive non-sticky dough for gluten-containing flours, such as wheat flour, as the most commonly used flour in baking. The gluten network, comprised of the two proteins gliadins and glutenin, gives the dough its characteristic rheological properties and cohesiveness and contributes to water absorption. Gliadin contributes to the extensibility, whereas glutenin relates to dough elasticity and cohesiveness. In the absence of salt, water–protein interactions are favoured over protein–protein interactions; thus the gluten network cannot develop, and a non-cohesive sticky dough is produced. The authors of a recent literature survey on the role of salt in bread making conclude that particle replacement of NaCl with KCl at a level of ≈25–50% would be acceptable in terms of the dough-handling properties,[30] although the taste may still be challenged (bitter/metallic). Magnesium chloride and calcium chloride are less successfully employed, causing unfavourable dough rheology. This is argued to be due their position in the Hofmeister series, which ranks ions based on their ability to cause protein aggregation or disassociation where Na^+ and K^+ are equivalent.[31] It is concluded that, instead of alternative salts, despite the lack of studies in low-sodium environments, enzymes with the ability to alter dough properties might overcome the problem of creating sticky doughs.[30]

18.4 Fat

Fat as part of our diet encompasses triacyl glyceride (TAG)–based lipid systems that may be crystalline, liquid, or a mixture thereof at the point of consumption. The degree of crystallinity or the solid fat content depends on the chemistry of the fatty acids esterified onto the glycerol backbone, their distribution within the TAG molecule, and then the variety of TAGs making up the lipid system. Strictly speaking, the term "fat" would relate to a (at least partially) crystalline system, whereas a non-crystalline lipids system would be referred to simply as oil.

Fat has the highest energy density among the nutrients that structure foods. Dietary fat recommendations surround reduction of total fat, in light of food energy intake, and saturated fat intake; the reader is referred to a recent review in light of communicating nutrient information to consumers.[32] Fats and oils contribute in a variety of ways to structuring foods, and fat replacement is all about the structure functionality and maintaining the mouthfeel or texture properties of the full-fat counterpart. The two most common alternative approaches include the introduction of a water- or air-based phase and substitution with an oleogel. The types of products that tend to be considered in fat reduction strategies are fat-based spreads, chocolate, water-in-oil emulsion-based products such as sauces and dressings but also ice cream, meat and baked products such as puff pastry. Table 18.2 provides an overview of fat replacers, while microstructure-based reduction approaches for sugar, salt, and fat are discussed in the following section.

18.5 Microstructure Approaches to Sugar, Salt, and Fat Reduction

18.5.1 Particles with Designed Morphology

Designed-particle–based salt reduction approaches include salt crystals of designed size and shape and hollow particles.

Table 18.2 Fat replacement in different product categories.

Product category	Fat replacer	Examples
Baked[33]	Complex carbohydrates	Inulin, maltodextrin, plant fibres
	Gums and gels	Pectins, oleogels, whey protein
	Whole foods or partial food matrices	Fruits (banana for high starch content, avocado for oil content), vegetables, legume (for high starch and protein contents) or cereal-based ingredients
	Combinations of the preceding	
Comminuted meat[34]	Carbohydrates	Pectin, carrageenan, inulin, citrus fibre
	Proteins	Soya protein, whey protein isolates (WPI)
	Combinations of the preceding	Preformed WPI and carrageenan gel with tapioca starch
Ice cream	Carbohydrates	Modified sweet potato starch,[35] guar gum and basil gum seed,[36] okra gum,[37] pectin[38]
	Mixed protein/ polysaccharide systems	Soya protein isolate/xanthan gum[39]

The process of fractionating ground salt into differently sized fractions and applying these to fried, sliced potato crisps has revealed highest saltiness perception from the smallest size fraction (<106 μm) and has been correlated to enhanced dissolution kinetics and thus the delivery rate of the sodium to the taste receptors.[40] Following the dissolution of commercial salts in artificial saliva with video-microscopy revealed that some salts fragmented during dissolution into pyramid-shaped crystals, whereas others remained one entity until complete dissolution.[41] The larger surface area of fragmented crystals led to faster dissolution and would thereby enhance saltiness perception, based on the observation reported for saltiness perception from different size fractions.[40] Morphology of salt varies across commercially available salts and enhanced saltiness perception were linked to highly agglomerated, flat, or pyramidal crystals, with low circularity.[41] This observation suggests that opportunity for salt reduction lies in the targeted design of specific morphologies of salt crystals and has triggered some interesting research activity, such as the crystallisation of salt in single microdroplets, creating dendric or a single cubic crystal depending on conditions,[42] and a fundamental study into enhancing the knowledge of nucleation and growth of sodium chloride crystals into morphologies with predicted faster dissolution.[43]

In the case of sugar, amorphous porous sugar particles to replace solid sugar crystals for the delivery of structure and sweetness in commercial confectionery products have been introduced in the patent literature.[44]

18.5.2 Bubbles as Tastant-excluding Fillers

Bubbles in water-based gels act as inert fillers concentrating the amount of added tastant in the continuous gel phase. Without reducing the overall amount of sugar or salt, an increase in sweetness or saltiness has been reported.[45] Hence, a reduction in added sugar or salt is possible for retaining the original sweetness and saltiness of the product. Developing this approach further by considering the presence of a congruent aroma volatile, the possibility of achieving an 80% reduction in total sodium with no loss of saltiness perception was reported to be possible.[46]

18.5.3 Droplets

Various types of emulsion design have been shown or suggested to be applicable to sugar, salt, and fat reduction. These include oil-in-water (o/w), water-in-oil (w/o), and water-in-oil-in-water (w/o/w) emulsions to lower the content of either of the three health-sensitive nutrients discussed in this chapter in aqueous-continuous emulsion-based foods, and water-in-water (w/w) emulsions, as well as so-termed air-filled emulsions (AFEs) specifically for fat reduction.

18.5.3.1 o/w Emulsions

Incorporating oil droplets into an aqueous phase continuous product is analogous to the bubbles-as-tastant excluding filler principle. The drawback of introducing oil droplets into a food system to reduce sugar or salt is the associated increase in the fat content. A partial solution to this dilemma is the introduction of water-in-oil-in-water (w/o/w) emulsions, *i.e.* replacing a proportion of the oil inside the droplets with water.[47] These emulsions can also be designed as oral tastant release systems as will be discussed.

18.5.3.2 w/o Emulsions

For fat reduction purposes, w/o emulsion is obviously a more suitable emulsion microstructure to adopt. However, these systems have received less attention in literature than their o/w counterparts, partly due to the much more limited toolbox at our disposal in terms of interfacial stabilisation. More recently, Pickering (particle-based) stabilisation approaches have emerged as a means to design interfacial stabilisation in emulsions, originally discovered in the early 1900s.[48] Beginning in the early 2000s, it has attracted much interest for the stabilisation of interfaces in food emulsions as alternative to surfactants, hydrophobically modified polysaccharides, and proteins. Designer Pickering particles offer enhanced microstructure stability and flexibility in formulation over, for example, proteins with limited tolerance to salt and pH conditions. The enhanced stability is attributed to the practically irreversible adsorption of the particles at the interface.[49]

Pickering stabilisation of w/o emulsions has been demonstrated on the basis of fat crystals, solid lipid particles, and lignin-containing particles. The fat crystal approach tends to be realised *in situ* comparable to, or by utilising processing equipment encountered in, the manufacture of margarine, specifically a surface-scraped heat exchanger and a pin stirrer.[50] The processing conditions determine the physical properties of the fat crystals that stabilise the interface; for cocoa butter as the bulk phase, the presence of the w/o interface in the system has been demonstrated to affect the cocoa butter polymorphism.[51] Solid lipid particles are particles that are typically created *via* an oil-in-water emulsification route, and therefore, while more suited to be applied to aqueous continuous emulsions, they could be also transferred into a fat continuous phase. The authors of a recent review on Pickering emulsions for foods[49] concluded that fat particles were outstanding stabilisers for w/o emulsions. Lignin-containing particles for food emulsion stabilisation, as noted on the back of research into understanding the Pickering functionality of cocoa particles,[52] would readily stabilise oil-in-water emulsion interfaces. Hydrothermal treatment of another candidate lignin-rich material, ground coffee waste, has then been demonstrated to enable the stabilisation of w/o emulsion interfaces.[52b]

18.5.3.3 w/o/w Emulsions

Emulsions characterised as w/o/w are aqueous continuous o/w emulsions with an internal aqueous phase emulsified into the oil phase. The fat reduction functionally of w/o/w emulsions is through the replacement of a fraction of the oil with water. While these systems are increasingly discussed in the literature, their application in commercial food application is scarce. This is most likely due to the challenge of formulating stable w/o/w emulsions, involving not only at least one additional ingredient but also an additional processing step. One of the biggest challenges is maintaining the process-induced microstructure, *i.e.* the locking of the dispersed aqueous phase into the oil droplets. If the water is mobile and an osmotic pressure difference exists between the two aqueous phases of w/o/w emulsions, the oil droplets frequently lose their load into the external aqueous phase. Conversely, the swelling of the oil droplets may be observed due to diffusion from the external to the internal aqueous phase. These emulsions may be produced in a one-step process,[53] but their final microstructure tends to be more controlled by first processing the internal w/o emulsion and then incorporating this as the oil phase into the aqueous continuous emulsion phase as a second processing step. The additional ingredient is required to stabilise the internal w/o emulsion. Except for the aforementioned interfacial stabilisation *via* fat crystals, long-term successful stabilisation of the internal interface in w/o/w emulsions is realistically limited to a very small selection of lipophilic surfactants with polyglycerol polyricinoleate applied most successfully. However, it is generally regarded as not label friendly and, as per the food additive directive 95/2/EC of the European Parliament, limited in application to, for example, 4 g/1 kg dressing. Alternative surfactants, applicable in the presence of salt in either product aqueous phase, a not uncommon scenario in food formulations, include sucrose ester and sorbitan monooleate.[54]

The successful stabilisation of the external interface in w/o/w emulsions with sodium caseinate, gelatin, BSA, whey protein isolate, gum arabic, and hydrophobically modified starch is mentioned in a highly cited review article on double emulsions stabilised by food biopolymers.[55] Other successfully applied emulsifiers include non-chemically modified waxy rice starch, lecithin combined with Tween 20 and sodium carboxymethyl cellulose[56] and Tween 20[57] and pea protein isolate;[58] this is not an exhaustive list of emulsifiers or references. It can be expected that the recently demonstrated stabilisation of simple o/w emulsion interfaces with dual modified waxy rice starch (enzymatically debranched and esterified with OSA)[59] can be successfully applied to w/o/w emulsions.

The stabilisation of w/o/w emulsions with starch as the external emulsifier brings opportunity for their application to sugar and salt reduction. Starch hydrolyses in contact with salivary amylase. As a result, the internal droplet phase highly concentrated in salt or sugar is released, delivering a change in tastant concentration to the taste receptors that is perceived as enhanced sweetness[61] or saltiness.[60] Application to salt reduction has been demonstrated not only for simple emulsion systems but also for a close-to-market tomato soup formulation.[61]

18.5.3.4 *w/w Emulsions*

The microstructure of w/w emulsions is that of a droplet phase dispersed in a continuous phase, although bicontinuous microstructures are often observed for roughly equal volumes of both aqueous phases. Their structure similarity to conventional emulsions, *i.e.* those containing oil, suggests their application in emulsion-based foods for fat reduction. Replacing all lipid components in a food formulation impacts on mouthfeel and flavour perception and is therefore typically avoided. These emulsions form through phase separation of two incompatible hydrocolloids in aqueous solution, governed by the phase diagram. These systems have been developed a few years ago now, have been described in the context of formulation engineering of food emulsions,[62] and also have been the subject of a recent review.[63] In presence of at least one gelling hydrocolloid in the w/w emulsion system, their rheology or compression stress/strain behaviour may resemble that of low-fat spreads.[64] Notably, the phase behaviour of maltodextrin with either ι-carrageenan or gelatine was studied in great detail in industry R&D.[65]

The design and properties of w/w emulsions and their interfaces continue to attract interest in the research community. Progress in particular has been made in the design of particles that will locate at the w/w interface, enhancing the opportunity for the process-induced trapped microstructure design of these systems.[63,66]

18.5.4 Gelled Particles

In the formulation of healthier foods, gelled particles are considered fat replacers. There are various ways of designing these on the basis of a gelling single- or mixed-hydrocolloid system. Microparticulated protein and fluid gel particles are the two most widely known microstructures.

Whey protein is the most frequently applied protein processed into microparticles[67] and is thus considered in sensory[68] and thermal stability studies,[69] although microparticles have also been developed base on egg white[70] and mung bean protein[71] microparticles, as well as in the presence of pectin.[72]

Fluid gel particles can be processed from a range of gelling polysaccharides through superimposing a network disrupting process, such as shear during gelation, not to be confused with the comminution of an already formed gel. The scientific principles are now reasonably well understood,[73] in addition to acting as rheology modifier to assume the role of fat, their functional properties as foam stabiliser, lubricant, and feedstock for additive manufacture are being investigated (Figure 18.4).[74]

18.5.5 Oleogels

The term "oleogel," or "organogel," denotes a vegetable oil, mineral oil, or organic solvent–based system that is structured by a self-assembled network by a structurant, the organogelator. Some aspects of these systems and how their structure development can be modelled are also discussed in Chapter 15 of this book. This section focuses on their use as replacers of saturated fats by healthier oils. Obviously, edible oleogels demand the use of vegetable oils and food-grade organogelators, and the materials are chosen with the final application in mind. Oleogelators have become of interest due to their ability to replace hydrogenated fat in order to structure high-lipid products, and progress has recently been reviewed.[76] The authors of this review stress the importance that oleogels have gained in foods due to the combined action of providing structure as well as health benefits, backed up by data on the steep increase in scientific publications and patents in the last 10 years.

The different types of oleogels that have been reported in literature include those structured by crystalline material, cross-linked or entangled polymers, or reverse spherical micelles (see the already cited review article for detail[76]). While only ethylcellulose is available as an edible polymeric oleogelator, among a wider choice of crystallising structurants, wax esters have received considerable interest, for example rice bran wax (RBW).[77] RBW is a natural wax extracted from rice bran, a by-product of rice milling. It is one of the oleogelators that has been applied in the oil droplets of o/w emulsions, for example for the stabilisation of the air bubbles in ice cream[78] where droplet destabilisation is part of the functionality. For RBW crystals to form just inside the oil droplets, not piercing the droplet interface, a viscoelastic hydrophilic emulsifier such as gum arabic is required. For targeted interfacial crystallisation, the presence of an oil-soluble amphiphile such as lecithin is beneficial (manuscript in preparation).

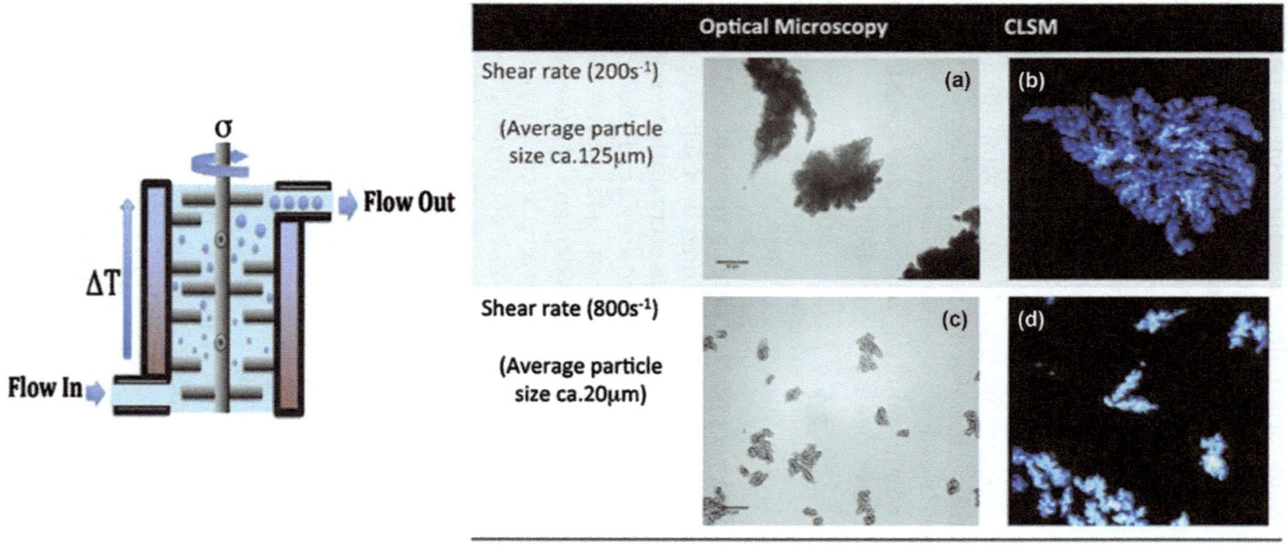

Figure 18.4 Whey protein fluid gel particles. Left: Schematic representation of a process set-up to set fluid gels using a temperature gradient to induce gelation whilst applying shear. Right: (a, c) and confocal light scanning micrographs (b, d) depicting whey protein fluid gel particles. Scale bar represents 50 μm.

18.6 Conclusions

In conclusion, replacement strategies for health-compromising tastants have probably been exhausted, and levels of reduction are limited by their technological functions. In particular, overcoming the challenges of designing foods with uncompromised structure and microstructure remains, in addition, of course, to ensuring an acceptable flavour profile. Undoubtedly future efforts to achieve further reductions of sugar, salt, and fat need to consider the manufacturing paradigm as a variable.

References

1. (a) M. Muller-Bolla and S. Domejean, *Cah. Nutr. Diet.*, 2018, **53**, 341–346; (b) D. J. Tybor, A. R. Beauchesne, R. J. Niu, M. M. Shams-White and M. Chung, *Curr. Dev. Nutr.*, 2018, **2**, 10.
2. D. Malta, K. S. Petersen, C. Johnson, K. Trieu, S. Rae, K. Jefferson, J. A. Santos, M. M. Y. Wong, T. S. Raj, J. Webster, N. R. C. Campbell and J. Arcand, *J. Clin. Hypertens.*, 2018, **20**, 1654–1665.
3. (a) H. E. Billingsley, S. Carbone and C. J. Lavie, *Nutrients*, 2018, **10**, 16; (b) E. Fattore and E. Massa, *Int. J. Food Sci. Nutr.*, 2018, **69**, 916–927.
4. A. Regan, M. P. Kent, M. M. Raats, A. McConnon, P. Wall and L. Dubois, *Nutrients*, 2017, **9**, 9.
5. M. W. J. Noort, J. H. F. Bult, M. Stieger and R. J. Hamer, *J. Cereal Sci.*, 2010, **52**, 378–386.
6. (a) A. C. Mosca, J. A. Rocha, G. Sala, F. van de Velde and M. Stieger, *Food Hydrocolloids*, 2012, **27**, 448–455; (b) A. Syarifuddin, C. Septier, C. Salles and T. Thomas-Danguin, *Food Qual. Prefer.*, 2016, **48**, 59–69.
7. M. Stieger and F. van de Velde, *Curr. Opin. Colloid Interface Sci.*, 2013, **18**, 334–348.
8. J. Busch, C. Tournier, J. E. Knoop, G. Kooyman and G. Smit, *Chem. Senses*, 2009, **34**, 341–348.
9. C. Morris, A. L. Koliandris, B. Wolf, J. Hort and A. Taylor, *Chemosens. Percept.*, 2009, **2**, 1–8.
10. K. R. Goldfein and J. L. Slavin, *Compr. Rev. Food Sci. Food Saf.*, 2015, **14**, 644–656.
11. H. R. McCain, S. Kaliappan and M. A. Drake, *J. Dairy Sci.*, 2018, **101**, 8619–8640.
12. R. Di Monaco, N. A. Miele, E. K. Cabisidan and S. Cavella, *Curr. Opin. Food Sci.*, 2018, **19**, 92–97.
13. S. Gibson, M. Ashwell, J. Arthur, L. Bagley, A. Lennox, P. J. Rogers and S. Stanner, *Perspect. Public Health*, 2017, **137**, 237–247.
14. K. A. Le, F. Robin and O. Roger, *Curr. Opin. Clin. Nutr. Metab. Care*, 2016, **19**, 310–315.
15. C. Kruger, in *Sugar and Bulk Sweeteners*, ed. M. Fowler, S. T. Beckett and G. Ziegler, John Wiley and Sons Inc., New York, 2017.
16. R. G. M. van der Sman and S. Renzetti, *Crit. Rev. Food Sci. Nutr.*, 2018, **59**, 1–15.

17. S. Struck, D. Jaros, C. S. Brennan and H. Rohm, *Int. J. Food Sci. Technol.*, 2014, **49**, 1963–1976.
18. E. Wilderjans, A. Luyts, K. Brijs and J. A. Delcour, *Trends Food Sci. Technol.*, 2013, **30**, 6–15.
19. J. W. Donovan, *J. Sci. Food Agric.*, 1977, **28**, 571–578.
20. C. S. Kim and C. E. Walker, *Cereal Chem.*, 1992, **69**, 206–212.
21. R. West and D. Rousseau, *Crit. Rev. Food Sci. Nutr.*, 2018, **58**, 1917–1936.
22. C. Clarke, *The Science of Ice Cream*, Royal Society of Chemistry, Cambridge, 2012.
23. T. Hutton, *Br. Food J.*, 2002, **104**, 126–152.
24. (a) U. Hoppu, A. Hopia, T. Pohjanheimo, M. Rotola-Pukkila, S. Makinen, A. Pihlanto and M. Sandell, *Foods*, 2017, **6**, 12; (b) A. Allison and A. Fouladkhah, *Foods*, 2018, **7**, 15; (c) A. Fouladkhah, D. Berlin and D. Bruntz, *Food Rev. Int.*, 2015, **31**, 341–354; (d) T. Mills and I. T. Norton, in *Salt Reduction in Food*, ed. I. T. Norton, P. Fryer and J. E. Norton, John Wiley & Sons, Ltd., Chichester, UK, 2013, pp. 163–173.
25. K. Cepanec, S. Vugrinec, T. Cvetkovic and J. Ranilovic, *Compr. Rev. Food Sci. Food Saf.*, 2017, **16**, 881–894.
26. E. S. Inguglia, Z. H. Zhang, B. K. Tiwari, J. P. Kerry and C. M. Burgess, *Trends Food Sci. Technol.*, 2017, **59**, 70–78.
27. I. Bae, J. H. Park, H. Y. Choi and H. K. Jung, *Korean J. Food Sci. Anim. Resour.*, 2017, **37**, 793–798.
28. L. L. Ferrao, E. B. Silva, H. L. A. Silva, R. Silva, N. Mollakhalili, D. Granato, M. Q. Freitas, M. C. Silva, R. S. L. Raices, M. C. Padilha, P. B. Zacarchenco, M. Barbosa, A. M. Mortazavian and A. G. Cruz, *Food Res. Int.*, 2016, **86**, 93–102.
29. N. J. Gaudette and Z. Pietrasik, *J. Sens. Stud.*, 2017, **32**, 10.
30. N. A. Avramenko, R. T. Tyler, M. G. Scanlon, P. Hucl and M. T. Nickerson, *Food Rev. Int.*, 2018, **34**, 204–225.
31. (a) R. A. Miller and R. C. Hoseney, *Cereal Foods World*, 2008, **53**, 4–6; (b) H. Salovaara, *Cereal Chem.*, 1982, **59**, 422–426; (c) K. R. Preston, *Cereal Chem.*, 1989, **66**, 144–148.
32. A. G. Liu, N. A. Ford, F. B. Hu, K. M. Zelman, D. Mozaffarian and P. M. Kris-Etherton, *Nutr. J.*, 2017, **16**, 15.
33. K. Colla, A. Costanzo and S. Gamlath, *Foods*, 2018, **7**, 12.
34. E. Varga-Visi and B. Toxanbayeva, *Acta Aliment.*, 2017, **46**, 181–186.
35. A. S. Babu, R. Parimalavalli and R. J. Mohan, *J. Food Meas. Charact.*, 2018, **12**, 2426–2434.
36. F. Javidi and S. M. A. Razavi, *J. Food Meas. Charact.*, 2018, **12**, 1872–1884.
37. N. S. Aziz, N. S. Sofian-Seng, S. M. Yusop, K. F. Kasim and N. S. M. Razali, *Food Sci. Technol. Res.*, 2018, **24**, 519–530.
38. H. Zhang, J. L. Chen, J. H. Li, C. Y. Wei, X. Q. Ye, J. Shi and S. G. Chen, *Molecules*, 2018, **23**, 11.
39. R. Liu, L. G. Wang, Y. Liu, T. Wu and M. Zhang, *Food Hydrocolloids*, 2018, **81**, 39–47.
40. R. Rama, N. Chiu, M. C. Da Silva, L. Hewson, J. Hort and I. D. Fisk, *J. Texture Stud.*, 2013, **44**, 338–345.

41. M. Quilaqueo, L. Duizer and J. M. Aguilera, *Food Res. Int.*, 2015, **76**, 675–681.
42. A. Utoft, K. Kinoshita, D. L. Bitterfield and D. Needham, *Langmuir*, 2018, **34**, 3626–3641.
43. J. Desarnaud, H. Derluyn, J. Carmeliet, D. Bonn and N. Shahidzadeh, *J. Phys. Chem. Lett.*, 2018, **9**, 2961–2966.
44. R. De Acutis, A. S. Whitehouse, L. Forny, V. D. M. Meunier, M. Dupas-Langlet, J. P. N. Mahieux, A. S. Witthaus and L. M. Dupas, in *Amorphous Porous Particles Having a Specified Closed Porosity, Useful in a Food Product e.g. Confectionery Product and Culinary Product and as Natural Low Calorie Sugar Alternative, Comprise Sugar, a Bulking Agent and a Surfactant*, Nestec Sa (Nest-C), 2016, pp. A3383191–A3383194.
45. S. M. Goh, B. Leroux, C. A. G. Groeneschild and J. Busch, *J. Food Sci.*, 2010, **75**, S245–S249.
46. N. Chiu, L. Hewson, N. Yang, R. Linforth and I. Fisk, *LWT–Food Sci. Technol.*, 2015, **63**, 65–70.
47. G. Muschiolik and E. Dickinson, *Compr. Rev. Food Sci. Food Saf.*, 2017, **16**, 532–555.
48. (a) S. U. Pickering, *J. Chem. Soc.*, 1907, **91**, 2001–2021; (b) R. Aveyard, B. P. Binks and J. H. Clint, *Adv. Colloid Interface Sci.*, 2003, **100**, 503–546.
49. C. Linke and S. Drusch, *Crit. Rev. Food Sci. Nutr.*, 2018, **58**, 1971–1985.
50. V. di Bari, J. E. Norton and I. T. Norton, *J. Food Eng.*, 2014, **122**, 8–14.
51. V. Di Bari, W. Macnaughtan, J. Norton, A. Sullo and I. Norton, *Food Struct.*, 2017, **12**, 82–93.
52. (a) J. Gould, J. Vieira and B. Wolf, *Food Funct.*, 2013, **4**, 1369–1375; (b) J. Gould, G. Garcia-Garcia and B. Wolf, *Materials*, 2016, **9**, 14; (c) J. M. Gould, S. Furse and B. Wolf, *Front. Chem.*, 2016, DOI: 10.3389/fchem.2016.00011.
53. A. T. Florence and D. Whitehill, *Int. J. Pharm.*, 1982, **11**, 277–308.
54. L. Zhang and B. Wolf, in *Alternatives to PGPR for the Stabilisation of w/o and w/o/w Emulsions*, 2018.
55. E. Dickinson, *Food Biophys.*, 2011, **6**, 1–11.
56. J. Teixe-Roig, G. Oms-Oliu, G. R. Velderrain-Rodriguez, I. Odriozola-Serrano and O. Martin-Belloso, *Food Bioprocess Technol.*, 2018, **11**, 2229–2241.
57. A. P. E. Pagano, N. Khalid, I. Kobayashi, M. Nakajima, M. A. Neves and E. L. Bastos, *Food Res. Int.*, 2018, **109**, 489–496.
58. M. Lad, L. Hewson and B. Wolf, *Flavour*, 2012, **1**, 13.
59. S. Jain, T. Winuprasith and M. Suphantharika, *Food Hydrocolloids*, 2019, **89**, 153–162.
60. (a) N. Chiu, L. Hewson, I. Fisk and B. Wolf, *Food Funct.*, 2015, **6**, 1428–1434; (b) N. Chiu, A. Tarrega, C. Parmenter, L. Hewson, B. Wolf and I. D. Fisk, *Food Hydrocolloids*, 2017, **69**, 450–458.
61. M. Kasprzak, P. Wilde, S. E. Hill, S. E. Harding, R. Ford and B. Wolf, *Food Funct.*, 2019, **10**, 4242–4255.
62. A. K. Pawlik, P. J. Fryer and I. T. Norton in *Formulation Engineering of Food Emulsions*, ed. J. E. Norton, P. J. Fryer and I. T. Norton, Wiley-Blackwell, Chichester, 2013, pp. 82–106.

63. J. Esquena, *Curr. Opin. Colloid Interface Sci.*, 2016, **25**, 109–119.
64. I. T. Norton, A. B. Norton, F. Spyropoulos, B. J. D. Le Reverend and P. Cox, in *Rheological Control and Understanding Necessary to Formulate Healthy Everyday Foods*, ed. I. T. Norton, F. Spyropoulos and P. Cox, Blackwell Publishing Ltd., 2011, pp. 219–253.
65. L. Lundin, I. T. Norton, T. J. Foster, M. A. K. Williams, A. M. Hermansson and E. Bergstrom, *Gums Stab. Food Ind.*, 2000, **10**, 167–180.
66. (a) T. Nicolai and B. Murray, *Food Hydrocolloids*, 2017, **68**, 157–163; (b) J. F. Chen, J. Guo, T. Zhang, Z. L. Wan, J. Yang and X. Q. Yang, *J. Agric. Food Chem.*, 2018, **66**, 4200–4207.
67. C. Chung, B. Degner and D. J. McClements, *Food Res. Int.*, 2014, **56**, 136–145.
68. (a) K. Liu, M. Stieger, E. van der Linden and F. van de Velde, *Food Hydrocolloids*, 2016, **60**, 186–198; (b) K. Liu, Y. J. Tian, M. Stieger, E. van der Linden and F. van de Velde, *Food Hydrocolloids*, 2016, **52**, 403–414.
69. E. Cakir-Fuller, *Food Hydrocolloids*, 2015, **47**, 41–50.
70. R. Liu, Z. J. Tian, Y. S. Song, T. Wu, W. J. Sui and M. Zhang, *Food Sci. Technol. Res.*, 2018, **24**, 817–827.
71. P. Sirikulchayanont, S. Jayanta, P. Pradipasena and O. Miyawaki, *Int. J. Food Prop.*, 2007, **10**, 621–630.
72. C. C. Sun, R. Liu, B. Liang, T. Wu, W. J. Sui and M. Zhang, *Food Res. Int.*, 2018, **108**, 151–160.
73. (a) I. T. Norton, D. A. Jarvis and T. J. Foster, *Int. J. Biol. Macromol.*, 1999, **26**, 255–261; (b) A. Gabriele, F. Spyropoulos and I. T. Norton, *Food Hydrocolloids*, 2009, **23**, 2054–2061; (c) A. Gabriele, F. Spyropoulos and I. T. Norton, *Soft Matter*, 2010, **6**, 4205–4213; (d) B. J. D. Le Reverend, I. T. Norton, P. W. Cox and F. Spyropoulos, *Curr. Opin. Colloid Interface Sci.*, 2010, **15**, 84–89; (e) A. Gabriele, PhD Thesis, *Fluid Gels: Formation, Production and Lubrication*, University of Birmingham, Birmingham, 2011.
74. (a) A. Lazidis, L. D. Parizotto, F. Spyropoulos and I. T. Norton, *Food Hydrocolloids*, 2017, **73**, 110–119; (b) A. L. Ellis, T. B. Mills, I. T. Norton and A. B. Norton-Welch, *Food Hydrocolloids*, 2019, **87**, 371–381; (c) A. L. Ellis, A. B. Norton, T. B. Mills and I. T. Norton, *Food Hydrocolloids*, 2017, **73**, 222–228; (d) S. Holland, C. Tuck and T. Foster, *Food Biophys.*, 2018, **13**, 175–185.
75. I. F. Farres, R. J. A. Moakes and I. T. Norton, *Food Hydrocolloids*, 2014, **42**, 362–372.
76. A. J. Martins, A. A. Vicente, R. L. Cunha and M. A. Cerqueira, *Food Funct.*, 2018, **9**, 758–773.
77. L. S. K. Dassanayake, D. R. Kodali, S. Ueno and K. Sato, *J. Am. Oil Chem. Soc.*, 2009, **86**, 1163–1173.
78. (a) D. C. Z. Botega, A. G. Marangoni, A. K. Smith and H. D. Goff, *J. Food Sci.*, 2013, **78**, C1334–C1339; (b) M. E. Moriano and C. Alamprese, *LWT–Food Sci. Technol.*, 2017, **86**, 371–376.

CHAPTER 19

Food Structure Development for Specific Population Groups

W. A. ALSANEI[a] AND J. CHEN*[b]

[a] King Abdulaziz University, Department of Food Science and Nutrition, Idara Street, Jeddah, 21589, Saudi Arabia; [b] Zhejiang Gongshang University, School of Science and Biotechnology, 18 Xuezheng Street, Hangzhou, Zhejiang 310018, China
*Email: jschen@zjgsu.edu.cn

19.1 Introduction

The world's population is ageing at a rate that has been continuously rising globally in the last few decades. It is anticipated that the ratio of older population (≥60 years) to youth will increase quickly in many countries in the near future.[1,2] A consequence of ageing is the deteriorated health conditions affecting appetite and functional ability to swallow, a symptom of dysphagia, which leads to altered oral food intake and malnutrition.[3] Therefore, the development of food and liquid structures (*i.e.* textures) has become an area of interest in order to meet the increased demands on specific food products that suit the needs of these population groups.

This chapter begins with an introduction on the controlling mechanisms of eating and swallowing, the main clinical characteristics of dysphagia (causes, prevalence, symptoms, and signs), and factors which affect one's eating and swallowing ability. This is followed by a short summary on the assessments and grading of eating and swallowing capability through instruments and sensory evaluation. The core part of the chapter is to review the key outcomes of some existing literature of texture standards of

Food Chemistry, Function and Analysis No. 18
Handbook of Food Structure Development
Edited by Fotis Spyropoulos, Aris Lazidis and Ian T. Norton

Published by the Royal Society of Chemistry, www.rsc.org

dysphagia management in countries such as the United States, the UK, Australia, and Japan, as well as the International Dysphagia Diet Standardisation Initiative (IDDSI). The chapter ends with prominent elements of some approaches to texture creation and modification, with special focus on the use of functional ingredients and innovative techniques.

19.2 Mechanisms Controlling Eating and Swallowing

Food oral processing is one of the most basic and important biological functions that comprises two actions: eating and swallowing. Both actions have complicated behaviors involving voluntary and involuntary neuromuscular responses.[4] These responses trigger three swallowing stages. First is the initial suckling or mastication in the oral that causes the creation and transportation of the bolus towards the pharynx while triggering the swallowing reflex. This is followed by the transportation of the bolus through the pharynx and finally the transfer from the esophagus to the stomach. A significant amount of studies in respect to food oral processing and oral physiological behavior have been reported in literature.[5,6] However, problems of eating and swallowing among dysphagia patients require food texture alteration and modification by either clinical, home care practitioners or food services for the health and well-being of these specific population groups. The core question is how and to what extent is the food texture going to be altered? What are the guiding rules for such texture modification? To be able to answer these questions, a basic understanding of the controlling mechanisms of eating and swallowing is essential.

The structural cavity of eating and swallowing comprises different types of tissues that have different anatomical and physiological functionalities when joined together during the food oral processing and oral propulsive phase of swallowing. The capability of these mechanisms is greatly affected by the bolus size and formation, the strength and coordination of oral muscles, and oral sensation. Numerous studies have focused on studying bolus swallowing in vulnerable populations in order to developing food texture through compensatory and rehabilitation strategies.[7–9]

19.2.1 Tongue Capability

One of the most obvious physiological factors in the oral stage of eating and swallowing is the tongue, especially its strength and capability for food manipulation.[10,11] Its critical role appears when a network of nerves and lingual muscle mass incorporate with tongue–palate interaction in order to transfer a food bolus posteriorly to the pharynx.[12] Such cooperation enables the tongue to be the primary cause of kinetic energy to initiate the swallowing of a bolus.[7] However, many patients with dysphagia and some elderly are characterized by weak tongue strength, thus causing impaired sensory feedback involving a premature leakage of the bolus into the pharynx. For this reason, clinical studies of swallowing have focused on the safe and pleasurable food oral intake through tongue rehabilitation, in particular its muscle strength.

19.2.2 Orofacial Muscle Activities

In addition to tongue capability, the activities and movements of orofacial muscles also contribute essentially to the transferring of ingested food through the oropharynx, facilitating masticatory process, and controlling the chewing efficiency. However, the function and feature of such muscles are found to change as a function of age.[13] Thus, a number of therapeutic procedures had been proposed principally to rehabilitate and strengthen the orofacial muscles in order to improve bolus sensory input prior to swallowing and to coordinate the movements of orofacial muscles.[14]

19.2.3 Temporo-spatial Coordination between Breathing and Swallowing Tracks

Based on the coordination between the orofacial muscles and tongue pressure generation, a bolus becomes ready to shift from the oral stage into the pharynx. During the swallowing process, a number of valves (soft palate, tongue base, and pharyngeal wall) must open and close involuntarily and then direct the bolus down the esophagus. At one time, the larynx must close to prevent the bolus from entering the lungs and windpipe. Thus, a precise temporal-spatial coordination between breathing and swallowing tracks during oral processing is necessary to ensure a safe bolus flow and to prevent complications of respiratory pneumonia.[13]

A significant number of dysphagia and non-dysphagia research studies have led to a better understanding of temporo-spatial coordination and complex physiologic function and manometric events of swallowing. Direct (videofluoroscopy) and/or indirect (electromyography, swallow sounds) bolus measurements were made possible by observing the volume, consistency, flow, and timing of the bolus in responses from individuals of different age, gender, as well as sensory responses to aspiration and bolus residue.[14–16] Results confirm that variations in the temporal and muscle activation patterns exist between individuals, whether normally or abnormally swallowing.

19.2.4 Bolus Formation Prior to Swallowing

Bolus formation and food disintegration occur simultaneously during mastication. Even though such a physical process appears a relatively simple, it involves a number of oral manipulations either simultaneously or sequentially performed by the teeth, tongue, and oral musculature. As a result, food (*i.e.* the solid) is broken down into small particle sizes, moistened by saliva, and formed into a cohesive mass that can be swallowed easily and safely. Food particle size reduction by mastication is crucial for bolus formation, along with saliva incorporation. Insufficient mastication leads to dysphagia.[17,18] Thus, it is generally agreed that food structure modification is the dominant method for controlling the risk of mastication impairment for patients with dysphagia.

19.3 Swallowing Disorders (Dysphagia)

Swallowing disorder (dysphagia) is characterized clinically as a serious health problem. It can affect any part of the swallowing stages and is a common problem among elderly populations, as well as in individuals who have suffered neurological injuries and in newborn, infant, or premature babies.

Dysphagia is a comorbid condition caused by a wide variety of health conditions including stroke, head and neck cancer, Parkinson's disease, amyotrophic lateral or multiple sclerosis, achalasia, diffuse esophageal spasm, esophageal cancer, esophageal stricture, myasthenia gravis, oeosinophilic oesophagitis, and xerostomia (dryness of oral mucosa).[19–32] All of these health complications, independently or collectively, can compromise one's overall well-being and quality of life. Accurate estimates of dysphagia's prevalence among these population groups cannot be made due to the wide variety of its causes.[19,33–39] However, it was reported in 2014 that dysphagia affects 1 in 25 adults in the United States,[34] with the rates of dysphagia's prevalence in the general population ranging from 1.7% to 16%, with prevalence much higher in elderly people.[40–42]

The most common symptom of dysphagia is the subjective sensation of difficulty during swallowing.[43] However, clinicians use a variety of symptoms and signs as indicators of swallowing dysfunction, including odynophagia (painful swallowing), coughing or gagging when swallowing, choking when eating, regurgitation, cyanosis with feeds, and delayed swallows.[44–46] In addition, signs such as perceived mouth coating, post-swallow residue, and pharyngeal coating, excessive saliva (drooling), recurrent aspiration pneumonia, and poor oral management can also cause and/or exacerbate swallowing disorders.[7,47–50] Moreover, dysphagia is often accompanied by unexplained weight loss, dehydration, and malnutrition, which are the predominant symptoms in most patients and may potentially lead to death.[51–53]

19.4 Assessments and Grading of Eating and Swallowing Capability

19.4.1 Videofluoroscopy

Dysphagia is best diagnosed by videofluoroscopy (VFS), which has become a gold standard assessment tool for radiological evaluation, in which a barium-contained bolus will be used for swallow monitoring.[54] The technique is commonly performed on the entire process of swallowing, not only to clinically confirm the presence of swallowing dysfunction (*i.e.* bolus movement) but also to assess the degree of dysphagia (*i.e.* aspiration).[15,55,56] During assessment of barium swallow, the effects of bolus volume and rheology (*i.e.* bolus viscosity and chewed bolus) on the physiology of swallowing in healthy participants as well as dysphagic patients were observed by the dynamic images from VFS video.[57] Semi-quantitative assessment can be made for a number of swallowing parameters, including bolus shape, bolus movement, number of swallows, oral transit time, pooling above the larynx, thin bolus penetration into the trachea, and direct aspiration of thin bolus into the lungs.[56,58–60] Once the pathophysiology is defined, a valuable information can be drawn for rehabilitation treatments and best practice to feed the individual patient.[61,62]

19.4.2 Surface Electromyography

Surface electromyography (sEMG) is another useful instrumental method for dysphagia assessment. It is a graphic record for the collective electrical activity controlled by the nervous system in the form of signals representing the anatomical and physiological properties during the orofacial and pharyngeal muscular performances.[63] It is categorized as a radiation free and reliable screening method for optimal patient management, but it is not the most reliable for a proper investigation in neurogenic dysphagia.[14,64] During oral processing and swallowing, recorded signals for an individual can be analyzed regarding peak frequency, peak length, amplitude, root-mean-square amplitudes, total work, and other parameters in order to correlate with oral processing behaviors. Importantly, several studies confirmed that the rheological characteristics (*i.e.* viscosity) of thin and thick boluses have an effect on swallow duration by sEMG technique.[65–67] More adhesive stimuli compared to liquid and thin paste showed an increase in the amplitude of sEMG activity.[68] Moreover, food hardness (*i.e.* physical properties) can be studied for its dissociation during mastication including number of cycles, sequence duration, and masticatory frequency by sEMG recording.[69,70]

19.5 Texture Standardization of Diets for Specific Population Groups

19.5.1 Needs of Texture-modified Diet

Texture-modified diet is an important strategic intervention for dysphagia management. This can be achieved by changing the fluid consistency and food texture to different grades using thickening agents and adjusting/developing the texture of foods.

Despite diet modification is seen as an effective therapeutic treatment of dysphagia for swallowing rehabilitation, a big knowledge gap still exists between scientific understanding of texture grading and its standardization at the international scale.[71,72] Unfortunately, incidences of mortality associated with swallowing disorders occur often when patients are served with wrongly labeled food with unsuitable textured consistencies.[72–74] Inconsistency and difference in the description of texture-modified foods are very common between or even within countries, among food-producing establishments, food providers, health professionals, and food-handling practices. Great progress had been achieved in some countries to develop national standards in classifying terminologies of texture-modified foods and drinks. Such development can globally contribute to improving patient safety and care.[74–78]

19.5.2 Published National Standards

19.5.2.1 U.S. National Dysphagia Diet (NDD)

Back to 2002, the first standard so-called the National Dysphagia Diet (U.S.) was published in the United States by the National Dietetic Association for specific population groups. This standard consists of three levels of textured-modified diets that are Dysphagia's Advance, Ground, and Pureed (Table 19.1). Criteria for this classification are characterized by the physical forces of food chewing and swallowing such as cohesiveness, adhesiveness, firmness, springiness, *etc.*, while fluid standard is divided into four levels based on fluid viscosity and consistency scales: *spoon-thick* (>1750 cP), *honey-like* (351–1750 cP), *nectar-like* (51–350 cP), and *thin* (1–50 cP) (Table 19.2).

If a patient is diagnosed with a difficulty of swallowing food or liquid, a special dietary regimen shall be used in order to ensure an adequate nutrition intake and body hydration. Such diet relies on modifying the consistency and altering the rheological properties of common foods and fluids. In many cases, food thickeners are used for such purposes. The choice of texture level is closely associated with the patient's health status (type and degree of dysphagia, tongue pressure, and orofacial weakness, fatigue, tooth loss, lack of saliva, poor mastication force, organic causes). It influences the amount of time spent on preparation and feeding. The texture choice becomes more complicated when considering the influences of temperature on food preparation (*e.g.* instant food thickeners and commercially packaged prethickened beverages during sensory swallowing test, in lab or factory), the use of different food-thickening agent (type, amount, ingredients), different types and amount of dispersing media, *etc.*[7,79–87] These factors have a great influence on the overall structure and consistency of diets in relation to the NDD guidelines.

Considerable inconsistencies commonly exist in the rheological properties among products thickened with instant food thickeners and commercially packaged prethickened beverages.[86,88–92] It should be pointed out that rheological information of food products for patients with dysphagia are filled under moderate deformation speeds (up to 30 s^{-1}) to comply with the requirements of the production line.[92] This rate is quite different from the shear rate in modeling of a natural swallow (estimated from 1 to 1000 s^{-1}), fast swallow (100 s^{-1}), or in the mouth of an individual with dysphagia who has difficulty in propelling the food.[92–94] Real oral shear rate of dysphagia patients is still unknown. Therefore, as a compromise, NDD uses 50 s^{-1} as a standard measure of shear rate. This, of course, requires further scientific validation.

Table 19.1 National Food Texture Scale. Adapted from United States, UK, Australia, and Japan.

Status	Country	Scaling (grading) code	Description of texture modification
Foods	USA	Regular	• All foods allowed (unmodified foods) • For patients able to eat an unrestricted diet
		NDD Level 3	• Dysphagia-Advance • 2.5 cm "bite-sized" • For patients with some ability to chew • Soft foods that require more chewing ability
		NDD Level 2	• Dysphagia-Mechanical Altered • 0.6 cm • For patients with some ability to chew • Cohesive, moist, semisolid foods, requiring some chewing
		NDD Level 1	• Dysphagia-Pureed • For patients with significant impairment in control • Homogeneous, very cohesive, pudding-like, requiring very little chewing
	UK	Regular	• All foods allowed (unmodified foods)
		Texture E	• Closest to regular textures • Fork Mashable Dysphagia Diet • 1.5 cm
		Texture D	• Per-mashable Dysphagia Diet
		Texture C	• Thick Puree Dysphagia Diet
		Texture B	• Thin Puree Dysphagia Diet
		Texture A	• Puree
	Australia	Regular	• All foods allowed (unmodified foods)
		Texture A	• Soft • Can be chewed with minimal effort • 1.5 cm
		Texture B	• Minced and Moist • Easily mashed with a fork • 0.5 cm
		Texture C	• Smooth Pureed
	Japan	Level 0 – Initial diet	• Severity: Severe
		Level 1 – Dysphagia Diet I	• Severity: Moderate
		Level 2 – Dysphagia Diet II	• Severity: Moderate
		Level 3 – Dysphagia Diet III	• Severity: Mild
		Level 4 – Transitional Diet	• Severity: Mainly (masticatory impairment)

19.5.2.2 UK National Descriptors for Texture Modification in Adults

In 2002, the British Dietetic Association (BDA), in cooperation with the Royal College of Speech and Language Therapists (RSCLT), developed their own

Table 19.2 National Viscosity Scale. Adapted from United States, UK, Australia, and Japan.

Status	Country	Scaling (grading) code	Description of consistency modification
Liquid	USA	Spoon-thick	>1750 cP
		Honey-like	351–1750 cP
		Nectar-like	51–350 cP
		Thin	1–50 cP
	UK	Thin Fluid	• Still as water
		Naturally Thick Fluid	• Product leaves a coating on an empty glass
		Thick Fluid Stage 1	• Fluid to which a commercial thickener has been added to thicken consistency, *but*: – Can be drunk through a straw – Can be drunk from a cup if advised or preferred – Leaves a thin coat on the back of a spoon
		Thick Fluid Stage 2	• Fluid to which a commercial thickener has been added to thicken consistency, *but*: – Cannot be drunk through a straw – Can be drunk from a cup – Leaves a thick coat on the back of a spoon
		Thick Fluid Stage 3	• Fluid to which a commercial thickener has been added to thicken consistency, *but*: – Cannot be drunk through a straw – Cannot be drunk from a cup – Needs to be taken with a spoon
	Australia	Regular	• Regular (unmodified fluids)
		Level 150	• Mildly thick • Nectar consistency • Fluid runs freely off the spoon but leaves a mild coating on it
		Level 400	• Moderately thick • Gravy consistency • Fluid slowly drips in dollops off the spoon's end
		Level 900	• Extremely thick • Custard consistency • Fluid sits on the spoon and does not flow off it
	Japan	Regular	• Regular (unmodified fluids)
		Level 1	• Mildly thick • 50–150 mPas • Line spread value: 36–43 mm
		Level 2	• Moderately thick • 150–300 mPas • Line spread value: 32–36 mm
		Level 3	• Extremely thick • 300–500 mPas • Line spread value: 30–32 mm

original descriptive standards for food and fluid texture modifications intended for a specific life stage and categorized as UK National Descriptors for Texture Modification in Adults.[95] In 2009, this standard was reviewed again by BDA, and an updated version entitled Diet Dysphagia Food Texture Descriptors was published in 2011 and was endorsed by the BDA, RCSLT, Hospital Caterers Association (HCA), and the National Nurses Nutrition Group (NNNG).[78] The UK standard is termed by *Texture*, followed by an English alphabetical letter to cover the whole scale of food texture (ranging from *E* for regular texture to *A* for pureed texture) (Table 19.1). For fluids and beverages, consistency is the key property for classification, divided into five categories from naturally thin and thick fluids to three further Stages of fluid consistency (Table 19.2).

It should be noted that the UK standard was an adult-oriented guideline. It may not suit preterm, newborns, and infants who suffer from swallowing difficulties. However, some studies reported that the thinner consistency can be described for infant babies as long as such material can easily flow through a nipple.[96,97] This description may involve risk of health complication because of the different interpretations in these critical periods of life. An appeal has been raised of the need to review the existing literature on infant's thickened liquid and their formula. Outcomes can help many newborn nursing caregivers who frequently report difficulty attempting to constitute ready-to-feed preterm and infant formulas to the prescribed thickness.[98,99] Another suggestion has emphasized the need of an accurate scale of the thickened consistency with attention to different types of thickeners and the types of milk used that can suit all these cases.[100,101]

19.5.2.3 Australian Texture-modified Foods and Fluids

In 2007, Australia published its own national standard of terminology and definitions for texture-modified food and thickened fluids by the Dieticians Association of Australia and Speech Pathology Australia.[74] This standard used the word "Texture" followed by English alphabetical letter for three grades of specific food textures (ranging from *A*, which refers to soft foods of size up to 1.5 cm that can be chewed with minimal effort to *C*, which refers to puree texture) (Table 19.1). Compared to the UK standard, the Australian standard has fewer texture grades (four compared to six grades in the UK standard). For fluids and beverages, Australia's standard also has fewer levels compared to the UK standard. It is divided into three Levels followed by numerical numbers (150, 400, and 900) to indicate increased viscosity (Table 19.2), where low-viscous liquids are used for patients with a mild dysphagia, and thicker liquids are used to manage more severe conditions of dysphagia.[102] Note that the numbers of 150, 450, and 900 in Australia standard are only indicative and are not intended for any real proportional meaning.

19.5.2.4 Japanese Modified Diet for Dysphagic Persons

Multiple standards are published in Japan, and they are working in parallel under supervision of the Japanese Society of Dysphagia Rehabilitation.

Food for Special Dietary Uses (FOSDU), published in 1994 by the Japanese Ministry of Health, Labour and Welfare, was the first standard in the country especially for elderly with difficulty in masticating or swallowing.[103] In 2009, significant developments in the Japanese standard were made when shear rate of 2–3 s^{-1} was introduced for the assessment of gel- and liquid-type (sol-type) foods.[104] However, such a shear rate value is much lower than that used by NDD (50 s^{-1}) for viscosity evaluation.[16]

In 2013, a clear classification of the Japanese standard was made by the Task Force of Modified Diet for Dysphagic Persons by the Japanese society.[105] It comprises of five grade classifications for food and three grade classifications of thickness for fluids and beverages. Food grade is categorized by word "Level" (from 0 to 4) followed by initial diet, dysphagia diet I, dysphagia diet II, dysphagia diet III, and transitional diet, respectively (Table 19.1). Grades for fluids and beverages are given by word "Level," followed by number (from 1 to 3),[106] with descriptive terms given for each level: mildly, moderately, and extremely thick consistency, respectively (Table 19.2). Similar to the NDD system, this standard also gives defined viscosity values for each level.

Over the last decade, rheological properties have been intensively studied in Japan to evaluate their suitability as care foods for dysphagic patients.[107] Patients with dysphagia were found to be easily aspirated by drinking water but not yogurt. High flow speed of water is believed to be the reason (water was measured as 0.5 m s^{-1} during swallowing, but yogurt was only 0.2 m s^{-1}).[108–110] Both viscosity and velocity through the pharynx were measured for various solutions prepared from commercial thickening agents with different concentrations (in the consistency of jam-like, yogurt-like, potage-like solutions) and compared with "yogurt as it is" and "fractured yogurt" as a model bolus of hand-mixed yogurt.[111] At a shear rate of 10 s^{-1}, the viscosity of yogurt-like thickened solution was lower than that of the so-called fractured yogurt. Measurements of the flow speed at pharynx by color spectra technique revealed that fractured yogurt was similar to that of the yogurt but slower than that of jam-like, yogurt-like, potage-like solutions. A major finding was that similar consistency of natural products (yogurt) and constituted products with various thickeners (yogurt-like consistency) does not necessarily give similar swallowing behavior.

19.5.2.5 International Dysphagia Diet Standardisation Initiative (IDDSI)

Over the past two decades, many institutions have adapted and implemented various texture standards, while many others display a reluctance due to inconsistency among these standards.[112–115] Clinicians and researchers are concerned about possible risks involved due to the diversity of these standards. Therefore, a clear internationally accessible classification system for food texture and liquid consistency for dysphagia management is urgently needed. For this reason, globally integrated teamwork was first initiated in

2013 and came to fruition in late 2015 with the International Dysphagia Diet Standardisation Initiative (IDDSI) framework. The framework was designed for dysphagia individuals of all the ages, with different cultural backgrounds, and from a diversity of health care settings.[116] IDDSI framework consists of two opposite-arranged pyramids with distinct eight levels (from 0 to 7) for foods and drinks (classified by numbers, names, and colors) (see Figure 19.1).[117] Based on the gravitational flow through a 10 mL slip-tip syringe for 10 s, Levels (0–4) refer to thin, slightly, mildly, moderately, and extremely thickened consistencies of drinks, respectively. While Levels (3–7) refer to liquidized, pureed, minced, and moist, soft and bite-sized and regular and easy to chew food textures verified by fork and/or spoon drip test.

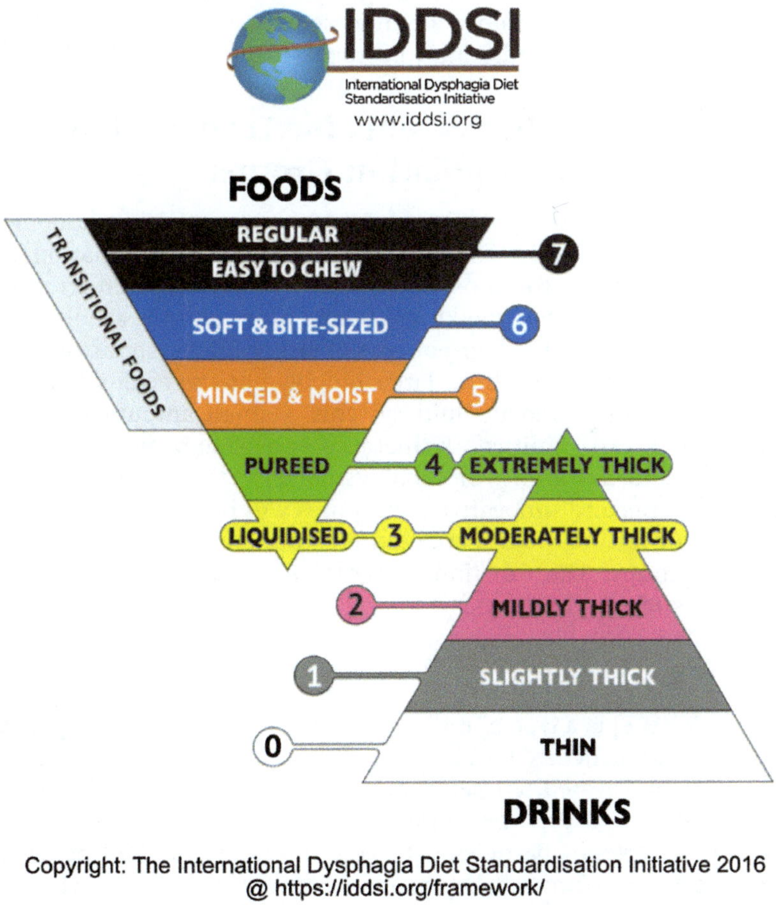

Figure 19.1 International Dysphagia Diet Standardisation Initiative (IDDSI) Framework categorizes texture-modified food and fluids into eight levels with numbers, term descriptions, and colors.

In this framework, there is an overlapped area within Level 3 and 4 zones where two different terms refer to the same texture and flow properties.

Compared to other national standards, IDDSI is most self-consistent, offering great advantages and benefits to food manufacturers, health care givers, and consumers. The framework combines solid foods and liquid beverages into a single continuous system. Grades of texture are represented by three parallel coding systems: number, term, and color. This gives great flexibility for food provision and frontline application. Grade assessment and characterization are simple to implement, involving no expensive instruments and no complicated and difficult to understand jargon. The framework is welcomed by international food companies, and its implementation is highly recommended across the world. The framework has been translated into many different languages and has been implemented by a number of professional bodies for endorsement and adoption in Germany, the UK, Canada, the United States, Portugal, Spain, and China.[118–122]

19.6 Functional Ingredients for Texture Modification for Specific Population Groups

The food industry actively seeks new techniques and methodologies in order to produce food for the need of these special population groups. Ingredients are always the most important factor for the creation of food structure and texture. Of all food ingredients, proteins and carbohydrates are the most effective structure components of either food systems or fluids and beverages. In solid foods, these large biopolymer molecules help to form structural matrixes, while in fluid systems a small amount of protein or polysaccharides can significantly increase consistency as well as regulate the rheology and flowability of such systems.[123,124] Here, the use of functional ingredients and innovative techniques will be highlighted as feasible approaches for texture modification and texture creation of food for specific population groups. The functional ingredients that follow are only a few examples of such applications.

19.6.1 Whey Proteins

Whey protein (WP) is a type of milk protein widely used in food applications. It has a globular structure and high solubility at neutral pH conditions. WP denatures at an elevated temperature when its structure unfolds capable of forming clusters and even networks. A decrease of pH will also lead to the aggregation of protein particles with very different colloidal behavior. WP can be in different forms: whey protein concentrate (WPC), whey protein isolate (WPI), and whey protein hydrolysate (WPH). Generally, WP is an ideal structure-forming and structure-controlling ingredient.

Yogurt is seen as a healthy food product offering important physiological benefits to human body. It is also a very important food for elderly and dysphagia patient due to its potential health and well-being impact but more importantly due to its better controlled swallowing. For example, the addition of WP to yogurt has been found to be beneficial to enhance yogurt's texture and sensory quality,[125] with an increased hardness, cohesiveness, elasticity, and water-holding capacity.[126,127] When WP is used in combination with casein in yogurt products, a much smoother texture was observed than its reference. Moreover, WP can also be an effective structural enhancer for many other gel-type foods.

19.6.2 Polysaccharides

Polysaccharides are nondigestable, flavorless, large-molecular-weight carbohydrates. They are probably the most important category of functional ingredients for food applications.

Xanthan gum is a popular polysaccharide that is used to thicken fluids. It is a natural high-molecular-weight polysaccharide produced by a fermentation process. Due to its unique rheological properties, it has been used as a very effective thickener and stabilizer for a number of commercial formulations aimed for dysphagia patients.[128] A very unusual rheological property of xanthan gum is its extensional behavior. A xanthan-based fluid has a much higher extensional viscosity than that of starch and other polysaccharide-based solutions.[129,130] It is not clear yet how the extensional viscosity affects swallowing, but researchers seem to agree that good bolus stretchability may help to maintain bolus cohesiveness during swallowing. Appropriate bolus cohesiveness could mean less aspiration. This, of course, requires further experimental validation.

Another good example of polysaccharides for dysphagia applications is the beta glucan (β-glucan) naturally found in the dietary fibers of cereal grains such as oats and barley.[131] It is considered a value-added functional ingredient in food products due to its functional and bioactive properties including water-holding capacity, gel formation, thickening, and stabilizing. Adding 1% barley β-glucans shows a most desirable effect on the yogurt's firmness, viscosity, consistency, and sensory attributes, including creaminess, mouthfeel, color, and appearance, superior to that observed for 2% addition of guar gum and inulin.[132] Moreover, such a type of functional ingredient not only contributes to improve the texture of various food systems, it also enhances the health benefit by increasing dietary fibers, which helps better regulation of the blood lipid levels and controlling of some chronic diseases among the elderly.[132–135]

19.6.3 Nanostructured Carriers

Strictly speaking, nanostructured carriers are not food ingredients but a novel technique for delivery functional ingredients (*i.e.* physiologically active food components) in optimizing desirable health benefits (*e.g.* an aging population).[136] During the past decade, there has been a dramatically increased activity towards the use of food-applicable nanocarrier systems (*e.g.* lipid-based carriers). These include nanoemulsions, nanoliposomes, polymeric nanoparticles, solid lipid nanoparticles.

Encapsulation is another novel technique for the texture modification of diets for specific population groups. For example, nanostructured lipid carriers (NLC) have been used as a functional ingredient in food systems both for texture purposes and for nutrition delivery.[137] NLC can be used as an effective carrier for some functional ingredients, such as vitamins A, D, K, and E. These nutrients are essential for health but difficult to deliver because of their insolubility in water. Therefore, NLC functionality acts as a nutraceutical delivery system by enhancing bioavailability, ensuring adequate stability of such nutrients and their protection from oxidation, pH, and enzymatic degradation.[138–141] The importance of NLC thus lies in its promising future as a food carrier for foods and beverages with essential micronutrients for specific population groups.

19.6.4 Incorporation of microbubbles

The potential applications of microbubble incorporation have recently gained a considerable interest in studying it as a promising food ingredient (or a novel technique) due to its significant influence on the development of food structures.[142–145] An aerated food system is comprised of micron-sized gas microbubbles substantially dispersed in the aqueous phase (*i.e.* aerated-gel foods). The presence of air bubbles in the gel weakens its structure,[146] leading to unique physical and sensory properties. The main controlling factors of bubbled food include the volume fraction of the incorporated air and the size distribution of the bubbles. Despite no report of the application of microbubble incorporation for dysphagia or elderly food, the potential of this novel ingredient (technique) is great. The incorporation of microbubbles in the food system leads to a wide range of structural and textural properties. Aroma components can be easily contained in the bubble phase that gives a burst release of food aroma during eating. More importantly, the air ingredient comes with no health risk at all to dysphagia patients.

19.7 Conclusions

The continuous growth of the elderly population occurs not only in developed countries but also in many developing countries. Weakened oral physiological functions mean that many elderly people have much reduced capability of eating and swallowing. Therefore, providing quality food and beverages for the health and well-being of these fast-growing population groups has become a big challenge for the food industry as well as for society. While the food industry is developing food products with suitable texture for safe eating and swallowing in close association with hospitals, nursing homes, and other health care settings, a big challenge is the assurance of an adequate texture standard of such products. For this reason, a number of standards have been developed in the last two decades in the

United States, the UK, Australia, Japan, and a few other countries. The latest addition to the texture standards of diets for dysphagia people is the International Dysphagia Diet Standardisation Initiative (IDDSI). The IDDSI framework was developed for global use for dysphagia individuals of all ages, with different cultural backgrounds, and in all health care settings. Texture modification can be achieved by many different approaches. The application of functional ingredients and novel processing techniques are the most promising approaches for the production of food and beverages with specific texture in order to suit the needs of these specific population groups.

References

1. W. Leslie and C. Hankey, *Healthcare*, 2015, **3**, 648–658.
2. C. Gilleard and P. Higgs, *Int. J. Ageing Later Life*, 2007, **2**, 13–30.
3. L. Watson, W. Leslie and C. Hankey, *Rev. Clin. Gerontol.*, 2006, **16**, 23–34.
4. J. A. Logemann, in *Evaluation and Treatment of Swallowing Disorders*, ed. J. A. Logemann, College-Hill Press, San Diego, 1983, pp. 11–36.
5. J. A. Logemann, *Curr. Opin. Otolaryngol. Head Neck Surg.*, 1998, **6**, 395–400.
6. K. Matsuo and J. B. Palmer, *Phys. Med. Rehabil. Clin. North America*, 2008, **19**, 691–707.
7. W. A. Alsanei and J. Chen, *J. Texture Stud.*, 2014, **45**, 1–12.
8. M. A. Nicosia, J. A. Hind, E. B. Roecker, M. Carnes, J. Doyle, G. A. Dengel and J. Robbins, *J. Gerontol., Ser., A*, 2000, **55**, M634–640.
9. J. A. G. Stierwalt and S. R. Youmans, *Am. J. Speech-Language Pathol.*, 2007, **16**, 148–156.
10. T. Park, Electronic Thesis or Dissertation in Speech-Language Pathology, *Effects of Oropharyngeal Strengthening Exercise (OSE) on Tongue Strength, Submental Muscle Activity, and Quality of Life in a Healthy Elderly Population*, Ohio University, 2015. Available at https://etd.ohiolink.edu/.
11. W. A. Alsanei, J. Chen and R. Ding, *Food Res. Int.*, 2015, **67**, 331–337.
12. I. J. Cook, W. J. Dodds, R. O. Dantas, M. K. Kern, B. T. Massey, R. Shaker and W. J. Hogan, *Dysphagia*, 1989, **4**, 8–15.
13. J. D. Subtelny and M. Sakuda, *Am. J. Orthod. Dentofacial Orthopedics*, 1966, **52**, 495–517.
14. M. Vaiman and E. Eviatar, *Head Face Med.*, 2009, **5**, 1–11.
15. E. Eisenhuber, W. Schima, E. Schober, P. Pokieser, A. Stadler, M. Scharitzer and E. Oschatz, *Am. J. Roentgenol.*, 2002, **178**, 393–398.
16. H. Kumagai, A. Tashiro, A. Hasegawa, K. Kohyama and H. Kumagai, *Food Sci. Technol. Res.*, 2009, **15**, 203–210.
17. J. Chen, *Trends Food Sci. Technol.*, 2015, **45**, 222–228.
18. S. C. Hutchings, K. D. Foster, J. E. Bronlund, R. G. Lentle, J. R. Jones and M. P. Morgenstern, *J. Food Eng.*, 2012, **109**, 736–744.

19. J. A. Y. Cichero and K. W. Altman, *Nestle Nutr. Inst. Workshop Ser.*, 2012, **72**, 1–11.
20. M. González-Fernández, L. Ottenstein, L. Atanelov and A. B. Christian, *Curr. Phys. Med. Rehabil. Rep.*, 2013, **1**, 187–196.
21. A. Schindler, N. Denaro, E. G. Russi, N. Pizzorni, P. Bossi, A. Merlotti, M. Spadola Bissetti, G. Numico, A. Gava, E. Orlandi, O. Caspiani, M. Buglione, D. Alterio, A. Bacigalupo, V. De Sanctis, G. Pavanato, C. Ripamonti, M. C. Merlano, L. Licitra, G. Sanguineti, J. A. Langendijk and B. Murphy, *Crit. Rev. Oncol. Hematol.*, 2015, **96**, 372–384.
22. J. G. Kalf, B. J. M. de Swart, B. R. Bloem and M. Munneke, *Parkinsonism Related Disord.*, 2012, **18**, 311–315.
23. S. Kawai, M. Tsukuda, I. Mochimatsu, H. Enomoto, Y. Kagesato and H. Hirose, *Stomato-pharyngology*, 2000, **12**, 373–379.
24. M. Poorjavad, F. Derakhshandeh, M. Etemadifar, B. Soleymani, A. Minagar and A. H. Maghzi, *Mult. Scler.*, 2010, **16**, 362–365.
25. J. E. Pandolfino and A. J. Gawron, *JAMA*, 2015, **313**, 1841–1852.
26. M. A. Khashab, A. A. Messallam, M. Onimaru, E. N. Teitelbaum, M. B. Ujiki, M. E. Gitelis, R. J. Modayil, E. S. Hungness, S. N. Stavropoulos, M. H. El Zein, H. Shiwaku, R. Kunda, A. Repici, H. Minami, P. W. Chiu, J. Ponsky, V. Kumbhari, P. Saxena, A. P. Maydeo and H. Inoue, *Gastrointest. Endosc.*, 2015, **81**, 1170–1177.
27. B. Sunde, J. Ericson, K. Kumagai, L. Lundell, J. A. Tsai, M. Lindblad, I. Rouvelas, S. Friesland, N. Wang and M. Nilsson, *Diseases of the Esophagus*, 2016, **29**, 442–447.
28. Y. Dai, C. Li, Y. Xie, X. Liu, J. Zhang, J. Zhou, X. Pan and S. Yang, *Cochrane Database Syst. Rev.*, 2014, **30**, CD005048.
29. E. Forti, G. Bonato, L. Dioscoridi, M. Cintolo, F. Pugliese, L. Cristoferi, A. Tringali, V. Caputo, V. Motta, M. D. Torre, S. Conforti and M. Mutignani, *Dysphagia*, 2018, **33**, 399–402.
30. S. Im, S. Suntrup-Krueger, S. Colbow, S. Sauer, I. Claus, S. G. Meuth, R. Dziewas and T. Warnecke, *Eur. J. Neurol.*, 2018, **25**, 1235–1242.
31. T. S. Gunasekaran, C. Chu, N. Ronquillo, R. Chennuri, B. Adley, K. Borgen, A. Schwartz, R. Newberry and J. Berman, *Can. J. Gastroenterol. Hepatol.*, 2017, **2017**, 1–6.
32. K. K. Takahashi, *Clin. Calcium*, 2012, **22**, 59–65.
33. M. Cabre, M. Serra-Prat, E. Palomera, J. Almirall, R. Pallares and P. Clavé, *Age Ageing*, 2010, **39**, 39–45.
34. N. Bhattacharyya, *Otolaryngol.–Head Neck Surg.*, 2014, **151**, 765–769.
35. N. Roy, J. Stemple, R. M. Merrill and L. Thomas, *Ann. Otol., Rhinol., Laryngol.*, 2007, **116**, 858–865.
36. L.-C. Lin, S.-C. Wu, H. S. Chen, T.-G. Wang and M.-Y. Chen, *J. Am. Geriatr. Soc.*, 2002, **50**, 1118–1123.
37. S. Teramoto, Y. Fukuchi, H. Sasaki, K. Sato, K. Sekizawa and T. Matsuse, *J. Am. Geriatr. Soc*, 2008, **56**, 577–579.
38. S. K. Rosenvinge and I. D. Starke, *Age Ageing*, 2005, **34**, 587–593.

39. C. DeMatteo, D. Matovich and A. Hjartarson, *Dev. Med. Child Neurol.*, 2005, **47**, 149–157.
40. D. F. Roden and K. W. Altman, *Otolaryngol. Clin. North America*, 2013, **46**, 965–987.
41. B. Kertscher, R. Speyer, E. Fong, A. M. Georgiou and M. Smith, *Dysphagia*, 2015, **30**, 114–120.
42. J. Robertson, D. Chadwick, S. Baines, E. Emerson and C. Hatton, *Intellect. Dev. Disabil.*, 2017, **55**, 377–391.
43. R. Martino, N. Foley, S. Bhogal, N. Diamant, M. Speechley and R. Teasell, *Stroke*, 2005, **36**, 2756–2763.
44. C. S. Hammond, *Lung*, 2008, **186**, 35–40.
45. K. L. Duffy, *Curr. Probl. Pediatr. Adolesc. Health Care*, 2018, **48**, 71–73.
46. C. O. Muller, L. Derycke, A. Kheniche, G. Garcia, S. Bernard, N. Teissier and A. Bonnard, *Pediatr. Surg. Int.*, 2018, **34**, 879–884.
47. N. Zijlstra, M. Mars, R. A. de Wijk, M. S. Westerterp-Plantenga and C. de Graaf, *Int. J. Obes.*, 2007, **32**, 676–683.
48. H. Adle-Biassette and J. A. Golden, in *Developmental Neuropathology*, John Wiley & Sons, 2nd edn, 2018, p. 560.
49. J. Murray, S. E. Langmore, S. Ginsberg and A. Dostie, *Dysphagia*, 1996, **11**, 99–103.
50. I. Tagashira, H. Tohara, Y. Wakasugi, K. Hara, A. Nakane, Y. Yamazaki, M. Matsubara and S. Minakuchi, *Arch. Gerontol. Geriatr.*, 2018, **74**, 106–111.
51. F. Zerbib and T. Omari, *Nat. Rev. Gastroenterol. Hepatol.*, 2015, **12**, 322–331.
52. J. Lagergren, E. Smyth, D. Cunningham and P. Lagergren, *Lancet*, 2017, **390**, 2383–2396.
53. M. Cabre, M. Serra-Prat, L. Force, E. Palomera and R. Pallares, *Med. Clin.*, 2008, **131**, 167–170.
54. K. M. Hiiemae and J. B. Palmer, *Crit. Rev. Oral Biol. Med.*, 2003, **14**, 413–429.
55. T. Ono, K. Hori, Y. Masuda and T. Hayashi, *Sensors*, 2010, **10**, 176–202.
56. P. Clavle, M. De Kraa, V. Arreola, M. Girvent, R. Farre, E. Palomera and M. Serra-Prat, *Aliment. Pharmacol. Ther.*, 2006, **24**, 1385–1394.
57. S. Nair, A. Surendaran, J. Menon, S. Sreedharan and P. Sylaja, *Ann. Ind. Acad. Neurol.*, 2016, **19**, 249–251.
58. E. Saitoh, S. Shibata, K. Matsuo, M. Baba, W. Fujii and J. B. Palmer, *Dysphagia*, 2007, **22**, 100–107.
59. E. R. Oommen, Y. Kim and G. McCullough, *Dysphagia*, 2011, **26**, 318–323.
60. K. Matsuo and J. B. Palmer, *Curr. Opin. Food Sci.*, 2016, **9**, 1–10.
61. M. G. Rugiu, *Acta Otorhinol. Ital.*, 2007, **27**, 306–316.
62. M. Gasparin, C. Schweiger, D. Manica, A. C. Maciel, G. Kuhl, D. S. Levy and P. J. C. Marostica, *Pediatr. Pulmonol.*, 2017, **52**, 41–47.
63. R. Merletti, A. Rainoldi and D. Farina, *Exercise Sport Sci. Rev.*, 2001, **29**, 20–25.

64. R. Ding, C. R. Larson, J. A. Logemann and A. W. Rademaker, *Dysphagia*, 2002, **17**, 1–12.
65. J. A. Logemann, B. R. Pauloski, A. W. Rademaker, L. A. Colangelo, P. J. Kahrilas and C. H. Smith, *J. Speech Hear. Res.*, 2000, **43**, 1264–1274.
66. R. Ding, J. A. Logemann, C. R. Larson and A. W. Rademaker, *J. Speech Hear. Res.*, 2003, **46**, 977–989.
67. L. Reimers-Neils, J. Logemann and C. Larson, *Dysphagia*, 1994, **9**, 101–106.
68. D. Inagaki, Y. Miyaoka, I. Ashida and Y. Yamada, *J. Oral Rehabil.*, 2008, **35**, 656–663.
69. I. Ashida, H. Iwamori, S.-Y. Kawakami, Y. Miyaoka and A. Murayama, *J. Texture Stud.*, 2007, **38**, 87–99.
70. M. A. Peyron, K. Maskawi, A. Woda, R. Tanguay and J. P. Lund, *J. Dent. Res.*, 1997, **76**, 789–795.
71. J. A. Hinchey, T. Shephard, K. Furie, D. Smith, D. Wang, S. Tonn and I. Stroke, Practice Improvement Network, *Stroke*, 2005, **36**, 1972–1976.
72. J. M. Garcia and E. I. Chambers, *Am. J. Nurs.*, 2010, **110**, 26–33.
73. A. Zamora Mur, C. Palacín Ariño, A. I. Guardia Contreras, A. Zamora Catevilla, E. Clemente Roldán and J. Santaliestra Grau, *Med. Familia. Semergen*, 2018, **44**, 168–173.
74. M. Atherton, N. Bellis-Smith, J. A. Y. Cichero and M. Suter, *Nutr. Diet.*, 2007, **64**, S53–S76.
75. U. T. Andersen, A. M. Beck, A. Kjaersgaard, T. Hansen and I. Poulsen, *e-SPEN J.*, 2013, **8**, e127–e134.
76. National Dysphagia Diet Task Force and American Dietetic Association, *National Dysphagia Diet: Standardization for Optimal Care*, American Dietetic Association, Chicago, IL, 2002.
77. Ministry of Health, Labour and Welfare, Food Safety Information: Food with Health Claims in Japan, 2009. Available at https://www.mhlw.go.jp/english/topics/foodsafety/.
78. National Patient Safety Agency RCoSaLT, British Dietetic Association, National Nurses' Nutrition Group and Hospital Caterers' Association, Dysphagia diet food texture descriptions, 2011. Available at https://www.rcslt.org/members/publications/dysphagia_diet_texture_descriptions.
79. T. J. Valenzano, A. A. Waito and C. M. Steele, *Dysphagia*, 2016, **31**, 598–609.
80. R. Momosaki, H. Yasunaga, H. Matsui, H. Horiguchi, K. Fushimi and M. Abo, *Geriatr. Gerontol. Int.*, 2015, **15**, 694–699.
81. L. Wright, D. Cotter and M. Hickson, *J. Human Nutr. Diet.*, 2008, **21**, 555–562.
82. P. K. Lieu, M. S. Chong and R. Seshadri, *Ann. Acad. Med. Singapore*, 2001, **30**, 148–154.
83. R. J. Dewar and M. J. Joyce, *Dysphagia*, 2006, **21**, 264–269.
84. J. M. Garcia, E. Chambers, Z. Matta and M. Clark, *Dysphagia*, 2008, **23**, 65–75.
85. P. Dodrill, in *Dysphagia*, ed. M. E. Groher and M. A. Crary, Mosby, St. Louis, 2 edn., 2016, pp. 325–350.

86. J. M. Garcia, E. Chambers, Z. Matta and M. Clark, *Dysphagia*, 2005, **20**, 325–335.
87. S.-G. Kim, W. Yoo and B. Yoo, *Prev. Nutr. Food Sci.*, 2014, **19**, 358–362.
88. J. A. Y. Cichero, O. Jackson, P. J. Halley and B. E. Murdoch, *J. Speech Hear. Res.*, 2000, **43**, 537–547.
89. C. M. Steele, P. H. H. M. Van Lieshout and D. H. Goff, *Dysphagia*, 2003, **18**, 182–195.
90. I. Germain, T. Dufresne and H. S. Ramaswamy, *J. Food Eng.*, 2006, **73**, 64–74.
91. G. Chi-Fishman and B. C. Sonies, *Dysphagia*, 2002, **17**, 278–287.
92. C. Payne, L. Methven, C. Fairfield and A. Bell, *Dysphagia*, 2011, **26**, 27–33.
93. L. Strowd, J. Kyzima, D. Pillsbury, T. Valley and B. Rubin, *Chest*, 2008, **133**, 1397–1401.
94. M. O'Leary, B. Hanson and C. H. Smith, *Int. J. Lang. Commun. Disord.*, 2011, **46**, 17–29.
95. Joint working party of the British Dietetic Association and the Royal College of Speech and Language Therapists, National Descriptors for Texture Modification in Adults, British Dietetic Association, 2002. Available at https://www.findusfoodservices.se/content/SE/Special%20Foods/Fakta%20&%20forskning/Konsistensanpassad%20mat/TMF_National_descriptors_UK.pdf.
96. J. Cichero, T. Nicholson and P. Dodrill, *Dysphagia*, 2011, **26**, 264–271.
97. S. Stuart and J. M. Motz, *Dysphagia*, 2009, **24**, 412–422.
98. M. M. Gosa and P. Dodrill, *Nutr. Clin. Pract.*, 2017, **32**, 238–244.
99. D. S. Levy, E. Osborn, K. A. Hasenstab, S. Nawaz and S. R. Jadcherla, *J. Parenter. Enteral Nutr.* 2019, **43**, 290–297.
100. S. Hines, J. McCrow, J. Abbey and S. Gledhill, *Int. J. Evidence-Based Healthcare*, 2010, **8**, 252–255.
101. M. B. M. Almeida, S. C. Gomes Junior, J. B. Silva, D. A. Silva and M. E. L. Moreira, *Revista CEFAC*, 2017, **19**, 683–689.
102. J. A. Y. Cichero, *Nutr. J.*, 2013, **12**, 54.
103. H. S. Park, D.-K. Kim, S. Y. Lee, K.-H. Park, *J. Texture Stud.*, 2017, **48**, 362–369.
104. Ministry of Health Labour and Welfare, Japan, Food Safety Information: Food with Health Claims, 2009. Available at http://www.mhlw.go.jp/english/topics/foodsafety/fhc/index.html.
105. K. Nishinari, K. Kohyama, H. Kumagai, T. Funami and M. C. Bourne, *Food Sci. Technol. Res.*, 2013, **19**, 519–521.
106. T. Funami, *J. Texture Stud.*, 2016, **47**, 313–322.
107. K. Nishinari, *Food Sci. Technol. Res.*, 2009, **15**, 99–106.
108. A. Hasegawa, A. Otoguro, H. Kumagai and F. Nakazawa, *Nippon Shokuhin Kagaku Kogaku Kaishi*, 2005, **52**, 441–447.
109. A. Hasegawa, F. Nakazawa and H. Kumagai, *Nippon Shokuhin Kagaku Kogaku Kaishi*, 2008, **55**, 330–337.
110. A. Hasegawa, F. Nakazawa and H. Kumagai, *Nippon Shokuhin Kagaku Kogaku Kaishi*, 2008, **55**, 541–548.

111. A. Tashiro, K. Ono, A. Hasegawa-Tanigome, H. Kumagai and H. Kumagai, *Japan J. Food Eng.*, 2010, **11**, 177–185.
112. R. Samuels and D. D. Chadwick, *J. Intell. Disabil. Res.*, 2006, **50**, 362–370.
113. A. C. F. Amaral, L. A. Rodrigues, R. M. M. M. Furlan, L. C. C. Vicente and A. R. Motta, *CoDAS*, 2015, **27**, 541–549.
114. D. Becker and L. Tews, in *A Comprehensive Guide to Geriatric Rehabilitation*, ed. T. L. Kauffman, R. Scott, J. O. Barr and M. L. Moran, Churchill Livingstone, Oxford, 3rd edn., 2014, pp. 409–413.
115. G. Hall and K. Wendin, *Ann. Nutr. Metab.*, 2008, **52**, 25–28.
116. C. M. Steele, W. A. Alsanei, S. Ayanikalath, C. E. A. Barbon, J. Chen, J. A. Y. Cichero, K. Coutts, R. O. Dantas, J. Duivestein, L. Giosa, B. Hanson, P. Lam, C. Lecko, C. Leigh, A. Nagy, A. M. Namasivayam, W. V. Nascimento, I. Odendaal, C. H. Smith and H. Wang, *Dysphagia*, 2015, **30**, 2–26.
117. International Dysphagia Diet Standardization Initiative (IDDSI), 2017. Available at www.iddsi.org.
118. P. Lam, S. Stanschus, R. Zaman and J. A. Cichero, *Br. J. Neurosci. Nursing*, 2017, **13**, S18–S26.
119. S. Pownall and C. Taylor, *Br. J. Neurosci. Nursing*, 2017, **13**, 260–268.
120. R. O. Dantas and L. Oliveira, *Revista CEFAC*, 2018, **20**, 382–387.
121. M. Su, G. Zheng, Y. Chen, H. Xie, W. Han, Q. Yang, J. Sun, Z. Lv and J. Chen, *J. Texture Stud.*, 2018, **49**, 2–10.
122. S. Perez, L. Andrea, C. Cuadrado Vives and B. Beltran de Miguel, *Nutr. Clin. Diet. Hosp.*, 2017, **37**, 89–97.
123. C. Ritzoulis and P. D. Karayannakidis, in *Modifying Food Texture*, ed. J. Chen and A. Rosenthal, Woodhead Publishing, 2015, pp. 51–69.
124. C. E. Jorgensen, R. K. Abrahamsen, E.-O. Rukke, A.-G. Johansen, R. B. Schuller and S. B. Skeie, *Int. Dairy J.*, 2015, **47**, 6–18.
125. E. A. Foegeding, J. P. Davis, D. Doucet and M. K. McGuffey, *Trends Food Sci. Technol.*, 2002, **13**, 151–159.
126. B. Delikanli and T. Ozcan, *Int. J. Dairy Technol.*, 2014, **67**, 495–503.
127. B. Delikanli and T. Ozcan, *J. Food Process. and Preserv.*, 2017, **41**, e13101.
128. B. Sezgin, D. Durusoy, M. S. Demirci, K. Ozturk, I. Kaya, S. Eyigor and S. Gode, *Eur. Arch. Oto-Rhino-Laryngol.*, 2018, **275**, 2997–3005.
129. E. K. Hadde and J. Chen, *J. Food Eng.*, 2019, **245**, 18–23.
130. M. Q. Waqas, J. Wiklund, A. Altskär, O. Ekberg and M. Stading, *J. Texture Stud.*, 2017, **48**, 507–517.
131. J. L. Buttriss and C. S. Stokes, *Nutr. Bull.*, 2008, **33**, 186–200.
132. N. Benkeblia and P. S. Thondre, in *Polysaccharides: Natural Fibers in Food and Nutrition*, ed. N. Benkeblia, CRC Press, 2014, pp. 233–258.
133. A. Ahmad, F. M. Anjum, T. Zahoor, H. Nawaz and S. M. R. Dilshad, *Crit. Rev. Food Sci. Nutr.*, 2012, **52**, 201–212.
134. A. Ahmad and M. Kaleem, in *Biopolymers for Food Design*, ed. A. M. Grumezescu and A. M. Holban, Academic Press, 2018, pp. 351–381.

135. D. El Khoury, C. Cuda, B. L. Luhovyy and G. H. Anderson, *J. Nutr. Metabol.*, 2012, 28.
136. C. M. Hasler, *Food Technol. Mag.*, 1998, **52**, 57–62.
137. M. Fathi, M. R. Mozafari and M. Mohebbi, *Trends Food Sci. Technol.*, 2012, **23**, 13–27.
138. Z. Fang and B. Bhandari, *Trends Food Sci. Technol.*, 2010, **21**, 510–523.
139. P. Zimet and Y. D. Livney, *Food Hydrocolloids*, 2009, **23**, 1120–1126.
140. M. Fathi, J. Varshosaz, M. Mohebbi and F. Shahidi, *Food and Bioprocess Technol.*, 2013, **6**, 1464–1475.
141. G. Y. Liu, J. M. Wang and Q. Xia, *Eur. Food Res. Technol.*, 2012, **234**, 391–398.
142. E. Dickinson, *Food Hydrocolloids*, 2018, **77**, 372–385.
143. F. L. Tchuenbou-Magaia, I. T. Norton and P. W. Cox, in *Gums and Stabilisers for the Food Industry*, Royal Society of Chemistry, 2009, pp. 113–125.
144. T. A. M. Rovers, G. Sala, E. Van der Linden and M. B. J. Meinders, *J. Texture Stud.*, 2016, **47**, 220–230.
145. G. M. Campbell and E. Mougeot, *Trends Food Sci. Technol.*, 1999, **10**, 283–296.
146. R. N. Zuniga, U. Kulozik and J. M. Aguilera, *Food Hydrocolloids*, 2011, **25**, 958–967.

Subject Index

References to figures are given in *italic* type. References to tables are given in **bold** type.

acusto optical tunable filter (AOTF), 291
Aero, 101
AFM. *See* atomic force microscopy (AFM)
agar, 326–327, *327*
agitation, 99
air-filled emulsions, 68, 83
alginate, **6**, 18, *19*, 20, 326–327, *327*, 395
AOTF. *See* acusto optical tunable filter (AOTF)
apparent viscosity, 8
atomic force microscopy (AFM), 375–378
- emulsion microstructure, visualization of, 77–78
- future perspective of, 378–379
- of lipids, 377
- of oleogels, 389–391, *390*
- plant-based foods, 378
- of polysaccharides, 376–377
- of proteins, 378, *378*
- setup, *376*

Australian standard for texture-modified foods, 467

baked foam foods, mechanical testing of, 210–212, *211*
baked products, salt in, 447–448
baking of dough and bread, 108
BDA. *See* British Dietetic Association (BDA)
bioactive compounds, encapsulation and delivery of, 259–280
- basic structures, 263–264, **263**
- core-shell-type encapsulates
  - coated melt extrudates for flavours, 276–277, *276*, *277*
  - core–(multi-)shell colloidal delivery systems, 280
  - fish oil, complex coacervates of, 277–280, *278*, *279*
- food products, 260–262
  - iron-fortified bouillon cube, 261
  - regulators of digestion, 261–262
- homogeneous core systems
  - colloidal lipid and protein-based particles, 267–270, *269*, *270*
  - emulsion-based delivery systems, 265, 267, *268*
  - micronutrient and nutraceutical-based colloidal particles, 264–265
  - particles based on associative complexes, 270–272, *271*, *272*

issues in, **260**
matrix-type encapsulates, 272–275
colloidal particle systems, 275, *275*
multiple emulsions, 273
spray-dried and melt extrudated flavours, 273–274, *273*
theobromine in lipid particles, 274–275, *274*
biochemical stresses
foods as time-dependent soft matter systems under, 240–241
biscuits, **153**, *157,* 442–443, *443*, *444*
bloom formation mechanisms, 145–148, *146*, *147*
bolus formation prior to swallowing, 461–462
bovine serum albumin (BSA), 331
branched polysaccharides, 15–16, *15*, *16*
bread
bubble structures in, 105–111
characterization of, 109–111
bubble fraction, 110–111
gas-free density, 110
gas volume fraction, 110–111
specific volume, 110
void fraction, 110–111
inclusion steps
baking, 108, *157*
mixing, 106–107
proving and fermentation, 107–108
physical and chemical controlling factors
flour, types of, 108
gluten, 108–109
water absorbance, 109
yeast/non-yeast, 109
porous heterogeneous system, 293–295, *293*, *294*, *295*
bread dough
microstructure studies of, 201–202, *202*
tensile test, *202*, 212
breathing and swallowing, temporospatial coordination between, 461
bright field microscopy, 289
British Dietetic Association (BDA), 465, 467
brittle baked wafer fracture, 213, *213*
BSA. *See* bovine serum albumin (BSA)
bubble fraction, 110–111
bubbles, in food structure development, 93–112
chocolate, 101–105
bubble inclusion methods, 102–103
influential factors in bubble inclusion, 103–105
operational variables, 105
dough and bread, 105–111
bubble inclusion, 106–108
characterization, 109–111
physical and chemical controlling factors, 108–109
milk, 94–100, *95*
foam processing methods, 98–100, *99*, **100**
process parameters, 95–98
as tastant-excluding fillers, 450

Cadbury, 101
cakes, *154*, 443, 445
calcium alginate. *See* sodium/calcium alginate
Camembert cheeses, 51–52
carbohydrate gelation, 393–395

carrageenan
 milk reactivity of, 23–24
 small-angle scattering analysis of, 326–327, *327*
casein-based gels
 cheeses, 47–52
 yogurt, 44–47, *45*, *46*
caseins, characteristics of, 33–34
CATA. *See* check all that apply (CATA)
cavitation flow regime, 71
CCP. *See* colloidal calcium phosphate (CCP)
cellular structures
 baked and extruded foam foods, mechanical testing of, 210–212, *211*
 brittle baked wafer fracture, finite element modeling of, 213, *213*
 cereal/snacks products, 159–162, *160*
 edible foam food products, manufacturing methods of, 209–210
 foam foods using XMT, microstructural characterization of, 212–213, *212*
cellulose, small-angle scattering analysis of, 322–324, **323**
cell wall structure, 378
cereal-based foods, 151–153, *154*, 240–241.
 biscuits, **153**, *157*, 442–443, *443*, *444*
 bread, 108, *157*, 293–295, *293*, *294*, *295*
 cakes, *154*, 443, 445
 dough, 295–296, *295*, *296*
 fibrous vegetable proteins, 296–297, *297*, *298*
 See also cereal/snack products
cereal/snack products, 151–169, *152*
 cellular structure, 159–162, *160*
 composition, *154*
 modeling approaches, 164, 166–167, *168*
 physical properties, **153**
 rheological properties, 162–164, *165*
 state diagrams, 156–158, *157*, **158**
 wheat components, 153–156
charged polysaccharides, 14
check all that apply (CATA), 236–237
cheese, 47–52
 Camembert, 51–52
 Emmental or Swiss-type, 50–51
 mechanical performance of, 204–205, *204*
 mozzarella, 48–50
 salt in, 446–447
chocolate
 bloom formation mechanisms, 145–148, *146*, *147*
 bubble structure. *See* chocolate, bubble structures in
 cooling model, 141–145, *142*, *143*, *144*, *145*
 food structure development in, 134–148
 formulation, 135–137, **136**
 future needs, 148
 light and confocal microscopy of, 301–303, *302*, *303*, *304*
 manufacturing process, 134–135, *135*
 mechanical performance of, 205
 polymorphism in cocoa butter, 139–148, **140**
  chocolate cooling model, 141–145, *142*, *143*, *144*, *145*
 scanning electron microscope of, 372–373, *373*
 structure control
  liquid chocolate mass making, 137–138, *138*
  tempering and forming, 139
 sugar in, 445
 See also chocolate, food structure development in

chocolate, bubble structures in, 101–105
inclusion methods, 102–103
extrusion, 103
gas sparging under pressure, 102
reverse phase aeration, 103
vacuum application, 102
influential factors in bubble inclusion, 103–105
extrinsic factors, 104
intrinsic factors, 103–104
system factors, 104–105
operational variables, 105
fluid and gas flow rate, 105
mixing head and speed in continuous process, 105
vibration effect, 105
CIJs. *See* confined impinging jets (CIJs)
CMC. *See* critical micelle concentration (CMC)
coalescence, 69, 83, 95, 102
droplet coalescence *versus* droplet breakup, 70–72
partial, 63, 69
coated melt extrudates, for flavours, 276–277, *276*, *277*
cocoa butter, polymorphism in, 139–145, **140**
chocolate cooling model, 141–145, *142*, *143*, *144*, *145*
cohesive zone model (CZM), 202–204, *203*
cold aeration, 99
cold gelation, 44
colloidal calcium phosphate (CCP), 33, 34, 44, 45
nanoparticles, 329
colloidal lipids, 267–270, *269*, *270*
colloid mills, 73
colloids, 80, *266,* 268
colloidal particle systems, 269–272, 275, *275*
definition of, 4
mechanical performance of, 201–205, *202*, *203*, *204*
micronutrient and nutraceutical-based, 264–265
computer-based molecular methods, 383–398
carbohydrate gelation, 393–395
docking simulations, 386–387
future perspective of, 397–398
MM force fields, 385–386, *386*
molecular dynamics simulations, 387–388
molecular modelling, 384–385, *384*
protein gelation, 396–397, *397*
triglyceride systems, structure formation in, 388–393
oleogels, 389–393, *390*, *391*, *392*, *393*
conching, 137–138
cone geometry, 177
confined impinging jets (CIJs), 74, *75*
confocal scanning laser microscopy (CSLM)
cereal-based porous and heterogeneous foods
bread, 293–295, *293*, *294*, *295*
dough, 295–296, *295*, *296*
fibrous vegetable proteins, 296–297, *297*, *298*
fat-rich foods
chocolate, 301–303, *302*, *303*, *304*
mayonnaise, 303–305, *304*
fat structure, characterization of, 121
food structure analysis using, 287–306
characterization methods, 290–292
future perspective, 305–306
plant-based foods, 297–301, *299*, *300*, *301*
of protein gels, 206

cooling rate, effect on fat structure, 127–128, *127*
core-(multi-)shell colloidal delivery systems, 280
core-shell-type encapsulates
coated melt extrudates for flavours, 276–277, *276*, *277*
core-(multi-)shell colloidal delivery systems, 280
fish oil, complex coacervates of, 277–280, *278*, *279*
CPD. *See* critical point drying (CPD)
creaming, 69
critical concentration of hydrocolloids, 9, *10*
critical micelle concentration (CMC), 64
critical point drying (CPD), 368
crowding effect, 207
cryogenic scanning electron microscopy (cryo-SEM), *147*, 368, *369*, 373, *374*
dehydration, 368–369, *369*
cryogenic transmission electron microscopy (cryo-TEM), 119, *272*
of bread dough, 201–202, *202*
of colloidal lipids, *268*
fat structure, characterization of, 119–120, *120*, *127*, *129*
cryo-SEM. *See* cryogenic scanning electron microscopy (cryo-SEM)
cryo-TEM. *See* cryogenic transmission electron microscopy (cryo-TEM)
crystallization
under shear, 128–129, *129*
temperature, effect on fat structure, 127–128, *127*
CSLM. *See* confocal scanning laser microscopy (CSLM)
cup geometry, 178, *179*
CZM. *See* cohesive zone model (CZM)

DH. *See* dry-heating (DH)
DIC. *See* differential interference contrast (DIC)
Diet Dysphagia Food Texture Descriptors, 467
differential interference contrast (DIC), 289
differential scanning calorimetry (DSC), 79, 140
of cereal/snacks products, 156–157, *157*
diffusion-limited cluster–cluster aggregation (DLCA), 125
dispersed phase volume fraction, 80
DLCA. *See* diffusion-limited cluster-cluster aggregation (DLCA)
DLS. *See* dynamic light scattering (DLS)
DMA. *See* dynamic mechanical analysis (DMA)
docking simulations, 386–387
double emulsions, 67, 82, 83
dough, 295–296, *295*, *296*. *See also* dough, bubble structures in
dough, bubble structures in, 105–111
inclusion steps
baking, 108
mixing, 106–107
proving and fermentation, 107–108
physical and chemical controlling factors
flour, types of, 108
gluten, 108–109
water absorbance, 109
yeast/non-yeast, 109
droplets
breakup *versus* droplet coalescence, 70–72
o/w emulsions, 450
w/o emulsions, 450–451
w/o/w emulsions, 451–452
w/w emulsions, 452
dry-heating (DH), 40–41
drying, scanning electron microscope, 368
DSC. *See* differential scanning calorimetry (DSC)

duplex emulsions, 67, 82, 83
dynamic light scattering (DLS), 78
dynamic mechanical analysis (DMA), 162, 164
dysphagia, 459, 462–463, 467. *See also* Diet Dysphagia Food Texture Descriptors; texture standardization of diets

eating
  capability
    surface electromyography, 463
    videofluoroscopy, 462–463
  controlling mechanisms of
    orofacial muscle activities, 461
    tongue capability, 460–461
edible foam food products, manufacturing methods of, 209–210
edible oils, 336
EDTA. *See* ethylenediaminetetraacetic acid (EDTA)
egg white protein-based gels, 41–44, *43*
electric field processing, 422–435
  novel perspectives toward food structure, 432–434, *433*
  research needs, 434
  thermal and nonthermal processing, **426**
    high-pulsed electric technologies, 425–428
    ohmic heating, 428–432
electromyography (EMG), 232
electron microscopy (EM)
  emulsion microstructure, visualization of, 77
EMG. *See* electromyography (EMG)
Emmental or Swiss-type cheeses, 50–51
EM. *See* electron microscopy (EM)
emulsification, 72–73
  microchannel, 74
emulsifiers
  in fats, addition of, 129
  influence on chocolates, 104
emulsion microstructure
  characterization of
    components, 78
    flow and thermal behavior, assessment of, 79
    visualization, 77–78
  complex, 66–69
    duplex emulsions, 67
    gelled emulsions, 67
    nanoemulsions, 66
    water-in-water emulsions, 67–68
  conventional, 60–66, *61*
    continuous phase, 60–62
    dispersed phase, 62
    interface, 62–66
  for encapsulation and release, 81–82
  fat, salt and sugar reduction, 82–84
  and oral processing, 80–81
  and performance, 79–84
  and rheology, 80
emulsion
  air-filled emulsions, 68, 83
  -based delivery systems, 265, 267, *268*
  double emulsions, 67, 82, 83
  filled-hydrogel particle emulsions, 68, 82, 83
  -filled protein gels, property studies of, 206–207
  food structure development in, 59–84
  gelled emulsions, 67, 81–82
  microstructure. *See* emulsion microstructure
  multiple emulsions, 273
  nanoemulsions, 66
  pickering emulsions, 65–66
  processing routes, 69–77

emulsion (*continued*)
mechanical methods, 70–76, *75*, *76*
nonmechanical methods, 77
scanning electron microscope of, 373
small-angle scattering analysis of, 334–335
systems, food structure development in, 59–84
future perspective, 84
thermodynamic aspects, 69
encapsulation, 340–341
emulsion microstructure and, 81–82
energy density, 72–73
energy efficiency, 72–73
environmental scanning electron microscope (ESEM), 369–371, *370*, *371*
of bloom formation mechanisms, 146, *147*
ESEM. *See* environmental scanning electron microscope (ESEM)
ethylenediaminetetraacetic acid (EDTA), 330
extruded foam foods, 210–212, *211*
extrusion, 103

fat, 448
in chocolate, 103–104
crystallization, 335
lubrication, control of, 191–192, *192*
reduction
bubbles as tastant-excluding fillers, 450
droplets, 450–452
emulsion microstructure and, 82–84
gelled particles, 453, *454*
oleogels, 453
particles with designed morphology, 448–449
replacement in product categories, **449**
-rich foods, light and confocal microscopy of
chocolate, 301–303, *302*, *303*, *304*
mayonnaise, 303–305, *304*
See also fat systems
fat systems, 115–132
characterization
cryogenic transmission electron microscopy, 119–120, *120*
fractal dimensions, 124–125
^{1}H spin diffusion NMR, 119
microstructure, 120–121, *122*
polymorphism, 116–117
Scherrer analysis, 117–119
small-angle X-ray scattering, 117–119, *118*
ultra-small-angle X-ray scattering, 121, 123–124, *124*
wide-angle X-ray scattering, 116–117
factors affecting
cooling rate, 127–128, *127*
crystallization temperature, 127–128, *127*
crystallization under shear, 128–129, *129*
emulsifiers, addition of, 129
TAG composition, 126–127
structure to functional properties, relating
oil-binding capacity, 130–131
rheological properties, 131–132
fermentation, 45, 47, 106, 107–108, 159
FFA. *See* free fatty acid (FFA)

fibers in foods, adding, 214
fibrils, 39–40
fibrous vegetable proteins, 296–297, *297*, *298*
filled-hydrogel particle emulsions, 68, 82, 83
finite element modeling
  of bread dough, 202, *203*
  of brittle baked wafer fracture, 213, *213*
fish oil, complex coacervates of, 277–280, *278*, *279*
flavours, coated melt extrudates for, 276–277, *276*, *277*
FLIM. *See* fluorescence lifetime microscopy (FLIM)
flocculation, 69
fluid flow rate, 105, 208
fluorescence lifetime microscopy (FLIM), 306
fluorescence recovery after photobleaching (FRAP), 292, 303
foaming temperature of milk, 97–98
foam processing methods
  agitation, 99
  cold aeration, 99
  steam injection, 98–99, *99*
  supersaturation, 100
Food for Special Dietary Uses (FOSDU), Japan, 468
Food Hydrocolloids Trust, 5
food oral processing (FOP), 169
  physiology, 231–234, *233*
foods, as time-dependent stimuli, 237
  biochemical stresses, 240–241
  mechanical stress, 239–240
  thermal stresses, 241
food structure design
  control of viscosity, 185–190, *187*, *188*, *189*
  fat lubrication, control of, 191–192, *192*
  yield stress, 190–191
FOP. *See* food oral processing (FOP)
FOSDU. *See* Food for Special Dietary Uses (FOSDU)
Fourier transform infrared (FTIR) spectroscopy
  of oleogels, 390
fractal aggregates of heating protein solution, 38–39
fractal dimensions, 124–125
FRAP. *See* fluorescence recovery after photobleaching (FRAP)
free choice profiling, 236
free fatty acid (FFA), 96
FTIR. *See* Fourier transform infrared (FTIR) spectroscopy
functional groups of hydrocolloids, 16–17

gas flow rate, **100**, 105
gas-free dough density, 110
gas sparging under pressure, 102
gas volume fraction, 110–111
GDL. *See* glucono-delta-lactose (GDL)
gelatine gels, *192*, 208–209, *208*, *209*
gelation of hydrocolloids, 11–13, *11*, **11**, **13**
gelled emulsions, 67, 81–82
gelled particles, 453
gels, property studies of, 205–209, *208*, *209*
globular proteins, characteristics of, 34–35
glucono-delta-lactose (GDL), 44
gluten, 108–109
  tensile test, 201, *202*
GPCRs. *See* G-protein-coupled receptors (GPCRs)
G-protein–coupled receptors (GPCRs), 227
gum arabic, 24–25, *25*
gums, small-angle scattering analysis of, 324–327

HCA *See* Hospital Caterers Association (HCA)
hedonic testing, 237
hemicellulose polysaccharides, 394
heteroprotein complex coacervates (HPCC), 35–37

high-energy mechanical methods, 73
high-methoxyl pectins, 21–22
high-molecular-weight emulsifiers, 64–65
high-pressure homogenizers, 73
high-pressure processing (HPP), 407–408, *408*
  and food structure, 412–414
high-pulsed electric technologies
  high-voltage electrical discharge, 425–427
  pulsed electric fields, 427–428
high-shear mixers, 73
high-voltage electrical discharge (HVED), 425–427
HLB. *See* hydrophilic lipophilic balance (HLB)
homogeneous core systems
  colloidal lipid and protein-based particles, 267–270, *269*, *270*
  emulsion-based delivery systems, 265, 267, *268*
  micronutrient and nutraceutical-based colloidal particles, 264–265, *266*, *267*
  particles based on associative complexes, 270–272, *271*, *272*
Hospital Caterers Association (HCA), 467
HPCC. *See* heteroprotein complex coacervates (HPCC)
HPP. *See* high-pressure processing (HPP)
^{1}H spin diffusion NMR, 119
HVED. *See* high-voltage electrical discharge (HVED)
hydrocolloids, 3–27
  branched and linear polysaccharides, 15–16, *15*, *16*
  charged and neutral polysaccharides, 14
  definition, 3
  examples
    carrageenan, milk reactivity of, 23–24
    gum arabic, 24–25, *25*
    high-/low-methoxyl pectins, 21–22
    methyl cellulose, 22–23
    sodium/calcium alginate, 18–21, *19*, **21**
    xanthan, 26
  functional groups, 16–17
  functionality, 5–13
  future perspectives, 26–27
  gelation, 11–13, *11*, **11, 13**
  molecular confirmation, 17–18
  molecular weight, 13–14
  molecular weight distribution, 13–14
  solubility, 6–7, **7**
  viscosity, 8–10, *8*, *10*
  See also interpenetrating hydrocolloid networks (IHNs)
hydrophilic lipophilic balance (HLB), 63

ice cream, 99, 274–275, 445–446, **449**
IDDSI. *See* International Dysphagia Diet Standardisation Initiative (IDDSI)
IGBT. *See* insulated gate bipolar transistor (IGBT) pulses
IHNs. *See* interpenetrating hydrocolloid networks (IHNs)
inherent viscosity. *See* intrinsic viscosity
insulated gate bipolar transistor (IGBT) pulses, 432
interfacial-driven flow regime, 70
interfacial rheology, 79
intermediate-energy mechanical methods, 74, *75*, *76*
International Dysphagia Diet Standardisation Initiative (IDDSI), 468–470, *469*
interpenetrating hydrocolloid networks (IHNs), 327–328
intrinsic viscosity, 9, *10*

ionic surfactants, 64
iron-fortified bouillon cube, 261

Japanese Modified Diet for Dysphagic Persons, 468
jelly formulation, **13**
JND. *See* just noticeable sensorial difference (JND)
just noticeable sensorial difference (JND), 235

kinetic jaw movement (KJM), 232
KJM. *See* kinetic jaw movement (KJM)
Kraft Foods, 101
Krieger–Dougherty model, 189

laminar flow regime, 70
LAOS. *See* large amplitude oscillatory shear rheology (LAOS)
large amplitude oscillatory shear rheology (LAOS), 180–181
lens aberrations, 367, *367*
light microscopy (LM)
    cereal-based foods
        bread, *293*, *294*, 293–295
        dough, 295–296, *295*, *296*
        fibrous vegetable proteins, 296–297, *297*, *298*
    emulsion microstructure, visualization of, 77
    fat-rich foods
        chocolate, 301–303, *302*, *303*, *304*
        mayonnaise, 303–305, *304*
    food structure analysis using, 287–306
        characterization methods, 288–290
        future perspective, 305–306
        plant-based foods, 297–301, *299*, *300*, *301*
linear polysaccharides, 15–16, *15*, *16*
lipids, 115
    atomic force microscopy of, 377
    small-angle scattering analysis of
        emulsions and emulsion gels, 334–335
        fat crystallization, 335
        oleogels, 336
        oxidation of edible oils, 336
liquid chocolate mass making, 137–138, *138*
LM. *See* light microscopy (LM)
long spacings, 117
low-methoxyl pectins, 21–22
low-molecular-weight surfactants, 63–64

magnetic resonance imaging (MRI), 154–155
Maillard reaction, 331
matrix-type encapsulates, 272–275
    colloidal particle systems, 275, *275*
    multiple emulsions, 273
    spray-dried and melt extrudated flavours, 273–274, *273*
mayonnaise, 303–305, *304*, 334
MC. *See* methyl cellulose (MC)
MD. *See* molecular dynamics (MD) simulations
meat proteins, small-angle scattering analysis of, 331–332
mechanical performance, food structure development for, 199–221
    cellular structures
        baked and extruded foam foods, mechanical testing of, 210–212, *211*
        brittle baked wafer fracture, finite element modeling of, 213, *213*
        edible foam food products, manufacturing methods of, 209–210
        foam foods using XMT, microstructural characterization of, 212–213, *212*
    colloids, 201–205, *202*, *203*, *204*
    future trends, 219–221

mechanical performance, food structure development for (*continued*)
gels, 205–209, *208*, *209*
particulate composites, 201–205, *202*, *203*, *204*
short fiber composites
fibers in foods, adding, 214
mechanical characterization of, 215–218, *215*, *216*, *217*, *218*
oral breakdown, stimulation of, 218–219, *219*
mechanical stresses
foods as time-dependent soft matter systems under, 239–240
MEF. *See* moderate electric fields (MEF)
melt extrudated flavours, 273–274, *273*
membrane emulsification, 74
membrane processing, 411–412, *412*
and food structure, 417
methoxyl pectins, 21–22
methyl cellulose (MC), 22–23, 393–394
microbubbles, 472
microchannel emulsification, 74
microgels, 38, 190–191
micronutrient-based colloidal particles, 264–265, *266*, *267*
microparticulated whey protein (MPWP), 193, *194*
microstructure of fat, 120–121, *122*
microwave processing. *See* radiofrequency and microwave processing
milk
foaming temperature, 97–98
pH, 97
proteins, 96
resonant soft X-ray scattering (RSoXS), 329
small-angle scattering analysis of, 328–331
See also milk, bubble structures in
milk, bubble structures in, 94–100, *95*
foam processing methods
agitation, 99
cold aeration, 99
steam injection, 98–99, *99*
supersaturation, 100
process parameters, 95–98
change of pH, 97
foaming temperature, 97–98
free fatty acid, 96
influence of milk heat treatment, 98
milk protein/lipid, 96
skim milk *versus* whole milk, 97
mini traction machine (MTM), 183, *183*, 186
mixing, in dough and bread, 106–107
MM. *See* molecular modelling (MM)
moderate electric fields (MEF), 430–432
molecular confirmation of hydrocolloids, 17–18
molecular dynamics (MD) simulations, 387–388
molecular modelling (MM), 384–385, *384*
force fields, 385–386, *386*
molecular weight of hydrocolloids, 13–14
distribution of, 13–14
mozzarella cheeses, 48–50
MPQ model, 188, 189, *189*
MPWP. *See* microparticulated whey protein (MPWP)
MRI. *See* magnetic resonance imaging (MRI)
MTM. *See* mini traction machine (MTM)
mucilages, 324–327
multiple emulsions, 273

nanoemulsions, 66
nanoparticles, 65–66

nanostructured carriers, 471–472
nanotubes, 37
NA. *See* numerical aperture (NA)
National Food Texture Scale, **465**
National Nurses Nutrition Group (NNNG), 467
National Viscosity Scale, **466**
NDD. *See* U.S. National Dysphagia Diet (NDD)
neutral polysaccharides, 14
neutron magnetic resonance (NMR), 78
neutron scattering (NS), 78
NMR. *See* neutron magnetic resonance (NMR)
NNNG. *See* National Nurses Nutrition Group (NNNG)
nonionic surfactants, 64
non-yeast, 109
nosespace proton transfer reaction–mass spectrometry (PTR-MS), 243
NS. *See* neutron scattering (NS)
numerical aperture (NA), 288–289, 292
nutraceutical-based colloidal particles, 264–265, *266*, *267*

ohmic heating (OH), 409–410, *410*, 428–432
    equipment, **424**
    and food structure, 416
    moderate electric fields, 430–432
    pulsed, 432
OH. *See* ohmic heating (OH)
oil-binding capacity of fats, 130–131
oil–water interfacial tension, 78
oleogels
    small-angle scattering analysis of, 336
    structure and complexation, 389–393, *390*, *391*, *392*, *393*
    sugar, salt and fat reduction in, 453
olfaction, 227
optimal sensory performance, 225–248, *226*
    chemosensory perception, 241–243, *242*
    complex design structure, mechanical perception of, 244–246, *247*
    foods, as time-dependent stimuli
        biochemical stresses, 240–241
        mechanical stress, 239–240
        thermal stresses, 241
    future avenues, 246, 248
    oral cavity
        food oral processing physiology, 231–234, *233*
        olfaction, 227
        sensory physiology, 226–227
        taste, 228
        trigeminal and texture, 228–231, *229*, *230*
    quantification and product performance, 235–239, *235*, *238*
    reverse engineer sensory performance, 234–239
oral processing, 80–81, 176, 182, 240–241, 460, 463
    physiology, 231–234
orofacial muscle activities, 461
oscillatory rheology, 179–181
Ostwald ripening, 69
o/w emulsions, 450
oxidation of edible oils, 336

packaging materials, 341
PALM. *See* photoactivated localization microscopy (PALM)
partial coalescence, 69
particulate composites, 200–205, *202*, *203*, *204*
PCW. *See* plant cell wall (PCW) polysaccharides

PDMS. *See* polydimethylsiloxane (PDMS)
PEF. *See* pulsed electric field (PEF)
pendant drop technique, 78
phase inversion, 69
phase inversion temperature (PIT) method, 77
pH of milk, 97
photoactivated localization microscopy (PALM), 306
pickering emulsions, 65–66
PIT. *See* phase inversion temperature (PIT) method
plane geometry, 177
plant-based foods
atomic force microscopy of, 378
light and confocal microscopy of, 297–301, *299*, *300*, *301*
scanning electron microscope of, 373–374, *374*
plant cell wall (PCW) polysaccharides, 322–324, **323**
plant proteins, 205, 332–333
PLM. *See* polarized light microscopy (PLM)
PNA. *See* protein network analysis (PNA)
POH. *See* pulsed ohmic heating (POH)
polarized light microscopy (PLM)
of fat-rich foods, 303, *303*
fat structure, characterization of, 120
polydimethylsiloxane (PDMS), 182–183
polymorphism
in cocoa butter, 139–145, **140**
effects on fat structure, 116–117
polyphenol–polysaccharide interactions, 337–338
polysaccharides
atomic force microscopy of, 376–377
branched, 15–16, *15*, *16*
cellulose, 322–324, **323**
charged, 14
gums, 324–327
interpenetrating hydrocolloid networks, 327–328
linear, 15–16, *15*, *16*
mucilages, 324–327
neutral, 14
plant cell wall, 322–324, **323**
seaweed, 326–327, *327*
small-angle scattering analysis of, 317–318, **318**
texture modification, functional ingredients for, 471
processed meat, 447
protein gels
molecular modelling and computer simulation approaches to gelation, 396–397, *397*
property studies of, 205–206
protein network analysis (PNA), 155
protein-polysaccharide coacervates, 338–339
proteins, 29–53
atomic force microscopy of, 378, *378*
food matrices, examples
casein-based gels, 44–52, *45*, *46*
egg white protein-based gels, 41–44, *43*
functional properties, factors affecting, *31*
future trends, in food structure development, 52–53
as high-molecular-weight emulsifier, 64–65
milk and egg white proteins, characteristics of, 32–35, **32**
caseins, 33–34
globular proteins, 34–35
small-angle scattering analysis of
meat proteins, 331–332
milk proteins, 328–331
other proteins, 333–334
plant proteins, 332–333

supramolecular assemblies obtained from protein environment, *36*, **37**
heteroprotein complex coacervates, 35–37
nanotubes, 37
supramolecular assemblies obtained on heating protein solution, 37–41
fibrils, 39–40
fractal aggregates, 38–39
microgels, 38
multistranded ribbons, 39–40
protein microparticles, 40–41
spherulites, 39–40
PTR-MS. *See* nosespace proton transfer reaction–mass spectrometry (PTR-MS)
pulsed electric field (PEF), 427–428
equipment, manufacturers of, **424**
processing, 410–411, *411*
and food structure, 416–417
pulsed ohmic heating (POH), 432

QCM. *See* quartz crystal microbalance (QCM)
QDA. *See* quantitative descriptive analysis (QDA)
QM. *See* quantum mechanics (QM)
quantification of sensory perceptions, 235–239, *235*, *238*
quantitative descriptive analysis (QDA), 236, 237
quantum mechanics (QM), 385
quartz crystal microbalance (QCM), 186

radiofrequency and microwave processing, 408–409, *409*
and food structure, 414–416
reaction-limited cluster–cluster aggregation (RLCA), 125
reedings, 210
relative viscosity, 9
resonant soft X-ray scattering (RSoXS), 314–315, 329
restructured foods, 20
retrogradation, 16
reverse phase aeration, 103
rhamno-galactorunans I (RG-I), 395
rheological properties
of cereal/snacks products, 162–164, *165*
of fats, 131–132
rheology
definition of, 175
dynamic measurement, 177–179, *178*, *179*
emulsion microstructure and, 80
future prospects of, 195
interfacial, 79
model of concentrated dispersions, 206
oscillatory, 179–181
in sensory perception, 192–195, *194*
theory, 176–177
thin film, 79, 181–182
See also rheological properties
rheometry, 240
RLCA. *See* reaction-limited cluster-cluster aggregation (RLCA)
Royal College of Speech and Language Therapists (RSCLT), 465, 467
RSCLT. *See* Royal College of Speech and Language Therapists (RSCLT)
RSoXS. *See* resonant soft X-ray scattering (RSoXS)

salt
in baked products, 447–448
in cheese, 446–447
in processed meat, 447
reduction
bubbles as tastant-excluding fillers, 450

salt (*continued*)
droplets, 450–452
emulsion microstructure and, 82–84
gelled particles, 453, *454*
oleogels, 453
particles with designed morphology, 448–449
SAOS. *See* small amplitude oscillatory shear rheology (SAOS)
SasView, 317
SAXS. *See* small-angle X-ray scattering (SAXS)
scanning electron microscope (SEM) imaging
of bloom formation mechanisms, 146, *147*
of coated melt extrudates, *276*
components of, 362
cryo-SEM, *147*, 368, *369,* 373, *374,* 368–369, 369
electron-specimen interactions, 365–366
foam foods, microstructural characterization of, 212
future perspective of, 378–379
high-resolution imaging, compromises of, 362–364, *363*, *364*
image formation, 365–366, *365*, *366*
lens aberrations, 367, *367*
limitations imposed by vacuum, overcoming, 367–374
cryo-SEM, 368–369, *369*
drying, 368
environmental SEM, 369–371, *370*, *371*
examples, 371–374
of parmesan cheese, *204*
of protein gels, 206
of short fiber composites, 216, *218*
Scherrer analysis of fat structure, 117–119
seaweed polysaccharides, 326–327, *327*
sedimentation, 69
SE-HPLC. *See* size exclusion high-performance liquid chromatography (SE-HPLC)
sEMG. *See* surface electromyography (sEMG)
SEM. *See* scanning electron microscope (SEM)
sensory perception, 192–195, *194*
sensory performance
definition of, 225–226
optimal, design structures for, 225–248
See also optimal sensory performance
sensory physiology, 226–227
SESANS. *See* spin-echo small-angle neutron scattering (SESANS)
SHI. *See* strain-hardening index (SHI)
short fiber composites
fibers in foods, adding, 214
mechanical characterization of, 215–218, *215*, *216*, *217*, *218*
oral breakdown, stimulation of, 218–219, *219*
short spacings, 116
sitosterol–oryzanol gels, 391
size exclusion high-performance liquid chromatography (SE-HPLC), 154
skim milk *versus* whole milk, 97
SLM. *See* solid lipid microparticles (SLM)
SLS. *See* static light scattering (SLS)
small amplitude oscillatory shear rheology (SAOS), 179–180
small-angle scattering, food structure characterisation using, 309–342
complex systems, 336–339
polyphenol–polysaccharide interactions, 337–338
protein-polysaccharide coacervates, 338–339

starch-lipid complexes, 337
encapsulation, 340–341
future perspectives of, 341–342
lipids
emulsions and emulsion gels, 334–335
fat crystallization, 335
oleogels, 336
oxidation of edible oils, 336
methods, 312–317, *314*
packaging materials, 341
polysaccharides, 317–318, **317**
cellulose, 322–324, **323**
gums and mucilages, 324–327, *327*
interpenetrating hydrocolloid networks, 327–328
plant cell wall polysaccharides, 322–324, **323**
starch, 318–321
proteins
meat proteins, 331–332
milk proteins, 328–331
other proteins, 333–334
plant proteins, 332–333
techniques, 310–312, *311*, **312**
whole foods, 339–340
small-angle X-ray scattering (SAXS), 123, 312, 320–341
fat structure, characterization of, 117–119, *118*
sodium/calcium alginate, 18–21, *19*, **21**
solid lipid microparticles (SLM), 207
solubility of hydrocolloids, 6–7, **7**
specific viscosity, 9
specific volume of dough, 110
spherulites, 39–40
spin-echo small-angle neutron scattering (SESANS), 335
spray-dried flavours, 273–274, *273*
starch-lipid complexes, 337
starch, small-angle scattering analysis of, 318–321
static light scattering (SLS), 78
steam injection, 98–99, *99*
STED. *See* stimulated emission depletion (STED)
stimulated emission depletion (STED), 306
stirred yogurt, 47
strain-hardening index (SHI), 163
Stribeck curves, 183–185, *184*
sugar, 441–446
in biscuits, 442–443, *443*, *444*
in cakes, 443, 445
in chocolate, 104, 445
in ice cream, 445–446
reduction
bubbles as tastant-excluding fillers, 450
droplets, 450–452
emulsion microstructure and, 82–84
gelled particles, 453, *454*
oleogels, 453
particles with designed morphology, 448–449
supersaturation, 100
supramolecular assemblies
obtained by tuning the protein environment, *36*, **37**
heteroprotein complex coacervates, 35–37
nanotubes, 37
obtained on heating protein solution, 37–41
fibrils, 39–40
fractal aggregates, 38–39
microgels, 38
multistranded ribbons, 39–40
protein microparticles, 40–41
spherulites, 39–40
surface electromyography (sEMG), 463

sustainable food processing systems, 405–418
  food structure
    high-pressure processing and, 412–414
    membrane processing and, 417
    ohmic heating and, 416
    pulsed electric field processing and, 416–417
    radiofrequency and microwave processing and, 414–416
  future perspectives of, 417–418
  high-pressure processing, 407–408, *408*
  membrane processing, 411–412, *412*
  ohmic heating, 409–410, *410*
  overview of, 406–407
  pulsed electric field processing, 410–411
  radiofrequency and microwave processing, 408–409, *409*
swallowing
  capability, assessments and grading of
    surface electromyography, 463
    videofluoroscopy, 462–463
  controlling mechanisms of
    bolus formation prior to swallowing, 461
    orofacial muscle activities, 461
    tongue capability, 460–461
  disorders, 462. *See also* dysphagia
  tracks and breathing, temporospatial coordination between, 461

TAGs. *See* triacylglycerols (TAGs)
Task Force of Modified Diet for Dysphagic Persons, 468
taste sensing, 228, 242
TCATA. *See* temporal check-all-that-apply (TCATA)
TDS. *See* temporal dominance of sensations (TDS)
TEM. *See* transmission electron microscopy (TEM)
tempering, chocolate manufacturing, 104, 139
temporal check-all-that-apply (TCATA), 237
temporal dominance of sensations (TDS), 237
texture modification,
  microbubbles, 472
  nanostructured carriers, 471–472
  polysaccharides, 471
  whey proteins, 470–471
texture-modified diet, needs of, 463–464
texture sensations, 228–231, *229*, *230*
texture standardization of diets
  published national standards
    Australian Texture-modified Foods and Fluids, 467
    International Dysphagia Diet Standardisation Initiative, 468–470, *469*
    Japanese Modified Diet for Dysphagic Persons, 467–468
    UK National Descriptors for Texture Modification in Adults, 465, 467
    U.S. National Dysphagia Diet (NDD), 464
  texture-modified diet, needs of, 463–464
theobromine in lipid particles, 274–275, *274*
thermal stresses
  foods as time-dependent soft matter systems under, 241

thermotransient receptor potential (TRP) channels, 230
thermotropic gels, 41
thin film rheology, 181–182
tongue capability, 460–461
transmission electron microscopy (TEM), *36*, *45*
  of colloidal lipids, *269*, *270*
  See also cryo-TEM
triacylglycerols (TAGs), 115–117, 128–131, 388
  composition, 126–127
tribology, 79, 240
  definition of, 175
  future prospects of, 195
  methods, 182–185
  in sensory perception, 192–195, *194*
  Stribeck curves, 183–185, *184*
  theory, 182–185
tribometry, 240
trigeminal sensations, 228–231, *229*, *230*
triglycerides, 115, 388–393
TRP. *See* thermotransient receptor potential (TRP) channels
turbulent flow, 71

UK National Descriptors for Texture Modification in Adults, 465, 467
ultra-small-angle X-ray scattering (USAXS), 121, 123–124, *124*, 127, 312, 335
ultrasonic homogenization, 73
uniaxial compression tests, 211, *211*
USAXS. *See* ultra-small-angle X-ray scattering (USAXS)
U.S. National Dysphagia Diet (NDD), 464

vacuum application, for bubble inclusion in chocolates, 102
vane geometry, 178, *179*
vibration, effect on bubble-containing chocolates, 105
videofluoroscopy, 462–463
viscosity
  apparent, 8
  of chocolates, 104–105
  control of, 185–190, *187*, *188*, *189*
  of hydrocolloids, 8–10, *8*, *10*
  intrinsic, 9, *10*
  relative, 9
  specific, 9
void fraction, 110–111

WA. *See* water absorbance (WA) of flour
wafer. *See* brittle baked wafer
water absorbance (WA) of flour, 109
water-in-water (w/w) emulsions, 67–68
water jelly formulation, **13**
Wattle blossom model of gum arabic, *25*
WAXS. *See* wide-angle X-ray scattering (WAXS)
Weber–Fechner law, 235, *235*
Weber's law, 235
wheat components, in cereal/snacks products, 153–156
whey protein microparticles (WPM), 41
whey proteins (WP), 34, *36*, **37**, 40–41, 45–46, *194, 454*
  in moderate electronic fields, 431
  texture modification, functional ingredients for, 470–471
whole foods, small-angle scattering analysis of, 339–340
whole milk *versus* skim milk, 97
wide-angle X-ray scattering (WAXS), 116–117, 335
w/o emulsions, 450–451
w/o/w emulsions, 451–452
WP. *See* whey proteins (WP)
WPM. *See* whey protein microparticles (WPM)
w/w emulsions, 452

xanthan, 26
X-ray diffraction (XRD), 117, 140
    chocolate cooling, validation of, 141, 143, *143*
X-ray microtomography (XRT), 159–160, *274*, *277*
    of brittle baked wafer fracture, 213
    of coated melt extrudates, 277
    foam foods, microstructural characterization of, 212–213, *212*
XRD. *See* X-ray diffraction (XRD)
XRT. *See* X-ray microtomography (XRT)
xyloglucan, 394–395

yeast, 109
yield stress, 190–191
yogurt, 44–47, *45*
    manufacture, 45–47, *46*
    stirred yogurt, 47

zwitteronic surfactants, 64